"*The Routledge Companion to Pragmatism* hits a sweet spot. Well organized and approachable, it offers the novice a fine introduction to this vital tradition. Innovative in its conception and comprehensive in its coverage, it contains much that will inform and challenge even the most knowledgeable specialists."

— **Robert Brandom**, *University of Pittsburgh*

THE ROUTLEDGE COMPANION TO PRAGMATISM

The Routledge Companion to Pragmatism offers 44 cutting-edge chapters—written specifically for this volume by an international team of distinguished researchers—that assess the past, present, and future of pragmatism. Going beyond the exposition of canonical texts and figures, the collection presents pragmatism as a living philosophical idiom that continues to devise promising theses in contemporary debates. The chapters are organized into four major parts:

- Pragmatism's history and figures
- Pragmatism and plural traditions
- Pragmatism's reach
- Pragmatism's relevance

Each chapter provides up-to-date research tools for philosophers, students, and others who wish to locate pragmatist options in their contemporary research fields. As a whole, the volume demonstrates that the vitality of pragmatism lies in its ability to build upon, and transcend, the ideas and arguments of its founders. When seen in its full diversity, pragmatism emerges as one of the most successful and influential philosophical movements in Western philosophy.

Scott Aikin is Associate Professor of Philosophy at Vanderbilt University. He specializes in pragmatism, epistemology, argumentation theory, and ancient philosophy. He is the author of *Epistemology and the Regress Problem* (2010) and *Evidentialism and the Will to Believe* (2014).

Robert B. Talisse is W. Alton Jones Professor of Philosophy and Professor of Political Science at Vanderbilt University. His research focuses on political philosophy, with an emphasis on democracy, equality, and justice. His most recent book is *Sustaining Democracy* (2021).

ROUTLEDGE PHILOSOPHY COMPANIONS

Routledge Philosophy Companions offer thorough, high quality surveys and assessments of the major topics and periods in philosophy. Covering key problems, themes and thinkers, all entries are specially commissioned for each volume and written by leading scholars in the field. Clear, accessible and carefully edited and organised, *Routledge Philosophy Companions* are indispensable for anyone coming to a major topic or period in philosophy, as well as for the more advanced reader.

Titles include:

THE ROUTLEDGE COMPANION TO FEMINIST PHILOSOPHY
Edited by Ann Garry, Serene J. Khader, and Alison Stone

THE ROUTLEDGE COMPANION TO PHILOSOPHY OF PSYCHOLOGY, SECOND EDITION
Edited by Sarah Robins, John Symons, and Paco Calvo

THE ROUTLEDGE COMPANION TO MEDIEVAL PHILOSOPHY
Edited by Richard Cross and JT Paasch

THE ROUTLEDGE COMPANION TO PHILOSOPHY OF PHYSICS
Edited by Eleanor Knox and Alastair Wilson

THE ROUTLEDGE COMPANION TO ENVIRONMENTAL ETHICS
Edited by Benjamin Hale, Andrew Light, and Lydia A. Lawhon

THE ROUTLEDGE COMPANION TO PHILOSOPHY OF PAINTING AND SCULPTURE
Edited by Noël Carroll and Jonthan Gilmore

THE ROUTLEDGE COMPANION TO PRAGMATISM
Edited by Scott Aikin and Robert B. Talisse

For more information about this series, please visit: https://www.routledge.com/Routledge-Philosophy-Companions/book-series/PHILCOMP

THE ROUTLEDGE COMPANION TO PRAGMATISM

Edited by Scott Aikin and Robert B. Talisse

Routledge
Taylor & Francis Group
NEW YORK AND LONDON

Cover image: © Getty Images

First published 2023
by Routledge
605 Third Avenue, New York, NY 10158

and by Routledge
4 Park Square, Milton Park, Abingdon, Oxon, OX14 4RN

Routledge is an imprint of the Taylor & Francis Group, an informa business

© 2023 Taylor & Francis

The right of Scott Aikin and Robert B. Talisse to be identified as the authors of the editorial material, and of the authors for their individual chapters, has been asserted in accordance with sections 77 and 78 of the Copyright, Designs and Patents Act 1988.

All rights reserved. No part of this book may be reprinted or reproduced or utilised in any form or by any electronic, mechanical, or other means, now known or hereafter invented, including photocopying and recording, or in any information storage or retrieval system, without permission in writing from the publishers.

Trademark notice: Product or corporate names may be trademarks or registered trademarks, and are used only for identification and explanation without intent to infringe.

ISBN: 978-1-138-55551-8 (hbk)
ISBN: 978-1-032-34770-7 (pbk)
ISBN: 978-1-315-14959-2 (ebk)

DOI: 10.4324/9781315149592

Typeset in Bembo
by codeMantra

CONTENTS

List of Contributors   *xi*

   Introduction   1
   Scott Aikin and Robert B. Talisse

PART I
Pragmatism's History and Figures   **5**

1 The Metaphysical Club   7
   Cheryl Misak

2 C.S. Peirce's Pragmatism   13
   Albert Atkin

3 William James   19
   Alexander Klein

4 John Dewey   26
   David L. Hildebrand

5 Jane Addams   35
   Núria Sara Miras Boronat

6 Alain Locke's Critical Pragmatism as Applied to Race and Culture   41
   Corey L. Barnes

7 Sidney Hook   47
   Robert B. Talisse

8	C.I. Lewis between Classical and Contemporary Pragmatism *Peter Olen*	53
9	Quine and American Pragmatism *Yemima Ben-Menahem*	58
10	Wilfrid Sellars and Pragmatism *Willem A. deVries*	63
11	Richard Rorty: Narrative as Anti-Authoritarian Therapy and as Cultural Politics *Susan Dieleman*	70
12	Hilary Putnam *Maria Baghramian and Matthew Shields*	75
13	Cornel West and Prophetic Pragmatism *Eduardo Mendieta*	81
14	Susan Haack and Worldly, Realist Pragmatism *Robert Lane*	89
15	Nicholas Rescher's Methodological Pragmatism *Michele Marsonet*	95
16	Robert Brandom *Chauncey Maher*	101

PART II
Pragmatism and Plural Traditions   107

17	Pragmatism's Family Feud *Henry Jackman*	109
18	One Hundred Years of Pragmatism at Harvard *Douglas McDermid*	118
19	Pragmatism in Britain and Italy in the Early 20th Century *Gabriele Gava and Tullio Viola*	128
20	Pragmatism and Analytic Philosophy *Henrik Rydenfelt*	138
21	Pragmatism and Continental Philosophy *Paul Giladi*	147

22 Prospects for "Big-Tent" Pragmatic Phenomenology   159
    J. Aaron Simmons

23 Pragmatism and Its Prospects   168
    Michael Bacon

PART III
Pragmatism's Reach   **177**

24 Pragmatism and Logic   179
    F. Thomas Burke

25 Pragmatism and Metaphysics   191
    Claudine Tiercelin

26 Peirce, James, and Dewey as Philosophers of Science   206
    Jeff Kasser

27 Pragmatism and Language   217
    David Boersema

28 Pragmatism in the Philosophy of Mind   226
    Aaron Zimmerman

29 Pragmatism and Cognitive Science   239
    Shaun Gallagher

30 Knowledge-Practicalism   252
    Stephen Hetherington

31 Pragmatism and Religion   264
    Sami Pihlström

32 Pragmatism and the Moral Life   276
    Diana B. Heney

33 Artworld Practice, Aesthetic Properties, Pragmatist Strategies   287
    Robert Kraut

34 Pragmatism and Political Philosophy   300
    Matthew Festenstein

35 Pragmatism and Metaphilosophy   311
    Scott Aikin

PART IV
Pragmatism's Relevance   **321**

36 Pragmatism and Philosophical Methods   323
    Andrew Howat

37 Pragmatism and Expressivism   331
    David Macarthur

38 Pragmatism and Naturalism   343
    James R. O'Shea

39 Pragmatist Theories of Truth   351
    Cornelis de Waal

40 Pragmatism and Insurrectionist Philosophy   358
    Lee A. McBride III

41 Latin American Philosophy, U.S. Latinx Philosophy, and Anglo-American
    Pragmatism   366
    Denise Meda Calderon and Andrea J. Pitts

42 Pragmatism and Race   375
    Jacoby Adeshei Carter

43 Meaning and Inquiry in Feminist Pragmatist Narrative   380
    Shannon Dea

44 Pragmatism and Environmental Philosophy   387
    Evelyn Brister

Index   *395*

CONTRIBUTORS

Scott Aikin is Associate Professor of Philosophy at Vanderbilt University. He specializes in epistemology, argumentation theory, ancient philosophy, and American philosophy. He is the author of *Epistemology and the Regress Problem* (2010) and *Evidentialism and the Will to Believe* (2014).

Albert Atkin (Macquarie University, Australia) is philosopher who works on Peirce, pragmatism, the philosophy of race, and the social and political dimensions of language and epistemology. He is the author of two books – *Peirce* (2016), *The Philosophy of Race* (2014) – and numerous articles.

Michael Bacon is Senior Lecturer in Political Theory in the Department of Politics, International Relations, and Philosophy at Royal Holloway, University of London. He has written various pieces on pragmatism, with an excessive focus on Rorty's work. He is currently completing a monograph on the writings of Ralph Waldo Emerson.

Maria Baghramian is Full Professor of Philosophy at UCD School of Philosophy and project leader of the Horizon 2020 European Commission Research Project *Policy Expertise and Trust in Action (PERITIA)*. She has published extensively on the problem of relativism, the philosophy of Hilary Putnam, and most recently on the social epistemology of trust and expertise.

Corey L. Barnes is Assistant Professor of Philosophy at the University of San Diego. Professor Barnes' research interests are the philosophy of race; social and political philosophy, particularly regarding cosmopolitanism; and Africana philosophy. In his current research, Professor Barnes aims to systematize many different parts of Alain Locke's philosophy (e.g., his philosophical anthropology, value theory, and religious thinking), arguing that what connects the parts of Locke's philosophy is his underlying commitment to establishing a cosmopolitan community. Further, Professor Barnes' current research examines questions regarding what race is, what we mean when we use the term, and the historical development of the concept of race.

Yemima Ben-Menahem is Professor (Emerita) of Philosophy at the Hebrew University of Jerusalem, specializing in the philosophy of science. Other interests of hers include American pragmatism and the philosophy of history. Her publications include papers on realism, explanation, causation, historical contingency, the foundations of quantum mechanics and statistical mechanics, as well as papers on Quine, Putnam, Wittgenstein, and Davidson. She is the author of *Conventionalism* (2006) and *Causation in Science* (2018).

Contributors

David Boersema is retired emeritus Distinguished University Professor of Philosophy at Pacific University (in Forest Grove, Oregon). Among his publications are *Pragmatism and Reference* (2008), *Philosophy of Science: Text and Readings* (2009), and *Philosophy of Art: Aesthetic Theory and Practice* (2013).

Núria Sara Miras Boronat is Associate Professor of Moral Philosophy at the Philosophy Department of the University of Barcelona. She wrote her Dissertation on *Wittgenstein and Gadamer: Language, Praxis, Reason* (2009). She has been a guest researcher and lecturer at the Humboldt-University of Berlin, the University of Leipzig, the University of Parma, and the Autonomous National University of Mexico. Her main research areas are the political philosophy, pragmatism, the philosophy of play, and feminism. She has published the chapter "Addams and Gilman: The Foundations of Pragmatism, Feminism and Social Philosophy" in the book *Pragmatism and Social Philosophy. American Contributions to a European Discipline* edited by Michael Festl (2021). Together with Michela Bella, she is editing the book *Women in Pragmatism: Past, Present, and Future* (2022).

Evelyn Brister is Professor of Philosophy at the Rochester Institute of Technology. Pragmatism guides her research examining ethical, epistemological, and political issues that are raised in conservation science. She is the editor, with Robert Frodeman, of *A Guide to Field Philosophy* (2020), a collection of essays examining collaborations between philosophers and policymakers.

F. Thomas Burke is Professor in the Department of Philosophy at the University of South Carolina. He is the author of *What Pragmatism Was* (2013) and *Dewey's New Logic* (1994). He is co-editor of *George Herbert Mead in the Twenty-First Century* (2013) and *Dewey's Logical Theory: New Studies and Interpretations* (2002). He pursued graduate studies in mathematics and philosophy way back in the late 20th century at the University of New Mexico and Stanford University, earning a PhD in philosophy from Stanford in 1992.

Denise Meda Calderon is PhD candidate in the Department of Philosophy at Texas A&M University. Her research engages Latinx philosophy, Mexican philosophy, and Chicanx philosophy to theorize the relationship between communal practices of resistance to legacies of colonialism and non-Western interpretations of death. Denise is also a Graduate Research Fellow with the Glasscock Humanities Center and a Race and Ethnicity Studies Institute Graduate Student Fellow at Texas A&M University. Previously, she was editorial assistant for the *Inter-American Journal of Philosophy*.

Jacoby Adeshei Carter is Associate Professor of Philosophy, Chair of the Department of Philosophy at Howard University, Director of the Alain Leroy Locke Society, and co-editor of the *African American Philosophy and the Diaspora Book Series* published by Palgrave/Macmillan. He is the author of *African American Contributions to the Americas' Cultures: A Critical Edition of Lectures by Alain Locke* (2016) and co-editor of *Philosophic Values and World Citizenship: Locke to Obama and Beyond* (2010). His forthcoming books include *Philosophizing the Americas: An Inter-American Discourse*, an anthology of African American, Latin American, and Caribbean philosophy, and *Insurrectionist Ethics: Radical Perspectives on Social Justice*, a collection of essays that address advocacy on behalf of oppressed groups to promote radical social transformation in the interest of justice.

Shannon Dea is Professor of Philosophy and Dean of Arts at University of Regina, Canada. The author of *Beyond the Binary: Thinking about Sex and Gender* (2016), she lives on Treaty 4 territory and the homeland of the Métis Nation.

Contributors

Willem A. deVries is Professor Emeritus at the University of New Hampshire and co-editor of Routledge Studies in American Philosophy. He writes on Sellars and Hegel with forays into related periods and topics.

Cornelis de Waal is Professor of Philosophy at Indiana University–Purdue University Indianapolis and Editor-in-Chief of the *Transactions of the Charles S. Peirce Society*. His most recent book is *Introducing Pragmatism: A Tool for Rethinking Philosophy* (2021). Currently, he is editing the *Oxford Handbook of Charles S. Peirce*.

Susan Dieleman is Assistant Professor of Philosophy at Southern Illinois University Edwardsville. She is co-editor of *Pragmatism and Justice* (2017) and *The Ethics of Richard Rorty: Moral Communities, Self-Transformation, and Imagination* (forthcoming) and has several published essays on Richard Rorty's thought.

Matthew Festenstein is Professor of Political Philosophy at the University of York, UK. His books include *Pragmatism and Political Theory: From Dewey to Rorty* (1997) and *Negotiating Diversity: Culture, Deliberation, Trust* (2005). His current research interests include John Dewey, trust, and emergency politics.

Shaun Gallagher is Lillian and Morrie Moss Professor of Excellence in Philosophy at the University of Memphis, and Professorial Fellow at the School of Liberal Arts, University of Wollongong. He held the Humboldt Foundation Anneliese Maier Research Fellowship (2012–2018). His publications include *Action and Interaction* (2020); *Enactivist Interventions: Rethinking the Mind* (2017); *Phenomenology* (2012); *The Phenomenological Mind* (3rd edition with Dan Zahavi, 2021); *How the Body Shapes the Mind* (2005); and as an editor, *Oxford Handbook of the Self* and *Oxford Handbook of 4E Cognition*. He's editor-in-chief of the journal *Phenomenology and the Cognitive Sciences*.

Gabriele Gava is Associate Professor of Theoretical Philosophy at the University of Turin. He has published articles in leading philosophical journals on Peirce, Kant, pragmatism, and epistemology. He is the author of *Peirce's Account of Purposefulness: A Kantian Perspective* (2014). His second book, *Kant's Method in the Critique of Pure Reason*, is forthcoming. He is Assistant Editor of the journal *Studi Kantiani*.

Paul Giladi is Senior Lecturer in Philosophy at Manchester Metropolitan University. He has published articles in leading philosophy journals and edited collections on Hegel, pragmatism, critical social theory, feminism, and contemporary Anglo-American philosophy. Dr. Giladi is also editor of *Responses to Naturalism: Critical Perspectives from Idealism and Pragmatism* (2019), editor of *Hegel and the Frankfurt School* (2020), as well as co-editor (with Nicola McMillan) of the forthcoming collection *Epistemic Injustice and the Philosophy of Recognition*.

Diana B. Heney is Assistant Professor of Philosophy at Vanderbilt University. Her research centers around ethics, bioethics, well-being, and American pragmatism. She is the author of *Toward a Pragmatist Metaethics* (2016).

Stephen Hetherington is Emeritus Professor of Philosophy, UNSW, Sydney. He was editor-in-chief of *Australasian Journal of Philosophy* (2013–2022) and is (for Cambridge University Press) editor-in-chief of *Elements in Epistemology*. His books include *Epistemology's Paradox* (1992), *Good Knowledge, Bad Knowledge* (2001), *How to Know* (2011), and *Knowledge and the Gettier Problem* (2016) – not forgetting *What Is Epistemology?* (2019). His edited books include *Epistemology Futures*

(2006), *The Philosophy of Knowledge: A History* (2019; four volumes, as general editor), *The Gettier Problem* (2019), and (with Nicholas D. Smith) *What the Ancients Offer to Contemporary Epistemology* (2020).

David L. Hildebrand is Professor of Philosophy at the University of Colorado Denver and President of the Society for the Advancement of American Philosophy. Besides two books on Dewey, he has contributed major entries on Dewey to the *Stanford Encyclopedia of Philosophy,* the *Cambridge Companion to Pragmatism,* and on Dewey and linguistic pragmatism to *The Oxford Handbook of Dewey* and *A Companion to Rorty.* Current research includes technology's influence upon aesthetics, education, and democratic problem-solving.

Andrew Howat is Associate Professor of Philosophy at California State University, Fullerton. His research focuses on pragmatist perspectives on metaphysics and philosophical methodology, with a particular emphasis on C.S. Peirce. His current work explores the implications of Peirce's commitment to scholastic realism for his conception of truth and approach to conceptual clarification. His website is andrewhowat.com.

Henry Jackman is Associate Professor of Philosophy at Toronto's York University and former president of the William James Society. In addition to William James, his work focuses primarily on contemporary theories of reference and representation. A selection of his papers on both topics can be found at www.jackman.org.

Jeff Kasser is Associate Professor of Philosophy at Colorado State University. He specializes in epistemology and the history of American pragmatism, with forays into philosophy of science and philosophy of religion. He is currently trying to reconcile Charles Peirce's doubt-belief theory of inquiry with his philosophy of statistical inference. He hopes to get back to his work on William James' expressivist approach to epistemic norms soon. He has published in such journals as *Erkenntnis, European Journal of Philosophy,* and *Transactions of the Charles S. Peirce Society.* He also wrote and delivered *Philosophy of Science,* a course produced for The Great Courses (formerly The Teaching Company).

Alexander Klein is Canada Research Chair and Associate Professor of Philosophy at McMaster University, where he is Director of the Bertrand Russell Research Centre and Head of the Digital Philosophy Laboratory. In addition to numerous journal articles and book chapters, he is the author of *Consciousness Is Motor: Warp and Weft in William James* and editor of the *Oxford Handbook of William James,* both of which are forthcoming.

Robert Kraut is Professor of Philosophy at The Ohio State University. He has held visiting positions at Pittsburgh and Rutgers, and was a Marta Sutton Weeks Fellow at the Stanford Humanities Center. His primary interests are metaphysics, aesthetic theory, and the philosophy of language; his *Artworld Metaphysics* (2010) explores the complexities of artwork interpretation, description, and evaluation from a neo-pragmatist perspective.

Robert Lane is Professor of Philosophy at the University of West Georgia. He is the author of *Peirce on Realism and Idealism* (2018), editor for Peirce submissions to the *Transactions of the Charles S. Peirce Society,* and associate editor of Susan Haack (ed.), *Pragmatism, Old and New* (2006).

David Macarthur is Associate Professor of Philosophy at The University of Sydney. He has published widely on liberal naturalism, metaphysical quietism, pragmatism, skepticism, common sense, perception, ordinary language, and philosophy of art, especially architecture, photography, and film. He has edited *Hilary & Ruth-Anna Putnam, Pragmatism as a Way of Life* (2017), and with Mario De Caro co-edited *Naturalism in Question* (2004); *Naturalism and Normativity* (2010); and *Hilary Putnam, Philosophy in an Age of Science* (2012).

Chauncey Maher is Associate Professor of Philosophy at Dickinson College, USA. Along with several articles, he has published *Plant Minds* (2017) and *The Pittsburgh School of Philosophy: Sellars, McDowell, Brandom* (2012).

Michele Marsonet is Professor of Philosophy at the University of Genoa, Italy, and Fellow of the Center for Philosophy of Science of the University of Pittsburgh. Among his books are *Idealism and Praxis* (2008) and the edited *The Problem of Realism* (2002).

Lee A. McBride III is Professor of Philosophy at The College of Wooster (Ohio). McBride specializes in American philosophy, ethics, political philosophy, and philosophy of race. He is the author of *Ethics and Insurrection: A Pragmatism for the Oppressed* (2021). He is the editor of *A Philosophy of Struggle: The Leonard Harris Reader* (2020) and co-editor (with Erin McKenna) of *Pragmatist Feminism and the Work of Charlene Haddock Seigfried* (2022).

Douglas McDermid is Professor of Philosophy at Trent University (Canada), where he has taught since 2002. He is the author of two books: *The Varieties of Pragmatism* (2006) and *The Rise and Fall of Scottish Common Sense Realism* (2018).

Eduardo Mendieta is Professor of Philosophy and Latina/o Studies at Penn State University. His most recent publications include editing Richard Rorty's *Pragmatism as Anti-Authoritarianism* (2021) and with Amy Allen, *Decolonizing Ethics: The Critical Theory of Enrique Dussel* (2011).

Cheryl Misak is University Professor and Professor of Philosophy at the University of Toronto. Her books include *Frank Ramsey: A Sheer Excess of Powers* (2020), *Cambridge Pragmatism* (2018), and *The American Pragmatists* (2013).

Peter Olen is Assistant Professor of Philosophy at Lake-Sumter State College. His research addresses issues at the intersection of epistemology, the philosophy of science, and the philosophy of language from a historical perspective. In his spare time, Peter promotes live events and concerts throughout Florida.

James R. O'Shea is Professor of Philosophy at University College Dublin (UCD). He edited *Kant's Critique of Pure Reason: A Critical Guide* (2017) and *Sellars and His Legacy* (2016) and is the author of *Kant's Critique of Pure Reason: An Introduction and Interpretation* (2014) and *Wilfrid Sellars: Naturalism with a Normative Turn* (2007). He also publishes articles on classical and contemporary pragmatism, history of analytic philosophy, Hume's philosophy, perceptual epistemology, and naturalism.

Sami Pihlström is Professor of Philosophy of Religion at the University of Helsinki, Finland. He has published widely on pragmatism, realism, philosophy of religion, ethics, metaphysics, and transcendental philosophy. His recent books include *Death and Finitude* (2016), *Kantian Antitheodicy* (with Sari Kivistö, 2016), *Pragmatic Realism, Religious Truth, and Antitheodicy* (2020), *Why Solipsism Matters* (2020), as well as the forthcoming volume, *Pragmatist Truth in the Post-Truth Age*.

Contributors

Andrea J. Pitts is Associate Professor of Philosophy at the University of North Carolina, Charlotte. She is the author of *Nos/Otras: Gloria E. Anzaldúa, Multiplicitous Agency, and Resistance* (2021), and co-editor, with Mark Westmoreland, of *Beyond Bergson: Examining Race and Colonialism through the Writings of Henri Bergson* (2019) and *Theories of the Flesh: Latinx and Latin American Feminisms, Transformation, and Resistance* with Mariana Ortega and José M. Medina (2020).

Dr. Henrik Rydenfelt is Docent of Practical Philosophy and Media and Communication Studies at the University of Helsinki, Finland. Since 2008, he has coordinated the Nordic Pragmatism Network. His most recent publications include "Realism without Representationalism," "Recent Problems of the Public," and "Democracy and Moral Inquiry."

Matthew Shields is Postdoctoral Fellow at University College Dublin with Horizon 2020 European Commission Research Project *Policy Expertise and Trust in Action* (*PERITIA*). His recent publications include "Conceptual Domination," "On Stipulation," and "Conceptual Change and Future Paths for Pragmatism."

J. Aaron Simmons is Professor of Philosophy at Furman University. He is the author of *God and the Other* (2011), co-author of *The New Phenomenology* (2013), and editor of numerous volumes including *Christian Philosophy* (2018), *Kierkegaard and Levinas* (2008), and *Phenomenology for the Twenty-First Century* (2016).

Robert B. Talisse is W. Alton Jones Professor of Philosophy at Vanderbilt University. He specializes in political philosophy, with an emphasis on democratic theory. He is the author of over 100 articles and 15 books.

Claudine Tiercelin is Professor at the College de France (Chair of Metaphysics and Philosophy of Knowledge). She is also member of the Institut de France and of the Academia Europea. Her fields of scholarship cover analytic metaphysics, epistemology, and classical and contemporary pragmatism (Peirce, James, Ramsey, Putnam).

Tullio Viola is Assistant Professor of Philosophy of art and culture at Maastricht University. Prior to his position in Maastricht, he worked as a postdoctoral fellow in Berlin and Erfurt. His research focus is pragmatism, post-Kantian philosophy, and the philosophy of culture and history. His first monograph, *Peirce on the Uses of History*, was published in 2020. In other works, he explored the connections between pragmatism and European philosophy (France and Germany in particular).

Aaron Zimmerman teaches philosophy at the University of California, Santa Barbara where he writes on the philosophy of mind and epistemology, with a focus on moral psychology and moral epistemology. He is the author of numerous articles and two books on these subjects: *Moral Epistemology* (2010) and *Belief: A Pragmatic Picture* (2018). He is also co-editor, with Karen Jones and Mark Timmons, of the *Routledge Handbook of Moral Epistemology* (2019).

INTRODUCTION

Scott Aikin and Robert B. Talisse

Carrying on an intellectual tradition is a Janus-faced enterprise. Like Janus, the Roman god of transitions, one must look both forward and backward. To start, a living tradition must have a causal continuity with its past. The texts and debates from its crucial figures must be carefully preserved and interpreted; distinctive themes must be kept alive, insights appreciated, and founding arguments clarified. Yet living traditions are not museum pieces. A viable tradition must be applicable to contemporary circumstances. Contemporary practitioners must demonstrate the relevance of their tradition by importing its characteristic principles and arguments into the fray of current debate. Naturally, this will occasion new challenges and problems for the tradition. Contemporary critics of the tradition will have innovative lines of objection, and material arrangements will change in ways that the founders did not anticipate.

Thus, a tradition's distinctive insights will need updating and revision. Sometimes, more drastic measures are necessary. Given that any tradition will have internal debates at its founding or in its early development, those who seek to carry on that tradition must be open to the possibility that there are errors in the tradition's inception – if the tradition's founders disagreed, *at least one* was wrong, and maybe *all* were. The formative moments of a tradition are animated by debates over how the program's most worthy insights can be developed and perfected. And what can be discarded. Accordingly, those who seek to keep a tradition alive must engage those debates anew, by addressing challenges from external critics and internal disputes among fellow practitioners. Without this forward-looking face, one open to transitions and revisions, the tradition degrades into dead dogma.

In editing this volume, we have sought to apply the Janus rule to *pragmatism*. Living by the Janus rule means one must keep the themes and insights of pragmatism's genesis and development fresh and relevant, yet one must also find new avenues for applying and revising those commitments. In other words, one must *look both ways*.

Pragmatism, as an approach not only to philosophical problems but also to philosophy itself, requires that we regularly ask: *Is this (still) working?* Consequently, we need to be open to the answer that our approaches do not work, even with respect to what seem the most powerful thoughts driving the tradition. "The Pragmatists" is an appellation all too regularly reserved for just a few (and often just three) men who lived at the end of the 19th and beginning of the 20th century. But we think that the insights of the pragmatist tradition extend well beyond what these initial figures had anticipated. The broad variety of current philosophical research captured by the term "pragmatist" shows that the tradition is very much alive and growing. Pragmatism, in this regard, is a poster child for successful philosophical programs. It is a viable and competitive approach in nearly every current philosophical debate; in fact, in almost any matter of theoretical import, once can find a pragmatist program among the live options. As with any philosophical approach, pragmatism has its faults and sites of contestation, but it falls to those carrying on the tradition to make corrections and answer the objections. We are *the pragmatists* now.

Pragmatism is the name of a philosophical movement founded in America at the dawn of the 20th century by Charles Peirce, William James, and John Dewey. Though developed by its founders as primarily a thesis concerning the meaning of terms and sentences, it quickly expanded to encompass views about truth, knowledge, experience, and value. Throughout the 20th century, pragmatism developed even further, eventually coming to be associated with distinctive claims regarding science, logic, nature, mind, philosophical method, and the limits of philosophy itself. As things stand today, "pragmatism" remains a widely used term, and several different views identify as pragmatist. One wonders, however, whether pragmatism today is anything other than a fashionable, but nonetheless meaningless, label. For a philosophical movement founded in an essay titled "How to Make Our Ideas Clear," this possibility is unlovely. Still, this grim possibility is a consequence of the forward-facing part of the Janus principle.

A regrettable reaction is foreseeable. A tempting response to the seemingly irreducible variety of uses of the term has been that of *retrenchment*. Retrenchers have devoted thousands of pages to the claim that *pragmatism proper* saw its culmination in the philosophies of Peirce, James, and Dewey. Pragmatism, the retrenchers contend, is strictly the set of philosophical ideas found in those formative thinkers' works and in the works of those who devote themselves to interpreting and restating what they said. Retrenchment's companion is a historical story according to which pragmatism proper was abandoned or "eclipsed" in the mid-20th century by the philosophical idiom of analytic philosophy that prevails to this day in the English-speaking world. This *eclipse narrative* allows for the easy dismissal of post-Deweyan philosophy in America as decidedly non-pragmatist. Accordingly, as the account concludes, the renewed interest in Dewey stimulated by Richard Rorty's work in the early 1980s and carried forward by the program broadly termed "analytic pragmatism" is fundamentally flawed, because it is rooted in a distorted version of Dewey's pragmatism. Those endorsing the tactic of retrenchment claim the term "pragmatism" for the "classical" pragmatists alone.

There are obvious difficulties with retrenchment. First is that the so-called "classical" pragmatists were divided over fundamental philosophical issues. That clear set of commitments is, in fact, a fractious collection. Second, the mainstream "analytic" philosophical idiom that the retrenchers identify as the force that pushed pragmatism to the margin of the profession is replete with theses, arguments, and ideas that have their source in (and are explicitly attributed to) classical pragmatist writings. With a clear view of the matter, one sees that the tradition of pragmatism in America extends continuously from the so-called "classical" pragmatists through much of the mid-century and contemporary work that the retrenchers identify as "analytic" (and as such non-pragmatist). That is, once we acknowledge the deep-rooted philosophical disputes among the "classical" figures, the pragmatist family drastically expands to include, in addition to Peirce, James, and Dewey, many of the giants of 20th-century philosophy in America: C. I. Lewis, Nelson Goodman, Wilfrid Sellars, W. V. O. Quine, Hilary Putnam, Donald Davidson, and Richard Rorty. And then we will also see pragmatism as a *continuing* philosophical movement being carried forward in the work of Susan Haack, Cornel West, Philip Kitcher, Christopher Hookway, Nicholas Rescher, Cheryl Misak, and Huw Price – a collection of philosophers who draw on a common idiom but in the end are not in agreement over central philosophical issues.

This volume proceeds from the premise that pragmatism has always been a motley collection of broadly empiricist, humanist, and naturalist views. The approach those views inform does not determine the results, since the program is at heart a commitment to properly run inquiry. More importantly, pragmatism has never been a doctrine or dogma but rather the site of the *development* of certain strands of naturalistic, empiricist, and humanistic investigation. Accordingly, today there are distinctively pragmatist views within each major area of philosophy. In any specific case, the precisely pragmatist element and the philosophical connection to ideas and arguments of classical pragmatism are up for grabs.

We have sought to produce a state-of-the-art report on the current arrangement of pragmatist theses in the most research active fields within philosophy. That is, unlike many of the current collections and readers on pragmatism, this volume is not organized around the central historical figures and their ideas. Of course, there are orienting chapters on the central figures (Part I). However, the lion's share of this Companion looks beyond the exegesis of the founding texts and distinctive theses of pragmatism. The chapters in Part II explore the historical trends affecting pragmatism's development and that of its competing philosophical traditions. Parts III and IV, respectively, feature chapters that examine the most important topical areas in contemporary philosophy and explore the content, nature, and prospects of the distinctively pragmatist theses at work within them.

In editing this volume, we took up the challenge of producing a document proving that pragmatism is a living tradition rather than a relic. To the extent that these chapters supply readers with questions, challenges, objections, and dilemmas concerning what pragmatism is, and what within it remains viable, we will have succeeded in our Janus-faced enterprise.

PART I
Pragmatism's History and Figures

1
THE METAPHYSICAL CLUB

Cheryl Misak

Pragmatism's Origin Story

A central insight of the tradition of pragmatism is that any domain of inquiry and belief—science, mathematics, ethics—is *human* inquiry and belief. Our philosophical concepts of truth, knowledge, justice, and so on should start with that insight, not with high metaphysics, a supernatural God, an all-powerful sovereign, or some other absolute ideal. Pragmatists, that is, turn away from what they argue is a fruitless quest to secure a grounding for our beliefs. The best of the pragmatists (in my view) argue that there are nonetheless objective standards. The fact that truth is in some sense relative to humans need not entail that "anything goes". Belief must be responsive to experience, which provides an external check, even if we can say nothing about that externality without bringing to it our concepts and interpretations. One of the two primary founders of pragmatism, Charles Sanders Peirce, put forward this position. A belief is a habit of expectation which is conditioned by the force of experience. A true belief is one that would fit with all experience and argument if we were to inquire as far as we could on the matter. The other primary founder of pragmatism, William James, sometimes agreed with Peirce and, at other times, seemed to suggest that truth is what works here and now, for some believer.[1]

Pragmatism has its roots in a short-lived reading group in Cambridge (Massachusetts) in 1872, called the Metaphysical Club. Louis Menand nicely captured the personalities, if not the philosophy, in his Pulitzer Prize winning *The Metaphysical Club* (2001). The club's participants included some who would become America's finest intellectuals—Peirce and James, as well as Oliver Wendell Holmes Jr. Peirce described the club in a colorful way:

> It was in the earliest seventies that a knot of us young men in Old Cambridge, calling ourselves, half- ironically, half-defiantly, "The Metaphysical Club,"—for agnosticism was then riding its high horse, and was frowning superbly upon all metaphysics—used to meet, sometimes in my study, sometimes in that of William James. It may be that some of our old-time confederates would today not care to have such wild-oats-sowings made public, though there was nothing but boiled oats, milk, and sugar in the mess. Mr. Justice Holmes, however, will not, I believe, take it ill that we are proud to remember his membership ... Nicholas St. John Green was one of the most interested fellows, a skillful lawyer and a learned one, ... Chauncey Wright, something of a philosophical celebrity in those days, was never absent from our meetings. I was about to call him our corypheus; but he will better be described as our boxing-master whom we—I particularly—used to face to be severely pummelled ...

> Wright, James, and I were men of science, rather scrutinizing the doctrines of the metaphysicians on their scientific side than regarding them as very momentous spiritually. The type of our thought was decidedly British. I, alone of our number, had come upon the threshing-floor of philosophy through the doorway of Kant, and even my ideas were acquiring the English accent.
>
> (CP 5. 12; 1907)

Shortly after this burst of fruitful philosophical discussion, Peirce would publish his two founding documents of pragmatism: the 1877 "The Fixation of Belief" and the 1878 "How to Make Our Ideas Clear". In these papers, the pillars of pragmatism are constructed. We must start our inquiries and our philosophy where we find ourselves, laden with a body of belief which we do not doubt. Those beliefs must be responsive to experience; otherwise they are not genuine beliefs or candidates for truth. When the force of experience upsets a belief, inquiry is sparked until we reach a settled belief. If a belief would be forever settled because it met future experience perfectly, then there is no truth higher than that.

During the same period, James was grappling with his own version of pragmatism. In 1875, he produced a draft of a paper that would be published as "The Will to Believe". In the draft, he argued that, given the dearth of evidence for or against the existence of God, if believing in God makes me happier, then I have a duty to believe in God: "any one to whom it makes a practical difference (whether of motive to action or mental peace) is in duty bound to ... it" (1987 [1875]: 293). James would improve on this position. But in 1875, we have a clear picture of the two general approaches of pragmatism still with us today. On the one hand, we have Peirce's idea that truth is what would best fit with all the experience the community of inquirers pushed into the indefinite future. On the other, we have James's idea that truth is what would best fit with an individual's experience.

The members of Metaphysical Club tended to be a self-destructive bunch. Chauncy Wright and Nicholas St. Green would meet their deaths early, after years of overdrinking and under-sleeping. Peirce managed to alienate himself from employers, so much so that, after a brief stint teaching at Johns Hopkins University, he was locked out of academia. James had periods of depression throughout his life. Holmes seems to be the only one who had robust mental health, despite a terrible civil war.

These were the men (alas, no women) of the Metaphysical Club who founded America's home-grown philosophical school. If we unpack Peirce's origin story, we find the seeds for the subsequent evolution of the various branches of pragmatism.

British Empiricism and Kantian Transcendentalism

When Peirce says, "The type of our thought was decidedly British", he means that pragmatism is aligned with the empiricism of John Locke, George Berkeley, Alexander Bain, and David Hume. The basic idea of empiricism is that our beliefs must be grounded in, verified by, or responsive to experience. Each term—grounded, verified, responsive—gives rise to a different kind of empiricism. Reductionism has it that our beliefs must be traced by and grounded in experience. Verificationism has it that all meaningful beliefs must be such that some experience can show them to be true or false.

Pragmatism is a version of empiricism that holds the connection between belief and experience to be less strict—a belief must be responsive to experience. It also takes from Bain the idea that belief and action are intimately connected. Peirce credits St. John Green with urging upon the Metaphysical Club Bain's definition of belief as that upon which we are prepared to act. This dispositional account of belief was to be a mainstay of Peirce's pragmatism, and it was picked up by one of the strongest pragmatists of the next generation, Frank Ramsey in Cambridge, England, who then figured out how to measure and assess partial belief (Ramsey 1926).

As Peirce suggests, it would be a mistake to think of pragmatism simply as a brand of empiricism. He knew Kant's *Critique of Reason* off by heart and was influenced by Kant's idea that there are some regulative assumptions of our practices that we must assume if we are to continue with those practices. For instance, we need to assume there is a truth of the matter to the question at hand; otherwise it would make no sense to inquire into it. But Peirce was clear that needing to assume something does not confer upon the assumption the status of necessity. Peirce wanted to naturalize Kant—do away with transcendental deductions and move to talking about the requirements of belief and inquiry. Peirce was "not one of those transcendental apothecaries ... —they are so skilful in making up a bill—who call for a quantity of big admissions, as indispensible Voraussetzungen of logic" (CP 2. 113; 1902). Clarence Irving Lewis, who taught a famous Kant course for decades at Harvard, was the pragmatist who followed Peirce in entering the threshing floor of philosophy through the doorway of Kant. His 1929 *Mind and the World Order* describes our body of beliefs as forming a pyramid, with the most comprehensive, such as those of analytic definition and logic, at the top, and singular judgments at the bottom. Everything in our body of knowledge is connected in some way to experience through this network of beliefs, and everything is revisable. If that sounds like Quine, it is because it is. Quine was Lewis's student.

William James was not tempted by even a naturalized Kant, and his pragmatism remained more resolutely empiricist. Since much will be written about James and Peirce in this volume, I will confine myself to the lesser known, but nonetheless important, founders of pragmatism—St. John Green, Wright, and Holmes.

Nicholas St. John Green

Green earned a B.A. from Harvard in 1851 and a law degree in 1861. At the time of the Metaphysical Club meetings, he was a law professor at Boston University. He died in 1876, shortly after the Metaphysical Club ceased operations. He is thus mostly remembered, when he is remembered at all, for Peirce crediting him with bringing Bain's dispositional account of belief to pragmatism.

That was indeed his most influential contribution. But it is important to note that Green also put forward a pragmatist account of causation. In 1870, he published "Proximate and Remote Cause" in the *American Law Review*, which was edited by Holmes. In it, he argued that there is no single cause or chain of causation for an event. There are always multiple causes:

> From every point of view from which we look at the facts, a new cause appears. In as many different ways as we view an effect, so many different causes ... can we find for it. The true, the entire, cause is none of these separate causes taken singly, but all of them taken together. ... There is no chain of causation consisting of determinate links ranged in order of proximity to the effect. They are rather mutually interwoven with themselves and the effect, as the meshes of a net are interwoven.
>
> (1870: 13)

We choose or pick out from this mesh of causes what we call the proximate cause. Our choice is based on our needs and interests.

Green offers us not a metaphysical account of causation, in which there are objective causes in the world, but a practical, agent-centered, pragmatist account:

> In this view of causation there is nothing mysterious. Common people conduct their affairs by it, and die without having found it beyond their comprehension. When the law has to do

with abstract theological belief, it will be time to speculate as to what abstract mystery there may be in causation; but as long as its concern is confined to practical matters it is useless to inquire for mysteries which exist in no other sense than the sense in which every thing is a mystery.

(1870: 13)

Green came to meetings of the Metaphysical Club already a pragmatist.

Chauncey Wright

Like Green, Wright was older than the others in the Metaphysical Club and died soon after it wound up.

After his Harvard B.A., Wright became a "computer" for the American Nautical Almanac. His heart was not in it. As Simon Newcomb, the eminent mathematician and astronomer who was a colleague at the Almanac, put it, he had:

> an abominable habit of doing his whole year's work in three or four months, during which period he would work during the greater part of the night as well as of the day, eat little, and keep up his strength by smoking. The rest of the year he was a typical philosopher of the ancient world, talking ... but never writing.
>
> *(Thayer 1971: 70)*

After devising calculating shortcuts that enabled him to dispense quickly with his Almanac work, Wright would do what he really loved—talking philosophy. He died in 1875, when he was only 45, his habits catching up to him.

He, like Green, was a huge influence on the other emerging pragmatists. He brought an evolutionary perspective to pragmatism. He was a correspondent of Darwin, who was impressed by Wright's scientific ability. We have seen Peirce calling him his boxing master. He also boxed James. Appalled by James's idea that we have a duty to believe what makes our lives go better, he lay in wait for an opportunity to have what he thought was a much-needed "duel" with his friend:

> I have carried out my purpose of giving Dr. James the two lectures I had in store for him. I found him just returned home on Wednesday evening. His father remarked in the course of talk, that he had not found any typographical errors in William's article. ... I said that I had read it with interest and had not noticed any typographical errors. The emphasis attracted the youth's attention, and made him demand an explanation, which was my premeditated discourse. ... He fought vigorously, not to say manfully; but confessed to having written under irritation ... On Friday evening I saw him again and introduced the subject of the "duty of belief" as advocated by him in the Nation. He retracted the word "duty". All that he meant to say was that it is foolish not to believe, or try to believe, if one is happier for believing. But even so he seemed to me to be more epicurean (though he hates the sect) than even the utilitarians would allow to be wise ... He quite agrees that evidence is all that enforces the obligation of belief, and that it does this only in virtue of its own force as evidence. Belief is only a matter of choice, and therefore of moral duty, so far as attending to evidence is a volitional act; and he agreed that attention to all accessible evidence was the only duty involved in belief.
>
> *(Madden 1963: 45)*

Wright died not long after this exchange, having convinced James to alter his position. When "The Will to Believe" was finally published 20 years later, he argued that one has a right to believe ahead of the evidence, not a duty. Wright also influenced Holmes, who said of him:

> Chauncey Wright a nearly forgotten philosopher of real merit, taught me when young that I must not say necessary about the universe, that we don't know whether anything is necessary or not. I believe that we can bet on the behavior of the universe in its contact with us. So I describe myself as a *bet*tabilitarian.
>
> *(Howe 1961: 252)*

Oliver Wendell Holmes Jr.

What Wendell Holmes learned from Chauncey Wright is a perfect encapsulation of pragmatism. There is no mysterious necessity in the universe. Rather, we human beings find and bet that certain things are necessary. We meet the future with our beliefs about what is necessary and the future will let us know whether we are right or not.

Holmes went on to become one of America's most famous Supreme Court justices and legal theorists. He did not identify himself with the pragmatist movement, perhaps because he was not engaged with philosophers but with English legal scholars and German legal scientists. Nonetheless, one can detect a strong current of pragmatism running through his thought. The ideas hammered out in the Metaphysical Club animated the view of law he developed during the late 1800s, a view that did not undergo significant revision during the rest of his long and illustrious career.

Holmes was interested in the common law—as Frederic Kellogg puts it—the bottom-up theory of law (2007: 19). He described the common law as starting with cases "and only after a series of determinations on the same subject matter" does it come, "by … induction to state the principle" (1995 [1870]: 213). Law is not something that is set in statue or in immutable moral truths. It is an evolving enterprise of inquiry that starts from precedent (from where we are) and then is driven by experience, conflict, and unanticipated problems: "The life of the law has not been logic: it has been experience" or "the felt necessities of the time" (1882: 1). The law does not consist of a fixed body of doctrines and syllogisms, but rather, it is an organic structure that has come together in response to experience. Holmes says that all theories that consider the law "only from its formal side" are failures (1882: 36–37). For, whatever code or set of principles might be adopted, "[n]ew cases will arise which will elude the most carefully constructed formula" and will have to be reconciled (1995 [1870]: 213).

Holmes's theory of law is an expression of Peirce's idea that belief is stable until prompted by some positive doubt, which then sparks inquiry. We can also detect Peirce's account of truth in Holmes: if law were to reach a stage where conflicts no longer arise, then its aim would be fulfilled. But he does not treat the attainment of this aim as a likely possibility. Its pursuit is supported by what Peirce would call a hope. Courts, for Holmes, provide the venue for inquiry and experience in law. They are engaged in a fallible search for the best answer we can come to, given the time in which we are living and the circumstances in which we find ourselves. Again, that is as good an expression of the pulse of pragmatism that one could hope for.

Note

1 See Misak (2013) for this argument and for the foundation of much of what follows.

References

Green, Nicholas St. John (1870). "Proximate and Remote Cause." *American Law Review*, 4/201.

Holmes, Oliver Wendell Jr. (1882). *The Common Law*. London: MacMillan.

——— (1995 [1870]). "Codes, and the Arrangement of Law." In *The Collected Works of Justice Holmes: Complete Public Writings and Selected Judicial Opinions of Oliver Wendell Holmes*. 3 vols. Vol i. Ed. Sheldon M. Novick. Chicago: University of Chicago Press, 212–221.

Howe, Mark DeWolfe (1961). *Holmes–Pollock Letters: The Correspondence of Mr. Justice Holmes and Sir Frederick Pollock, 1874–1932*. Cambridge: Belknap Press.

Kellogg, Frederic R. (2007). *Oliver Wendell Holmes, Jr.: Legal Theory, and Judicial Restraint*. Cambridge: Cambridge University Press.

Lewis, Clarence Irving ([1956]1929). *Mind and the World Order: Outline of a Theory of Knowledge*. New York: Charles Scribner's Sons.

Madden, Edward (1963). *Chauncey Wright and the Foundations of Pragmatism*. Seattle: University of Washington Press.

Menand, Louis (2001). *The Metaphysical Club: A Story of Ideas in America*. New York: Farrar, Straus, and Giroux.

Misak, Cheryl (2013). *The American Pragmatists*. Oxford: Oxford University Press.

Peirce, Charles Sanders (1931–1966). *Collected Papers of Charles Sanders Peirce*. 8 vols. Eds. Charles Hartshorne and Paul Weiss (vols. I–VI), A. Burks (vols. VII–VIII). Cambridge: Belknap Press. (References to this work are in standard form: (CP vol. number: paragraph number.)

Ramsey, Frank Plumpton (1926). "Truth and Probability" in 1990. *Philosophical Papers*. Ed. D. H. Mellor. Cambridge: Cambridge University Press, 52–95.

Thayer, James Bradley (1971 [1878]). *Letters of Chauncey Wright: With Some Account of His Life*. New York: Lennox Hill.

2
C.S. PEIRCE'S PRAGMATISM

Albert Atkin

Introduction

Charles Sanders Peirce (1839–1914) (pronounced "purse") was an enigmatic and polymathic philosopher, credited by his friend and fellow philosopher William James as the founding figure of pragmatism. Peirce was born into a well-connected intellectual family in Cambridge, Massachusetts – his father was the famous Harvard mathematician Benjamin Peirce, his maternal grandfather was a state Senator, and his older brother James followed their father into a Mathematics Professorship at Harvard. His home was also frequently visited by leading intellectuals in the arts and sciences, and this intellectual environment encouraged Peirce to develop and explore a series of philosophical and scientific interests which occupied him for the rest of his life.

Peirce only held an academic position briefly, at Johns Hopkins University, between 1879 and 1884, but worked for leading scientific institutions such as the Harvard Observatory and the US Coastal and Geodetic Survey. Early in his career, he published widely in philosophy, mathematical logic, semiotics, chemistry, and photometric research, and it was during this time that he published the work for which he is now most famous – his account of pragmatism and the pragmatic elucidation of truth.

Peirce was, by all accounts, an irascible figure, and he frequently courted controversy. A difficult divorce from his first wife, a series of poor financial investments, and acrimonious departures from his posts at Johns Hopkins, the Geodetic Survey, and the Harvard Observatory meant that Peirce increasingly retreated into isolation at his home in rural Pennsylvania, and although he continued to write on topics, including pragmatism, his published works largely petered out. At the time of his death from bowel cancer in 1914, he left around 80,000 pages of unpublished work.

Clear and Distinct Ideas

Peirce's pragmatism is most distinct in that it is intended as a maxim for clarifying concepts for the purpose of scientific and philosophical inquiry rather than as a broader statement of philosophical methodology. Of course, Peirce's early work on his pragmatic maxim influenced William James and F.C.S. Schiller to develop pragmatist philosophies of their own, but for Peirce, pragmatism was primarily a tool for clarifying meanings. It is helpful then to understand just how Peirce came to see pragmatism this way.

In a series of five papers published in 1877 and 1878 in the *Popular Science Monthly*, we find the earliest clear statements of Peirce's pragmatic maxim. For Peirce, it is important for the proper conduct of science and inquiry that we have a full and clear grasp of the concepts we use. Adopting some Cartesian and Rationalist terminology, Peirce notes that we can have a "clear and distinct" apprehension of any given concept or thing. Clearness, which Peirce calls the first grade of apprehension, is an unreflective familiarity such that "acquaintance with the idea" leads us "to have lost all hesitancy in recognizing it in ordinary cases" (EP1: 124–125 [1878]). Distinctness, or the second grade of apprehension, is to understand a concept such that we can "give a precise definition of it, in abstract terms" (EP1: 125 [1878]). So, for instance, I might clearly demarcate foodstuff from non-foodstuff in my daily life and ordinary interactions with the world yet be unable to provide an abstract definition of "food". This suggests that we have a clear, but not distinct, understanding of the concept. If, however, I can give some additional abstract definition of "food" – perhaps mentioning broad nutritional classes and the notion of consumption for sustenance or some such – then I could be rightly said to have both a clear and a distinct apprehension of the concept of "food".

The notions of clearness and distinctness as modes of apprehension, however, for Peirce, are inadequate. Clearness may very well involve a lack of hesitation or reflection, but it seems obvious that I can engage with some concept without hesitation even though that concept may well be meaningless. For instance, I may refuse to pick up a knife I have dropped or decline to accept scissors when they are directly passed to me, but this unreflective, non-hesitant behavior doesn't suggest there is any meaning to my behavior beyond simple superstition. Moreover, I can meet the requirements of "distinctness" by providing an abstract definition without this being a guarantee of meaningfulness either. We can, for example, give precise definitions for absurd objects, mythical creatures, and empty references, but this doesn't show that we have a deeper understanding or clearer comprehension than if we had simply failed to engage with an empty concept at all. All of which brings us to the crux of Peirce's pragmatism and, if standard histories are to be believed, the origins of American pragmatism.

The Pragmatic Maxim

For Peirce, what is needed to address the inadequacies of clearness and distinctness is an additional, third grade of clearness, which later became known as the "pragmatic maxim":

> Consider what effects, which might conceivably have practical bearings, we conceive the object of our conception to have. Then, our conception of these effects is the whole of our conception of the object.
>
> *(EP1: 132 [1878])*

Whilst this now-famous paragraph is widely considered to be the core statement of Peirce's pragmatism, the maxim we find here is worth exploring further to get a clear sense of what Peirce's pragmatism amounts to. In particular, we can see it as containing two elements: a pragmatic account of meaning or conceptual "elucidation" and a pragmatic guide to meaningful inquiry. We'll explore these in order.

To see how the maxim gives an account of meaning, we can turn to one of Peirce's own examples. Directly after introducing the maxim, Peirce discusses what we mean when we call an object "hard". We might have an unreflective understanding of hardness – we use a hammer when we need something hard to strike a tack. We might also have some rudimentary definition for what we mean by calling a thing hard. But for us to have a full understanding of the concept, we must be able to say what the "conceived effects" would be – what should we expect to experience from our practical interactions with "hard" objects?

These expectations and practical interactions are often characterized as a list of conditional statements (Atkin 2016: Ch 2; Misak 1991: Ch 1) that exhaust our understanding of any given concept. For "hard", we might expect the list of classificatory conditionals to include such statements as if X is hard, then X will not be easily scratched; if X is hard, then X will be highly resistant to surface indentation; and so on. For any concept, on Peirce's third grade of clearness, then, an exhaustive list of conditional statements such as these, that capture our practical experiences with a concept, gives us the complete picture of that concept.

As an account of pragmatic meaning or elucidation, we should note that there are two elements to the picture that Peirce suggests here. First, the pragmatic maxim, as an account of meaning, focuses on understanding. As should be clear, for Peirce, the meaning of any concept is only fully realized when we have an elucidation of what it would mean for us to understand the experiential effects of holding that concept to be the case. It is, then, through our understanding as inquirers and users of concepts that we come to a full pragmatic elucidation and thus a complete account of meaning. As Cheryl Misak notes (1991: 3 fn2), this bears similarities to the Manifestationist accounts of meaning found in Michael Dummett's work, and at its core, meaning is explained in the behaviors, expectations, and understanding of concept users – "Our idea of anything is our idea of its sensible effects" (EP1: 132 [1878]).

The second and related strand of Peirce's account of pragmatic meaning is that it focuses on use. Our understanding of the sensible effects of any given concept is manifest in our use and application of those concepts to our practical interactions with the world around us. When we compile our lists of expected sensible effects for, say, the concept "hard", we are not simply asking questions about the concept in abstract. Instead, we are applying that concept and developing our pragmatic sense of sensible effects by asking of particular objects if the concept applies to them. Is this diamond hard? Would it resist scratching or surface indentation? Would it fulfill a purpose I have for it in some experimental or even mundane domestic role? These are questions, answers, and sensible experiential outcomes obtained by using these concepts in our practical interactions with the world.

In terms of understanding Peirce's pragmatic maxim as a tool for meaningful inquiry, we see some relation to the theory of meaning implicit in his account. For Peirce, by providing a third grade of clearness, the pragmatic maxim can play an important role in delineating fruitful inquiries from empty pursuits and in keeping us honest about our intellectual pursuits. In particular, Peirce suggests that the pragmatic maxim can help us to do at least three things in our inquiries: first, to see where apparent disputes are merely illusory and so block fruitful inquiry; second, to see where the object of inquiry is in fact empty or beyond useful investigation; and third, to show where we are guilty of pursuing unnecessary metaphysical categories and entities.

In the first instance, merely apparent disputes come about where opposing sides on any given debate fancy there to be conceptual differences between them which, when subject to pragmatic elucidation, turn out to be empty. Peirce's famed example here is disagreement between Catholics and Protestants about the sacraments.

> It is foolish for Catholics and Protestants to fancy themselves in disagreement about the elements of the sacrament, if they agree in regard to all their sensible effects, here or hereafter.
> *(EP1: 132 [1878])*

Under Peirce's maxim, Catholics and Protestants would compile the same list of sensible effects and expectations from the taking of sacrament. Protestants take bread and wine to be symbolic of the body and blood of Christ; Catholics take them to be literally the flesh and blood of Christ, yet both agree that the experiential sensible effects of taking sacrament are to taste bread and wine.

From Peirce's position, whether Protestants argue that this is because of the symbolism of the sacrament and Catholics argue that the sacrament retains the "accidental" qualities of bread and wine during transubstantiation is irrelevant – in pragmatic terms there exists no difference between them. Their debate is empty.

The second useful element of the pragmatic maxim as a tool of inquiry is that it shows us where claims and positions are empty. For Peirce, for instance, the emptiness of the debate between Catholics and Protestants about the sacrament comes about because of the pragmatic emptiness of the Catholic notion of transubstantiation – if the sensible effects of taking bread and wine before and after transubstantiation are pragmatically indistinguishable, then the concept of transubstantiation makes no practical difference at all and so is empty. For Peirce, many pursuits in what he calls "ontological" (or a priori) metaphysics are shown to be similarly empty in pragmatic terms:

> It [pragmatism] will serve to show that almost every proposition of ontological metaphysics is either meaningless gibberish, – one word being defined by other words, and they by still others, without any real conception ever being reached, – [...] all such rubbish being swept away, what will remain of philosophy will be a series of problems capable of investigation by the observational methods of the true sciences[.]
>
> *(EP2: 338 [1905])*

For Peirce, then, by showing where concepts and theories lack practical and observable consequences, the pragmatic maxim gives us the means for delineating meaningful inquiry and philosophy from pseudo-problems.

The final function that Peirce sees for the pragmatic maxim as a tool of inquiry is that it places a clear "ontological" stop on the types of question that we are inclined to pursue. Again, one of the examples that Peirce uses to illustrate his third grade of clearness is instructive here. Peirce asks, what do we mean by "force"? His own elaboration is lengthy and circumlocutious, but he settles on the idea that we conceive of force in terms of our experience of change in motion. The crucial point though is that judging by Peirce's pragmatic maxim, this is the whole of our conception, and our inquiries are bounded by the answers to what conceivable practical and sensible effects we can expect to fall under the concept of force. If we ask any further questions about what force is, we are misleading ourselves:

> Whether we ought to say that a force is an acceleration, or that which causes an acceleration, is a mere question of propriety of language [...] Yet it is surprising to see how this simple affair has muddled men's minds. In how many profound treatises is not force spoken of as a "mysterious entity", which seems to be only a way of confessing that the author despairs of ever getting a clear notion of what the word means!
>
> *(EP1: 136 [1878])*

Peirce's pragmatic maxim, then, is a tool for clarifying the meaning of the terms we use – it gives us a third and final grade of clarity. Further, it serves as a guide to meaningful and worthwhile inquiry – it highlights empty debate, demarcates meaningful inquiry from pseudo-problems, and places a positivist boundary on our explication of concepts.

The Development of Pragmatism

So far, we have seen Peirce's seminal statement of pragmatism as a maxim of inquiry from the late 1870s. What is notable though is that at no point in these early statements did Peirce ever call his account "pragmatism". This title was not to arrive until William James, in an 1898 address to Berkeley University, named his own philosophy "pragmatism" and identified Peirce's 1878 maxim as its source. Peirce was, for the most part, ready to adopt the label and noted that he had presented a paper to the Metaphysical Club in the early 1870s, "expressing some of the opinions I have been urging all along under the name pragmatism" (EP2: 400 [1907]). Moreover, the similarity between James' canonical statements of his own pragmatism and Peirce's maxim is striking. For instance:

> To attain perfect clearness in our thoughts of an object, then, we need only consider what effects of a conceivably practical kind the object may involve – what sensations we are to expect from it, and what reactions we must prepare. Our conception of these effects, then, is for us the whole of our conception of the object, so far as that conception has positive significance at all.
>
> *(James 1907/1975: 29)*

But the surface similarities here hide important and significant differences between the pragmatism of Peirce and James. We shall not engage in any kind of exhaustive analysis here, but there are some notable and instructive divergences in the two views of pragmatism that arise from these positions. First, James' interest in such a maxim is geared strongly towards individual reaction and conduct to the sensations that arise from our thought and interaction with an object. This may seem like a simple matter of emphasis, but the impacts are interesting. For instance, James' own take on the question of transubstantiation is enlightening here:

> Since the accidents of the wafer don't change in the Lord's supper, and yet it has become the very body of Christ, […] a tremendous difference has been made, no less a one than this, that we who take the sacrament, now feed upon the very substance of divinity.
>
> *(James 1907/1975: 46–47)*

For Peirce, the sensible effects of a pragmatic elucidation do not raise above the empirical sense experiences of eating sacraments. For James, the individual psychological and behavioral experiences count as part of the effects. This changes the resultant account of meaning (and of course truth) and even shifts what counts as meaningful inquiry. For Peirce, scientific and empirical observations aide in marking the boundaries of proper philosophical inquiry – this would rule out much of religion, metaphysics, and even ethics. For James, however, there are effects and sensations to be had in these domains, and thus, much of religion, metaphysics, and ethics would count as domains of worthwhile philosophical inquiry. Indeed, we can see here that, for Peirce, the pragmatic maxim is simply a tool for clarifying philosophical inquiry in the manner of science, but that, for James, pragmatism becomes a methodological approach for approaching philosophical questions at the precise point where scientific observation gives out.

The move towards pragmatism as a methodology that expands our approach to philosophical questions that we find emerging in James' work here was not lost on Peirce. He was, of course, encouraged by the opportunity to reengage with questions of pragmatism and mark out his own contribution clearly, but he was nonetheless adamant that in James' (and others) hands, pragmatism had run beyond his original aspirations for the maxim. He criticized other pragmatists for acting as though they were:

> carrying about them the master key to all the secrets of metaphysics. … I make pragmatism to be a mere maxim of logic instead of a sublime principle of speculative philosophy.
>
> *(EP2: 134 [1903])*

Peirce made numerous attempts to restate and refashion his pragmatism in light of work from fellow pragmatists and to deal with what he felt were shortcomings in his earlier work (see Atkin 2016: Ch 2; Hookway 2004). However, even in these later statements, some written only a year before his death, Peirce still maintained that the pragmatic maxim of 1878 was, in its essentials, the correct reading of his pragmatism.

References

Atkin, Albert. (2016). *Peirce*. Abingdon: Routledge.
Hookway, Christopher. (2004). "The Principle of Pragmatism: Peirce's Formulations and Examples." *Midwest Studies in Philosophy* 28 (1): 119–136.
James, William. (1907/1975). *Pragmatism: A New Name for Some Old Ways of Thinking*. Cambridge, MA: Harvard University Press.
Misak, Cheryl. (1991). *Truth and the End of Inquiry: A Peircian Account of Truth*. Oxford: Clarendon Press.

3
WILLIAM JAMES

Alexander Klein

Imagine a photograph, taken straight on from above, of a human left hand resting palm down on a table. We might think of the photograph as representing the hand in virtue of *resembling* it. In the image, the index finger appears shorter than the middle finger, and it appears to the left of the thumb. The hand appears to be on top of the table. All the various relations in the image between the fingers, between the hand and the table—left-right, long-short, over-under—match the corresponding relations between the real objects, from the perspective of the camera that originally shot the image. Perhaps the colors match as well.

Now consider the idea of the same hand in the photographer's mind. Her idea of the hand consists (let us suppose) of a visual image of the hand as pictured from the same perspective as the camera. We might think that the idea also represents the hand in virtue of resembling it, much in the way the photograph might be thought to represent the real hand in virtue of resembling it.

Alternatively, we might think the idea was *generated* through a neuro-psychological process that is somehow akin to the way light reflected off the hand causes an irradiation pattern in the film or CCD and that the idea represents the hand in virtue of having the right kind of causal history. More recently, so-called causal theorists like Dretske and Fodor (Dretske 1981; Fodor 1987) have made just this kind of suggestion that a mental state like the one in my example represents the hand not in virtue of resembling the hand but in virtue of having been appropriately caused by it. *Resembling* and being causally *generated* by an object are not mutually exclusive, of course, and older copy theorists like Hume long incorporated causal generation into their view as well (Garrett 2006, 307).

At the center of William James's pragmatism was a conception of mental representation that staunchly denied both copy- and object-generation style theories of the sorts sketched above. For James, ideas do not represent in virtue of either resembling or having been caused by their objects. Instead, James developed a voluntaristic, forward-looking approach that explained representation in terms of causal consequences. In this chapter, I will offer a brief examination of James's approach to mental representation, explaining what makes his view "voluntaristic," what makes it "forward-looking," and how it sets the tenor for his general approach to philosophy.

By "mental representation," I shall mean the familiar phenomenon of a mental state's *picking out* or *being about* some other object, whether that object is physical or mental. And I take it a *theory* of mental representation (or of "meaning," as I shall also say) should identify the feature or property of mental states in virtue of which those states count as representations (whenever they do so count).[1] James's theory, very roughly, says that an idea represents an object in virtue of affording an ability to navigate to the object and, so to speak, to transact one's business with the object.

DOI: 10.4324/9781315149592-5

The view is voluntaristic in that mental representation requires *having* some "business" to transact with an object in the first place, and just what business one might have with an object is a matter that is hatched—sometimes, but not always—by the agent herself. We shall therefore see that his psychological work on willing helps bring the philosophical view about meaning into focus.

James first achieved intellectual celebrity for his work in the nascent field of empirical psychology, especially for a series of pioneering essays that culminated in his classic 1890 *Principles of Psychology*. He styled psychology as a marriage between physiology (a field in which he had had extensive training) and philosophy.[2] During the subsequent decade, he turned his attention to what he called "popular philosophy," evidently making a play on the more familiar idea of "popular science" practiced by people like Ernst Mach and especially Herbert Spencer and his legion of followers.[3] A central interest explored throughout James's philosophical output was the question of whether science and religion are compatible, and when he began discussing pragmatism publicly (first in an 1898 lecture at UC Berkeley called "Philosophical Conceptions and Practical Results"), he expressly sought to apply pragmatism to this problem.[4]

By James's own description, a "pivotal" part of his pragmatism was his account of "truth" (James 1909/1978, 3), and the account of truth itself is but a short step from his account of representation. He says his "first statement" (James 1909/1978, 4) of the account of truth had been given in an 1884 essay, "The Function of Cognition," and that essay in fact devotes considerable attention to representation as well.

It is natural to assume that *resemblance* is somehow involved with representation, especially if we think of ideas as mental images, as per the example in my opening paragraphs. But "The Function of Cognition" raises objections that suggest that resemblance at least cannot be *sufficient* for representation. Much like (but also much before) more recent causal accounts of content (such as Kripke 1972; Stampe 1977), James called attention to the symmetric nature of resemblance, which appears to be at odds with the asymmetric nature of representation. James offered the example of two toothaches that may "resemble each other, but do not on that account represent, stand for, or know each other" (James 1909/1978, 21). Contrast this with the photograph in my example. If the photograph resembles the hand, the hand must, in turn, resemble the photograph, but since the hand doesn't *represent* the photograph, resemblance cannot be sufficient for representation. One can run the same argument for the relationship between the hand and the visual idea of the hand.

So we have one reason for doubting at least the sufficiency of resemblance for establishing representation. What about object generation? Why not think that ideas represent objects in virtue of having been caused by them?

More recently, causal generation has been regarded as a factor that can help address the symmetry problem with copy theories discussed above. In a classic paper, Stampe offers the example of a photograph of one of a pair of identical twins (Stampe 1977). The photograph resembles each twin equally, we may suppose, and yet only represents one twin. Why? Stampe's answer helped inspire a generation of causal theorists: the image was created (caused) by light that bounced off one twin rather than the other, and the twin that is appropriately causally connected with the image is the twin represented by the image (and mutatis mutandis for an idea of a twin).

James recognized this very problem.

> Suppose, instead of one q [the object of some particular thought], a number of real q's in the field. If the gun shoots and hits, we can easily see which one of them it hits. But how can we distinguish which one the feeling knows?
>
> *(James 1909/1978, 21)*

Instead of appealing to the thought's causal *history* to answer this question, James proposed that an idea is in fact quite like a gun. Just as the gun shows what reality it's pointing at "by breaking it," the idea refers to some reality in virtue of conferring a "power of interfering" with it (James 1909/1978, 21–22). Like the bullet, what matters is where the idea *leads*.

Isn't it simpler or more intuitive to think of representation in terms of where an idea *comes from*? One factor that might explain why James wasn't more tempted by appeals to causal history is his general approach to scientific reasoning, which was broadly inspired by Comtean positivism (Klein 2015; Pearce 2015). Comte was an early advocate of *hypothetical* reasoning (Laudan 1981, ch.9). On this approach, our scientific theories need not arise inductively from data we have collected in the past—they can stem from *hypotheses*, or educated guesses about nature, and a good guess can come from a creative flash as much as from a natural extension of what we have already observed. In other words, the epistemic value of a hypothesis is not derived from what *generated* it but rather from whether or not it is borne out by future experience. Here is one way James put the point. As empiricists,

> [o]ur great difference from the scholastic lies in the way we face. The strength of his system lies in the principles, the origin, the *terminus a quo* of his thought; for us the strength is in the outcome, the upshot, the *terminus ad quem*. Not where it comes from but what it leads to is to decide. It matters not to an empiricist from what quarter an hypothesis may come to him: he may have acquired it by fair means or by foul; passion may have whispered or accident suggested it; but if the total drift of thinking continues to confirm it, that is what he means by its being true.
>
> *(James 1897/1979, 24)*

Strikingly, for James, empiricists do *not* find either the meaning or justification of every idea in past experience. He views empiricism as an attitude central to the sciences (James 1897/1979, 21), one that says that we can attain truths by reasoning hypothetically, by postulating new ideas, and then checking how well they conform to *future* experience. Like scientific hypotheses, our everyday ideas do not gain legitimacy from their source, according to Jamesian empiricism—they get their legitimacy from where and how well they lead.[5]

To be sure, in the above passage James is making an epistemological rather than a semantic point. But statements of his similarly forward-looking semantic commitments are plentiful. For example:

> the same thought may be clad in different words; but if the different words suggest no different conduct, they are mere outer accretions, and have no part in the thought's meaning. If, however, they determine conduct differently, they are essential elements of the significance. "Please open the door," and "*Veuillez ouvrir la porte*," in French, mean just the same thing; but "D—n you, open the door," although in English, *means* something very different. Thus to develop a thought's meaning we need only determine what conduct it is fitted to produce: that conduct is for us its sole significance.
>
> *(James 1898/1975, 259)*

This is one of James's early statements of the so-called pragmatic maxim.[6] As he understands it, the maxim says that two thoughts are different if and only if they are "fitted" to produce different "conduct." James attributes his awareness of this principle to an article by his friend Charles Sanders Peirce (Peirce 1878). But he had also independently developed a sophisticated psychological account of just *how* thoughts produce conduct—quasi-mechanically—in his *Principles of Psychology* (James 1890/1981). I will examine the "how" of it, below. First we should ask this question: in virtue of *what* do two sequences of bodily motion constitute different "conducts"?

The answer is that like a bullet, conduct has a *target*, and so James individuates conduct on the basis of the goal to which it is directed. We can see this in another classic illustration of his forward-looking account of representation:

> Suppose me to be sitting here in my library at Cambridge, at ten minutes' walk from "Memorial Hall," and to be thinking truly of the latter object. My mind may have before it only the name, or it may have a clear image, or it may have a very dim image of the hall, but such an intrinsic difference in the image makes no difference in its cognitive function. Certain *extrinsic* phenomena, special experiences of conjunction, are what impart to the image, be it what it may, its knowing office.
>
> For instance, if you ask me what hall I mean by my image, and I can tell you nothing; or if I fail to point or lead you towards the Harvard Delta; or if, being led by you, I am uncertain whether the Hall I see be what I had in mind or not; you would rightly deny that I had "meant" that particular hall at all, even tho my mental image might to some degree have resembled it. The resemblance would count in that case as coincidental merely, for all sorts of things of a kind resemble one another in this world without being held for that reason to take cognizance of one another.
>
> On the other hand, if I can lead you to the hall, and tell you of its history and present uses; if in its presence I feel my idea, however imperfect it may have been, to have led hither and to be now *terminated*; if the associates of the image and of the felt hall run parallel, so that each term of the one context corresponds serially, as I walk, with an answering term of the other; why then my soul was prophetic, and my idea must be, and by common consent would be, called cognizant of reality.
>
> That percept was what I *meant*, for into it my idea has passed by conjunctive experiences of sameness and <u>fulfilled intention</u>. Nowhere is there jar, but every later moment continues and corroborates an earlier one.
>
> *(MT 1904, 62–63, italics original, underline added)*

Notice that James presents the Memorial Hall example as illustrating an instance of "knowing," or of being "cognizant of a reality." And he seems to use these two phrases as roughly synonymous with "thinking truly" so that the example in effect concerns both what the thought of Memorial Hall "meant" (that's the "thinking" part) and whether or not *what* was meant was accurate (that's the "truly" part). It is not entirely clear how to disambiguate the semantic and epistemic considerations at play here, but James apparently sees meaning (and truth) as grounded in a thought's capacity to help the agent *navigate to* and *talk about* an object in ways that "fulfill" an "intention," and to do so in ways that the agent's peers would accept. The intention specifies, in effect, a goal. When I am thinking of Memorial Hall, I am thinking of it as a target of some purposive action sequence, for James. For example, I might be thinking of it as the target of a physical action sequence, perhaps as a place I would like to go, sit, and work on campus. Or it might be the target of an intellectual sequence, perhaps as a focus of discussion on the history of department locations at Harvard.

We can distill James's view this way. A thought successfully represents, he thinks, just in case the thought would aid in fulfilling the intended action sequence through the actual, contingent environment (physical or intellectual) in a manner that would satisfy not only me but also my rational peers.

Notice from this passage that James is an externalist about mental representation in that what a thought means is not purely contained in the thought itself but depends on the environment in which the thought is embedded. So, on offer is a special kind of causal theory, where ideas represent in virtue of aiding in achieving a goal in the actual, contingent environment.

One standard problem traditional causal theories of representation (i.e., object-generation theories) have faced is how *misrepresentation* is possible.[7] My idea of water may be triggered both by real water and by mirages, but we don't want to say that my idea of water means either water or mirages. So the backwards causal theorist needs a way to disambiguate representation-conferring from accidental causes.

Though James's causal account is forward-looking, he might be thought to face a similar problem, one that Wittgenstein once raised against Bertrand Russell (in a related context).[8] Suppose we understand an object that a mental representation picks out as whatever would terminate my navigation or intervention action. A punch in the nose (Wittgenstein notes) might terminate my navigation just as much as actually finding Memorial Hall. How can a forward-looking account avoid this problem?

James has theoretical resources, found in his psychological work on will, for addressing this problem. A brief overview of these resources should give readers a sense of how James might account for misrepresentation and also how distinctively voluntaristic James's general approach to mental representation is.

On the view sketched above, an idea is about Memorial Hall just in case it supports an intentional *action* of a certain kind—namely, the intentional navigation to or intervention with the relevant object. At their most fundamental level, actions involve two components, for James. First, the agent hatches a goal for herself. In the paradigmatic case, hatching a goal means framing what he calls an "anticipatory image"—an idea of what it will have felt like to make some bodily movement (James 1890/1981, 1111–1112).[9] For example, an archer might hatch the goal of shooting an arrow at a target by thinking of what it will have felt like to have performed the relevant motions. This anticipatory feeling is essentially a goal representation (and different sense modalities can be involved in such representations). I submit that in the more philosophical passages quoted above, where James is discussing an "intention" I seek to "fulfill," as when I intend to navigate to Memorial Hall, the "intentions" are just goal representations in the sense articulated in his earlier psychological work.

The second component of intentional action—in the paradigmatic case of physical actions—is a sequence of muscular contractions that are naturally (as an evolutionary-physiological matter; see Klein Forthcoming) caused by the conscious awareness of the goal representation. For example, when I think of raising my hand, I am thinking of what it will feel like to perform this action, and James argues that this thought will naturally tend to trigger the arm-raising motion, provided no rival thoughts occur (such as the thought that I am at an auction and do not want to place a bid). This, in a small nutshell, is James's ideo-motor theory of action.[10]

Now suppose I have an idea of the coffee cup on my desk. My idea is about the coffee cup in virtue of affording me aid in performing a relevant goal-directed action (perhaps grabbing the cup and taking a swig). I need not actually perform the action for my idea to represent the cup—what matters is that the cup *affords* help in performing the action *were* I to perform it (James 1909/1978, 63). Thus, not just any old (actual or possible) interaction with an object will establish reference—*my interaction must be in accord with my goal representation*.

In the paradigmatic case—representation of external objects[11]—the world about which I am thinking is in effect populated by loci of practical activities. The cup I am thinking about is the thing from which I can sip my coffee, and the table is the place I can rest the receptacle.[12] I call this approach "voluntaristic" because goal representations are often endogenously generated. I decide what business I have to transact with the coffee cup, and thus the very content of mental representations depends in part on my own subjective interests.

The Jamesian solution to the problem of misrepresentation can quickly be illustrated by contrasting the case of my veridical idea of the cup on my desk with my hallucinatory idea of water in the desert. In the latter case, my idea *misrepresents* because, when I try to transact my business with the mirage, my aim (slaking my thirst) is thwarted. This explains the phrase I underlined in the Memorial Hall passage, above—the notion that reference involves an actual or possible "fulfilled intention."

Finally, one should emphasize that there is a social dimension to reference, as James conceives it. While I may be the most important agent who gets to decide what business I might *legitimately* have with a represented object, I am not the *only* agent who has a say, on James's view. Refer again to the Memorial Hall passage—one requirement of having a thought that successfully refers, in his example, is that I can converse coherently *with others* about this or that aspect of the Hall ("if *you* ask me what hall I mean…"). Thus, if I am so severely dehydrated that I hallucinate a feeling of satisfaction when imbibing the imaginary water in the desert, James can deny that I have had an idea of water on grounds that onlooking "critics"[13] would not agree.[14] James ultimately articulated a forward-looking, voluntaristic form of pragmatism. In this chapter I have sketched the voluntaristic, forward-looking account of mental representation on which his pragmatism was built.

Notes

1 My use of "representation" loosely follows Cummins (1989, 12), who uses the term "mental content" in a similar way. I prefer "representation" because James came (late in his career) to be wary of the distinction between consciousness and content (Klein 2020) and because James uses the term "representation" himself (e.g. at James 1909/1978, 21).
2 See the letter to Charles Eliot, December 2, 1875 (CWJ 4.527), where James specifically positioned himself professionally as someone who could develop psychology at Harvard by uniting these "two 'disciplines'"—physiology and philosophy—"in one man." I discuss James's physiological background in Chapter 1 of Klein (Forthcoming).
3 For a discussion on the impact of Spencerian popular science on various pragmatists, see Pearce (2020, ch. 3). On Mach's dedication of one volume of his own popular scientific writings to James, see Stadler (2017).
4 For more on James's pragmatist approach to religion, see Bush (2020) and Klein (2019).
5 An account from cognitive science that draws on James's forward-looking empiricism about representation is Buckner (Forthcoming). My own reading of James in this paper is stimulated by Buckner's account, with which I am broadly sympathetic.
6 For a helpful discussion of James's distinctive take on this maxim, see Jackman (2022).
7 Fodor calls this the "disjunction problem" (Fodor 1987, 101–102). For a discussion, see Cummins (1989, 40, 57). Buckner also discusses ways in which a James-style, forward-looking account can skirt the disjunction problem in Buckner (Forthcoming).
8 Though he always rejected the pragmatist theory of truth, by the 1921 *Analysis of Mind*, Russell would adopt something like James's philosophy of "pure experience" (which Russell called "neutral monism"). Wittgenstein had objected to Russell's neutral-monist account of the content of a desire: "if I wanted to eat an apple and someone punches me in the stomach so that I lose my appetite, then the punch was the thing I originally wanted." Russell himself had made a similar objection to James's account of representation, years earlier; for a discussion, and for the Wittgenstein quotation, see (Griffin 2015, 11).
9 The term more frequently used in psychology today is "response image."
10 James's ideo-motor theory of action has a fascinating history, and it has come in for a revival in contemporary cognitive science; see Stock and Stock (2004) and Shin et al. (2010). James introduces the term "ideo-motor action" at James (1890/1981, 1130).
11 Since James often treats the representation of external, physical objects as paradigmatic, readers may wonder whether and how such an account can be extended to more abstract ideas such as those at play in mathematics or logic. For an ingenious solution to this problem on James's behalf, see Jackman (2020).
12 The resemblance to Gibson's later ecological psychology is unlikely to be an accident—for James's influence on Gibson (primarily via James's student E. B. Holt), see Heft (2001, 2002).
13 James gave an ineliminable, theoretical role to an idealized "critic" even in his earliest articulation of this forward-looking account of mental representation; see James (1909/1978, 16).
14 For helpful critical feedback on an early draft of this paper, I thank Henry Jackman.

References

Buckner, Cameron. Forthcoming. "A Forward-Looking Theory of Content." *Ergo*.
Bush, Stephen. 2020. "James and Religion." In *The Oxford Handbook of William James*, edited by Alexander Klein. New York: Oxford University Press. doi:10.1093/oxfordhb/9780199395699.013.31.

Cummins, Robert. 1989. *Meaning and Mental Representation*. Cambridge: MIT Press.
Dretske, Fred I. 1981. *Knowledge and the Flow of Information*. Cambridge: MIT/Bradford Press.
Fodor, Jerry A. 1987. *Psychosemantics: The Problem of Meaning in the Philosophy of Mind, Explorations in Cognitive Science*. Cambridge: MIT Press.
Garrett, Don. 2006. "Hume's Naturalistic Theory of Representation." *Synthese* 152 (3):301–319.
Griffin, Nicholas. 2015. "Russell's Neutral Monist Theory of Desire." *Russell: The Journal of Bertrand Russell Studies* n. s. 35 (1):5–28.
Heft, Harry. 2001. *Ecological Psychology in Context: James Gibson, Roger Barker, and the Legacy of William James's Radical Empiricism*. Mahwah, NJ: L. Erlbaum.
———. 2002. "Restoring Naturalism to James's Epistemology: A Belated Reply to Miller & Bode." *Transactions of the Charles S. Peirce Society* 38 (4):559–580.
Jackman, Henry. 2020. "James, Intentionality, and Analysis." In *The Oxford Handbook of William James*, edited by Alexander Klein. New York: Oxford Univesity Press. doi:10.1093/oxfordhb/9780199395699.013.13.
———. 2022. "James's Pragmatic Maxim and the 'Elasticity' of Meaning". In *The Jamesean Mind*, edited by Sarin Marchetti. Abingdon, 274–283. Oxon: Routledge.
James, William. 1890/1981. *The Principles of Psychology*. Edited by Frederick H. Burkhardt, Fredson Bowers and Ignas K. Skrupskelis, *The Works of William James*. Cambridge: Harvard University Press.
———. 1897/1979. *The Will to Believe, and Other Essays in Popular Philosophy*. Edited by Frederick H. Burkhardt, Fredson Bowers and Ignas K. Skrupskelis, *The Works of William James*. Cambridge: Harvard University Press.
———. 1898/1975. "Philosophical Conceptions and Practical Results." In *Pragmatism*, edited by Fredson Bowers and Ignas K. Skrupskelis, 257–271. Cambridge: Harvard University Press.
———. 1909/1978. *The Meaning of Truth*. Edited by Frederick H. Burkhardt, Fredson Bowers and Ignas K. Skrupskelis, *The Works of William James*. Cambridge, MA: Harvard University Press.
Klein, Alexander. 2015. "Science, Religion, and 'the Will to Believe.'" *HOPOS: The Journal for the International Society for the History of Philosophy of Science* 5 (1):72–117.
———. 2019. "Between Anarchism and Suicide: On William James's Religious Therapy." *Philosophers' Imprint* 19 (32):1–18.
———. 2020. "The Death of Consciousness? James's Case against Psychological Unobservables." *Journal of the History of Philosophy* 58 (2):293–323.
———. Forthcoming. *Consciousness Is Motor: Warp and Weft in William James*. New York: Oxford University Press.
Kripke, Saul A. 1972. *Naming and Necessity*. Cambridge, MA: Harvard University Press.
Laudan, Larry. 1981. *Science and Hypothesis: Historical Essays on Scientific Methodology*. Dordrecht: D. Reidel.
Pearce, Trevor. 2015. "'Science Organized': Positivism and the Metaphysical Club, 1865–1875." *Journal of the History of Ideas* 76 (3):441–465.
———. 2020. *Pragmatism's Evolution: Organism and Environment in American Philosophy*. Chicago, IL: University of Chicago Press.
Peirce, Charles Sanders. 1878. "How to Make Our Ideas Clear." *Popular Science Monthly* 12 (January):286–302.
Shin, Yun Kyoung, Robert W. Proctor, and E. J. Capaldi. 2010. "A Review of Contemporary Ideomotor Theory." *Psychological Bulletin* 136 (6):943–974.
Stadler, Friedrich. 2017. "Ernst Mach and Pragmatism — the Case of Mach's Popular Scientific Lectures (1895)." In *Logical Empiricism and Pragmatism*, edited by Sami Pihlström, Friedrich Stadler and Niels Weidtmann, 3–14. Cham, Switzerland: Springer International.
Stampe, Dennis W. 1977. "Toward a Causal Theory of Linguistic Representation." *Midwest Studies in Philosophy* 2 (1):42–63.
Stock, Armin, and Claudia Stock. 2004. "A Short History of Ideo-Motor Action." *Psychological Research* 68 (2–3):176–188.

4
JOHN DEWEY

David L. Hildebrand

Introduction

John Dewey (1859–1952) was a founder of American pragmatism, a prominent American intellectual in the first half of the 20th century, and an educator whose ideas have had global reach.

Dewey was a systematic thinker, and his views on human nature, conduct, and experience informed his work in logic, ethics, epistemology, metaphysics, aesthetics, and religion. Though he was not keen on labels, he preferred "cultural naturalism" over "pragmatism" or "instrumentalism." In broad strokes, Dewey's aim was metaphilosophical as he sought to criticize and reconstruct philosophy to render it consistent with a Darwinian worldview committed to contemporary life and its problems (Dewey 1916a: 3). Here, Dewey was following James' repudiation of philosophy's overly technical and intellectualistic character. While Dewey's work was an approach typical of philosophy—working at more general removes than most sciences—he explicitly directed it toward socially critical work, including helping to connect different theoretical areas with one another. (Dewey refers to this role for philosophers as "liaison officers") (Dewey 1925: 306).

Typical of other classical pragmatists, Dewey criticizes historic dualisms. He places an arguably greater emphasis upon reconstruction. For example, he criticizes the typical "mind/body" dichotomy and suggests an alternate account called, in *Experience and Nature*, "body-mind." Such dualism-busting work is evident in many of his titles—e.g. *School and Society, Child and Curriculum*, and *Democracy and Education*. The saliency of Dewey's critical work on dualisms is most evident in the enormous range of theorists *outside* of philosophy who continue to use his work to escape their own field's constrictive dualisms. Dewey's critiques have been utilized in areas concerned with art, education, environment, journalism, politics, public administration, medicine, psychiatry, sociology, and more.

Biographical

Dewey was born October 20, 1859 in Burlington, Vermont to Archibald Dewey and Lucina Rich Dewey. Growing up in Burlington, Dewey was raised in the Congregationalist Church and attended public school and the University of Vermont (ages 15–19). In 1882, after teaching high school for two years in Oil City, PA, Dewey enrolled in Johns Hopkins' new graduate program in philosophy. He graduated in two years, studying logic with Charles S. Peirce, history of philosophy with George Sylvester Morris, and psychology (physiological and experimental) with Granville Stanley Hall. Dewey's dissertation (now lost) criticized Kant from an Idealist standpoint ("The Psychology of Kant").

Dewey taught and led departments at a number of universities (shorter periods at the University of Michigan and the University of Minnesota, a decade at the University of Chicago, and the last 26 years at Columbia University). Dewey married Harriet Alice Chipman in 1886; they had six children and adopted another. Two children died young. Chipman significantly influenced Dewey's advocacy for women, as did Jane Addams.[1]

Dewey's career spanned six decades and produced an extraordinary body of work: 40 books and approximately 700 articles in over 140 journals. Dewey wrote for many different audiences, within and outside academia. He was active politically, helping to lead, support, or found organizations such as the American Civil Liberties Union, the American Association of University Professors, the American Philosophical Association, the American Psychological Association, and the New School for Social Research. He traveled widely, including Turkey, Mexico, the Soviet Union, and South Africa; he traveled and lived abroad in Japan and China for two years.

In 1946, almost two decades after his first wife's death (1927), Dewey married Roberta Lowitz Grant; they adopted two children. He died of pneumonia at home in New York City on June 1, 1952.

Intellectual Influences: Hegel, Darwin, and James

Dewey read widely and his intellectual influences were complex; I will make just a few comments to help orient readers.

First, while Peirce was one of Dewey's professors in graduate school, there was no significant influence at the time. (Years later, after reading Peirce, Dewey attributed important credit to his influence.) Dewey's main early influences were Darwinian biology, Wundtian experimental psychology, and Neo-Hegelian idealism. These views raised for Dewey a question as to whether the world was fundamentally creative and spiritual or biological and material; his attempt to reconcile these motivated his early efforts. He saw in both sides the idea of "organism" and responded by trying to merge experimental psychology with idealism as a "new psychology."

Hegel, and Hegelianism, was another important influence on Dewey, in both philosophical and personal ways. Dewey writes,

> Hegel's synthesis of subject and object, matter and spirit, the divine and the human, was, however, no mere intellectual formula; it operated as an immense release, a liberation. Hegel's treatment of human culture, of institutions and the arts, involved the same dissolution of hard-and-fast dividing walls, and had a special attraction for me.
>
> *(Dewey 1930: 153)*

While the degree to which Dewey's work was enduringly "Hegelian" is debated, it is clear that his early enthusiasm for Hegelian views is reflected in lifelong attempts to integrate experience's various dimensions (bodily and psychical, practical and theoretical, etc.) rather than keep them separate.

Darwin profoundly influenced Dewey in many ways. Most generally, Dewey imbibed a Weltanschauung, a view of reality as a natural complex of changing, transactional processes, none aiming at fixed ends. Such framing allows Dewey to reconceive the traditional account of experience; no longer a detached mental "stuff," experience comes to signify the doing and undergoing of organisms in environments, the "functions and habits, of active adjustments and readjustments, of coordinations and activities, rather than of states of consciousness" (Dewey 1910: 5). Dewey develops this conception of experience throughout his career, and it had a profound impact upon every area of his philosophy.

Along with Darwin, William James' influence on Dewey was gargantuan. James' *Principles of Psychology* (1890) showed how psychology might account for consciousness unified and intelligent character without appealing to anything transcendent. Dewey credited James' work with giving his thinking "a new direction and quality" (Dewey 1930: 157). Indeed, it was James' rejection of old account of mind (as discrete, elementary states) for a new one (as a biologically based "stream of consciousness") that, Dewey said, "worked its way more and more into all my ideas and acted as a ferment to transform old beliefs" (Dewey 1930: 157). Additionally, James' approach to empiricism ("radical empiricism") taught Dewey that appeals to absolutes never inform practical action; such guidance actually comes from a "study of the deficiencies, irregularities and possibilities of the actual situation" (Dewey 1922: 199). Part of this actual situation, for philosophers, is their starting point—why and how they, as individuals, choose and approach the philosophical issues they do. James' focus on "temperament" in his 1907 *Pragmatism* germinates in Dewey's genealogical approach to understanding various historical philosophies.

Pragmatism

Dewey seldom labeled his view "pragmatism," preferring "cultural naturalism," "humanism," and "experimentalism." He spent little energy dickering over labels; instead, he spent his time arguing that meanings (of objects, concepts, terms, propositions, and theories) can have direct and existential consequences in inquiry and action (of some kind or another). Dewey makes this especially clear in "What Pragmatism Means by Practical" (Dewey 1908b); there, he distinguishes the meaning of an *object* from the meaning of an *idea*. He cites, with approval, Peirce's view that the meaning of an *object* is to be found in its conceptual content and whatever practical effects the object requires of us (or commits us to, logically). But some assessments of meaning start, instead, with an *idea* rather than the object. Dewey explains this difference, writing,

> Pragmatism will, of course, look to future consequences, but they will clearly be of a different sort when we start from an idea as idea, than when we start from an object. For what an idea as idea means, is precisely that an object is *not* given. The pragmatic procedure here is to set the idea "at work within the stream of experience."
>
> *(Dewey 1908a: 103)*

What is important to note, here, is how committed Dewey is to the existential (even "realistic") consequences, which flow from the meaning of ideas, as "ideas are essentially intentions (plans and methods), and that what they, as ideas, ultimately intend is *prospective*—certain changes in prior existing things" (Dewey 1908a: 99). Dewey is not disagreeing with James but fore-fronting James' emphasis upon what he calls the "concrete." Dewey does this by invoking Peirce, writing,

> But all of these ideas are colored and transformed by the dominant influence of experimental science: the method of treating conceptions, theories, etc., as working hypotheses, as directors for certain experiments and experimental observations. Pragmatism as attitude represents what Mr. Peirce has happily termed the "laboratory habit of mind" extended into every area where inquiry may fruitfully be carried on.
>
> *(Dewey 1908a: 100)*

That last point—about carrying pragmatism into "every area"—limns another signal element of pragmatism. Too often, Dewey argues, what is "practical" is associated with the "merely" personal, with phenomena dismissively subjective. But this undermines the importance of knowledge as the *interactive conduct* of sentient creatures with environments—to make a difference. Making a difference in this real and overt sense is to escape that "species of confirmed intellectual lock-jaw called epistemology" (Dewey 1908b: 140, n. 6) and instead seek "relevancy and bearing in the generating ideas of its contemporary present" (Dewey 1908b: 143). Here, again, Dewey hews close to Peirce's direction for pragmatism to only take up "real and living doubt"—doubts which, as Dewey put it, comprise the "actual crises of life" (Dewey 1917: 43).

Another misconstrual often made by critics (of pragmatists' emphasis upon the practical) is to read it as the abject *subordination* of thought to action. This, too, is unfair. While pragmatists sought to divorce philosophy from scholasticism to better redress problems of everyday life, they understood that some problems may, at present, be largely conceptual, long-term, or both. The problem of staffing a hospital and that of figuring out why a basement is flooded both require concepts and actions; pragmatism take no stance on what proportion of conception-to-action is correct, overall. But in either case, the proof of the process will emerge in the resolution of the problem.

As with Peirce and James before him, inquiry for Dewey is always situated; it emerges from the needs and goals of certain people, at certain times and places. No inquiry is ever motivated by pure wonder, nor is it ever done in a vacuum. Every inquiry draws upon certain histories (comprising previous inquiries, events, etc.) and is done by actual inquirers with whichever attributes (cultural, personal, emotional, bodily) they possess at that time. Recognition of some of these factors can help inquirers strive for a more "neutral" perspective—in other words, one which seeks to counterweigh some of the biases implicit in the conditions of inquiry—but notice: this is a *strategic* move, meant to improve the results. It is not seeking, in Hilary Putnam's phrase, a "view from nowhere" or pure objectivity. Those goals, pragmatists argue, are bankrupt. "So far as the question of the relation of the self to known objects is concerned," Dewey writes, "knowing is but one special case of the agent–patient, of the behaver–enjoyer–sufferer situation" (Dewey 1910: 120).

Contextualized by situations, pragmatic tests of meaning become relevant to inquiry; a successful inquiry is one which works, where "works" means the ability to use its solutions, conclusions, and judgments to transform the initial problematic situation into one which is satisfactory, however provisionally. Satisfactions with staying power render their methods as "true," which neopragmatist Richard Rorty called a "rhetorical pat on the back."

In sum, the most important takeaway for Dewey's pragmatism is that it orients philosophy toward inquiry able to navigate *us* toward the *future*. Knowing is not spectatorship of what "really" exists; it is the tool of clever critters to predict and shape future consequences. "Knowing," Dewey wrote, "is literally something which we do; that analysis is ultimately physical and active; that meanings in their logical quality are standpoints, attitudes, and methods of behaving toward fact, and that active experimentation is essential to verification" (Dewey 1917: 367). This tool should "be applied as widely as possible; and to things as diverse as controversies, beliefs, truths, ideas, and objects" (Dewey 1908a: 101).

Beyond Empiricism, Rationalism, and Kant

Because Dewey's pragmatism characterizes the various elements of knowing (ideas, theories, telescopes, laboratories) in terms of their function, it is best understood as beyond or beside the older dualistic traditions of realism/idealism or empiricism/rationalism. Very crudely, classical empiricists argued that knowledge arises from sensory experience, whereas rationalists claimed that the certainty endemic to knowledge could only come from sources neither material nor merely probable: the mind.

Kant, of course, sought to split the difference by refusing an exclusively original place to either percepts or concepts; for him, mind and world are together determinative of what could be. Kant called his strategy a "Copernican Revolution." Dewey rejects this characterization, arguing that Kant "edited a new version of old conceptions about mind and its activities in knowing, rather than evolved a brand new theory" (Dewey 1929: 232). Pragmatism (or experimentalism) is far more revolutionary because in it,

> an idea in experiment is tentative, conditional, not fixed and rigorously determinative. It controls an action to be performed, but the consequences of the operation determine the worth of the directive idea; the directive idea does not fix the nature of the object.
>
> *(Dewey 1929: 231)*

Summarizing a key difference between pragmatism and Kant, Dewey writes that unlike the contemporary experimental approach he is espousing, "There is nothing hypothetical or conditional about Kant's forms of perception and conception. They work uniformly and triumphantly; they need no differential testing by consequences" (Dewey 1929: 231).

Older divisions between mind and matter, concept and percept, are also undermined by Dewey's functionalist approach. In its logical processes, Dewey writes,

> the datum is not just external existence, and the idea mere psychical existence. Both are modes of existence—one of *given* existence, the other of *possible*, of inferred existence. ... In other words, datum and ideatum are divisions of labor, cooperative instrumentalities, for economical dealing with the problem of the maintenance of the integrity of experience.
>
> *(Dewey 1903: 339–340)*

Inquiry

Inquiry is so central to Dewey's pragmatism that despite other central tenets in his work—notably, experience—I devote the remainder of this essay to it. What is inquiry? Why did it become so important for his philosophy? And how does it inform nearly every other area of his philosophical work?

One must begin with Peirce and James. Dewey's "Some Stages of Logical Thought" (1900) and his collaborative volume, *Studies in Logical Theory* (1903), echo Peirce's well-known 1877–1878 papers on doubt and belief (Peirce 1877, 1878). There, Peirce located knowing in a biological and natural frame, arguing that reflective inquiry arises when habitual conduct is interrupted. Such interruptions Peirce calls "doubt," and their satisfactory resolution is "belief." We can see Dewey's debt to Peirce in his characterization of the development of knowledge as a "doubt-inquiry" function (or process) and his wider understanding that, for Peirce,

> habit on his view is first a cosmological matter and then is physiological and biotic—in a definitely existential sense. [H]abit ... operates in and through the human organism, but that very fact is to him convincing evidence that the organism is an integrated part of the world in which habits form and operate. ... In so many words he says "Logic is rooted in the social principle."
>
> *(Dewey 1946: 151)*

These general ideas were also, of course, in James' work as well; Dewey especially drew upon James' 1890 *Principles of Psychology* and his 1898 lecture, "Philosophical Conceptions and Practical Results" (James 1978). After Dewey's 1903 Chicago School volume was published, James' positive review gave Dewey great pleasure; in a letter, he wrote to James: "I have simply been rendering back in logical vocabulary what was already your own" (Dewey in Perry 1935: 526).

Early Work in Psychology

Dewey's earlier work in psychology was crucial to his theory of inquiry. His seminal paper "The Reflex Arc Concept in Psychology" (1896) both applauded and criticized the "reflex arc" model. On the one hand, the model innovated a way for the emerging school of physiological psychology to escape from mysterious "psychic entities" of then-regnant introspectionist psychology. However, Dewey found the reflex-arc model to be too flawed for uncritical acceptance, a "a patchwork of disjointed parts, a mechanical conjunction of unallied processes" (Dewey 1896: 97). Dewey's (now-celebrated) move was to portray individual acts as integrated in a wider, ongoing field of "sensori-motor coordinations," circuits in which environment and organism dynamically adjust and reconstitute. "What Dewey proposes," Thomas Alexander writes, "is to start with the idea of the organism already dynamically involved with the world and aiming toward unified activity" (Alexander 1987: 129).

This model of behavior (and, indeed, the entire organism-environment picture) applies to inquiry directly; inquiry is conduct, something we do. We do not inquire (or aspire to inquire, as Descartes had) from a detached conceptual standpoint, abstracted away from involvements in the world. Just as the child (in Dewey's example) interacts with the candle stimulus by *already* bringing a set of past experiences and present desires, the adult inquirer, too, comes upon a stimulus (or problem) replete with present engagements. This erases a common assumption that philosophical inquiry is categorically different from commonsense ones. *Any* inquiry starts in the midst of life, funded willy-nilly by concepts, tools, habits, etc.

Inquiry's Pattern

As a fundamental tool for human survival and growth, inquiry exists in innumerable venues and takes many forms. In works such as "Analysis of Reflective Thinking" (Dewey 1933) and *Logic: The Theory of Inquiry* (Dewey 1938), Dewey describes prevalent and recurrent patterns of inquiry. In contrast to much in the philosophical tradition, inquiry begins with (1) a unique *feeling* of something problematic; this quality pervades the coming inquiry and provides a directing focus. That initial feeling, however unique, is conceptually indeterminate and so (2) a *problem* must be carefully formulated; this formulation remains open to later revision. Next, (3) a preliminary *hypothesis* is constructed to organize relevant percepts and ideas and to integrate them for eventual experiments. Prior to experimentation, inquiry (4) imagines possible *implications* of the meanings involved and the possible consequences of the proposed experiment; this stage spots errors, contradictions, and potential obstacles. Finally, (5) there is *action*; the hypothesis is tested and evaluated in terms of how well it addresses the initial problem. Inadequate results may require returning and revising earlier phases of inquiry.

The pattern of inquiry described is schematic; Dewey drew it from exemplary forms of inquiry, such as those in the empirical sciences. Following this basic approach, he argues, could improve a wide variety of methods for solving problems, not least in fields of ethics and social science, as well as in philosophy.

Inquiry Applied: Education and Democracy

Author of numerous influential books and articles in education, Dewey was also active in many other ways. He taught high school before going to graduate school (in Oil City, PA), devised curricula, and created the famous Laboratory School at the University of Chicago, administering it with his wife, Alice. Dewey both developed his theory of inquiry from work in psychology and education and turned around and applied and tested his theory in education and philosophy.

(N.B. Dewey decried the disciplinary distinction of education and philosophy as artificial. See Dewey 1916b: 338.)

What does inquiry have to do with schooling? Schools are where the personalities, desires, needs, and character of local children are developed with education's (more) abstract subject matters; curricula are supposed to equip students to thrive as individuals and as contributors to their community and wider society. The problem is that society is constantly changing, and schools must adapt to change without becoming pawns of more powerful societal forces (e.g. industry). If, as Dewey believed, each child is individual—in background, personality, creativity, and motivations—then schools respecting that child have to balance their needs with the pressures from the wider society. Indeed, preparation of children for later life is in effect empowering them to *transform* society so *it* best suits their situations. "The life of all thought," he wrote, "is to effect a junction at some point of the new and the old, of deep-sunk customs and unconscious dispositions, that are brought to the light of attention by some conflict with newly emerging directions of activity" (LW3: 6). Put another way, "inquiry" is relevant not only as a method taught to students but also as something educational institutions *themselves* need to do to create autonomy for the school and the students.

Utilizing the theory of inquiry as articulated above—functional, directed by present needs, tested in future experience—Dewey developed a philosophy of education that navigated between fierce debates among educational "traditionalists" and "romantics." From the traditionalists, Dewey accepted the notion that there were inheritances (norms, histories, technical discoveries, values) which could and should be part of the curriculum; no inquiry can begin in a vacuum, and no culture simply begins anew with its youngest generation. From romantics (progressives), Dewey accepted the immense importance of the present child's perspective; inquiry must be innovative in its hypotheses and experiments, ever-sensitive to the specific inquirer's *felt* sense of the problem.

This work on inquiry helped propel Dewey toward a new model of pedagogy. Once the metaphor of the child as "empty cabinet" waiting to be filled with information is discarded and replaced by an understanding of children as sensitive and experimental agents, a new approach is possible. "The question of education," Dewey writes, "is the question of taking hold of [children's] activities, of giving them direction" (Dewey 1900: 25).[2]

Democracy

In concert with education, effective inquiry was at the heart of Dewey's vision of democracy. The connections should already be obvious. Our native disposition is to inquire, to seek out solutions to problems or to devise new paths toward aesthetic experiences. Schools help refine and extend these dispositions, empowering children toward greater comprehension of extant society, more effective criticism, and (hopefully) the courage to experiment and effect change as they see fit. Done right, schooling incorporates but does not indoctrinate the child into a wider social world. Thus, the school's challenge is also that of democratic society: how can the norms, institutions, laws, and machineries of democracy incorporate citizens while nevertheless remaining flexible—hypothetical, if you will—to future directions not yet determined? How can a plurality be made flexible, fallible, and experimental?

The answer is complicated, but here I will tie it back just to inquiry. Democracy is a form of life constituted by and conducive to experimental inquiry. "Democracy," Dewey writes, "is not an alternative to other principles of associated life [but] the idea of community life itself" (Dewey 1927: 328). This means that the "core" of democratic life is not *content*, some *specific* set of norms, institutions, procedures, etc., but is *how* we conduct inquiry to solve problems. Training of democratic habits begins in schools, which must also model a democratic ethos. Democratic habits endure insofar as they shape citizens' expectations about what government is for and how it should work; habits must also be enacted by citizens' own participation in inquiries.

Most important, the habits of democratic inquiry reinforce citizens' faith in self-rule, the notion that experience *itself* provides the ultimate test of experiments done freely. The test of experience must never be extinguished or subordinated to any personality, fiat, dogma, or tradition. As this is a fitting note on which to end this chapter, I'll give Dewey the last word:

> Democracy is the faith that the process of experience is more important than any special result attained, so that special results achieved are of ultimate value only as they are used to enrich and order the ongoing process. Since the process of experience is capable of being educative, faith in democracy is all one with faith in experience and education. All ends and values that are cut off from the ongoing process become arrests, fixations. They strive to fixate what has been gained instead of using it to open the road and point the way to new and better experiences.
>
> *(Dewey 1939: 229)*

Notes

1 Dewey and his biographer/daughter Jane Dewey credit Jane Addams with helping Dewey advance his thinking on education, democracy, and philosophy as well. The very significant influence Addams had on Dewey is still being uncovered. See Dewey (1939), Seigfried (1999), Fischer (2013).
2 Dewey only became more emphatic about this over the course of his life. Forty-four years later, he wrote:

> There will be almost a revolution in school education when study and learning are treated not as acquisition of what others know but as development of capital to be invested in eager alertness in observing and judging the conditions under which one lives. Yet until this happens, we shall be ill-prepared to deal with a world whose outstanding trait is change.
>
> (Dewey 1944: 463)

References

Readers interested in inquiry are advised to consult the following chapters within this volume: Peirce, James, Logic, and Truth.

Works by John Dewey

Citations to John Dewey's works are to the 37-volume critical edition *The Collected Works of John Dewey, 1882–1953*, edited by Jo Ann Boydston (Southern Illinois University Press, 1969–1991). The series includes:

Dewey, J. (1967) *The Early Works*, 1882–1898, 5 vols., J.A. Boydston (ed.), Carbondale: Southern Illinois University Press. (Abbreviated *EW*)

Dewey, J. (1976) *The Middle Works*, 1899–1924, 15 vols., J.A. Boydston (ed.), Carbondale: Southern Illinois University Press. (Abbreviated *MW*)

Dewey, J. (1981) *The Later Works*, 1925–1953, 17 vols., J.A. Boydston (ed.), Carbondale: Southern Illinois University Press. (Abbreviated *LW*)

Specific Works Cited Here

(1896) "The Reflex Arc Concept in Psychology," EW5.
(1900) "Some Stages of Logical Thought," MW1.
(1903) *Studies in Logical Theory*, MW2.
(1908a) "Does Reality Possess Practical Character?" MW4.
(1908b) "What Pragmatism Means by Practical," MW4.

(1910) "A Short Catechism Concerning Truth," MW6.
(1916) "Brief Studies in Realism," MW6.
(1916a) "The Influence of Darwin on Philosophy," MW4.
(1916b) *Democracy and Education*, MW9.
(1917) "The Need for a Recovery of Philosophy," MW10.
(1922) "The Good of Activity," MW14.
(1925) *Experience and Nature*, LW1.
(1927) *The Public and Its Problems*, LW2.
(1929) *The Quest for Certainty: A Study of the Relation of Knowledge and Action*, LW4.
(1930) "From Absolutism to Experimentalism," LW5.
(1933) "Analysis of Reflective Thinking," LW8.
(1938) *Logic: The Theory of Inquiry*, LW12.
(1939) "Creative Democracy: The Task Before Us," LW14.
(1944) "Between Two Worlds," LW17.
(1946) "Peirce's Theory of Linguistic Signs, Thought, and Meaning," LW15.

Works by Others

Alexander, T.M. (1987) *John Dewey's Theory of Art, Experience, and Nature: The Horizons of Feeling*, Albany: State University of New York Press.
Dewey, J.M. (1939) "Biography of John Dewey," in P.A. Schilpp (ed.), *The Philosophy of John Dewey*, New York: Tudor Publishing Co, 3–45.
Fischer, M. (2013) "Reading Dewey's Political Philosophy through Addams's Political Compromises," *American Catholic Philosophical Quarterly*, 87 (2): 227–243.
James, W. (1978/1898) "Philosophical Conceptions and Practical Results," in J. McDermott (ed.), *The Writings of William James: A Comprehensive Edition*, Chicago, IL: University of Chicago Press, 345–362.
James, W. (1979/1907) *Pragmatism*, Cambridge, MA: Harvard University Press.
James, W. (1981/1890) *The Principles of Psychology*, Cambridge, MA: Harvard University Press.
Peirce, C.S. (1992/1877) "How to Make our Ideas Clear," in N. Houser and C.J.W. Kloesel (eds.), *The Essential Peirce, Volume 1, Selected Philosophical Writings*, (1867–1893), Indianapolis: Indiana University Press, 124–141.
Peirce, C.S. (1992/1878) "The Fixation of Belief," in N. Houser and C.J.W. Kloesel (eds.), *The Essential Peirce, Volume 1, Selected Philosophical Writings*, (1867–1893), Indianapolis: Indiana University Press, 109–123.
Perry, R.B. (1935) *The Thought and Character of William James*, vol. 2, New York: Little, Brown and Co.
Seigfried, C.H. (1999) "Socializing Democracy: Jane Addams and John Dewey," *Philosophy of the Social Sciences*, 29 (2): 207–230.

Further Reading

Alexander, T.M. (1987) *John Dewey's Theory of Art, Experience, and Nature: The Horizons of Feeling*, Albany: State University of New York Press. (Exceptional work on Dewey's metaphysics and aesthetics, which makes an excellent case for why aesthetic experience is critical to understanding Dewey's project.)
Eldridge, M. (1998) *Transforming Experience: John Dewey's Cultural Instrumentalism*, Nashville: Vanderbilt University Press. (Focuses on Dewey's instrumentalism, with emphasis on how it applies to Dewey's social, political, and religious philosophies.)
Hickman, L.A. (1990) *John Dewey's Pragmatic Technology*, Indianapolis: Indiana University Press. (Original interpretation of Dewey's approach to pragmatism and inquiry utilizing the concept of technology and setting Dewey's project into that wider arena.)
Hildebrand, D.L. (2003) *Beyond Realism and Antirealism: John Dewey and the Neopragmatists*, Nashville: Vanderbilt University Press. (Extended analysis of Dewey's pragmatism in contrast both to the early debates between realists and idealists and as taken up by contemporary neopragmatists Richard Rorty and Hilary Putnam.)
Pappas, G. (2008) *John Dewey's Ethics: Democracy as Experience*, Indianapolis: Indiana University Press. (Connects Dewey's extensive moral philosophy with his theory of experience, explaining both the originality and power of Dewey's ethical thinking while contrasting it with historical and contemporary thinkers.)

5
JANE ADDAMS

Núria Sara Miras Boronat

Early years: education and social reform

Together with John Dewey (1859–1952), Jane Addams is probably the most well-known pragmatist philosopher of the progressive era. She was also a tireless activist and social reformer who contributed toward the foundation of several organizations in defense of civil rights, women's enfranchisement, pacifism, and trade unionism. She was awarded the Nobel Peace Prize in 1931 (which she shared with Nicholas Murray Butler),[1] after being nominated 91 times (Shields 2017: 3).[2] Despite her popularity and relevance, her philosophical work remains largely unknown for both philosophical traditions to which she contributed most: pragmatism and feminism. This lack of recognition might be due to the fact that the first generation of women pragmatists barely had access to college education (Seigfried 1996: 60). Addams did attend the Rockford Seminary, an educational institution for female students, and managed to write hundreds of articles and ten books. One of her most controversial pieces, "A Modern Lear" (1912), was declared to be "one of the greatest things I ever read, both as to its form and its ethical philosophy" (Westbrook 1991: 89) by her dearest friend John Dewey. Another of her closest friends, the philosopher George Herbert Mead (1863–1931), tried to grant Addams an honorary doctorate from the University of Chicago, but the petition was denied by the administration because she was a woman. Eventually, she received an honorary degree from Yale University in 1910, becoming the first woman to receive such an accolade from this university.

Addams's key role in founding the pragmatist school of thought has been rediscovered since the 1990s thanks to pragmatist feminist scholars (Deegan 1989; Seigfried 1996; Whipps & Lake 2016), who have reread the canon and reintegrated the contribution of other women of the progressive era to the pragmatist genealogy. This historical revision has also called into question the traditional division between men's theoretical work in sociology and women's involvement and social work as separate projects. In fact, women such as Mary Whiton Calkins (1863–1930), Elsie Ripley Clapp (1879–1965), and Mary Parker Follett (1868–1933) played an active role in the new advancements of the American philosophy and science.

Before Jane Addams became a civic heroine, she had to overcome many of the limitations that women experienced at that time. Born in 1868 in Cedarville, Illinois, she was the eighth child of the mill owner John Huy Addams and his wife, Sara. Jane tragically lost her mother when she was two years old, as Sara was fatally injured when she fell as she was running to help a neighborhood woman give birth. Jane was raised by her elder sister Mary until her father married another woman. John Addams was himself a philanthropist and had an extensive library. He did his best to give his children a good education and set them an example of moral conduct. Jane inherited her father's admiration for Abraham Lincoln, Shakespeare, Plutarch, Giuseppe Mazzini, and Charles Dickens. Jane dreamed of becoming a writer like Louisa May Alcott and her character Jo in *Little Women* (1868). She failed to get into the prestigious Smith College, of which her father had also not approved, considering it too far from home. She was accepted to study Medicine at Rockford and was an excellent student, but, after the unexpected death of her father, to whom she was very close, during a visit in Michigan, she fell into deep depression that lasted for almost two years. As she recovered and graduated, she got permission to travel to Europe with a college friend, Ellen Gates Starr (1859–1940). As Jane Addams explains in her autobiographical book *Twenty Years at Hull-House* (1910), the trip to Europe changed her forever. She finally had the chance to see Stonehenge, the cathedral of Notre Dame, the Acropolis, the Sistine Chapel, and all manner of historical wonders. However, it was during a bullfight in Madrid in 1888 that the very simple plan of the settlement, which she had long been forming in her mind, really took shape (Addams 1998: loc. 86). Being exposed to brutality made her realize that it was time to stop dreaming and turn her ideas into action.

Building community: Hull-House and the social settlement movement

During her travels in Europe, Jane had discussed at length with Ellen the possibility of starting a philanthropic project together. Addams took her inspiration from Toynbee Hall, one of the first social settlements founded by Samuel and Henrietta Barnett in East London in 1884. Jane was aware that the word "settlement" was likely to sound bad to American ears, as she recalls that the word still had connotations that imply

> migrating from one condition of life to another ... We are not willing, openly and professedly, to assume that American citizens are broken up into classes, even if we make that assumption the preface to a plea that the superior class has duties to the inferior.
> *(Addams 1998: loc. 51)*

Assuming their duty toward the most disadvantaged, Jane and Ellen moved to an old house in Halsted Street. The house was donated by Charles Hull, one of the first inhabitants of Chicago in the 19th century. The area was one of the poorest and most troubled neighborhoods in America. When Ellen and Jane opened Hull-House, the directory of the district "listed nine churches and 250 saloons" (Menand 2001: 308). People from 18 different nationalities lived and worked together in insalubrious apartments and unequipped streets.

However, Hull-House gained the favor of the neighbors very quickly. The social settlement provided education and rooms for social clubs governed by the different social and national groups. The educational program was developed in line with the interests and needs of the neighbors, with the aid and sponsorship of the residents. The house accommodated long-term residents who met regularly to decide on social interventions in cooperation with the local authorities and other cultural institutions. Hull-House had classes, lectures, a museum of arts, a kindergarten, a playground, a nursery, and theater groups. It was visited by thousands of people every week. The House had some very eminent residents, such as Beatrice and Sydney Webb, Peter Kropotkin, Emma Goldman, Frank Lloyd Wright, and Charlotte Perkins Gilman, among others. Alice Hamilton (1869–1970) went on to be the leading expert in toxicology and occupational health while residing at Hull-House from 1897 to 1919. The social settlement also established a fruitful partnership with the University of Chicago. John Dewey and James Hayden Tufts were regular lecturers at the Working-People's Social Science Clubs. The House also welcomed renowned thinkers and activists such as the suffragette Susan B. Anthony and the anarchist Peter Kropotkin.

Hull-House was therefore more than a charitable institution; it was a "sociology laboratory" (Menand 2001: 306) and a "*feminist* think tank" (Hamington 2009: 25). The residents of Hull-House produced genuine social knowledge through methodological innovation. One of the most significant contributions is the collective work *Hull-House Maps and Papers. A Presentation of Nationalities and Wages in a Congested District of Chicago, Together with Comments and Essays on Problems Growing Out of the Social Conditions* (1895). The project was mainly coordinated by Addams and the resident Florence Kelley, who had arrived at Hull-House with her three little children fleeing from a violent husband. Kelley had translated the works of Friedrich Engels into English. The project consisted of a social survey inspired by Charles Booth's London Poverty Maps.[3] The new feature of the residents' work, however, was that they collected data not only on family wages but also on nationalities, backgrounds, and details of the households. The superposition of the nationality and wages data collected is represented visually with a color code and scale that individualizes each household, which differs significantly from Booth's mapping of average income per street. This shows that the residents who conducted the survey and then published a collection of essays interpreting the maps had a very complex and advanced understanding of cartography and its use as a means of diagnosis for social intervention (Font-Casaseca 2016).

Despite having certain philosophical concerns with respect to Karl Marx's economic determinism and its dependence on the theory of class consciousness, Addams was a great supporter of the labor movement. Hull-House provided spaces for union meetings, and it gained a reputation for radicalism. According to Addams, during the 1890s, Chicago

> seemed divided in two classes; those who held that "business is business" and who were therefore annoyed at the very notion of social control, and the radicals, who claimed that nothing could be done to really moralize the industrial situation until society should be reorganized.
> *(Addams 1998: loc. 163)*

The conflict between these two fronts broke out in 1893 with the so-called panic of 1893, a deep economic recession that led to violent riots in Chicago. Addams was asked to act as a mediator and peacemaker, but she severely criticized the repressive attitude of the local government and the factory bosses, especially George Pullman, to whom she dedicated the essay "A Modern Lear", which was considered so incendiary that it could be not published at the time.[4] In the midst of the chaos in the city, Addams received sad news from home. When the demonstrators heard that Addams needed to be with her family, they helped her to get a car to reach Kenosha, Wisconsin, where her beloved sister Mary died shortly afterward (Knight 2005: 316–322).

Bread and peace: nurturing radical democracy

The constant exchange with the residents and visitors at Hull-House inspired most of Addams's writings. She combined social activism with intellectual work, and she was acquainted with the scientific and political debates of the era when her first book *Democracy and Social Ethics* (1902) came out. Formed of essays that Addams had conceived separately, her book is an interesting sample of the progressivism inspired by Darwin's writings (Fischer 2019). Addams's social philosophy is based on the notion that a society is a living organism, the evolution of which depends on the qualitative changes that occur over time. She uses the terms "feudalism", "industrialism", and "association" to refer to distinct phases of the historical process. "Social ethics" describes both the stage of social relations at each phase and the normative dimension of democracy that enables us to enter an age of association, leaving individualistic industrialism behind. In this book, Addams advocates a theory of democracy that goes beyond the mere idea of government. For Addams, the democratic effort is connected to sympathy and a diversified human experience (Addams 2002a:

7). There are two important concepts that emerge from Addams's early social philosophy: the concepts of sympathetic knowledge and lateral progress. Sympathetic knowledge is produced in our interactions with others. It presupposes overcoming the false dichotomy between reason and emotions. Lateral progress, as the goal of social evolution, means focusing social action toward a goal other than economic distribution and welfare. It involves pushing the whole community forward, discovering what people really want and providing "the channels in which the growing moral force of their lives shall flow". In this respect, lateral progress might be slower, because it is not the result of an individual striving but rather it is "underpinned and upheld by the sentiments and aspirations of many others" (Hamington 2009: 44–45).

Addams's essays addressed topics such as charitable effort, education, political reform, work, domestic service, play, and the arts. In *The Spirit of the Youth and the City Streets* (1909), Addams denounces the lack of urban spaces for children and teenagers, thereby contributing to the Playground Movement and its vindication of play spaces in urban environment as key to democratic communal living. *A New Conscience and an Ancient Evil* (1912) reveals society's hypocrisy with respect to prostitution. In all her writing, Addams demonstrates her concern for the humblest and most vulnerable members of society: children, immigrants, and women. Moreover, she provides imaginative solutions that often go beyond the creative powers of her contemporaries.

Newer Ideals of Peace (1906) was the first of a series of philosophical writings devoted to peace in relation to city government, in opposition to the growing patriotism and militarism in the years before the First World War. She soon saw that a wave of bigotry and belligerence was expanding around the globe. In a conversation with William James around 1899, she had already mentioned the idea of a "moral equivalent of war" that inspired James's essay of 1910 (Hamington 2009: 98). However, Addams's philosophical arguments were removed from James's idea of reorienting human bellicosity toward noble social ends. Addams's solutions for warfare involved the inclusion of immigrants and women in the democratic processes and joining a growing internationalist movement that advocated embracing values other than individualism and competitiveness.

When war was declared in Europe, Addams joined an international pacifist movement mainly composed of women. The first International Congress of Women took place in 1915 in the Hague. She attended the congress as one of the founders of the Woman's Peace Party and drafted the resolutions with Emily G. Balch and Alice Hamilton. *Women at the Hague: The International Congress of Women and Its Results* (1915) expresses the protest "against the madness and the horror of war, involving as it does a reckless sacrifice of human life and the destruction of so much that humanity has labored through centuries to build up". Women are doubly victimized under war conditions, not only as civilians but also because "the horrible violation of women which attends all war" (Addams, Balch & Hamilton 2003: 72). Members of the Women's International League for Peace and Freedom,[5] of which Addams was elected President, sought to persuade world leaders to cease war, without success. Addams compiled her experiences in Europe visiting the civilians and soldiers at the front in *Peace and Bread in Time of War* (1922, 2002b). However, many of her fellow citizens in the United States considered Addams's efforts to bring peace unpatriotic. She was the subject of a smear campaign orchestrated by the most conservative sectors of public opinion, and she was prosecuted by the FBI, being considered one of the "Most Dangerous Women in America".[6]

Addams's legacy

Addams's life was intense, and many people have regarded her as a moral example of personal integrity. One of her last works, *The Long Road of Woman's Memory* (1916), deals precisely with memory and its important "role in interpreting and appeasing life for the individual ... and its activity as a selective agency in social reorganization" (Addams 2002c: 5). The two functions are not mutually exclusive and support each other. Addams's reflection of memory arose from a very popular episode at Hull-House that brought masses of people to Halsted Street, as it was reported that the residents were "harboring within its walls a so-called 'Devil Baby'" (Addams 2002c: 7). Addams used this incident to elaborate on the possibilities of immigrant women to survive isolation and brutality, some of them in the poorest conditions or with abusive husbands, and to generate myths and tales to pass down a wealth of female knowledge from mothers to daughters over the centuries. The Devil Baby gave the women of the neighborhood the opportunity to create a network of sorority and to build on their traumas. Because of her interest in the folklore and the cultural heritage of immigrant women, Addams is considered a predecessor of cultural feminism.

Hull-House was a place governed mainly by women, and it was also very tolerant in respect to love and sexual orientation. "Boston marriages", or, in other words, a domestic partnership of two women, were commonly accepted during the progressive era (Simmons 2009). The partnership may be romantic in nature, but not necessarily so. Addams was involved in a romantic relationship with two women: firstly, with Ellen Gates Starr, the co-founder of Hull-House, and from 1892 onward, with Mary Rozet Smith, who was Jane's partner until Mary's death in 1934, only some months before Addams's passing in 1935. Thus, Addams is to be acknowledged as a key figure for the LGBTQIA+ community.

Addams's works were rediscovered during the 1990s (Deegan 1998; Seigfried 1996). Her critical reception is, therefore, still *in the making*, to use a pragmatist concept. She gave a definite impulse to social work and was also influential in the founding of social and political pragmatist philosophy. Her enduring legacy will surely provide inspiration for creative ways of thinking to the coming generations of pragmatists.

Notes

1 See the arguments for awarding her the prize on the Nobel Prize website: https://www.nobelprize.org/prizes/peace/1931/addams/facts/.
2 Addams became the second woman to win the Nobel Peace Prize, after Baroness Berta Sophie Felicitas von Suttner (1843–1914).
3 See Charles Booth's maps on the website of the London School of Economics: https://booth.lse.ac.uk/map/14/-0.1174/51.5064/100/0.
4 The text was refused by the *Review of Reviews, Forum*, and the *Atlantic Monthly*. Survey published it 15 years later, in 1912 (Fischer 2019: 78–80).
5 The Women's International League for Peace and Freedom still exists today: www.wilpf.org.
6 The Federal Bureau of Investigation's files on Jane Addams are available here because of her historical interest: https://vault.fbi.gov/Jane%20Addams.

References

Addams, J. (1998) *Twenty Years at Hull House; with Autobiographical Notes*. Available at: http://www.gutenberg.org/ebooks/1325.
Addams, J. (2002a) *Democracy and Social Ethics*. Urbana and Chicago: University of Illinois Press.
Addams, J. (2002b) *Peace and Bread in Time of War*. Urbana and Chicago: University of Illinois Press.
Addams, J. (2002c) *The Long Road of Woman's Memory*. Urbana, Chicago, and Springfield: University of Illinois Press.
Addams, J., Balch, E. G. and Hamilton, A. (2003) *Women at the Hague. The International Congress of Women and Its Results*. Urbana and Chicago: University of Illinois Press.
Font-Casaseca, N. (2016) 'Mapas contra la injusticia urbana: La utopía prágmatica de la Hull House en Chicago a finales del siglo XIX', in *XIV Coloquio Internacional de Geocrítica. Las utopías y la construcción de la sociedad del futuro*. Barcelona, pp. 1–19.
Menand, L. (2001) *The Metaphysical Club*. New York: Farrar, Straus, and Giroux.
Seigfried, C. H. (1996) *Pragmatism and Feminism. Reweaving the Social Fabric*. Chicago and London: The University of Chicago Press.

Shields, P. M. (2017) *Jane Addams: Progressive Pioneer of Peace, Philosophy, Sociology, Social Work and Public Administration*. Mosbach: Springer.
Simmons, C. (2009) *Making Marriage Modern. Women's Sexuality from the Progressive Era to World War II*. Oxford: Oxford University Press.
Westbrook, R. B. (1991) *John Dewey and American Democracy*. Ithaca and London: Cornell University Press.
Whipps, J. and Lake, D. (2016) *Pragmatist Feminism, Stanford Encyclopedia of Philosophy*. Available at: https://plato.stanford.edu/entries/femapproach-pragmatism/.

Further Reading

Deegan, M. J. (2017) *Jane Addams and the Men of the Chicago School, 1892–1918, Jane Addams and the Men of the Chicago School, 1892–1918*. Abingdon: Taylor and Francis.
(Deegan's book on Addams's contribution to sociology was the beginning of the "revival of Jane Addams" in the 1990s.)
Fischer, M. (2019) *Jane Addams's Evolutionary Theorizing. Constructing 'Democracy and Social Ethics'*. Chicago: The University of Chicago Press.
(Fischer introduces the context of production of Addams's most widely read work, linking the essays to the theoretical debates of the era.)
Hamington, M. (2009) *The Social Philosophy of Jane Addams*. Chicago and Urbana: University of Illinois Press.
(Comprehensive introduction to Addams's social philosophy, Hamington sees the fundamentals of care ethics in Addams.)
Knight, L. W. (2005) *Citizen Jane Addams and the Struggle for Democracy*. Chicago and London: The University of Chicago Press.
(Knight's extraordinary biography of Jane Addams reconstructs the life and works of an impressive civic heroine.)

6
ALAIN LOCKE'S CRITICAL PRAGMATISM AS APPLIED TO RACE AND CULTURE

Corey L. Barnes

Introduction

Pragmatists reject philosophical approaches and claims to knowledge that have no practical application. For pragmatists, knowledge must be useful. Thus, an ideology's or proposition's truth need not depend on whether a referent corresponds to its subject but on whether it: (1) works to solve problems and therefore (2) progresses community in some way. And so the pragmatist is able to commit to only those problems or inquiries that affect community, where the answers have some usefulness to the world. Pragmatists often claim that this approach works for those interested in keeping philosophy's focus on communities. Even though pragmatism claims this virtue, one might wonder whether it is susceptible to the uselessness that it seeks to reject. This chapter presents Alain Locke's critical pragmatism as a response to pragmatism's limitation for solving communal problems. In fleshing out Locke's critical pragmatism, I appeal to his thought on race and culture.

Critical Pragmatism

Jacoby Carter (2016: 107) tells us that pragmatism is a great source if one is looking for theoretical justifications for the practical engagement of social problems.

> However, if one wants not just a practical philosophy that has turned away from the problems of philosophers, but one that engages concrete social problems from a philosophical perspective that uses philosophical engagement not as an end in itself but as a tool for social uplift and increased democracy, the bulk of pragmatism is less well-suited to the task.
>
> *(2016: 107)*

Though it rejects mere abstract problems with no cash value for solving communal problems, the critique is that pragmatism stops short of application. The philosopher who sees theoretical justifications as requiring application must either appeal to a different method or supplement pragmatism in a way that preserves much of what makes it valuable. Locke's critical pragmatism supplements pragmatism such that it is more applicable to community.

A critical pragmatist is "one that accepts some of the substantive and methodological tenets of pragmatism while at the same time remaining critical of that philosophical school" (2016: 107). Pragmatism is a damaged but salvageable tool that requires a clear understanding of its deficiencies. In addressing certain of its deficiencies, the critical pragmatist supplements pragmatism with other traditions. Locke's critical pragmatism appealed to African American traditions, and this appeal undoubtedly explains his recognition of pragmatism's failures.

Carter (2016) claims that Locke's pragmatism was critical for five reasons. First (*1*) "is its engagement with concrete social problems from a philosophical perspective—one that frames the issue in terms of the relevant values, principles, rationales, and goals germane to the situation" (Carter 2016: 108–109). Locke philosophized from within a particular community, having understood what was at play therein. Second (*2*), Locke's critical pragmatism was attentive to history and the unfolding of its problems, solutions that failed or were successful, and the problems that each created. Third (*3*), for Locke, philosophical engagement does not end with mere theory but with application to problems of the world. Fourth (*4*), because Locke was attentive to history and its problems, he understood that solutions generate problems. And so Locke was dissatisfied with a mere solution to a social problem. He sought to think through problems that might derive from solutions. Fifth (*5*) "is the critique of progress that is often piecemeal and incremental and recognition that recent or past gains may be partially eroded before they are further advanced" (Carter 2016: 108–109). Here, the claim is that Locke rejected incrementalism in favor of more radical change. For the remainder of the chapter, I discuss Locke's thinking on race and culture that engages each reason except (*5*) and offer reasons for why Locke should be understood as an incrementalist.

Critical Pragmatism, Race, and Culture

Though I think all of Locke's philosophical explorations support four of Carter's reasons, I will restrict my discussion to his thought on race and culture. Locke seems to be an early progenitor of the view that debates about what race is are less important than debates about its usefulness for solving racism—which he took to be America's most fundamental problem (Locke 1944). Rather consistently throughout his life, and motivated by the critical pragmatist requirement of practical implementation as the end of theory, Locke eschewed racial skepticism and eliminativism because of its uselessness. And so he refused to accept that there were no races because it had no anthropological (biological) backing. Instead, Locke thought that race was real in part because race and race-thinking affected the world and that appeal to them can be useful.

In a series of early lectures on race, Locke attempted to understand race contacts from a specific American vantage-point. These lectures have been described as "meditations on the worsening state of race relations in America" (Locke 1992: xxi). More than that, however, his early lectures were an attempt to situate race in a larger context of a world history that it had shaped. And he sought to understand both its manifoldness and potential for improved race contacts in America. The lectures were structured to move from the manifold meanings of race, through how race began, grew, developed, flourished, and the basis of their contact of one with another, through "laws" about race contacts, through a survey of contemporary creeds of race and an explanation/condemnation of their fallacies, to how progress in America is possible given this history.

On progress, Locke's goal—consistent with African American political thought at the time—was to end with a program. He argued for a kind of conformity to a civilization type, by which he meant an endorsement of common institutions (legal and social). However, mere social conformity (imitation) would not progress society. Social conformity requires a counter-doctrine by "alien" groups that he called "race-consciousness." Race-consciousness leads to both solidarity and expressive culture, and is a necessary condition for self-respect that is both inherently valuable and a necessary condition of social respect (Locke 1992: 99). Here we see the pragmatic point of rejecting skepticism and eliminativism. What is the point of saying that race does not exist? For Locke, the greatest consequence is the elimination of an implication of race—race-consciousness—that would improve group life and help to solve racial tension. Locke's pragmatic thought was that eliminating race would not solve racism. It would merely eliminate a tool for solving it.

In a later treatment of race, Locke attempted to answer a long-standing impasse regarding the relationship between races and cultures. On one side, anthropologists took culture to depend on race such that different races invariably produce different cultures. On the other side, anthropologists denied any relationship between race and culture. Locke's goal was to refute both and to explain that races are social phenomena deriving from cultures. This strategy in some sense completed W.E.B. Du Bois' project of preserving benefits of tying cultures to race, namely race-pride, solidarity, and combating stereotypes for oppressed groups, but without appeal to Du Bois' essentialism (Du Bois 2015).

For Locke, races develop historically and socially. Historical and social conditions section off people in ways that produce varying degrees of proclivities toward certain values, ways of seeing the world, modes of expression, etc. These conditions lead people to produce varying traits of culture-types. What binds members of groups together is a likeness in culture-types. Race, then, becomes the idea (a sort of name under which groups understand themselves) that better binds group members together, and thereby becomes a value that unifies culture-traits by causing a robust sense of solidarity among those who understand themselves as a unified group with a specific culture-type. It is culture-heredity, and "[i]nstead therefore of regarding culture as expressive of race, race by this interpretation is regarded as itself a culture product" (Locke 1989d: 193). Locke committed to a type of racial category constructionism, where the creation and persistence of race are brought about by cultural practices that represent a certain race as the particular one that it is. For Locke: "Race operates as tradition, as preferred traits and values. ... Race, then ... seems to lie in that peculiar selective preference for certain culture-traits and resistance to certain others" (Locke 1989d: 195; see also Locke 1989a).

Now we've seen how Locke's conception of race satisfies Carter's third reason for classifying him as a critical pragmatist. It preserves something useful for social uplift. And Locke applied it concretely in his engagement with the Harlem Renaissance, his thoughts on the duty of African Americans to be cultured, and his philosophy of education. Moreover, Locke's conception seems to allow race to be scientifically studied in a way that a biological notion cannot. One can study race scientifically if it is culture-heredity. But why *1*, *2*, and *4*?

With *1*, though the above illustrates Locke's engagement with race in world history, his primary target was a particular community, namely America. His interest in engaging a lengthy racial history in his lectures was the belief that discussing the particularities of race in America requires a connection between it and race throughout world history. And Locke was particularly careful to situate American-specific race contacts in a broader history of the world because he did not want them to appear as aberrations, those unrelated to the world, or completely alien to human nature. His careful consideration is elucidated by *2* and *4*.

With *2* and *4* Locke understood race problems in America to stem from various modes of contact throughout world history. If we think that race contacts have progressed to some degree, then we have good reason to hope that American race problems can be alleviated. Further, he rejected the idea that careful attention to the unfolding of the race-sense teaches us nothing about both the problems of and possible solutions to hostile race contacts in America. And when discussing the practical program of subordinate groups cultivating a race-consciousness, Locke was aware that such a solution would receive initial reactionary pushback. Still, Locke saw it as the soundest way to progress toward social equality and racial harmony.

Now Locke's engagement with race tended to go hand in hand with the larger category of culture, such that culture was implicated in both race and race problems (Locke 1992: 99–100). Locke's early conception took culture to be a condition of understanding, appreciating, and living up to the best and most representative forms of human expression (Locke 1989c: 177).

Locke understood culture to relate to both the mind and the senses. There is a reciprocal relationship between them such that cultivating the one by understanding, appreciating, and living up to the best requires cultivating the other. "The [woman and] man of culture is the [woman and] man of trained sensibilities, whose mind expresses itself in keenness of discrimination and, therefore, in the cultivated interests and tastes" (Locke 1989c: 177). Training ultimately develops a personality. And this personality requires development of two basic aspects, namely "the great amateur arts of personal expression—conversation and manners" (Locke 1989c: 181). Expressions are not merely artistic but regard use of language, mood, posture, and perhaps even nonverbal cues such as facial expressions and gestures. It regards the kinds of conversation in which one engages and the ways that one engages in them.

Additionally, culture regards acting in a way that is socially acceptable and respectful to others. So for Locke, understanding, appreciating, and living up to the best and most representative forms of human expression do not merely regard creative expressions. One understands, appreciates, and lives up to the best human character, and this involves ethical activity. The organization of one's life, along with how one fulfills obligations to others, is implicated in culture. And so culture regards the development of a personality with a refined disposition and particular traits of character capable of performing actions representative of the best.

With the additional requirement of a kind of personality, Locke seems to have thought that culture is attained only given a particular psychological disposition. One needs to develop attention to detail, courage, and even something like wit as character traits. Culture develops with or is equivalent to the whole of one's character in the same way that Aristotle thought of the virtuous agent. "Culture likewise is every inch representative of the whole personality when it is truly perfected" (Locke 1989c: 183). Like Aristotle's notion of *eudaemonia* as living and doing well, or living in accordance with virtue, Locke's cultured person is one who exemplifies excellence in personality and pursuits for the purpose of living well. This can only result in a life lived excellently *via* activity.

Like Aristotle, Locke thought we have an ethical duty to be cultured (Locke 1989c: 176). Locke took this duty to be self-imposed. Though he did not specify why we have this duty, the reason appears to be rather obvious. The duty to culture is a duty to self that results from our basic aim of flourishing and culture's role therein. But how does this progress society?

Living up to the best and most representative forms of human expression implies an engagement with and sharing of diverse cultural elements throughout the current and historical world. And so culture implies cultural reciprocity. Locke understood cultural reciprocity to be "a general recognition of the reciprocal character of all contacts between cultures and of the fact that all modern cultures are highly composite ones" (Locke 1989b: 73). Cultures do not arise from the work of one group to the exclusion of others. Though a group may be taken as representative of a culture *via* thematic tendencies, common idioms, styles, and forms, investigation into them reveals their composite natures.

Recognition that cultures are composite serves a number of critical pragmatic interests. First, it cures cultural imperialism and chauvinism. No group can take complete credit for its own culture or "Culture" more broadly. Second, it is necessary for maintaining cultural production. Locke encouraged cultural participation and cultural merging because these keep cultures from becoming stale. Third, the recognition leads to a desire to share with and participate in different cultures. This desire motivates different races and cultural groups being in contact with each other in a way that alleviates animosity. Finally, the duty to culture motivates respect for diversity without requiring uniformity.

Finally, Locke's early view of culture illustrates a commitment to critical pragmatism when we understand why Locke gave his lecture on culture, and to whom he was speaking. Locke gave "The Ethics of Culture" to Howard University freshmen in 1923. Locke would have been concerned with the perception of African Americans as contributors to and appreciators of culture. For someone who thought that art and culture was an essential way to garner respect as equal citizens, lecturing on the duty to culture would have been important because of the need to combat negative stereotypes of African Americans.

Conclusion

To conclude, what should we make of 5? I read Locke as an incrementalist, and his incrementalism as aligning with his pragmatic commitments.[1] His incrementalism views social progress as requiring stages and is a combination of three pragmatist traits, namely fallibilism, an appeal to experience as a guide for theory, and meliorism.

For Locke, solving concrete social problems should be guided by actual experiences and not mere proposals given abstract moral claims. This is particularly true for Locke given human fallibility. Given human fallibility, we should form certain beliefs as hypotheses that are undergirded by, and yet adjustable in light of, new experiences. Experiences give our beliefs new and better evidence at each step.

Radical change implies larger leaps, and thereby lessens the possibility that our beliefs are guided by our experiences. These disallow new evidence and experiences to guide and adjust social programs. Further, radical social change tends to derive from abstract moralizing. Abstract moralizing fails to consider both the world in which we all live and the world as it will be after change occurs. In this way, Locke thought that avoidable disaster was much more likely a result of radical change. And he thought that adjusting afterward would be much more difficult than adjustments made after incremental changes.

Still further, Locke thought that radical change would more likely suffer from reaction. Locke considered the variety of mindsets affected by radical social change. And he seemed aware that many of those mindsets were unequipped to handle it, whether because of bigotry, blind appeals to custom, or comfort with familiarity. And so he was concerned that social instability would be the likely result of radical change. For Locke, changes in mindsets need cultivation during stages of social progress. And time is needed to cultivate newer, more progressive mindsets via education, art, and religion. In any event, Locke's engagement with philosophy shows him to have been a critical pragmatist who understood that philosophy's greatest value was its ability to guide life and be applied to social problems, and not merely as a theoretical exercise about life and social problems.

Note

1 See "The Mandate System: A New Code of Empire" (2012) for Locke's most forceful endorsement of incrementalism.

References

Carter, J.A. (2016) '"Like Rum in the Punch": The Quest for Cultural Democracy,' in Carter, J.A. (ed.) *African American Contributions to the Americas' Cultures: A Critical Edition of Lectures by Alain Locke.* New York: Palgrave Macmillan, pp. 107–174.

Du Bois, W.E.B. (2015) 'The Conservation of Races,' in Chandler, N.D. (ed.) *The Problem of the Color Line at the Turn of the Twentieth Century: The Essential Early Essays*. New York: Fordham University Press, pp. 51–65.

Locke, A.L. (1944) 'The Negro Group,' in MacIver, R.M. (ed.) *Group Relations and Group Antagonisms*. New York: Institute for Religious Studies, Harper & Brothers, pp. 43–70.

Locke, A.L. (1989a) 'The Contribution of Race to Culture,' in Harris, L. (ed.) *The Philosophy of Alain Locke: Harlem Renaissance and Beyond*. Philadelphia: Temple University Press, pp. 202–208.

Locke, A.L. (1989b) 'Cultural Relativism and Ideological Peace,' in Harris, L. (ed.) *The Philosophy of Alain Locke: Harlem Renaissance and Beyond*. Philadelphia: Temple University Press, pp. 69–78.

Locke, A.L. (1989c) 'The Ethics of Culture,' in Harris, L. (ed.) *The Philosophy of Alain Locke: Harlem Renaissance and Beyond*. Philadelphia: Temple University Press, pp. 176–185.

Locke, A.L. (1989d) 'The Concept of Race as Applied to Social Culture,' in Harris, L. (ed.) *The Philosophy of Alain Locke: Harlem Renaissance and Beyond*. Philadelphia: Temple University Press, pp. 188–199.

Locke, A.L. (1992) *Race Contacts and Interracial Relations*. Stewart, J. (ed.). Washington, D.C.: Howard University Press.

Locke, A.L. (2012) 'The Mandate System: A New Code of Empire,' in Molesworth, C. (ed.) *The Works of Alain Locke*. New York: Oxford University Press, pp. 509–527.

7
SIDNEY HOOK

Robert B. Talisse

Though a major intellectual in his day, Sidney Hook (1902–1989) is largely ignored by contemporary philosophers working in the pragmatist idiom. When Hook is mentioned at all, he is presented as a tragic figure: the heir apparent to his teacher, who ultimately betrayed Deweyan pragmatism for an awkward combination of analytic philosophy and conservative politics (Capps 2003; Phelps 1997). Hook is taken to show that by mid-century, pragmatism was in "deep crisis" (West 1989: 124).

Nonetheless, throughout his career, Hook described himself as a committed pragmatist. He remained a staunch defender of naturalism and experimentalism in ethics, metaphysics, and epistemology; additionally, he developed uniquely pragmatist views in the philosophy of law, philosophy of history, and philosophy of education (Sidorsky and Talisse 2018).

This chapter focuses on what I see as the core of Hook's pragmatism, his conception of democracy. Hook develops a novel epistemic and deliberative conception that improves upon Deweyan democracy and provides a basis for ongoing work in pragmatist democratic theory.

Hook's Conception of Democracy

Hook begins from an ordinary formula of democracy: "a democratic society is one where the government rests upon the freely given consent of the governed" (1938: 285). But Hook does not define democracy strictly in these terms. He contends that this simple formula is incomplete and must be supplemented by a normative analysis of the idea of freely given consent.

According to Hook, a government rests upon the consent of the governed when (1) political mechanisms exist by which the governed may at regular intervals register their approval and disapproval of proposed government action and policy, and (2) the government acknowledges a prima facie duty to conduct itself according to the consent of the governed (1938: 285). That is, a basic condition for consent is participation on the part of the governed in the processes of government. This much is commonplace; but what about the requirement that the consent of the governed be freely given?

There are many conditions to be met if consent is to be given freely. Beginning with the most obvious of these, Hook observes:

> An election held in the shadow of bayonets, or in which one can vote only "Yes," or in which no opposition candidates are permitted is obviously one which does not register freely given consent.
>
> *(1959: 54)*

Minimally, then, there must be no procedural obstructions to participation if consent is to be free. There are, however, more subtle forms of interference. As "there are few things to which a starving man will not consent" (Hook 1959: 32), we also may speak of economic obstructions to free consent. For example:

> A threat to deprive the governed of their jobs or means of livelihood, by a group which has the power to do so, would undermine a democracy.
>
> *(1938: 286)*

This consideration may seem a variation on the foregoing examples of procedural obstacles; however, Hook goes beyond cases of blatant economic manipulation. Hook notes that:

> Differences in economic power make it possible for the more powerful economic group to exercise a much greater influence upon decisions that affect public welfare than their numbers or deserts warrant.
>
> *(1959: 54–55)*

Hook fears that unchecked economic power will be employed to "render nugatory even legislative action" (1959: 55). Moreover, the economically powerful enjoy "greater advantages in mobilizing resources to influence public opinion and consent" (1959: 55). Hook concludes:

> Where the political forms of democracy function within a society in which economic controls are not subject to political control, there is always a standing threat to democracy.
>
> *(1938: 286)*

That is, democracy requires not only protection of the general populace from the direct domination of the economically powerful; it requires that steps be taken to ensure that economic power is not employed to control or undermine democratic processes.

There are also epistemological impediments to free consent. Hook writes that "Even in the absence of physical and economic coercion, consent is not free if it is bound or blinded by ignorance" (1959: 38). The operative epistemological principle is elementary: One's consent to a political proposal is free just in the degree to which one does not misunderstand the meaning and relevant implications of the proposal. Access to relevant sources of information is certainly not a sufficient condition for understanding a proposition, but whatever the sufficient conditions are, understanding the meaning of a political proposal requires information. Thus, democracy requires that the governed have unrestricted access to relevant kinds of information:

> The expression of consent by the majority is not free if it is deprived of access to sources of information, if it can read only the official interpretation, if it can hear only one voice in the classroom, pulpit, and radio.
>
> *(Hook 1938: 287)*

Hook elaborates,

> If one is kept ignorant of alternatives, denied access to information, deprived of the opportunity to influence and be influenced by the opinions of others, consent is not free.
>
> *(1959: 54)*

The commitment of a democratic society to freely given consent requires that there is no hindrance to access to relevant sources of information and to the agencies of critical discussion. Without such minimal provisions, consent is not free since "the individual has no more freedom of action when his mind is deliberately tied by ignorance than when his hands are tied with rope" (1938: 287).

According to Hook, then, the very concept of freely given consent demands that we broaden our conception of democracy beyond the proceduralist formula (1944: 50; 1959: 62). That is, we must acknowledge that the mere existence of democratic procedural devices such as open elections and periodic referenda is necessary but not a sufficient condition for democracy.

We have so far focused upon negative conditions for democracy, namely, the absence of obstructions to free consent. These conditions are by themselves insufficient. Even in the absence of the kind of impediments identified above, democracy may not be realized. Hook holds that an election in which every negative condition has been satisfied may yet fail to register free consent if the governed are generally illiterate or incapable of open and critical discourse. Under conditions such as these, manipulation of public opinion to produce artificial consent is easy; there is democracy in name only. Thus, we must also identify the positive conditions for democracy.

Just as there are economic and epistemological obstructions to democracy which must be guarded against, there are economic and epistemological requirements for democracy which must be provided for. We cannot examine Hook's views about "economic democracy" here, but note that Hook sees the positive economic requirements for democracy to be in the service of fuller and more rational participation on the part of the governed; consequently, Hook sees as primary the positive epistemic requirements for democracy.

A positive condition for democracy is general literacy among the governed. A democratic society must therefore provide public education. But basic literacy, command of a language, is not enough (1959: 109). The principle that persons must discern the meaning and implications of a proposal before they can freely consent to it requires something beyond literacy. One's consent to proposal P is free just in the degree to which one has examined P in some critical way; that is, one needs to have considered implications of P, entertained objections, and evaluated alternative proposals. Examining a proposal in this way requires free public discourse. It follows that "democratic society cannot exist without free discussion" (1980: 117) since free discussion is a necessary condition for free consent. Hence, Hook sees free consent and free discussion as epistemic matters. More precisely, for there to be free consent, and thus democratic legitimacy, certain epistemic conditions must be satisfied; in particular, citizens must be able to inquire. Accordingly, Hook proposes what is today called an epistemic conception of democracy.

Hook on the Ethics of Discourse

Yet Hook sees that "wherever discussion flourishes, controversy is sure to arise" (1980: 118). Amid controversy and conflict, free discussion can be as great a hindrance to free consent as any overt obstruction. Certain modes of discourse tend to silence debate, generate confusion, discredit dissenters rather than address dissenting views, suppress relevant information, encourage dogmatism, and establish on the basis of an appeal to loyalty to tradition that which cannot be established by an appeal to evidence and analysis; that is, "Some kinds of discussion tend to undermine democratic society" (1980: 117). Democracy therefore requires that public discourse be conducted in an epistemically responsible way.

In an essay titled "The Ethics of Controversy," Hook identifies the core of his conception of epistemically responsible discourse in terms of ten "ground rules" for democratic discussion:

> Nothing and no one is immune from criticism.

> Everyone involved in a controversy has an intellectual responsibility to inform himself of the available facts.
>
> Criticism should be directed first to policies, and against persons only when they are responsible for policies, and against their motives or purposes only when there is some independent evidence of their character.
>
> Because certain words are legally permissible, they are not therefore morally permissible.
>
> Before impugning an opponent's motives, even when they legitimately may be impugned, answer his arguments.
>
> Do not treat an opponent of a policy as if he were therefore a personal enemy of the country or a concealed enemy of democracy.
>
> Since a good cause may be defended by bad arguments, after answering the bad arguments for another's position present positive evidence for your own.
>
> Do not hesitate to admit lack of knowledge or to suspend judgment if evidence is not decisive either way.
>
> Only in pure logic and mathematics, not in human affairs, can one demonstrate that something is strictly impossible. Because something is logically possible, it is not therefore probable. 'It is not impossible' is a preface to an irrelevant statement about human affairs. The question is always one of the balance of probabilities. And the evidence for probabilities must include more than abstract possibilities.
>
> The cardinal sin, when we are looking for truth of fact or wisdom of policy, is refusal to discuss, or action which blocks discussion.
>
> *(1980: 122)*

Taken together, these "rules" suggest the epistemic character of Hook's view of democratic theory. According to Hook, democratic discourse is an agonistic clash of argument conducted according to percepts that are designed to keep discourse epistemically responsible by enjoining us to attend to arguments and reasons, not rhetoric and power. In this way, Hook sees democratic discourse as involving,

> a perpetual invitation to sit down in the face of differences and reason together, to consider the evidence, explore alternative proposals, assess the consequences, and let the decision rest – when matters of human concern are at stake – with the consent of those affected by the proposals.
>
> *(1959: 38)*

Hook and Deweyan Democracy

Hook employs many of the characteristically Deweyan tropes. He refers to democracy as a "way of life" guided by the "method of intelligence" (Hook 1938). Yet his view is not a species of Deweyan democracy but an improvement on it.

Dewey saw the democratic way of life and the method of intelligence as deeply nested in his own comprehensive philosophical system. For Dewey, citizenship in a properly democratic society is not only a necessary condition for a flourishing life; it is also sufficient for it. This is because Dewey identifies democracy with the "the one, ultimate, ethical ideal of humanity" (EW1: 248) and with "a truly human way of living" (LW11: 218).[1] Indeed, Dewey proposed that a criterion for democracy is whether a society's institutions, customs, and laws "[set] free individual capacities" (MW5: 431). He, hence, defines democracy as the social and political manifestation of human "growth," which he identifies as "the only moral end" (MW12: 181). In short, on Dewey's view, democracy is the way of life that is necessary for human flourishing, because it is identical with it.

By contrast, Hook sees the democratic way of life in strictly epistemic terms. On Hook's view, democratic politics is not aimed at realizing a vision of human flourishing; in fact, unlike Dewey, Hook proposes no comprehensive theory of human flourishing. Rather, he sees democracy as aimed at satisfying certain epistemic conditions for responsible collective decision and self-government. Accordingly, Hook does not build into his conception of democracy a conception of "the one, ultimate, ethical ideal of humanity" (EW1: 248). Hook recognizes that since we live in an "age of conflict" (Hook 1975) where legitimate but incommensurable goods compete for realization (Hook 1960), any attempt to ground our democratic commitments in deep moral ideals will fail. He understood that the suggestion that the whole of society or policy should be aimed at the realization of some contestable moral ideal – even a pragmatist ideal – will fail to win the free consent of the populace, which is necessary for legitimacy (Hook 1960).

Although Hook agrees with Dewey that democracy presents a "task before us," he disagrees with Dewey over the nature of this task. Dewey saw the project of democracy as that of human moral perfection. For Hook, it is the project of human liberation. The democratic aspiration, as Hook understands it, is to cultivate a society whose collective decisions and modes of public discourse are increasingly more intelligent; for Hook, the criterion for democracy is the role that collective inquiry plays in processes of political decision. Whether a society that satisfies that criterion also necessarily cultivates flourishing human beings is, on Hook's account, an open question. His focus is on the conditions under which politics can be driven more by reasons and less by power.

Hook's conception of democracy suggests a conception of democratic citizenship. The epistemic and agonistic view of democracy calls for a multidimensional public square, a variety of public institutions, organizations, and forums devoted to political and civic discourse. Accordingly, it also encourages democratic participation and engagement. Hook thus offers a distinctive conception of deliberative democracy.

Hook and Contemporary Pragmatist Democratic Theory

Hook resisted Dewey's impulse to build into the conception of democracy a conception of the good life for human beings, focusing instead on the more modest epistemic proposal that a democratic society is one whose institutions, norms, and practices aim to direct political power by reasons and arguments rather than money and other forms of influence. In this way, Hook's conception of democracy represents an early formulation of the recent "Peircean" turn in pragmatist democratic theory, as found primarily in the works of Cheryl Misak (2000) and Robert Talisse (2007, 2009).[2]

According to this view, democratic social and political arrangements are the prerequisites for cultivating the epistemic capacities that we already embrace. As Peirce observes in "The Fixation of Belief," it is a "mere tautology" to assert of a belief that one holds that it is true (CP5.375).[3] This conceptual tie between belief and truth-aspiration provides the basis for a constitutive norm governing our beliefs (Misak 2004: 12): to assess a belief that we hold as false is typically to undermine it as a belief.

A pragmatic elucidation of the idea of assessing a belief as true suggests a second norm governing our beliefs, namely, we must take our beliefs to be in line with our best evidence. Taking our beliefs to be in line with – or at least not contradicted by – our best evidence is the way we assess them as true. We thus can identify two constitutive norms of belief: truth-aspiration and reason-responsiveness (Misak 2004: 12; Talisse 2007: 62).

The Peircean democrat now follows Hook in observing that in order to take ourselves to be responding properly to our best evidence, we must have access to a social epistemic system that encourages the free exchange of information and ideas. In other words, we cannot regard our beliefs as proper unless we can also regard ourselves as having access to reliable sources of information.

From there, the Peircean builds an argument for familiar social norms or free expression, open inquiry, and protected dissent. This then provides the basis for a defense of distinctively democratic norms, including freedom of the press, compulsory public education, the protection of public space, and, ultimately, accountable and responsive government.

Thus, a robust conception of democracy and democratic life can be built without appeal to controversial theories of human flourishing. The Peircean view, like Hook's position, draws strictly on epistemic values and aims. Of course, much more would need to be said to present the Peirce-Hook strain of pragmatist democratic theory in full. The objective here has been to simply indicate the general direction of that work.

Notes

1 References to Dewey's work will be keyed to the 37-volume *Complete Works* and will employ the standard formula: (volume number: page number).
2 See also Misak and Talisse (2014 and 2021) and Aikin and Talisse (2018).
3 Citations to Peirce's writing will refer to the *Collected Papers* and follow the standard formula: (volume number. paragraph number).

Works Cited

Aikin, Scott F. and Robert B. Talisse. 2018. *Pragmatism, Pluralism, and the Nature of Philosophy.* New York: Routledge.
Capps, John. 2003. "Pragmatism and the McCarthy Era." *Transactions of the C. S. Peirce Society* XXXIX, no. 1: 61–76.
Dewey, John. 1969–1991. *The Collected Works of John Dewey: The Early Works, The Middle Works, The Later Works.* 37 vols. Jo Ann Boydston, ed. Carbondale: Southern Illinois University Press.
Hook, Sidney. 1938. "The Democratic Way of Life." In *Sidney Hook on Pragmatism, Democracy, and Freedom,* Robert B. Talisse and Robert Tempio, eds. Amherst: Prometheus Books.
———. 1944. "Naturalism and Democracy." In *Naturalism and the Human Spirit,* Yervant H. Krikorian, ed. New York: Columbia University Press.
———. 1959. *Political Power and Personal Freedom.* New York: Collier Books.
———. 1960. "Pragmatism and the Tragic Sense of Life." In *Sidney Hook on Pragmatism, Democracy, and Freedom,* Robert B. Talisse and Robert Tempio, eds. Amherst: Prometheus Books.
———. 1975. "The Role of Reason in an Age of Conflict." In *Sidney Hook on Pragmatism, Democracy, and Freedom,* Robert B. Talisse and Robert Tempio, eds. Amherst: Prometheus Books.
———. 1980. *Philosophy and Public Policy.* Carbondale: Southern Illinois University Press.
Misak, Cheryl. 2000. *Pragmatism, Truth, Democracy.* New York: Routledge.
———. 2004. "Making Disagreement Matter." *The Journal of Speculative Philosophy* 18, no. 1: 9–22.
Misak, Cheryl and Robert Talisse. 2014. "Pragmatist Epistemology and Democratic Theory." *Journal of Political Philosophy* 22: 366–376.
———. 2021. "Pragmatism, Truth, and Democracy." *Raisons Politiques* 81: 11–27.
Peirce, Charles Sanders. 1931–1958. *The Collected Works of Charles Sanders Peirce.* 8 vols. Cambridge: Harvard University Press.
Phelps, Christopher. 1997. *Young Sidney Hook.* Ithaca: Cornell University Press.
Sidorsky, David and Robert Talisse. Winter 2018 Edition. "Sidney Hook," *The Stanford Encyclopedia of Philosophy,* Edward N. Zalta (ed.), https://plato.stanford.edu/archives/win2018/entries/sidney-hook/.
Talisse, Robert B. 2007. *A Pragmatist Philosophy of Democracy.* New York: Routledge.
———. 2009. *Democracy and Moral Conflict.* Cambridge: Cambridge University Press.
West, Cornel. 1989. *The American Evasion of Philosophy.* Madison: University of Wisconsin Press.

8
C.I. LEWIS BETWEEN CLASSICAL AND CONTEMPORARY PRAGMATISM

Peter Olen

Introduction

Clarence Irving Lewis serves as a transitional point for American pragmatism. While his early influences were founders or direct descendants of classical pragmatism, much of his research was shaped by pressures different from those of 19th-century philosophy. Lewis's major works developed during the early stages of the revolution in formal logic, the professionalization of philosophy, and the linguistic turn. Far from making only a minor difference, the emergence of analytic philosophy in North America changed the direction of philosophy by shifting the dominant question from "How do you know?" to "What do you mean?" (Lewis 1934: 125). This is not to say that Lewis eschewed hallmark commitments of classical pragmatism: exploring human experience, a fallibilist conception of knowledge, a pragmatic conception of meaning, and a close connection between thought and action were views endorsed and developed by Lewis. What complicates Lewis's place in the history of pragmatism is not his pragmatist credentials; behind John Dewey, Lewis was arguably the mid-20th century's most influential pragmatists. Yet, Lewis's complicated relationship to the linguistic turn and analytic philosophy (specifically logical positivism) makes his historical placement an issue of overlapping influences and commitments.

This is not to align Lewis with later linguistic pragmatists, such as Richard Rorty, who argued for the elimination of experience as a philosophically rich concept. Especially in his earlier works, Lewis was concerned to locate a place for experience outside of language. Even when developing a more linguistically grounded epistemology (as found in his seminal work, *An Analysis of Knowledge and Valuation*), Lewis was clear that knowledge must always be traced back to a firm footing in ineffable and potentially incommunicable experience. Although our knowledge is of objects and things, our experiences are those perceptual presentations and feelings that constitute the flux of human experience. Lewis is an important but overlooked transitional figure in the pragmatist tradition, with one foot in the action-based, value-driven world of experience that constitutes the foundation of classical pragmatism and the other in the formal and epistemologically focused future of professionalized philosophy.

Pragmatism

Lewis is best known for his notion of the pragmatic a priori, his defense of material implication in formal logic, and his conception of givenness in epistemology. How much of this can be classified as *directly* related to pragmatism is debatable. Surely, his conception of the pragmatic a priori plays a central role in his thinking (though this is more directly traced to Josiah Royce's influence). The necessity of extra-logical considerations for formal systems (as embodied in his arguments for material implication) seems like a pragmatist point, but his praise of rigorous formal methods in philosophy is somewhat out of step with classical pragmatism. One might trace Lewis's stress on the material and experiential, as opposed to formal, aspects of inference back to pragmatism. Yet, Lewis seems to attribute these views not to William James or Dewey but to other contemporary influences (such as his close reading of Bertrand Russell's and Alfred North Whitehead's *Principia Mathematica* or then-recent developments in physics). While Lewis does consider Charles Sanders Peirce's views as "consonant" with his own, this places him further away from James and Dewey and does not constitute a direct line of influence (Lewis 1930/1970: 12). Does this make Lewis's reflections on material implication part of his pragmatist leanings?

Despite these historiographical complexities, Lewis is a crucial step in the development of pragmatism. Having studied under William James, Josiah Royce, and Ralph Barton Perry at Harvard, Lewis's pragmatist credentials can be traced back to the movement's originators and direct descendants. Lewis himself characterized his relationship to other figures as such:

> It finally dawned upon me, with some surprise, that as nearly as my own conceptions could be classified, they were pragmatic; somewhere between James and the absolute pragmatism of Royce; a little to one side of Dewey's naturalism and what he speaks of as 'logic'.
>
> *(Lewis 1968: 38)*

Lewis is describing himself as a combination of influential pragmatists: there are idealist or Kantian elements in Lewis's thought that are absent in James and Dewey (though found in Peirce and Royce), a voluntarism and humanism that indebts him to James and Dewey, but a wariness of what Lewis saw as Dewey's surrendering of logic to psychology, as well as James's subjectivism that, while representing the importance of choice in our experiences, fails to account for the objective character of knowledge and truth.

While somewhat skeptical of identifying a set of beliefs held by all pragmatists, Lewis argues that Peirce, James, and Dewey shared a commitment to the pragmatic method (albeit in different formulations). In Lewis's hands, the pragmatic maxim morphs into a conversation about the verification or verifiability of empirical meanings in experience. In shifting philosophy from the question, "How do you know?" to "What do you mean?", Lewis interprets the pragmatic maxim as carving a practical line in our conceptions that brings us close to the verification criterion found in some stages of logical positivism (to be discussed in the following section). Knowledge and meaning begin and end in human experience, but they are not relegated to immediate or subjective categories of first-person experience (Lewis 1934/1970: 267–268). For Lewis, the practical difference a concept makes turns on *all* conceivable experience. This allows for a wider interpretation of verification, empirical meaning, and the pragmatic maxim. From Lewis's standpoint, insofar as one can conceivably entertain a difference one might find in *some* experience, we have found a difference that will both potentially make a difference and constitute something meaningful. This broad notion of verification and meaning, though grounded in practical concerns and outcomes, allows Lewis to make room for presentational experiences, conceive of meaningful experiences as occurring prior to (or separate from) language, and explain the relationship between subjective experiences and objective knowledge.

Following James and other pragmatists, Lewis thought of theoretical pursuits as subservient to human interests, passions, and needs. This is found in Lewis's arguments that value functions as a form of empirical knowledge, with judgments of truth always presupposing some sense of value (Lewis 1946: 365–367). Non-normative judgments of fact, those mythologized in eliminativist or reductionist understandings of science or universal laws, obscure our actual practices of inquiry and judgment. Any pursuit of knowledge or rational action requires values. And in the flux of experience, subjective and objective aspects of the world appear as the same (Lewis 1936/1970: 154).

This subservience of concepts and reasoning to human interests comes out most explicitly in Lewis's conception of the pragmatic a priori. Here, one finds Lewis arguing for an understanding of concepts and a priori categories as malleable to our practical interests. Instead of positing a rigid categorical structure that exists utterly independent of experience, Lewis's conception of the a priori reflects the importance of understanding conceptual change on pragmatic grounds. Our *knowledge* of the world is always conceptual, even if our experience may contain a non-conceptual element. Yet, this is not a rigid sense of "conceptual" (i.e. it is not required in the sense that one might think of Kant's forms of intuition). Depending on our purposes and goals, we can choose to apply or use different concepts to interpret our experiences. This is not to say we always pick concepts prior to our experience; we must, as Lewis says, learn which concepts are applicable for us (Lewis 1923: 174). Experience, on Lewis's reading, comes with all real and unreal aspects of the world. Our conceptual choices help structure how we interpret that experience. Yet, we must start with some definite conceptual structure (as there would be no other way to interpret our experiences or know about the world) that then changes based on our needs. There are even some concepts, such as those of science or mathematics, which we might see as near universal. But even here, Lewis argues, claims of universality bend to the needs of experience. If these ingrained concepts are successfully used to interpret our experience relative to our goals, there is no reason to think one might change them. Nonetheless, "continued failure to render experience intelligible in such terms might result eventually in the abandonment of that category altogether" (Lewis 1923: 175).

Even with an emphasis on human ends and needs, this view does not align Lewis with what he saw as an overly psychologistic conception of logic (as found in Dewey's work).[1] Lewis, as opposed to some of the earlier pragmatists, endorsed formal aspects of logic, especially in his development of alternative and intensional logics.[2] From Lewis's standpoint, Dewey's rejection of abstractions or formal notions is too rash; insofar as scientific concepts are operationally or pragmatically defined (or, at least, one imagines, traceable back to some conceivable difference in experience), formalizations are excellent concepts for making our ideas clear. Difficulties between intensional and extensional logics, for example, concern the meaning of implication operative in both logics as to how we should think of implication in various forms of reasoning. Lewis argued that extra-logical considerations are always relevant for understanding one's sense of implication and related concepts, as well as the role of inference within logical systems as a whole. The adequacy and meaning of inference and other concepts are not divorced from human experience but subservient to our practical concerns at some point. As Lewis puts it, "Goodness in a concept is not the degree of its verisimilitude to the given, but the degree of its effectiveness as an instrument of control" (Lewis 1930/1970: 85). While abstractions may remove a concept's direct connection to immediate experience, this does not mean such concepts are untethered from our practical concerns.

Lewis's pragmatism is most directly seen in his conception of the pragmatic a priori, his formulation of the pragmatic maxim, and his historical placement within the movement. While Lewis did diverge from some individual pragmatists' commitments (as is true for all pragmatists), these divergences constitute pragmatism's rich legacy as a "movement" centered on a loose set of commitments.

Pragmatism and Logical Positivism

Lewis was initially identified by Herbert Feigl as an ally of logical positivism (Blumberg and Feigl 1931). This is not to say that immigrating logical positivists opposed James or Dewey, but Lewis's work in logic and epistemology was more congenial to the positivist's own interests. Given this, it is not surprising he would appear as an intellectual ally for Rudolf Carnap, Hans Reichenbach, and others. Lewis was not only engaged in philosophical projects that were complimentary to views found in Carnap's *Der logische Aufbau der Welt* but also worked on formal aspects of philosophy. While leaving some room for philosophy to operate in broader contexts, Lewis treated the sciences as exemplary instances of human knowledge. So, there were plenty of reasons to think that Lewis's conceptualistic pragmatism would be closely aligned with logical positivism as philosophy headed into the mid-20th century.

Despite Feigl's initial assessment, Lewis saw fundamental differences between his pragmatism and logical positivism. Lewis himself characterized these disagreements not as minor "refinements of theory" but as directly related to a foundational understanding of meaning and philosophy itself (Lewis 1941/1970: 95). These disagreements are found as early as 1933 in Lewis's American Philosophical Association Presidential Address but are crystalized around four main points in a later article: the relevant conception of empiricism, the scope of science, the significance of metaphysics, and the status of normative and evaluative judgments (Lewis 1941/1970: 93). While pragmatism has a more active, engaged conception of empiricism than one finds in logical positivism, both movements constitute forms of empiricism. There are minor disagreements here, and Lewis occasionally remarks that Carnap's earlier empiricism is much more congenial to his own, but there is not a large enough difference to think of pragmatism's and logical positivism's conceptions of empiricism as being opposed to each other.

Lewis notes that pragmatism has always been broadly pluralistic, viewing science as an important method for getting at truth and fixing belief but not as the *only* method for doing so nor as the determiner of significance (Lewis 1941/1970: 99–100). Lewis chalks up this difference to pragmatism's emphasis on the scientific *method*, as opposed to its content, but this obscures the sense in which the two different movements' conceptions of meaning undercut the possibility of substantial agreement. This is partially due to logical positivism's narrow conception of verification and meaning. Restricting discussion of meaning and significance to the formal mode of speech simply renders philosophy unrecognizable for Lewis. Especially given pragmatism's commitment to the richness of experience, the positivist understanding of verification embodies a form of "extreme naturalism," eliminating the possibility of any meaningful metaphysics, of knowledge and truth, and of normative values (Lewis 1936/1970: 152).

It is not that Lewis was fundamentally opposed to logical positivism, and he retained a fondness for the movement into his later years, but their conception of philosophy was much narrower than his own.[3] Even if we find Lewis and Carnap in agreement on numerous issues, whether metaphysics and ethics count as meaningful subjects is a fairly large point of contention. Lewis's work in logic and epistemology was certainly more in step with the trajectory of professionalized philosophy, but these substantive differences cannot be papered over.

But perhaps this is wrong.[4] Perhaps Lewis focused on differences while ignoring commonalities that unite pragmatism and logical positivism against the seemingly last vestiges of rationalism and non-naturalism in American philosophy. Does this common enemy force us to view pragmatism and logical positivism as somewhat united? We should avoid what Daniel Dennett calls *hysterical realism*. There is no exact dividing line between those who count as pragmatists or positivists; we are not carving the world at its joints when identifying group membership. Lewis himself makes this point when thinking through the relationship between pragmatism and logical positivism:

> The attempt to characterize any philosophical movement is a somewhat dubious enterprise, and the comparison of two such is doubly dubious. A movement is to an extent a fiction: there are only the individuals thinkers agreeing in certain respects, presumably fundamental, and disagreeing in others; and to say anything important about their agreements without continual qualify references to their divergences is almost inevitably to be inaccurate to some degree.

(Lewis 1941/1970: 92)

While Lewis held some sympathies with logical positivism, his career is indicative of an oscillating interest in the movement. While more or less sympathetic at various times, we should understand Lewis as harboring deep disagreements with certain logical positivists. In terms of analytic philosophy's progression, Lewis's pragmatism has been overlooked, mostly eclipsed by W.V.O. Quine's and Wilfrid Sellars's categorization as pragmatists more closely aligned with analytic philosophy's progression in North America.[5]

Lewis's transitional role obscures his importance in histories of pragmatism and contemporary conversations within the tradition. This may have less to do with Lewis and more to do with a turn in recent pragmatist thinking. While pragmatism may have been founded as a far-reaching method, as well as a loose set of complimentary commitments, the current family resemblance is strained. Lewis did write on social, political, and ethical concerns, but he is known for his work in epistemology and logic. While contemporary pragmatism generally focuses on social and political issues, Lewis's work in this area (much as during his lifetime) has been ignored. Even debates about the relationship between experience and language, where Lewis's transitional place in 20th-century philosophy is clearly relevant, have ignored Lewis. This is a shame, not just because of the philosophical value of his arguments but also because of Lewis's place in the history of pragmatism. Many of Lewis's essays offer a considered articulation not just of pragmatism but also of the clear differences between its classical exponents. These were contemporaries for Lewis (at least for a good portion of his career), and thus his work can help clarify pragmatism's relationship not only to logical positivism but also to idealism and realism—and to itself.

Notes

1 See Dewey (1929: 153–155).
2 Peirce is an obvious exception to this point. But Peirce's forays into logic occur prior to the publication of *Principia Mathematica* and develop under different considerations than Lewis.
3 Lewis is generally concerned with Carnap's interpretation of logical positivism (though Reichenbach's work on probability constitutes a well-known disagreement with Lewis). Whether Lewis's interpretation of Carnap is correct is not addressed here.
4 See Richardson (2003) and Misak (2013) for alternative histories. My view is, I believe, more aligned with Murphey's biography of Lewis. See Murphey (2005).
5 Lewis's influence on Quine and Sellars constitutes a large issue that cannot be covered here. Both philosophers grapple with Lewis's conception of the given and the a priori at various points in their respective careers.

Bibliography

Dewey, J. (1929). *The Quest for Certainty*. New York: Minton, Balch & Company.
Goheen, John and Mothershead, John Jr., eds. (1970) *Collected Papers of Clarence Irving Lewis*. Stanford, CA: Stanford University Press.
Lewis, C. I. (1923). A Pragmatic Conception of the A Priori. *The Journal of Philosophy* 7: 169–177.
———. (1929). *Mind and the World Order*. New York: Dover Publications Inc.
———. (1930/1970). Logic and Pragmatism. In Goheen and Mothershead Jr. 1970: 3–19.
———. (1934). Experience and Meaning. *The Philosophical Review* 43: 125–146.
———. (1936/1970). Judgements of Value and Judgements of Fact. In Goheen and Mothershead Jr. 1970: 151–161.
———. (1941). Logical Positivism and Pragmatism. In Goheen and Mothershead Jr. 1970: 92–112.
———. (1946). *An Analysis of Knowledge and Valuation*. LaSalle: Open Court Press.
Misak, C. (2013). *The American Pragmatists*. Oxford: Oxford University Press.
Murphey, M. (2005). *C. I. Lewis: The Last Great Pragmatist*. Albany: State University of New York Press.
Richardson, A. (2003). The Fate of Scientific Philosophy in North America. In Gary L. Hardcastle and Alan W. Richardson (eds.), *Logical Empiricism in North America*. Minneapolis: University of Minnesota Press: 1–24.

9
QUINE AND AMERICAN PRAGMATISM

Yemima Ben-Menahem

Quine's standing as one of the greatest 20th-century analytic philosophers is unquestionable. His standing as a leading pragmatist, however, as suggested by his place in this volume, is more ambiguous and deserves a close examination of his pronouncements and, occasionally, his silences. This ambiguity is the focus of this chapter.

Unlike some of his contemporaries, such as Putnam, Rorty, and Davidson, Quine did not portray himself as a pragmatist. Indeed, he was sometimes quite dismissive about the meaning and importance of pragmatism as a distinct philosophical position: "it is not clear … what it takes to be a pragmatist. …the term 'pragmatism' is one we could do without" (1981b: 23). As we will see, Quine maintained that pragmatism is hardly distinguishable from empiricism and was therefore only willing to endorse it as such. Why then does his name come to mind when thinking of influential descendants of American pragmatism? One reason is the famous poetic ending of "Two Dogmas":

> Carnap, Lewis and others take a pragmatic stand on the question of choosing between language forms, scientific frameworks; but their pragmatism leaves off at the imagined boundary between the analytic and the synthetic. In repudiating such a boundary I espouse a more thorough pragmatism. Each man is given a scientific heritage plus a continuing barrage of sensory stimulation; and the considerations which guide him in warping his scientific heritage to fit his continuing sensory promptings are, where rational, pragmatic.
>
> *(Quine 1953: 46)*

By pragmatic considerations, Quine means considerations that are not imposed on us by either logic or experience, considerations that may involve norms and with regard to which we have discretion. That there is room for discretion and norm-guided choice is an immediate result of the underdetermination of theory by observation, one of the core tenets of "Two Dogmas" and Quine's philosophy in general.[1] By analogy with underdetermination in algebra (where, when there are fewer equations than variables, one can choose the values of some variables by fiat and solve the equations for the rest), the underdetermination of theory implies considerable amount of freedom with regard to theory choice. Underdetermination is illustrated by Quine's web of belief, whose inner parts can be variously connected with experience, creating the different options we can choose from. It is a seamless web, with no sharp boundary between analytic and synthetic truths, that is, in principle every component of the web could be revised to achieve better correspondence with experience. The web is not subject to rigid rules of change and modification;

when problems come up, there are only "soft" guidelines for choice among empirically equivalent alternatives. Primary among such choice-guiding considerations is, according to Quine, his "maxim of minimum mutilation" (1970: 85), which instructs us to keep most of the current web of belief intact, making changes only where they are necessary. It is noteworthy that Quine sees such pragmatic considerations as rational, implying that there could be other, less rational, considerations (e.g. authority), which occasionally affect our choice but which he would not sanction.

Quine represents his argument as an extension of pragmatism and yet, his referring to a logical positivist like Carnap in this context speaks against a direct association between the pragmatic argument espoused here and American pragmatism. What Quine means by pragmatism in this passage is epistemic discretion, which is indeed common to his own and Carnap's positions. It is doubtful whether, on the basis of this schematic evocation of pragmatism, we can see Quine's philosophy as rooted in the philosophical school bearing that name, for it is possible that what Quine has in mind is basically an "empiricism without dogmas" as suggested by the title of the final section of "Two Dogmas." We should therefore distinguish the more common use of the term "pragmatic," which can be linked quite naturally to epistemic discretion, from the rubric of pragmatism when associated more directly with the tenets of American pragmatism. To find out whether there are in fact such closer associates between Quine's philosophy and his pragmatist predecessors, one must conduct a more thorough search.

In 1975, Quine was invited to deliver a paper at a conference on "The Sources and Prospects of Pragmatism." His paper appeared in 1981 in two versions: the full version, entitled "The Pragmatists' Place in Empiricism," in a volume containing the conference papers (Quine 1981b), and an abridged version, under the name "Five Milestones of Empiricism" in *Theories and Things* (Quine 1981a). The latter contains only part of the former, excluding any discussion of pragmatism. The fact that in his collection Quine chose to include only the sections on the five milestones of empiricism indicates more than anything he said explicitly that he ascribed little significance to the impact of pragmatism on his own thought and was rather suspicious of the importance of pragmatism in general. The reader gets the same impression from what Quine does say about pragmatism in the unabridged version of the paper. His main complaint is, as we saw above, that it is not sufficiently clear what philosophical positions and commitments pragmatism involves. What pragmatists share, according to Quine, is empiricism, even if not the specific brand of empiricism he commends. Hence, the "five points where empiricism has taken a turn for the better" (1981b: 23), constituting the milestones on the road to Quine's own empiricism: (1) the shift from ideas to words, (2) the shift from terms to sentences, (3) holism—the shift from sentences to systems of sentences, (4) no analytic-synthetic dualism, and (5) naturalism—no prior philosophy.

Notably, none of the five points is attributed by Quine to the American pragmatists! Quine ascribes the first of these transitions to John Horne Tooke's critique of Locke and the second to Jeremy Bentham. The founding fathers of analytic philosophy, Frege, Russell, and Wittgenstein, as well as the logical positivists, are also referred to by Quine as promoting these two insights. The remaining three transitions are characteristically Quinean, although he deemphasizes his role as their proponent; for example, he cites August Comte as the originator of naturalism. In the ensuing discussion, Quine mentions various disagreements with the positions of American pragmatists on his five points. He criticizes Peirce for vacillating between words and ideas and between beliefs and sentences (although ultimately settling for sentences) and for not being sufficiently outspoken about holism. He criticizes James for being kind to wishful thinkers and both James and Dewey for declining (immanent) realism. Finally, he disagrees with Lewis about the analytic-synthetic distinction. Quine mentions in passing a couple of points of agreement with the pragmatists: fallibilism, the repudiation of Cartesian doubt, and the recognition of Darwinism as a key to understanding the human mind and its conceptual categories. There are two further points for which Quine credits the pragmatists somewhat more willingly: the man-made nature of truth and the social character of meaning. To the latter he refers as "behavioristic semantics," stressing that it was Dewey, rather than Wittgenstein, who first insisted "that there is no more to meaning than is to be found in the social use of linguistic forms" (1981b: 36–37).[2] The concluding lines of this rather critical paper on pragmatism are a bit more generous than its opening ones: Although he repeats the complaint that he "found little in the way of shared and distinctive tenets," he goes on to say: "The two best guesses seemed to be behavioristic semantics, which I so heartily approve, and the doctrine of man as truth-maker, which I share in large measure" (1981b: 37). Despite this acknowledgment, it remains a fact that Quine saw none of the five advances in empiricism as initiated by pragmatism and that he omitted the entire discussion of pragmatism from the version included in his collection.

Quine's somewhat dismissive attitude notwithstanding, he was apparently more deeply rooted in the American pragmatist tradition than his paper, as both the full and the abridged versions recognize. Most surprising, perhaps, are the similarities between Quine and James, a philosopher Quine criticizes in the full version of the pragmatism paper and hardly mentions elsewhere. Here are some examples. To begin with, James is committed to empiricism, a commitment evident not only in *Pragmatism* (1955) [1907] but also in *The Principles of Psychology* (James 1890), his pioneering attempt to turn psychology into an empirical science. His perception of pragmatism as continuous with empiricism is reflected in his portrayal of pragmatism as "a new name for some old ways of thinking" (the subtitle of *Pragmatism*). The principal epistemic desideratum, in his view, is conformity with experience. "But this all points to direct verifications somewhere, without which the fabric of truth collapses, like a financial system with no cash-basis whatever. ... Beliefs verified concretely by somebody are the posts of the whole superstructure" (1955 [1907]: 52). And again, "But all roads lead to Rome, and in the end and eventually, all true processes must lead to the face of directly verifying sensible experiences somewhere" (1955 [1909]: 141).[3]

Second, like Quine, James opposed the analytic-synthetic dichotomy. What traditional philosophers saw as eternal and incorrigible truths were for James only "the dead heart of the living tree" (1955 [1907]: 53). Not completely dead, however, for "how plastic even the oldest truths nevertheless really are has been vividly shown in our day by the transformation of logical and mathematical ideas, a transformation which seems even to be invading physics" (1955 [1907]: 53). The theorems of logic and mathematics and even some of the laws of nature were once conceived as representing "the eternal thoughts of the Almighty. His mind also thundered and reverberated in syllogisms. He also thought in conic sections, squares and roots and ratios, and geometrized like Euclid" (1955 [1907]: 48). In fact, however (James argues), all of these laws "are only a man-made language, a conceptual shorthand ... in which we write our reports of nature, and languages, as is well known, tolerate much choice of expression and many dialects" (1955 [1907]: 49). James's reason for seeing the system as man-made is the same as Quine's, namely, that language, with its categories and classifications, is a human creation. There is no privileged language that can be singled out as a true description of reality, no language that nature should have used to describe itself, so to speak. "The trail of the human serpent is thus over everything" (1955 [1907]: 53). The similarity with Quine is also manifest in James's account of the dual traffic between theory and experience. Whereas we typically change theory to accommodate recalcitrant experiences, both Quine and James also countenance the reverse process, whereby we sacrifice an observation sentence (or reinterpret it) in order to save parts of our theory. The feasibility of this option speaks against a simplistic picture of observation as a secure basis to which every theoretical sentence can be reduced. "New truths thus are resultants of new experiences and of old truths combined and mutually modifying one another" (1955 [1907]: 113).

Finally, James is explicit about underdetermination, discretion, and the criteria that are involved in the decision on the preferred theory.

> Yet sometimes alternative theoretic formulas are equally compatible with all the truths we know, and then we choose between them for subjective reasons ... taste included, but consistency both with previous truth and with novel fact is always the most imperious claimant.
>
> *(1955 [1909]: 142)*

When faced with a problem, a tension discovered within the existing system or a new experience that seems hard to fit into it, one should try to save what one can from the old system "for in this matter of belief we are all extreme conservatives" (1955 [1907]: 50). The reasonable method thus aims at "a minimum of disturbance," or "a minimum of modification" (Ibid.). The idea is not only identical with that of Quine in terms of substance but also uses the same terminology. "New truth ... marries old opinion to new fact so as ever to show a minimum of jolt, a maximum of continuity. We hold a theory true just in proportion to its success in solving this 'problem of maxima and minima'" (1955 [1907]: 50–51).

As far as I know, Quine nowhere mentions the similarity between his ideas and those of James. According to his scientific autobiography, however, James's *Pragmatism* was one of the only two philosophy books Quine read as a teenager. "I read them compulsively and believed and forgot all" (1986a: 6).[4]

The most important difference between Quine and the founding fathers of American pragmatism pertains to the notion of truth. Ironically, Quine's position on truth is not derived from any of his pragmatist predecessors, and yet, it is here that his pragmatism receives its clearest manifestation. Quine notes (1981b: 31) that we have no way of comparing theories in terms of their similarity to one another or in terms of their distance from the truth, a comparison he takes to be presupposed by Peirce's definition of truth. He certainly rejected the correspondence theory of truth (whatever that title could mean). On these points James and Quine are once more in agreement. But Quine was just as opposed to James's ideas about truth as he was to Peirce's. He characterizes his own concept of truth as immanent: "we are always talking within our going system when we attribute truth; we cannot talk otherwise" (1981b: 34).[5] It is our best scientific theory that tells us what is true and what is real. To accept this theory and still refuse to acknowledge its truth (or the reality of the entities it invokes) is senseless, according to Quine. The immanence of truth goes hand in hand with "unregenerate realism, the robust state of mind of the natural scientist who has never felt any qualms beyond the negotiable uncertainties internal to science" (1981b: 28). Consequently, for Quine, "physical objects are real, right down to the most hypothetical of particles, though this recognition of them is subject, like all science, to correction" (1981b: 33). Quine's immanent conception of truth is succinctly characterized in *Word and Object* as the denial of a "cosmic exile," namely, "a vantage point outside the conceptual scheme that he takes in charge" (1960: 275). But Quine is concerned that this immanent notion of truth "has the ring of cultural relativism" (1975: 327), a position he finds paradoxical:

> Truth, says the cultural relativist, is culture bound. But if it were, then he, within his culture, ought to see his own culture-bound truth as absolute. He cannot proclaim cultural relativism without rising above it, and he cannot rise above it without giving it up.
>
> *(1975: 327–328)*

Over the years, as a result of his preoccupation with the problems of underdetermination (of theory) and indeterminacy (of translation), Quine vacillated on the question of truth, weighing what he calls the ecumenical view, which grants the truth of "foreign" alternatives to "our own" theory, and the sectarian view that is more in harmony with immanent truth.[6] At the end of this winding path, truth and a kind of pragmatic realism consonant with it win the day. "Science is seen as pursuing and discovering truth rather than as decreeing it. Such is the idiom of realism, and it is integral to the semantics of the predicate 'true'" (1995: 67). The immanent concept of truth is a huge innovation. It allows Quine to decline skepticism while embracing fallibilism and to reject a naïve word-world correspondence while embracing (a modest conception of) truth and reality. Whether or not Quine saw himself as a pragmatist, in enriching the pragmatist tradition with this new synthesis of a number of its central tenets, Quine's position is thoroughly pragmatic.

Notes

1 See, however, Chapter 6 of *Conventionalism* (Ben-Menahem 2006) for a discussion of Quine's second thoughts on underdetermination.
2 Quine had already credited Dewey with the position usually ascribed to Wittgenstein in "Ontological Relativity" (Quine 1969: 27). It is a matter of debate, however, whether Wittgenstein's conception of meaning is in fact as close to that of Dewey as Quine suggests. See, for example, Glock (1996: 376ff).
3 The distance between this firm commitment to empirical support and the common image of James as sanctioning irresponsible make-believe should be obvious but has been repeatedly overlooked.
4 There are of course also significant differences between Quine and James. Unlike Quine, James was deeply engaged with moral philosophy and, despite his empiricist bent, sought to make room for metaphysics and religious belief.
5 The connection between Quine's naturalism and his immanent concept of truth is at the center of Verhaegh (2018).
6 For substantiation of this claim, see Gibson (1986), Quine's reply (1986b), and my work (2006).

References

Ben-Menahem, Y. (2006) *Conventionalism* (Cambridge: Cambridge University Press).
Gibson, R.F. Jr. (1986) 'Translation, Physics and Facts of the Matter', in L.A. Hahn and P.A. Schilpp (eds.), *The Philosophy of W.V. Quine, The Library of Living Philosophers* XVIII (LaSalle: Open Court), pp. 139–154.
Glock, H.J. (1996) *A Wittgenstein Dictionary* (Oxford: Blackwell).
James, W. (1890) *The Principles of Psychology* (New York: Holt).
James, W. (1955) [1907] [1909] *Pragmatism and Four Essays from The Meaning of Truth* (New York: Meridian Books).
James, W. (1956) [1897] *The Will to Believe and Other Essays in Popular Philosophy* (New York: Dover).
Quine, W.V. (1953) [1951] 'Two Dogmas of Empiricism', in *From a Logical Point of View* (Cambridge, MA: Harvard University Press), pp. 20–46.
Quine, W.V. (1960) *Word and Object* (Cambridge, MA: MIT Press).
Quine, W.V. (1969) 'Ontological Relativity', in *Ontological Relativity and Other Essays* (New York: Columbia University Press), pp. 26–68.
Quine, W.V. (1970) *Philosophy of Logic* (London: Prentice-Hall).
Quine, W.V.O. (1975) 'On Empirically Equivalent Systems of the World', *Erkenntnis* 9, 313–328.
Quine, W.V. (1981a) 'Five Milestones of Empiricism', in *Theories and Things* (Cambridge, MA: Harvard University Press), pp. 67–72.
Quine, W.V. (1981b) 'The Pragmatists' Place in Empiricism', in R.J. Mulvaney & P.M. Zeltner (eds.), *Pragmatism: Its Sources and Prospects* (Columbia: University of South Carolina Press), pp. 21–39.
Quine, W.V. (1986a) 'Autobiography', in L.E. Hahn & P.A. Schilpp (eds.), *The Philosophy of W.V. Quine* (La Salle: Open Court), pp. 3–46.
Quine, W.V. (1986b) 'Reply to Roger F. Gibson Jr.', in L.E. Hahn & P.A. Schilpp (eds.), *The Philosophy of W.V. Quine* (La Salle: Open Court), pp. 155–158.
Quine, W.V.O. (1995) *From Stimulus to Science* (Cambridge, MA: Harvard University Press).
Verhaegh, S. (2018) *Working from Within: The Nature and Development of Quine's Naturalism* (New York: Oxford University Press).

10
WILFRID SELLARS AND PRAGMATISM

Willem A. deVries

During Wilfrid Sellars' (1912–1989) life, he seemed to be an arm's length removed from Pragmatism, but, post-mortem, his relation to Pragmatism looks much closer. His apparent distance from Pragmatism is rooted in his family history. In his early years, Sellars viewed much of philosophy through his father's eyes.[1] Roy Wood Sellars was himself an accomplished philosopher who taught at the University of Michigan for 40 years. *Critical Realism* (1916) and *Evolutionary Naturalism* (1922) established him as a leading American philosopher of the day, but they also declared his independence from both the idealism still dominant in American universities and the Pragmatism that was, at the time, still ascendent. According to Wilfrid, Roy Wood regarded Pragmatism "as shifty, ambiguous, and indecisive. One thinks in this connection of Lovejoy's 'thirteen varieties,' though that, my father thought, would make too tidy a picture" (Sellars 1979 [NAO], "Introduction" ¶1).[2]

There was another criticism of Pragmatism imbibed from his father:

> "Time is unreal." "Sense data are constituents of physical objects." "Mind is a distinct substance." "We intuit essences." These are issues you can get your teeth into. By contrast, Pragmatism seemed all method and no results.
>
> *(Sellars 1979 [NAO], "Introduction" ¶2)*

Roy Wood did not shy away from metaphysics, but he did not see positive views coalescing out of Pragmatism's initial methodological prescriptions.

Though Pragmatism was not influential from his beginnings, it entered Wilfrid's thought fairly early on, before he began publishing. He encountered it as his "thought began to crystallize" (Sellars 1979 [NAO], "Introduction" ¶4). Dewey

> caught me at a time when I was moving away from 'the Myth of the Given' (antecedent reality?) and rediscovering the coherence theory of meaning. Thus it was Dewey's Idealistic background which intrigued me the most. I found similar themes in Royce and later in Peirce. I was astonished at what I had missed.
>
> *(Sellars 1979 [NAO], "Introduction" ¶4)*

(A life-long fascination with Idealism is also an important thread in Sellars' thinking.)

Sellars never outright declared allegiance to Pragmatism, nor is it mentioned in his "Autobiographical Reflections," where he reviews his early development. But pragmatist views show up

DOI: 10.4324/9781315149592-12

at critical junctures in Sellars' arguments. Because of this, after Sellars' death, he is increasingly described as a pragmatist.[3] Indeed, Sellars is treated as a canonical pillar on which the recent neo-Pragmatism of Richard Rorty, Robert B. Brandom, and Huw Price is erected.

In the following, we will look at several themes, concepts, and arguments that Sellars shares with Pragmatism.[4]

The priority of the practical in the constitution of meaning

The initial spark igniting Pragmatism was Peirce's maxim: "Consider what effects, that might conceivably have practical bearings, we conceive the object of our conception to have. Then, our conception of these effects is the whole of our conception of the object" (Peirce, CP 5.402). Our concepts of things concern the practical role those things can play in the economy of the world and especially our interactions with them. Correspondingly, our concept of a concept is exhausted by the practical bearings of concepts: they function as intermediaries modulating the responses made to inputs received as we encounter and cope with the world.

In the 20th century, this view (not always cognizant of its pragmatist heritage) has crystallized in functionalist theories of semantics. Some proponents of such theories emphasize the role of inference, for one "practical bearing" of a concept is its contribution to the valid inferences its users draw, whether in the context of belief acquisition and change or the context of planning action. This "inferentialism" has been defended and elaborated by Robert Brandom as central to his neo-Pragmatism.[5] While Brandom draws significant inspiration from Sellars (especially Sellars' 1953 essay "Inference and Meaning" [IM]), "inferentialism" seems too narrow in its connotations to capture Sellars' version of semantic functionalism.

Sellars certainly believes that a great deal of the semantic content of a concept is determined by the role it plays in valid inferences. This is patently true of the "logical words" themselves, which can be defined in terms of the inferences they license. But in Sellars' view, inferential role is significant for *every* concept. He distinguishes between formal and material inferences, where material inferences, although formally invalid, "determine the descriptive meaning of the expressions of a language within the framework established by its logical transformation rules" (Sellars 1953 [IM]: 336). For example, the inference from "X is red" to "X is colored" is formally invalid but still a perfectly good inference. It is not, according to Sellars, an enthymeme; there is a material rule of inference licensing this inference in English that is partially constitutive of the meaning of "red" and "color." Material inferences vary in kind. Besides the genus-species connection of the "red"-"colored" example, empirically well-verified connections among concepts (e.g., familiar causal connections) tend to become ensconced in language as material rules of inference ("It's raining, so the roads are wet.").

Inferences, however, are connections among items with propositional structure, but other connections are important to the semantic content of a concept or thought. Sellars also recognizes language-entry and language-exit transitions. Observation reports are his standard example of language-entry transitions, but perhaps greetings such as "Hello" or inquiries such as "What time is it?" count also as language-entry transitions. Sellars' examples of language-exit transitions are statements of intention immanently acted upon, e.g., "I'm going now," followed by the speaker's exit. Austinian performatives ("I now pronounce you man and wife") would, I believe, also count as language-exit moves, since extra-linguistic facts change as a result.

In "Some Reflections on Language Games," Sellars criticizes some pragmatists for thinking that viewing language as an instrument affords an *analysis* of "meaning" or "truth." But he subsequently remarks,

> if the pragmatist's claim is reformulated as the thesis that the language we use has a much more intimate connection with conduct than we have yet suggested, and that this connection

is intrinsic to its structure as language, rather than a "use" to which it "happens" to be put, then Pragmatism assumes its proper stature as a revolutionary step in Western philosophy.
(Sellars 1954 [SRLG] ¶49 in SPR: 340)[6]

Sellars echoes pragmatist thinking in his treatment of truth as well as meaning. His views on truth are not Peirce's, but they are Peirce-inspired. The predicate "is true" licenses a *performance*, namely the assertion of the statement to which truth is attributed:

> for a proposition to be true is for it to be assertible, where this means not *capable* of being asserted (which it must be to be a proposition at all) but *correctly* assertible; assertible, that is, in accordance with the relevant semantical rules, and on the basis of such additional, though unspecified, information as these rules may require. ... 'True', then, means *semantically* assertible ('S-assertible') and the varieties of truth correspond to the relevant varieties of semantical rule.
> (Sellars 1967 [SM], IV ¶26: 101)

Thus, for Sellars, truth is not so much what we're *destined* to believe as what it would be *ideal* to believe, given the constraints of our conceptual framework and the best possible functioning of our linguistic (and epistemological) faculties. As in Tarski, assessments of truth always require reference to a language or framework. Moreover, Sellars thinks we can develop the notion of the truth of a proposition in one framework F1 with respect to a second framework F2. But,

> however many sophisticated senses of 'true' may be introduced, and however important they may be, the connection of truth with *our current conceptual structure* remains essential, for the cash value of S-assertibility is assertion by us *hic et nunc*.
> (Sellars 1967 [SM], V ¶53: 134)

Thus, in this sense, Sellars does not see an essential reference to the future in attributions of truth.

However, the possibility of error, even systemic error, always haunts human endeavor. Sellars provides not only for belief change—recognizing that what one thought was true is not (or vice versa)—but also for conceptual change, even categorial change. There are several drivers of conceptual change. One is the changing environment: natural disasters, changing climates, changing social relations, etc. may drive belief change but can also drive conceptual change as well. The internal pressure to adapt ever better to even a relatively stable environment, to improve one's understanding of the world by the development of new concepts that better modulate our interaction with the world, is perhaps even more significant as a driver of conceptual change. Because of this, we can recognize that truth-as-we-see-it-now—even if it is an ideal—need not coincide with the truth-we-can-project as the ideal end of all inquiry. We not only revise our beliefs, we revise the conceptual frameworks that enable us to have beliefs, creating the possibility of new and better *kinds* of beliefs. Thus, the Peircean notion that truth emerges out of an ongoing process re-emerges in Sellars' thought.[7]

As Peircean as Sellars' treatment of meaning and truth may be, he departs from Peirce and some other pragmatists in his belief that any empirically meaningful language must also contain utterances via which language users *picture* their world.[8] For Sellars, neither meaning nor truth is a *relation*; meaning statements and truth claims neither *describe* nor *attribute relations* between words and the world; they *classify* expressions functionally or *evaluate* their status in the framework. But Sellars does believe that in any empirically meaningful language, there must be some basic sentences that, on some occasions of their use, realize a complex *picturing* relation between the utterance, construed now as a natural (phonetic, graphic, or whatever) object and the state of affairs it represents.[9] Such a picturing relation is, indeed, a natural relation, an isomorphism or co-variation through which the linguistic superstructure is knit into the causal structure of the world. This element in Sellars' thought differs from classical forms of Pragmatism, and consciously so.

> Peirce himself fell into difficulty because, by not taking into account the dimension of "picturing," he had no Archimedeian point outside the series of actual and possible beliefs in terms of which to define the ideal or limit to which members of this series might approximate.
> (Sellars 1967 [SM], V ¶75: 142)

The measure of the overall adequacy of a conceptual framework is its ability to permit construction of arbitrarily fine-grained and accurate pictures of tracts of space-time and the arrangement of objects in them. Science is the methodologically rigorous refinement of a framework that permits a maximally adequate picturing relationship.

Against the given

Another theme Sellars shares with Pragmatism is opposition to the notion that our knowledge of the world involves any kind of foundational *given*. By a given, I mean cognitive states that possess positive epistemic status independent of the epistemic status of any other cognitive state. The notion of a given has traditionally served to avoid both regress and circularity in the justificational structure of knowledge: some cognitive states were taken to be, as it were, absolute beginning points, foundation stones underlying the edifice of our knowledge. It was also traditionally believed that such states had to be *certain, incorrigible,* or even *infallible*, because, first, nothing could impugn them and, second, putatively, if nothing is certain, nothing can even be probable. While the pragmatist C. I. Lewis bought this latter idea, other pragmatists did not (consider, e.g., Dewey's *The Quest for Certainty*). One consequence of rejecting the given is accepting fallibilism, which is common among the pragmatists.

Sellars' classic essay "Empiricism and the Philosophy of Mind" announces itself as an attack on "the entire framework of givenness." Some have (mistakenly) construed this only in terms of an attack on the certainty of the foundations of our knowledge, but his critique reaches much deeper than that.[10]

The most important dimension of the attack on the given is the insistence that no cognitive state, all on its own, possesses any determinate content or epistemic status independently of its epistemic relations to other cognitive states. Thus, Sellars defends not only a coherence theory of conceptual content but also a kind of coherence theory of justification that is incompatible with any *given*. Sellars' epistemic holism is more systematic than that of Quine—a still more "thoroughgoing pragmatism."[11] It should be noted at once, though, that the kind of coherentism Sellars endorses for justification is not the simple idea that beliefs that hang together suffice for knowledge.

> [E]mpirical knowledge, like its sophisticated extension, science, is rational, not because it has a foundation but because it is a self-correcting enterprise which can put any claim in jeopardy, though not all at once.
> (Sellars 1956 [EPM] in SPR §38: 170)

The echoes of Peirce and Dewey are unmistakable here: knowledge can be made sense of only as the upshot of an ongoing process in which our cognitive behaviors are successively attuned to the world and our purposes. There is room for "input" from the world, and such "input" is sometimes feedback from the "output" crystallized in action. But no belief faces the "tribunal of experience" alone and independent; it couldn't make sense apart from the framework.

This second, more profound dimension of the abandonment of the given reflects back on the first: Sellars' fallibilism is profound, for he believes that the Manifest Image, the conceptual framework developed over millennia in terms of which humans came to understand themselves as persons amid a world populated with physical objects and other animals, will ultimately prove to be *false* and the world so constituted to be (in the Kantian sense) merely phenomenal. That framework will be replaced by a much more thoroughly elaborated and empirically well-supported Scientific Image. Sellars often packaged himself as a latter-day Kantian in analytic clothing, but he clearly departed from Kant in espousing scientific realism: as the self-correcting enterprise of empirical knowledge proceeds, especially in its cleaned-up, methodologically rigorous, and instrumentally sophisticated extension, science, it leads us to know what things in themselves are.

Sociality

The last doctrine common to Pragmatism and Sellars I will discuss here (given space limitations) is the sociality of thought and inquiry. Construed narrowly, it is a point about the structure of inquiry, but it reaches far deeper, for it is a rejection of the individualistic and methodologically solipsistic Cartesian understanding of the mind that dominated modern Western thought after the 17th century. How and why the pragmatists endorsed a thoroughly social view of humanity is treated in other essays in this volume, so I will leave that. But note that this is a theme Pragmatism shares with post-Kantian German Idealism, and thus is doubly attractive to Sellars.

Unlike the German Idealists, the pragmatists and Sellars are consciously post-Darwinian philosophers: they know that, however cloudy the origins of self-conscious, partially rational, language-using humanity may be, it is a layered story about the evolution of a social species that developed a sophisticated communication system that enabled a new level of enculturated learning that transformed its behavioral possibilities by vastly increasing both the kinds of behaviors (now, *actions*) available and the forms of control over them.

It is only within a linguistic community that a rich and determinate system of meanings can develop, and it is a communal enterprise to refine the system of meanings employed in order to better achieve the epistemic and practical goals of the community. We cannot understand knowledge or action in terms of an atomism that regards each person as an independent locus only accidentally joined with others in the pursuit of their goals. Concepts are communal possessions. Sociality is thus essential to the possibility of conceptuality. Perhaps even more important, it is essential to normativity in general. "To think of a featherless biped as a person is to think of it as a being with which one is bound up in a network of rights and duties ... to think of oneself and it as belonging to a community" (Sellars 1962 [PSIM] ¶112, in SPR: 39, in ISR: 407–408). Communities, for Sellars, are *practical* at their core:

> the fundamental principles of a community, which define what is "correct" or "incorrect," "right" or "wrong," "done" or "not done," are the most general common *intentions* of that community with respect to the behaviour of members of the group.
> *(Sellars 1962 [PSIM] ¶113, in SPR: 39, in ISR: 408)*

Sellars recognizes pressures that push us toward recognizing the maximal community, "the 'republic' of rational beings (cf. Kant's 'Kingdom of Ends')" (Sellars 1962 [PSIM] ¶112, in SPR: 39, in ISR: 407), governed by a universal set of rules.

> "Why one set of rules rather than another? How is the adoption of a set of rules itself to be justified?" I should like to be able to say that one justifies the adoption of rules pragmatically,

and, indeed, this would be at least a first approximation to the truth. The kinship of my views with the more sophisticated forms of pragmatism is obvious.

(Sellars 1949 [LRB] ¶43: 314)

As we've now seen, Sellars is at least kin to Pragmatism—if not a direct son, at least a close cousin, and a sire to one branch of a complex family tree. One of the complaints about the neo-pragmatists who claim descent from Sellars voiced by the not-so-neo is the neglect of *experience* in favor of its linguistic expression.[12] In Sellars, however, the nature of experience and its contribution to thought and action are constant themes. Many of the other tensions that exist within Pragmatism, say, between realism and forms of verificationism, or a certain cloudiness about truth, emerge within Sellars' own thinking as well and have spawned divergent readings of Sellars, just as there are divergent readings of the pragmatists.

Notes

1. See Sellars (1979 [NAO]), "Introduction" ¶1. Since publishing dates can be hard to remember (especially when the piece has been reprinted several times), I include the now industry-standard abbreviations for Sellars' works in citations.
2. Sellars' reference is to Lovejoy (1908a, 1908b).
3. For a different view, see Olen (2015).
4. See also O'Shea (2020).
5. See Brandom (1994, 2000).
6. The version of Sellars (1954 [SRLG]) reprinted in ISR is the earlier version from *Philosophy of Science* 21 (1954). It does not include substantial additions made by Sellars for the SPR version. This passage occurs there in ¶34, page 40.
7. For more on Sellars on truth, see Shapiro (2019).
8. Sellars' notion of picturing is based on the notion Wittgenstein deployed in his *Tractatus Logico-Philosophicus*, but with some significant differences. For instance, Wittgenstein's picturing relation is a relation between *facts*, whereas Sellars' is a relation between *objects*. Sellars' distinction between meaning and picturing is also similar to Huw Price's distinction between i-representation and e-representation in his Tilburg lectures: (Price 2013). For further argument that picturing fits well with Pragmatism, see Sachs (2018).
9. He also occasionally speaks of it as "mapping" relation. See, for instance, Sellars (1979 [NAO], chapter 5).
10. See deVries and Triplett (2000: xxvi–xxx) for a review of this literature.
11. See, for instance, "Sellars and Quine: Compare and Contrast" in Rosenberg (2007).
12. See, for example, Levine (2019).

Works cited

Brandom, R. B. (1994) *Making It Explicit*. Cambridge, MA: Harvard University Press.
———. (2000) *Articulating Reasons: An Introduction to Inferentialism*. Cambridge, MA: Harvard University Press.
deVries, W. A. and T. Triplett. (2000) *Knowledge, Mind, and the Given: A Reading of Sellars' "Empiricism and the Philosophy of Mind."* Indianapolis, IN: Hackett Publishing.
Dewey, J. (1960) *The Quest for Certainty: A Study of the Relation of Knowledge and Action*. New York: Capricorn Books.
Levine, S. (2019) *Pragmatism, Objectivity, and Experience*. Cambridge: Cambridge University Press.
Lovejoy, A. O. (1908a) "The Thirteen Pragmatisms. I." *The Journal of Philosophy, Psychology, and Scientific Methods*, 5, 1: 5–12. www.jstor.org/stable/2012277.
———. (1908b) "The Thirteen Pragmatisms. II." *The Journal of Philosophy, Psychology and Scientific Methods*, 5, 2: 29–39. www.jstor.org/stable/2011563.
Olen, P. (2015) "The Realist Challenge to Conceptual Pragmatism." *European Journal of Pragmatism and American Philosophy*, 7: 152–167. https://journals.openedition.org/ejpap/413.
O'Shea, J. R. (2020) "How Pragmatist was Sellars? Reflections on an Analytic Pragmatism." In *Wilfrid Sellars and Twentieth-Century Philosophy*, edited by Stefan Brandt and Anke Breunig. New York: Routledge, 174–206.

Peirce, C. S. (1965) *Collected Papers of Charles Sanders Peirce*. Volume V. *Pragmatism and Pragmaticism*. Edited by Charles Hartshorne and Paul Weiss. Cambridge, MA: Harvard University Press.

Price, H. (2013) *Expressivism, Pragmatism and Representationalism*. Cambridge: Cambridge University Press.

Rosenberg, J. (2007) "Sellars and Quine: Compare and Contrast." In *Wilfrid Sellars: Fusing the Images*. Oxford: Oxford University Press, 33–46.

Sachs, C. B. (2018) "'We Pragmatists Mourn Sellars as a Lost Leader': Sellars's Pragmatist Distinction between Signifying and Picturing." In *Sellars and the History of Modern Philosophy*, edited by Luca Corti and Antonio M. Nunziante. New York: Routledge, 157–177.

Sellars, R. W. (1916) *Critical Realism: A Study of the Nature and Conditions of Knowledge*. Chicago, IL: Rand-McNally and Co.

———. (1922) *Evolutionary Naturalism*. Chicago, IL: Open Court.

Sellars, W. S. (1949) [LRB]. "Language, Rules and Behavior." In *John Dewey: Philosopher of Science and Freedom*, edited by Sidney Hook. New York: The Dial Press, 289–315.

———. (1953) [IM]. "Inference and Meaning." *Mind*, 62: 313–338.

———. (1954) [SRLG]. "Some Reflections on Language Games." *Philosophy of Science*, 21: 204–228. Expanded and reprinted in Sellars 1963 (SPR); original reprinted in Sellars 2007 (ISR).

———. (1956) [EPM]. "Empiricism and the Philosophy of Mind." In *Minnesota Studies in the Philosophy of Science*, Vol. I, edited by Herbert Feigl and Michael Scriven. Minneapolis: University of Minnesota Press, 253–329. Reprinted in Sellars 1963 (SPR).

———. (1962) [PSIM]. "Philosophy and the Scientific Image of Man." In *Frontiers of Science and Philosophy*, edited by Robert Colodny. Pittsburgh: University of Pittsburgh Press, 35–78. Reprinted in Sellars 1963 (SPR) Sellars 2007 (ISR).

———. (1963) [SPR]. *Science, Perception and Reality*. London: Routledge and Kegan Paul. Reissued by Ridgeview Publishing Company in 1991.

———. (1967) [SM]. *Science and Metaphysics: Variations on Kantian Themes*. London: Routledge and Kegan Paul. Reissued by Ridgeview Publishing Company in 1992.

———. (1979) [NAO]. *Naturalism and Ontology: The John Dewey Lectures for 1974*. Atascadero, CA: Ridgeview Publishing Company.

———. (2007) [ISR]. *In the Space of Reasons*. Edited by Kevin Scharp and Robert B. Brandom. Cambridge, MA: Harvard University Press.

Shapiro, L. (2019) "Sellars, Truth Pluralism, and Truth Relativism." In *Wilfrid Sellars and Twentieth-Century Philosophy*, edited by Stefan Brandt and Anke Breunig. New York: Routledge, 174–206.

Further reading

Koons, J. (2009) *Pragmatic Reasons: A Defense of Morality and Epistemology*. London: Palgrave Macmillan. (Defends a sophisticated version of Pragmatism, resting on a novel account of strategy-based cooperative rationality.)

Kraut, R. (2010) "Universals, Metaphysical Explanations, and Pragmatism." *The Journal of Philosophy*, 107, 11: 590–609. (An interpretation of Sellarsian nominalism with an eye on Pragmatism.)

Maher, C. (2012) *The Pittsburgh School of Philosophy: Sellars, McDowell, Brandom*. New York: Routledge. (A coherent presentation of Sellars and his epigones.)

Williams, M. (2016) "Pragmatism, Sellars, and Truth." In *Sellars and his Legacy*, edited by James R. O'Shea. Oxford: Oxford University Press, 223–259. (Argues that Sellarsian picturing cannot do what Sellars wants it to and that while pragmatists can take a lot from Sellars, his views about truth are not among them.)

11
RICHARD RORTY: NARRATIVE AS ANTI-AUTHORITARIAN THERAPY AND AS CULTURAL POLITICS

Susan Dieleman

The diversity of traditions with which Richard Rorty engaged, the number of thinkers he invoked, and the sheer volume of writing he produced, can make approaching and engaging with his work a daunting task. Yet as Rorty himself points out, he had "really only one idea: the need to get beyond representationalism, and thus into an intellectual world in which human beings are responsible only to each other" (Rorty 2010: 474).[1] Or, as he puts it in an interview, "I think of my work as trying to move people away from the notion of being in touch with something big and powerful and nonhuman" (Mendieta 2006: 49). These passages suggest there are two moves that constitute Rorty's "one idea": (1) articulating the need to get beyond the notion that we have a responsibility to know by accurately representing what is fixed and eternal "out there," a move captured by the term "anti-authoritarianism," and (2) showing that responsibility to others is what remains after the representationalist project is abandoned, which he, late in his career, dubs "cultural politics."[2]

Insofar as Rorty has a method for making these moves, it is the use of stories, or narrative. But for Rorty, narrative is not just a method. It is simultaneously *how* he tries to persuade his readers as well as *what* he wants to persuade his readers of. Narrative is both the means (narrative as anti-authoritarian therapy) and the end (narrative as cultural politics), and these two aspects of his thought are connected in complex and sometimes confusing ways. He not only uses narrative as a form of therapy to dialectically dissolve the philosophical problems that give shape to and grow out of the representationalist paradigm but also, at the same time, commends narrative as the sort of activity philosophy should engage in once it has climbed up and abandoned the ladder of representationalism.

Adopting a new self-image of philosophy as cultural politics (to which I return below) is the metaphilosophical end at which Rorty aims. Yet to achieve this end requires that philosophy abandon the self-image it has adopted since at least the time of Descartes, a self-image of philosophy as capable of unmasking and accurately representing the nature of Reality. Rorty's strategy for accomplishing this anti-authoritarian goal, and promoting the self-image of philosophy as cultural politics, is a narratival form of therapy. By telling stories, and stories about the historical development of various features of the philosophical paradigm of representationalism in particular, Rorty aims to disabuse us of our apparent need to answer to an external authority. His goal is to locate the dialectic that leads us through and beyond the representationalist paradigm to what he will ultimately call cultural politics.

Rorty's therapeutic, anti-authoritarian narrative is first offered in *Philosophy and the Mirror of Nature* (1979; hereafter *PMN*) and the essays collected in *Consequences of Pragmatism* (1982; hereafter *CP*).[3] In the introduction to *PMN*, Rorty notes that the book is "therapeutic rather than constructive" (7); it aims to dissolve so-called perennial philosophical problems rather than propose new solutions to them. Rorty's particular therapeutic strategy involves narrating a story, one that – in a phrase he borrows from Wilfrid Sellars – attempts to see "how things, in the largest sense of the term, hang together, in the largest sense of the term" (*CP*: 29). In *PMN*, Rorty weaves a dialectical narrative about the development of modern philosophical inquiry, a paradigm according to which philosophy has a "special understanding of the nature of knowledge and of mind" and can therefore legitimize all other areas of knowledge and culture (*PMN*: 3). Beginning with Descartes through to (then) contemporary analytic philosophy, it narrates how philosophy of mind ("our glassy essence") and epistemology ("mirroring") came to dominate the modern philosophical landscape and how the foundationalist (read: authoritarian) urge exemplified by these fields has shaped philosophical inquiry since. *PMN* concludes with a commendation of hermeneutics, not to take the place of foundationalist philosophical inquiry but as "an expression of hope that the cultural space left by the demise of epistemology will not be filled" (*PMN*: 315). This hope, if not the particular expression of it as hermeneutics, can be found in the three heroes identified in *PMN*: Heidegger, Wittgenstein, and Dewey, each of whom, in their later works, abandoned foundationalism and focused on "warning us against those very temptations to which he himself had once succumbed" (*PMN*: 5).

Heidegger, Wittgenstein, and Dewey, as well as thinkers who are influenced by them, make regular appearances throughout Rorty's oeuvre, and each is put to work in service of Rorty's "one idea."[4] For example, Rorty also finds the anti-authoritarian impulse in the group of thinkers he labels "textualists" – literary critics and post-structuralists who sometimes "take their point of departure from Heidegger" (*CP*: 139). They have an aim similar to Rorty's, namely, to abandon the project and hope of discovering how things *really* are. In turn, textualists, he claims, "start from the pragmatist refusal to think of truth as correspondence to reality" (*CP*: 151).[5] Rorty regularly suggests that the anti-authoritarianism he recommends is best modeled for us by pragmatism, which he dubs "the chief glory of our country's intellectual traditions" (*CP*: 160). According to Rorty, pragmatism – and Dewey's pragmatism in particular, which he thinks "will have the greatest utility in the long term" (Rorty 1999: 12) – is clearly anti-authoritarian. This is a point Rorty makes explicitly in his 1999 paper, "Pragmatism as Anti-authoritarianism," when he writes, "As Dewey saw it, whole-hearted pursuit of the democratic ideal requires us to set aside *any* authority save that of a consensus of our fellow humans" (Rorty 1999: 7). Moreover, he lauds and emulates Dewey's use of narrative in service of this goal. He has, Rorty claims, an imaginative ability to tell "great sweeping stories about the relation of the human present to the human past" (14), stories that make it possible to set aside an external authority to which we owe allegiance and to place hope instead in cooperation with our fellow citizens and human beings.

It is, in part, the pragmatists' embrace of a Darwinian naturalism that enables them to adopt this anti-authoritarian position, and Rorty likewise adopts a naturalism that sees the evolution of language and culture as continuous with biological evolution. In *Contingency, Irony, and Solidarity* (1989; hereafter *CIS*), he recommends an evolutionary view of language in place of the view of language as a medium capable of doing a better or worse job of representing. Language does not represent the external world or internal beliefs but rather is a tool that can be used to cope in better and worse ways with our environment. And, because it is a tool, language is the sort of thing that evolves. As Rorty puts it, "Our language and our culture are as much a contingency, as much a result of thousands of small mutations finding niches (and millions of others finding no niches), as are the orchids and the anthropoids" (*CIS*: 16). This same naturalistic, and ultimately anti-authoritarian, insight is applied to the self. In *Truth and Progress* (1998; hereafter *TP*), Rorty writes, "it behooves us to give the self-image Darwin suggested to us a whirl, in the hope of having fewer philosophical problems on our hands" (*TP*: 48). Rather than understanding the human self as having a core essence knowledge of which philosophical inquiry makes possible, we should understand humans simply as "slightly-more-complicated-animals" (*TP*: 48).

Rorty's "one idea" can be detected in his later work as well, though it sometimes goes by other names. In *Philosophy as Cultural Politics* (2007; hereafter *PCP*), he narrates a story that introduces the concept of "redemptive truth," a concept that for Rorty captures the idea that there exists "a set of beliefs which would end, once and for all, the process of reflection on what to do with ourselves" (*PCP*: 90). The possibility of obtaining such a set of beliefs is what Rorty claims has, throughout history, oriented religious, philosophical, and literary activity. Yet the story he narrates advocates moving from philosophy, where "true belief is of the essence," to literature, which "offers redemption through making the acquaintance of as great a variety of human beings as possible" (*PCP*: 91). In a literary culture, the only authority that remains is other people, and thus redemption is found "in non-cognitive relations to other human beings, relations mediated by human artifacts such as books and buildings, paintings and songs" (*PCP*: 93). In short, with the move from philosophy to literature, we abandon the attempt to define ourselves in relation to something like "the true nature of reality" and instead define ourselves in relation to and, importantly, *with* other human beings. Literature, he claims, is the best way to engage in this project of figuring out who we are and what we want to be.

It is in this sense that Rorty is an anti-philosophy philosopher. He encourages an end to the philosophical paradigm that has shaped the field since at least Descartes and reaches its apotheosis in analytic philosophy of language. Yet Rorty did not see *PMN*, where this therapeutic narrative is first offered, as an attack on analytic philosophy, as many of his critics did and some continue to do. Instead, he saw it as "an attempt to carry out the postivists' original program" by seeing where the dialectical plot of his narrative would lead (Mendieta 2006: 20). This does not mean, however, that Rorty sees philosophy actually coming to an end. Rather, he commends a different understanding of what philosophy should become. In reflecting on the end of *PMN*, he later claims that he should have tried to "make a transition from philosophy as a discipline to a larger and looser activity" (Mendieta 2006: 21).

The idea of philosophy as a "larger and looser activity" is philosophy as cultural politics, a narratival approach to doing philosophy. As noted above, narrative is both *how* Rorty tries to persuade his readers (narrative as anti-authoritarian therapy) and *what* he wants to persuade his readers of (narrative as cultural politics). We can see in the *content* of *PMN* and the essays of *CP* Rorty's admonition that we move beyond the representationalist paradigm. But in the *style* of this work, we can see his commendation of a new philosophical self-image. Philosophy would become "a study of the comparative advantages and disadvantages of the various ways of talking which our race has invented" (*CP*: xl). In *CP*, he dubs this new method "culture criticism"; it later becomes "cultural politics."

"Cultural politics," Rorty stipulates, "covers, among other things, arguments about what words to use" (*PCP*: 3). It includes debates about what words and topics to include in, and what words and topics to excise from, our shared final vocabulary. Cultural politics is about shifting vocabularies in small and significant ways, projects to which philosophers can contribute by, as Rorty himself does, telling stories that either block or open up paths of inquiry. Such projects will go better, Rorty contends, if they consider the costs and benefits of including or excising particular words and topics and, thus, particular paths of inquiry. The changes to final vocabularies brought about by cultural politics are what drive cultural progress. It is because cultural politics expands logical space by saying things never said before that the radically new becomes possible. As Rorty puts it, "Cultural politics is the least norm-governed human activity. It is the site of generational revolt, and thus the growing point of culture – the place where traditions and norms are all up for grabs at once" (*PCP*: 21).

For Rorty, the political backdrop best suited for and supported by cultural politics is liberalism. He contends that "A liberal society is one whose ideals can be fulfilled [...] by the free and open encounters of present linguistic and other practices with suggestions for new practices" (*CIS*: 60).

In fact, he claims the whole purpose of a liberal society is to grant the freedom required for individuals to engage in cultural political activity, where new words and topics and paths of inquiries are proposed and either adopted or rejected. Rorty's hope, of course, is that cultural politics will be used in support of more freedom, of greater richness and variety, of new possibilities. Rorty's liberalism is a feature of his thought much explored by critics, sympathetic or otherwise, one that he knows he cannot defend in any non-circular way. Like other thinkers influenced by pragmatism, Rorty believes "that a circular justification of our practices … is the only sort of justification we are going to get" (*CIS*: 57). Yet he defends it nonetheless as that set of political ideals that we, in the post-industrial West, are lucky enough to have inherited. No other political system guarantees the freedoms required to engage in cultural politics, to make life freer, richer, and fuller.

Importantly, Rorty remains committed throughout to the idea that narrative is the best way to achieve the kind of consensus on which liberal democracy depends. As he notes in *Philosophy and Social Hope* (1999; hereafter *PSH*),

> The appropriate intellectual background to political deliberation is *historical narrative* rather than philosophical or quasi-philosophical theory. […] Social and political philosophy usually has been, and always ought to be, parasitic on such narratives.
> *(PSH: 231–232; emphasis added)*

This prescription – along with his narratival method – is applied to the U.S. context in *Achieving Our Country* (1998; hereafter *AOC*) where Rorty tells a story that connects the proclivity to engage in philosophical theorizing about the country with the tendency to abandon the patriotic hope required to make the country into an improved version of itself. What he calls the "cultural Left" has prioritized knowledge over hope; "Hopelessness," he claims, "has become fashionable on the left – principled, theorized, philosophical hopelessness" (*AOC*: 37). Better, he claims, to engage in "real politics" by working to end sadism, which results in identity-based inequalities, and selfishness, which results in material inequalities.

Unsurprisingly, Rorty thinks that narratives – and novels that present us with "sad and sentimental stories" (*TP*: 185) in particular – are the best tool at our disposal for minimizing both sadism and selfishness. Novels offer compelling stories that sensitize us to the pain of others; they show how "they" are like "us" in a way required for achieving solidarity: each of us is capable of suffering as a result of others' sadism and selfishness. Coming to understand others' ability to suffer, sometimes at our own hands, helps broaden the scope of our moral community. Thus, the narratives that novels and other similar forms of writing offer make solidarity – and thus liberal democracy – more likely to succeed.

Notes

1 The essay "Philosophy as a Transitional Genre" is also published in Rorty's 2007 collection of essays, *Philosophy as Cultural Politics*. However, the first sections of the essay, from which this passage is taken, are excerpted from that earlier publication.

2 This entry does not engage with Rorty's earliest work in analytic philosophy, which is well represented in the volume *Mind, Language, and Metaphilosophy: Early Philosophical Papers* (2014; hereafter *MLM*). The editors of the volume rightly comment, "given that Rorty's thinking never did radically change direction, there is […] plenty to be found here for those interested in the mature Rorty" *(MLM*: 1).

3 Arguably, these anti-authoritarian themes are also present in some of Rorty's earliest papers, recently collected in *On Philosophy and Philosophers: Unpublished Papers, 1960–2000* (2020; hereafter *OPP*), such as "Philosophy as Spectatorship and Participation."

4 Rorty notes that his first volume of collected papers, *Objectivity, Relativism, and Truth* (1990; hereafter *ORT*), focuses on themes and figures in the analytic tradition, primarily Quine, Sellars, and Davidson. His second volume of collected papers, *Essays on Heidegger and Others* (1991; hereafter *EHO*), focuses

on themes and figures in the continental tradition, primarily Heidegger, Derrida, and Foucault. Both volumes overlap, in particular in the way that each puts these analytic or continental thinkers in conversation with pragmatism.

5 It's worth noting that this essay, "Nineteenth-Century Idealism and Twentieth-Century Textualism," is where Rorty discusses the Bloomian notion of "strong misreading." Rorty in his own work largely adopts – to his critics' chagrin – the view that "[t]he strong misreader doesn't care about the distinction between discovery and creation, finding and making. [...] He is in it for what he can get out of it, not for the satisfaction of getting something right" (*CP*: 152).

Works Cited

Mendieta, E. (ed.) (2006) *Take Care of Freedom and Truth Will Take Care of Itself: Interviews with Richard Rorty*, Stanford: Stanford University Press.
Rorty, R. (1979) *Philosophy and the Mirror of Nature (PMN)*, Princeton: Princeton University Press.
———. (1982) *Consequences of Pragmatism (CP)*, Minneapolis: University of Minnesota Press.
———. (1989) *Contingency, Irony, and Solidarity (CIS)*, Cambridge: Cambridge University Press.
———. (1990) *Objectivity, Relativism, and Truth (ORT)*, Cambridge: Cambridge University Press.
———. (1991) *Essays on Heidegger and Others (EHO)*, Cambridge: Cambridge University Press.
———. (1998) *Truth and Progress (TP)*, Cambridge: Cambridge University Press.
———. (1999) "Pragmatism as Anti-Authoritarianism," *Revue Internationale de Philosophie* 53–207: 7–20.
———. (2007) *Philosophy as Cultural Politics (PCP)*, Cambridge: Cambridge University Press.
———. (2010) "Philosophy as a Transitional Genre," in C. J. Voparil and R. J. Bernstein (eds.), *The Rorty Reader*, Malden: Wiley-Blackwell, 473–486.
———. (2014) *Mind, Language, and Metaphilosophy: Early Philosophical Papers*, S. Leach and J. Tartaglia (eds.), Cambridge: Cambridge University Press.
———. (2020) *On Philosophy and Philosophers: Unpublished Papers, 1960–2000*, W. P. Malecki and C. Voparil (eds.), Cambridge: Cambridge University Press.

Further Reading

Auxier, R. E. and Hahn, L. E. (eds.) (2010) *The Philosophy of Richard Rorty*, McLean: Open Court Publishing. (A collection of 28 essays on Rorty's thought, written by pragmatist scholars, each of which is followed by a reply written by Rorty.)
Brandom, R. B. (ed.) (2000) *Rorty and His Critics*, Malden: Blackwell Publishers. (A selection of essays on Rorty's thought, written by key interlocutors, and including replies by Rorty.)
Festenstein, M. and Thompson, S. (eds.) (2001) *Richard Rorty: Critical Dialogues*, Malden: Blackwell Publishers. (A series of essays focusing on Rorty's ethical and political thought, with replies by Rorty.)
Janack, M. (ed.) (2010) *Feminist Interpretations of Richard Rorty*, University Park: The Pennsylvania State University Press. (A collection of essays considering Rorty's views from a feminist perspective, including key essays by Rorty and prominent feminist critics.)
Rondel, D. (ed.) (2021) *The Cambridge Companion to Rorty*, Cambridge: Cambridge University Press. (An overview covering major elements of Rorty's thought, written by contemporary Rorty scholars.)

12
HILARY PUTNAM

Maria Baghramian and Matthew Shields

Introduction

American, Harvard-based philosopher Hilary Putnam (1926–2016) began a sustained and rich engagement with pragmatism in the latter part of his career. This turn to pragmatism was part of a larger rethinking of his own earlier philosophical positions. The early Putnam was well known for highly original and influential work on realism in philosophy of science and mathematics, functionalism in philosophy of mind, and semantic externalism in philosophy of language.[1] According to the latter position, human language and thought aim primarily to represent a language and thought-independent reality. Although Putnam never completely abandoned his commitment to semantic externalism nor to realism, he began to question this general picture of the relationship between language/thought and the world. In developing this new outlook in the 1980s, Putnam began to invoke the ideas of the first-generation pragmatists, John Dewey and William James in particular, and, to a lesser extent, Charles Sanders Peirce.

Putnam's first exposure to pragmatism came in his undergraduate days in Pennsylvania, where his first teacher in the philosophy of science, C. West Churchman, was a pragmatist (a student of E. A. Singer Jr., who was, in turn, a student of William James). During his graduate studies, he also took a course at UCLA with Abraham Kaplan, who was strongly influenced by Dewey (Bella, Boncompagni, and Putnam 2015). But Putnam credits his "conversion" to pragmatism in large part to his wife Ruth Anna Putnam, a renowned scholar of the pragmatist tradition (Putnam and Putnam 2017: 18).

Putnam's Approach to Pragmatism

The label "neo-pragmatism" or "new pragmatism" has been applied to the work of a varied group of philosophers, chief among them W. V. O. Quine, Richard Rorty, Robert Brandom but also Joseph Margolis, Nicholas Rescher, Richard Bernstein, Susan Haack, and Huw Price. Even Wittgenstein and Habermas have been characterized as pragmatists by some commentators. But the questions of what exactly neo-pragmatism consists in and what connections it has with its ancestors remain open. For Putnam, pragmatism's primary appeal lies in its advocacy of a set of "fundamental theses" or interrelated conceptual and philosophical sensibilities rather than in narrowly circumscribed doctrines (1994: 152).[2] In fact, with respect to specific pragmatists' views, such as those of James and Dewey on truth or fictionalism about theoretical entities, Putnam is adamant that he does not endorse them (e.g. 2012 and 2017). For Putnam, pragmatism should be approached "not as a movement ... but as a way of thinking that I find of lasting importance, and an option (or at least an open question) that should figure in present-day philosophical thought" (Putnam 1995: xi).

In its negative mode, pragmatism amounts to a rejection of what Putnam dubs "metaphysical realism", a philosophical orientation that takes there to be "a set of 'ultimate' objects, the furniture of the world ... whose 'existence' is *absolute,* not relative to our discourse at all, and a notion of truth as 'correspondence' to these ultimate objects", which philosophers can supposedly observe by occupying a God's-eye view on the way the world ultimately is (1989: 231).[3] In contrast to this influential metaphysical realist orientation, the positive vision that Putnam endorses from the pragmatist tradition consists in the following general commitments:

1 *Anti-reductionism and pluralism about what there is* – views he found representatively (though not satisfactorily) defended by James:

> [W]e must avoid the common philosophical error of supposing that the term *reality* must refer to a single superthing instead of looking at the ways in which we endlessly renegotiate – and are *forced* to renegotiate – our notion of reality as our language and our life develop.
>
> *(1999: 9)*

At the same time, Putnam rejects James's specific brand of pluralism for suggesting that the world is a product of our own making. But Putnam agrees with the pragmatists that reality should not be cannibalized into a "single superthing":

> The world cannot be completely described in the language game of theoretical physics, not because there are regions in which physics is false, but because, to use Aristotelian language, the world has many levels of forms, and there is no realistic possibility of reducing them all to the level of fundamental physics.
>
> *(2012: 65)*

Whether we consider our everyday language of middle-sized dry goods or the more rarified languages of ethics, mathematics, the social and human sciences, or literary criticism, all are examples of languages that cannot be fully reduced or translated into the language of fundamental physics; the meaning of what we say in these domains would be lost by such reductions, along with knowledge we have gained. The domain of fundamental physics does not have a monopoly over meaningful, truth-apt, and reality-involving inquiry. Even within fundamental physics, we find a wealth of examples of the ongoing process of human inquiry expanding and revising our notion of reality:

> [Q]uantum mechanics is a wonderful example of how with the development of knowledge our idea of what counts as even a possible knowledge claim, our idea of what counts as even a possible object, and our idea of what counts as even a possible property are all subject to change.
>
> *(1999: 8)*

These expansions of and revisions to our languages, however, do not mean we are not constrained by an external reality – a point addressed in the next pragmatist thesis Putnam highlights.

2 *Realism.* Despite the many changes in Putnam's views during his philosophical career, one of his permanent preoccupations is the search for a satisfactory account of realism. Putnam defends what he calls "natural realism" in the last decades of his life, drawing directly from James. To articulate a satisfactory realism, we must "overcome the disastrous idea that has haunted Western

philosophy since the seventeenth century, the idea that perception involves an interface between the mind and the 'external' objects we perceive" (1999: 43). Most traditional and even contemporary accounts of perception assume that perception is always mediated by sense data or sensations: external objects impinge on us, which prompts sense data (or some equivalent notion), which we then interpret. On these views, we do not perceive tables or chairs; rather, we perceive sense data that we then interpret as tables or chairs. For Putnam, this philosophical picture is a dead end because it leaves mysterious how we are able to come into contact with the world or objects *at all*, prompting, in turn, a constant philosophical back-and-forth between the untenable positions of metaphysical realism and an idealist antirealism. Following James, Putnam's natural realism holds that "successful perception is a *sensing* of aspects of the reality 'out there' and not a mere affectation of a person's subjectivity by those aspects" (1999: 10). In other words, the natural realist accepts our perceptual reports and truth claims just as we articulate them in our everyday lives: we perceive tables and chairs, "not immaterial intermediaries" or sense data that we then interpret as tables or chairs (2017: 144). Citing James, Putnam explains that his position vindicates the "natural realism of the common man" (1999: 38).

3 *Rejection of dualisms.* Inspired by Dewey, Putnam repudiates various unhelpful and pernicious philosophical dualisms, including, most importantly, the subjective-objective dualism and its close relative, the fact-value dichotomy (e.g. Putnam 2002). Description and evaluation, according to Putnam, should not be seen as "two separate watertight boxes in which statements or uses of statements can be put. All description presupposes evaluation (although not necessarily moral evaluation) and all evaluation presupposes description" (2012: 70). But the interpenetration of fact and value does not open the floodgates to relativism. For Putnam, in order to make any claim about the way the world is, we need a sense of what is important and what not, what to pay attention to and what to pass over, what the relevant grounds for one's judgments are, how much weight they have, and some indication of how they are to be assessed or criticized. In short, we need a sense of realism in *all* of our inquiries, theoretical or practical, and this realism is necessarily bound up with and enabled by, rather than cordoned off from, values.

4 *Fallibilism conjoined with anti-skepticism.* Following the first-generation pragmatists, Putnam endorses the view that all beliefs are open to revision and all interpretations and methods of inquiry have a provisional authority only (1989, 1994). But he also draws the lesson from Peirce that doubt itself requires justification. While any claim can be questioned for the pragmatist, this questioning must be earned: it must be shown that we have genuine reason to doubt, a burden that, in turn, prevents pragmatism from collapsing into skepticism. In fact, Putnam believes that the very notion "that one can be both fallibilistic *and* antiskeptical is perhaps *the* unique insight of American pragmatism" (1994: 152). The stakes of this rejection of skepticism also implicate our political and moral lives because Putnam points out that "it is an open question whether an enlightened society can avoid a corrosive moral scepticism without tumbling back into moral authoritarianism" (1995: 2).

5 *The primacy of the agent point of view.* Following the pragmatists, Putnam argues that philosophy should not be in the business of attempting to indict, eliminate, or reduce concepts such as truth, reason, or meaning that play a fundamental role for the kinds of speaking and thinking creatures we are:

> The heart of pragmatism, it seems to me – of James's and Dewey's pragmatism, if not of Peirce's – was the insistence on the supremacy of the agent point of view. If we find that we must take a certain point of view, use a certain 'conceptual system', when we are engaged in practical activity, in the widest sense of 'practical activity', then we must not simultaneously advance the claim that it is not really 'the way things are in themselves'.
>
> *(1987: 70)*

Putnam's point is not that we should therefore return to traditional philosophical accounts of these concepts – where, for example, truth consists in correspondence to the way the world absolutely is. Instead, Putnam wants us to locate the role these foundational concepts play in our lives, what they allow us to do that we could not do otherwise. For example, Putnam argues that "[t]here is no eliminating the normative" in our linguistic and conceptual lives because our sense that utterances and thoughts can be correct or incorrect, true or false, is necessary for us to take ourselves to be talking and thinking about a shared world with others (1989: 246). As he puts the point elsewhere:

> [T]he pragmatists urged that the agent point of view, the first-person normative point of view, and the concepts indispensable to that point of view should be taken just as seriously as the concepts indispensable to the third-person descriptive point of view.
>
> *(2017: 168)*

We take up this line of thought in the following section.

Putnam versus Rorty on Pragmatism

Putnam's endorsement and deployment of the above theses are often particularly apparent in his engagement with the work of Richard Rorty, who also claims the heritage of pragmatism for his project. Although Putnam shares elements of Rorty's critique of their contemporaries' philosophical views – in particular, a repudiation of a God's-eye view on the world and a corresponding metaphysical realism – Putnam nonetheless rejects many of Rorty's specific views and Rorty's claims that his views have a pragmatist lineage. Following Dewey, both Putnam and Rorty argue that our justifications and standards for warrant for our beliefs are shaped by our historically situated interests and values. Epistemic standards are therefore never absolute or immutable but plural and evolve over time (Putnam 1990). Putnam and Rorty disagree strongly, however, on whether "in ordinary circumstances, there is usually a fact of the matter as to whether the statements people make are warranted or not" and "whether a statement is warranted or not is independent of whether the majority of one's cultural peers would *say* it is warranted or unwarranted" (1990: 21). From Rorty's perspective, whether a belief counts as warranted is exclusively a question about whether our fellow speakers and thinkers would count the belief as warranted; warrant or justification is therefore always fundamentally a social matter (because, in his view, there is no sense to be made of the world "telling us" how it ought to be made sense of). In turn, Rorty believes that we should move "everything over from epistemology and metaphysics to cultural politics, from claims to knowledge and appeals to self-evidence to suggestions about what we should try" (Rorty 1998: 57). For Putnam, however, the answerability of our beliefs and warrants for those beliefs to a world over and above our peers and a "reality not of our own invention plays a deep role in our lives and is to be respected" (Putnam 1999: 9).

If we assume, for example, that other speakers and thinkers have a final say over what counts as right or wrong, we would never be able to coherently disagree with a communal consensus, rendering impossible, in turn, the very notion of our making rational changes to our categories for understanding the world. Furthermore, our very practices of agreement and disagreement require that we take ourselves to be answerable to a world over and above our respective, idiosyncratic, or culturally specific commitments. If we did not, then we would have no reason to engage speakers whose semantic and epistemic commitments differ from our own. When, for example, other speakers utter declarative sentence "S" and we utter "~S", such utterances would then not be treated as contradictory but rather as products of hermetically sealed ways of talking and thinking about the world. But, for Putnam, pragmatists should aim to do justice to deeply ingrained practices such as those of agreement and disagreement rather than ignoring or eliminating them: "[M]y knowledge that *I disagree with what you just said,* is also knowledge that is as sure as any that we have. Such knowledge must be taken seriously by philosophers, not treated as an illusion to be explained away" (1994: 322). Once we center the agent point of view and, to take a specific example, our practices of agreement and disagreement even for speakers whose beliefs otherwise differ substantially from our own, Putnam points out that we discover that "*as* thinkers we are committed to there being *some* kind of truth, some kind of correctness which is substantial" (1989: 246). In other words, it is crucial to how we make sense of ourselves and the world that we do take our thoughts and utterances to be answerable to a world whose intelligibility our peers do not have a decisive ability to determine, but that exists over and above any communal consensus, no matter how resilient or robust.

Putnam's pragmatist approach to and preservation of the notions of truth, reality, reason, and meaning, however, should not be confused with traditional metaphysical realism. His position is that we should locate and analyze the crucial role these concepts play in our lives rather than resurrecting an illusory God's-eye view or attempting to excise them in the service of our philosophical theories. To fail to do so – in the ways that, on Putnam's reading, Rorty does – would be a betrayal of the pragmatist tradition and its "insistence on the supremacy of the agent point of view".

Conclusion

In addition to Putnam's endorsement of a pragmatist orientation to specific philosophical questions, he also shares the pragmatists' view that philosophy can and should matter to our moral and spiritual lives (1995; Putnam and Putnam 2017). This connection between philosophy and life as it shows up for agents (rather than as reimagined in a desiccated form from the philosopher's armchair), as we have seen, works in both directions: "[I]f there was one great insight in pragmatism, it was the insistence that what has weight in our lives should also have weight in philosophy" (1999: 70). In their co-authored collection of papers, both Putnams describe their hope that a pragmatist orientation will shape what philosophy will "look in the twenty-first century and beyond" (2017: 18).

In the previous sections, we sketched how in Putnam's own work, he borrowed from classical pragmatism in navigating the contested terrain concerning the nature of realism, truth, and rationality, using the resources of this tradition to defend views of these concepts that do not succumb either to traditional accounts that ignore the ways we are situated within history and culture or to revisionary approaches, such as Rorty's, that aim to rid philosophers of these concepts altogether. The pragmatist future for philosophy the Putnams imagined had an even broader scope, extending these insights into our moral and politics lives, into "reflection on our ways of living, and especially on what is wrong with those ways of living" (2013: 34).

Acknowledgments

The work on this publication has been made possible through funding from the European Union's Horizon 2020 research and innovation program under grant agreement No. 870883. The information and opinions contained herein are those of the authors and do not necessarily reflect those of the European Commission.

Notes

1 See, for example, the papers collected in Putnam (1975).
2 The positions numbered below in this section draw from and also expand on Putnam's own list in his 1994 work (152).

3 When we refer to "metaphysical realism" in this chapter, we have in mind how Putnam characterizes this view in the cited passage. For discussion of Putnam's changing views of metaphysical realism, see Putnam (2015).

Bibliography

Bella, M., Boncompagni, A., & Putnam, H. (2015) "Interview with Hilary Putnam," *European Journal of Pragmatism and American Philosophy [Online]*, VII-1. http://journals.openedition.org/ejpap/357.
Putnam, H. (1975) *Mind, Language and Reality: Philosophical Papers*, Volume 2, Cambridge: Harvard University Press.
———. (1989) *Realism and Reason: Philosophical Papers*, Volume 3, Cambridge: Cambridge University Press.
———. (1990) *Realism with a Human Face*, Cambridge: Harvard University Press.
———. (1994) *Words and Life*, Cambridge: Harvard University Press.
———. (1995) *Pragmatism: An Open Question*, Cambridge: Blackwell Publishers.
———. (1999) *The Threefold Cord*, New York: Columbia University Press.
———. (2002) *The Collapse of the Fact/Value Dichotomy and Other Essays*, Cambridge: Harvard University Press.
———. (2012) *Philosophy in an Age of Science: Physics, Mathematics, and Skepticism*, Cambridge: Harvard University Press.
———. (2013) "From Quantum Mechanics to Ethics and Back Again," in M. Baghramian (ed.), *Reading Putnam*, London and New York: Routledge, 19–35.
———. (2015) "Intellectual Autobiography," in R.E. Auxier, D.R. Anderson, and L.E. Hahn (eds.), *The Philosophy of Hilary Putnam*, Chicago: Open Court Publishing Company, 3–109.
Putnam, H., & Putnam, R.A. (2017) *Pragmatism as a Way of Life: The Lasting Legacy of William James and John Dewey*, Cambridge: Harvard University Press.
Rorty, R. (1998) *Truth and Progress: Philosophical Papers*, Volume 3, Cambridge: Cambridge University Press.

Further Reading

Goodman, R.B. (2008) "Some Sources of Putnam's Pragmatism," in M.U.R. Monroy, C.C. Silva, and C.M. Vidal (eds.), *Following Putnam's Trail*, New York: Rodopi B.V., 125–140.
Hildebrand, D. (2000) "Putnam, Pragmatism, and Dewey," *Transactions of the Charles S. Peirce Society* 36, 109–132.
MacArthur, D. (2008) "Putnam, Pragmatism and the Fate of Metaphysics," *European Journal of Analytic Philosophy* 4, 33–46.
Misak, C. (2013) "Hilary Putnam (1926 –)," in *The American Pragmatists*, Oxford: Oxford University Press, 238–244.
Pihlström, S. (2004) "Putnam and Rorty on their Pragmatist Heritage: Re-Reading James and Dewey," in E. Khalil (ed.), *Dewey, Pragmatism and Economic Methodology*, London: Routledge, 39–60.

13
CORNEL WEST AND PROPHETIC PRAGMATISM

Eduardo Mendieta

Cornel West (born 1953) is without question the most prominent and impactful African American philosopher, social activist, public, and organic intellectual of the last century (see Yancy 2001: 1–16, for one of the best overviews of his work). He is a self-identified pragmatist, Black Christian, Democratic Socialist, humanist, art and literary critic, i.e. a "bluesman in the life of the mind, and a jazzman in the world of ideas" to use one of his self-descriptions (West 2009). West is Du Bois, I. D. Wells, Dewey, Niebuhr, and Rorty rolled up into a unique towering intellect and a vast corpus. He is a Du Bois, because he has devoted most of his intellectual energies to celebrating, studying, archiving, and showcasing the achievements of African Americans; he is a Wells, because he has kept his focus on the systemic and racist violence that is ceaselessly performed on Black and Brown bodies and the war on the poor that is waged by capitalism and neoliberalism; he is a Dewey, because at the core of his philosophy is the unwavering commitment to the democratic project, what he calls "democratic *paideia*—the critical cultivation of an active citizenry," which for West demands "democratic *parrhesia*," the courage to speak truth to power (West 2004: 39); he is a Niebuhr, because his philosophical and literary outlook, his own vision of the American democratic project, is tempered by a tragic-comic vision that recognizes the fragility of the democratic project and hope, which is also informed by the prophetic Christian tradition; he is a Rorty, because his works, whether they be philosophical, religious, theological, historical, are informed by a voracious, encyclopedic, and profound understanding of the sources of America's distinctive philosophical tradition (pragmatism), and how it relates to European, or so-called continental, philosophy.

As jejune as these comparisons may appear, they are meant to signal how complex and encompassing West's *oeuvre* is. West himself has provided us with a succinct overview of his work:

> my writing constitute a perennial struggle between my African and American identities, my democratic socialist convictions and my Christian sense of the profound tragedy and possible triumph in life and history. I am a prophetic Christian freedom fighter principally because of distinctive Christian conceptions of what it is to be human, how we should toward one another and what we should hope for.
>
> *(West 1999: 13)*

What West calls the "perennial struggle between" his "African" and "American" identities is at the core of West's contributions to American pragmatism. West's prophetic pragmatism cannot be understood without a reference to the confrontation of this Black identity with an America that has enslaved, vilified, demeaned, lynched, segregated, and criminalized African Americans over its history while at the same time acknowledging that what America has become, especially in the 20th century, which West and Henry Louis Gates, Jr. called "the African American Century," is in part an accomplishment of African Americans (West and Gates 2000). In what follows, however, I will focus exclusively on West's appropriations, contributions, and challenges to pragmatism. I will do so by providing a "genealogy" of West's prophetic pragmatism while highlighting his own critical assessments.

The point of entry for understanding West's engagement with pragmatism is his 1982 book, *Prophesy Deliverance! An Afro-American Revolutionary Christianity*, a book that is as much a manifesto as a majestic entry into the life of letters in American culture. In the introduction, titled "The Sources and Tasks of Afro-American Critical Thought," West (1982: 15–24) names two major sources of such "critical thought": prophetic Christian thought and American pragmatism. The latter is quickly identified with the names of C. S. Peirce, William James, G. H. Mead, and J. Dewey. Here, Dewey is the pragmatist par excellence because he recognized that "philosophy is inextricably bound to culture, society, and history" (West 1982: 20). West argues that pragmatism shows us how the normative function of philosophy, as it may contribute to the project of philosophizing the Afro-American experience, is the "critical reflection and interpretation of a people's past for the purpose of solving specific problems" of members of a given culture face (Ibid.: 20). West is also quick to point out the shortcomings of pragmatism: its relative lack of attention to individuality, its failure to take class struggle seriously, and its idolatry of scientism (Ibid.: 21). Most notably, because this provides us with further clues to understand West's prophetic pragmatism, he identified five tasks in the face of two major challenges Afro-Americans face: self-image and self-determination. The tasks are to advance an "overarching interpretative framework" to deal with the two challenges; second, to engage in a "genealogical inquiry" into the sources of white supremacy; third, to reconstruct and evaluate Afro-American responses to white supremacy; fourth, to articulate a dialogical encounter between Afro-American Christian thought and Marxist-inspired social-political-economic analysis; fifth, to provide political prescriptions that address the present historical struggles for liberation (Ibid.: 23). These two challenges and the tasks they impose are critical for understanding West's development of his own version of pragmatism.

However, before I discuss further West's arguments for why pragmatism is one of the sources of Afro-American radical thought, I must note that already in an essay published in 1997 (when he was still a graduate student at Princeton), West identified pragmatism as a major source of Afro-American philosophy. The essay, titled "Philosophy and the Afro-American Experience" (West 1977–1978, reprinted in Lott & Pittman 2002: 7–32. I will cite from this later reprint), begins with the confession of an "antipathy" to the "ahistorical character of contemporary philosophy and the paucity of illuminating diachronic studies of the Afro-American experience" (Lott & Pittman 2002: 7). The second paragraph, however, is crucial, for there Dewey is identified as one of the key figures that can provide the requisite philosophical tools to make sense of the Afro-American experience in America. West writes:

> The philosophical techniques requisite for an Afro-American philosophy must be derived from a lucid and credible conception of philosophy. The search for such a conception requires us to engage in a metaphilosophical discourse, in philosophical reflections on philosophy. The most impressive metaphilosophical formulations of our age are those of Martin Heidegger, the later Wittgenstein, and John Dewey.

(Ibid.)

At the end of this sentence there is endnote 2, where we read:

> In the following exposition of metaphilosophy, I follow closely the brilliant work of Prof. Richard Rorty of Princeton University. Most of his work remains unpublished. I am confident his writing on metaphilosophical issues will be of paramount importance for thinkers in the future.
>
> (Ibid.: 27; see also West 1999: 26, 143)

Thus, we are justified in claiming that West began as a Rortyan pragmatist, or at the very least, that Rorty enabled him to relate in liberating ways to philosophy, in general, and pragmatism, in particular.

It is noteworthy that in this 1977 essay, West assesses the virtues and demerits of Heidegger and the late Wittgenstein, noting what Afro-American philosophy can take from each as it must reject what is not useful while championing the metaphilosophical contributions of Dewey. As he writes: "John Dewey's metaphilosophical views serve, for our purposes, as a form of synthesis and corrective for those of Heidegger and Wittgenstein. ... Philosophy is inextricably bound to culture, society, and history" (Ibid.: 11). Then, after the discriminating appropriations of the three great metaphilosophers of the 20th-century philosophy, West offers the following definition of Afro-American philosophy: "*Afro-American philosophy is the interpretation of Afro-American history, highlighting the cultural heritage and political struggles, which provide desirable norms that should regulate responses to particular challenges presently confronting Afro-Americans*" (Ibid.: 11, italics in original). West then proceeds to identify four distinct Afro-American traditions that seek to give philosophical voice to the Afro-American experience: the vitalist, the rationalist, the existentialist, and the humanist. This typology is extremely useful, synthetic, and panoramic. West has identified the currents, tensions, claims, and ideals of each one of these traditions, coupling each to several of the towering figures in Afro-American thinking over the past two centuries. West concludes his essay deeming the humanist tradition within Afro-American thinking as the most desirable because it is the one that celebrates individuality and democratic self-determination that promote "personal development, cultural growth and human freedom" (Ibid.: 27). This essay appears to be the genesis of Chapter 3 of *Prophesy Deliverance*, in which West has dispense with the discussions of Heidegger and Wittgenstein while moving his approbation of Dewey to the introduction. Additionally, West has modified the names of the four traditions he has identified. Now they are identified as exceptionalist, assimilationist, marginalist, and humanist traditions (West 1982: 70–71).

There are three other texts that need to be highlighted as key in the genealogy of West's version of pragmatism. The first is West's 1981 essay "Nietzsche Prefiguration of Postmodern American Philosophy" (West 1981: 241–269; reprinted in West 1999: 188–210). This essay is a close textual analysis of Friedrich Nietzsche's *The Twilight of the Idols* and *The Will to Power* and how key ideas in these two texts are echoed in the works of W. V. Quine, Nelson Goodman, Wilfred Sellars, Thomas Kuhn, and Richard Rorty. The aim of this cross-textual readings is to show how Nietzsche's philosophical revolutions achieve their most elaborate and explicit formulations in the anti-realism in ontology; the demythologization of the myth of the given, which West calls also anti-foundationalism in epistemology; and the detranscendentalization of the Cartesian concepts of the subject, which jettisons the "mind" as sphere of inquiry, which informs what here West calls "postmodern American philosophy." The goal of the essay, however, is to also argue that these moves of anti-realism, anti-foundationalism, and detranscendentalization lead to a "paralyzing nihilism and ironic skepticism" that must be corrected and tempered by a "countermovement" that can lead to their overcoming (West 1981: 242; see West 1999: 189). In West's analysis, none of the "American" postmodern philosophers offers such a "countermove." In fact, West judges them to be part of a culture permeated "by the scientific ethos, regulated by racist, patriarchal, capitalist and pervaded by debris of decay" (West 1981: 265; 1999: 210).

The second text that is relevant in our genealogy of West's prophetic pragmatism is his 1981 review essay of Richard Rorty's then recently published *Philosophy and the Mirror of Nature*. It should be noted that in this essay, West reveals not only how thoroughly he knows Rorty's works, beyond the book itself, including a slew of essays published before and right after his book, but also the literature on which Rorty relies. It should also be noted that West sees Rorty's work not simply as a rejection of epistemologically and linguistically focused analytic philosophy but also as a sustained reflection on the relationship between pragmatism and metaphilosophy, i.e. a reflection on how pragmatism had anticipated the transformations of philosophy once we make the linguistic turn, abandon epistemology-obsessed philosophy, and jettison the Cartesian philosophy of the subject. West celebrates the ways in which Rorty has made clear and advanced the consequences of Quine's holism (or pan-relationalism), Sellar's anti-foundationalism, Goodman's pluralism, and Kuhn's leveling of the ground between the sciences and the humanities. West also affirms Rorty's dethroning of philosophy as the Queen science that serves as both judge and prosecutor in all intellectual disagreements. Yet, West is also pointed and critical of Rorty's failure of nerve when it comes to following through on what his "transformation of philosophy" entails. In words that anticipate what West is going to write in *The American Evasion of Philosophy*, he chides Rorty:

> Yet, Ironically, Rorty's project, through pregnant with rich possibilities, remains polemical and hence barren. It refuses to give birth to the offspring it conceives. Rorty leads philosophy to the complex world of politics and culture, but does not permit it to get its hands dirty.
> *(West 1981–1982: 183, and also see Cormier 2010)*

As Rorty deconstructs philosophy, the more he retreats into its tower, refusing to enter the world that is post-philosophical (to use Rorty's expression).

The third text is West's contribution to the volume he co-edited with John Rajchman, *Post-Analytic Philosophy* (Rajchman and West 1985. See also West 1999: 183–187), titled "The Politics of American Neo-Pragmatism." Here, West takes a larger canvas and locates pragmatism and its resurgence in the 80s and 90s as part of the confrontation with the crisis of European/Western culture. West reviews once again the precursors of the pragmatist resurgence and coins new terms to describe what they rejected and demonstrated to be untenable, i.e. sentential atomism, phenomenalist reductionism, and analytical empiricism (Rajchman & West 1985: 260). West also returns to his assessment of Rorty's work, which expands on his earlier affirmations. West now foregrounds how Rorty's neo-pragmatism has two major consequences: First, Rorty's work leads to the collapse of the rigid distinction between the *Naturwissenschaften* (the natural sciences) and the *Geisteswissenschaften* (the human sciences), which is now to be understood in a Kuhnian sense, i.e. as their differences are not in methods or objects of study but in terms of the relative stability of their "normal" vocabularies. Second, since philosophy can no longer be conceived as a "tribunal of pure reason," and since the distinctions among the "sciences" is not in kind but of degrees, not of ontological realms but of vocabularies, philosophy has become just one more voice in the square of public deliberation. Philosophy's voice, one among other voices, is dubbed by West as "that of the informed dilettante or polypragmatic, Socratic thinker—among others in a Grand Conversation" (Rajchman and West 1985: 265). Yet, in more acerbic language, West expands his critique of Rorty's neo-pragmatism, which he now sees as a "form of ethnocentric posthumanism" that echoes the anti-humanism of Heidegger, Derrida, and Foucault, while also domesticating them within the narrow and safe confines of bourgeois capitalist humanism (Ibid.: 267). West argues that it is "impossible to historicize philosophy without partly politicizing (in contrast to vulgarly ideologizing), it" (Ibid.: 269). If philosophy can no longer remain in the tower of academia, and claim for itself some rhetorical and rational primacy, it must recognize that it is not simply one more voice in the public sphere and that it must play a role there, along with the other sciences, in educating the public and citizenry.

Thus far the focus has been on those early texts by West in which he explicitly discusses pragmatism; the resurgence of pragmatism, what he at times refers to as "postmodern American philosophy"; and the limits of neo-pragmatism because in these texts he swings between approbation and affirmation, on the one end, and criticism and rejection, on the other. These texts also document the depth and sustained attention to American philosophy during West's intellectual development. What makes West's intense engagement with pragmatism particularly striking is how he reads pragmatism and neo-pragmatism as opening a horizon of tasks, which require resources beyond philosophy. This is amply illustrated in another important landmark in West's corpus, namely his *Prophetic Fragments* from 1998, a book that gathers his writing from the 1980s from a bewildering array of publishing venues. The book is divided into three parts: "religion and politics," "religion and culture," and "religion and contemporary culture." The texts that make up this book were speeches, essays, commentaries, and mostly book reviews. Among the many things that can be said about this book, two are evident: West sure read a lot, and West sure was focused on the problems: inheritance, challenge, and promise of religion in American culture and the Afro-American experience (see also West 1993). The book opens with the text of a speech West gave on Martin Luther King, Jr. as a prophetic Christian organic intellectual, transverses the landscape of Afro-American culture during the 1970s and 1980s, and closes with a short story that West wrote as a young man in 1975, tellingly titled "Sing a Song." That the book closes with West's early, and we now know the story is semi-autobiographical, is important, as what holds together *Prophetic Fragments* together as a book is the focus on the entwinement between religion and the Afro-American musical creativity. Later West will describe this early work of fiction thusly:

> I finally wrote and submitted this autobiographical short story. I had hit on the motif of my work to come—to sing in spoken word and write texts like Duke Ellington played and Sarah Vaughan sang, to swing, to create an intellectual performance that had a blues sensibility and a jazzlike openness, to have the courage to be myself and find my voice in the world of ideas and the in the life of the academy.
>
> *(West 1999: 34)*

Prophetic Fragments, however, is suffused by reflection on what West means by "prophetic," and thus, this book is indispensable for understanding what he means by *prophetic* pragmatism. There is no better passage in this book that makes it clear what guides his vision than this:

> My prophetic outlook is informed by a deep, historical consciousness that accents the finitude and fallenness of all human beings and accentuates an international outlook that links the human family with a common destiny: an acknowledgement of the inescapable yet ambiguous legacy of tradition and the fundamental role of community; a profound sense of the tragic character of life and history that generates a strenuous mood, a call for heroic, courageous moral action always against the odds; and a biblically motivated focus on and concern for the wretched of the earth that keeps track of the historic and social causes for much (though by no means all) of their misery.
>
> *(West 1988: xi; see also 113, 120, and 267–280)*

West's *The American Evasion of Philosophy: A Genealogy of Pragmatism* (1989), however, is without question his sustained, profound, original, and still relevant engagement with pragmatism. It is without question the best historical reconstruction of the emergence, coming of age, decline, and eventual resurgence of American pragmatism. As Rorty put it on his blurb for the book: "I believe that *The American Evasion of Philosophy* ... may well become the standard account of the role of pragmatism in American thought." The book is also a bold interpretation of the origins of the basic themes of pragmatism in the work of Ralph Waldo Emerson, whose impact has been either neglected or unacknowledged. What West calls Emerson theodicy (which is optimistic, moralistic, and activistic) rests on three premises: first, the world is "congenial" and "supportive" of moral progress; second, the world, nature, is in flux and thus incomplete, and thus is open to human experimentation and creativity; third, human experimentality and its impact on the world have not been properly understood in the modern world (West 1989: 14–16). These three premises of Emerson's theodicy, according to West, can be reduced to three key assertions: that the only sin is limitation, that sin can be overcome, and that it is "beautiful and good" that sin exists so that it can be overcome. West proceeds to track Emerson's theodicy in the works of some usual figures in the pantheon of pragmatism: Peirce, James, and Dewey. What makes this book insightful and bold is how West includes among the major pragmatist figures that at first blush don't appear pragmatist, nor have been so considered by other pragmatists, such as Sidney Hook, C. Wright Mills, W. E. B. Du Bois, Reinhold Niebuhr, and Lionel Trilling, all of whom embodied in one way or another the dilemmas of the pragmatist intellectual during the middle of the 20th century. John Dewey appears in this genealogy as the height of pragmatism, because it is with Dewey that pragmatism achieves "intellectual maturity, historical scope, and political engagement" (Ibid.: 6). Quine and Rorty are identified with the decline and resurgence of pragmatism, respectively.

Notwithstanding the internal heterogeneity of the pragmatist movement, West identifies the common denominator that gives it coherence:

> ...its common denominator consists of a future-oriented instrumentalism that tries to deploy thought as a weapon to enable more effective action. Its basic impulse is a plebeian radicalism that fuels antipatrician rebelliousness for the moral aim of enriching individuals and expanding democracy. This rebelliousness, rooted in the anticolonial heritage of the country, is severely restricted by an ethnocentrism and a patriotism cognizant of the exclusion of people of color, certain immigrants, and women yet fearful of the subversive demands these people might make and enact.
>
> *(West 1989: 5)*

This leads West to claim that pragmatism's rejection of epistemology-centered philosophy results in the conception of philosophy as a form of cultural criticism that aims to intervene in the production of meanings about "America" as responses to distinct social and cultural challenges and crises. More concretely:

> American pragmatism is less a philosophical tradition putting forward solutions to perennial problems in the Western philosophical conversation initiated by Plato and more a continuous cultural commentary or set of interpretations that attempt to explain America to itself at a particular historical moment.
>
> *(Ibid.)*

Thus, the "evasion of philosophy" in the title means, more specifically, the evasion of a certain type of philosophy, the philosophy that Rorty ceaselessly called Platonic, Cartesian, Kantian, and the affirmation of philosophy as cultural critique with a practical intent. Karl-Otto Apel, an important pragmatic philosopher, and also one of the best interpreters of pragmatism, talked of the "transformation of philosophy" by pragmatism (Apel 1973, 1980). Analogously, we can say that when West writes "evasion," he also means "transformation."

West refers to his work as a "social history of ideas," as a "political interpretation of pragmatism" (Ibid.: 6), as well as a "critical self-inventory" at the service of a "historical, social, and existential" self-situation of West as an intellectual activist and a human being (Ibid.: 7). The subtitle of the book, however, is a "genealogy of pragmatism." Some have criticized West for using this term for it allegedly goes counter to what West does in the book (see Gooding-Williams 1991, who accuses West of vitalism and organicism, and misunderstanding not just Nietzsche and Foucault but also Du Bois). It is clear, nonetheless, why West calls his book a genealogy, at the very least because already in his book *Prophesy Deliverance!* West had clarified in Chapter 2, how he understands genealogy, namely as the study of the "emergence" (*Entstehung*) and "moment of emergence" (*Ursprung*), and not birth (*Geburt*) or descent/origin (*Herkunft*) (see West 1982: 48–49). The philosopher Colin Koopman in *Genealogy as Critique: Foucault and the Problems of Modernity* (Koopman 2013) makes a perspicacious and useful distinction among three types of genealogy or genealogical inquiry: the subversive, the vindicatory, and the problematizing (see also Allen 2013). West's genealogy of American pragmatism, it can be argued, should be read as subversive, because of the audacious ways in which new figures are incorporated in the roster of pragmatism; it is vindicatory, because it aims to show how the key virtues of pragmatism can be useful in our struggles for racial and social justice, and it is problematizing, in that it asks how does pragmatism contribute to our transforming and transformative understanding of "America." These three modalities of genealogy are in evidence in the last chapter of *The American Evasion of Philosophy* in which West's own prophetic pragmatism is juxtaposed to postmodernism and postmodern neo-pragmatism. In the last paragraph of the book, West states it succinctly and clearly:

> Prophetic pragmatism rests upon the conviction that the American evasion of philosophy is not an evasion in serious thought and moral action. Rather such evasion is a rich and revisable tradition that serves as the occasion for cultural criticism and political engagement in the service of an Emersonian culture of creative democracy.
>
> *(Ibid.: 239)*

Rorty's assessment in his blurb has turned out to be true, as West's book has become the "standard" account but also subversively generative critique of pragmatism. It has taught us to read pragmatism as being part of the project of imagining and forging a new America.

References

Allen, A. (2013) "Having One's Cake and Eating It Too: Habermas's Genealogy of Postsecular Reason," in C. Calhoun, E. Mendieta, and J. VanAntwerpen (eds.), *Habermas and Religion*. Cambridge: Polity, 132–152.
Apel, K.-O. (1973) *Transformation der Philosophie*, 2 Vols. Frankfurt am Main: Suhrkamp Verlag. This was partly translated in 1980 as *Toward a Transformation of Philosophy*. London, Boston, MA and Henley: Routledge & Kegan Paul.
Cormier, H. (2010) "Richard Rorty and Cornel West on the Point of Pragmatism," in R. E. Auxier and L. E. Hahn (eds.), *The Philosophy of Richard Rorty*. Chicago, IL and La Salle, IL: Open Court, 88–90.
Gooding-Williams, R. (1991) "Evading Narrative Myth, Evading Prophetic Pragmatism: Cornel West's 'The American Evasion of Philosophy.'" *The Massachusetts Review*, Vol. 32, No. 4 (Winter): 517–542.
Koopman, C. (2013) *Genealogy as Critique: Foucault and the Problems of Modernity*. Bloomington, IN: Indiana University Press.
West, C. (1977–1978) "Philosophy and the Afro-American Experience." *The Philosophical Forum*, Vol. 9, No. (2/3) (Winter-Spring): 117–148. Reprinted in T. Lott and J. P. Pittman (eds.) *A Companion to African-American Philosophy*. Malden, MA: Blackwell Publishing.
———. (1981) "Nietzsche Prefiguration of Postmodern American Philosophy." *Boundary 2*, Vol. 9/10, Vol. 9, no. 3–Vol. 10, no. 1 (Spring–Autumn), Why Nietzsche Now? A Boundary 2 Symposium: 241–269.

———. (1981–1982) "Book Review: *Philosophy and the Mirror of Nature*, Richard Rorty; Princeton: Princeton University Press, 1979, 401 pages, $ 20." *Union Seminary Quarterly Review*, Vol. XXXVII, Nos. 1&2: 179–185.

———. (1982) *Prophesy Deliverance!* Philadelphia, PA: The Westminster Press.

———. (1985) "The Politics of American Neo-Pragmatism," in J. Rajchman and C. West (eds.), *Post-Analytic Philosophy*. New York: Columbia University Press, 259–272.

———. (1988) *Prophetic Fragments*. Grand Rapids, MI: William B. Eerdmans Publishing Company.

———. (1989) *The American Evasion of Philosophy: A Genealogy of Pragmatism*. Madison, WI: The University of Wisconsin Press.

———. (1993) *Keeping Faith: Philosophy and Race in America*. New York and London: Routledge.

———. (1999) *The Cornel West Reader*. New York: Basic Civitas Books.

———. (2004) *Democracy Matters: Winning the Fight against Imperialism*. New York: The Penguin Press.

West, C. and Gates, Jr. H. L. (2000) *The African-American Century: How Black Americans Have Shaped Our Country*. New York: The Free Press.

West, C. with David Ritz (2009) *Brother West: Living Out Loud. A Memoir*. New York: Smiley Books.

Yancy, G. (2001) *Cornel West: A Critical Reader*. Malden, MA: Blackwell Publishers Inc.

Further Reading

Cormier, H. (2010) "Richard Rorty and Cornel West on the Point of Pragmatism," in R. E. Auxier and L. E. Hahn (eds.), *The Philosophy of Richard Rorty*. Chicago, IL and La Salle, IL: Open Court, 88–90. (This is the best article comparing and contrasting the similarities and differences between West and Rorty's neo-pragmatism, which is also followed by Rorty's thoughtful but doubled-down response to keep culture and politics separate.)

Johnson, C. S. (2003) *Cornel West & Philosophy: The Quest for Social Justice*. Routledge: New York and London. (This is an excellent monograph on West's overall philosophical work, not simply pragmatism.)

Putnam, H. (2001) "Pragmatism Resurgent: A Reading of *The American Evasion of Philosophy*," in G. Yancy (ed.), *Cornel West: A Critical Reader*. Malden, MA: Blackwell Publishers Inc, 19–35. (This is a very incisive and insightful reading of West's book by a former professor and colleague and co-teacher.)

Rorty, R. (1991) "The Professor and the Prophet." *Transition*, No. 52 (1991): 70–78. (Rorty's gracious as well as exasperated review of West's *The American Evasion of Philosophy*. What is fascinating is how far Rorty traveled in the direction of West's work, especially when one reviews Rorty's work from the late 90s and early 2000s.)

Wood, M. (2000) *Cornel West and the Politics of Prophetic Pragmatism*. Urbana, IL: University of Illinois Press. (This work is focused on West as a political philosophy.)

Wood, M. (2021) "Cornel West and the Prophetic Tradition," in M. L. Rogers and J. Turner (eds.), *African American Political Thought: A Collected History*. Chicago, IL and London: The University of Chicago Press, 705–729. (Very good overview and update of Wood's 2000 book.)

14
SUSAN HAACK AND WORLDLY, REALIST PRAGMATISM

Robert Lane

In his *Philosophical Investigations*, Wittgenstein urged the following as a method for philosophy: "[D]on't think, but look!" (§66). It's easy to imagine that his *Tractatus* resulted from the opposite approach: Think, but don't look!

Susan Haack understands that neither thinking nor looking is sufficient on its own. Like Peirce, Haack believes that philosophy should rely on "the method of experience and reasoning" (2005: 244–245) and that the relevant experience is not "the kind of specialized, recherché experience on which the empirical sciences call, but … familiar, everyday experience" (2016: 40–41; she cites Peirce, CP 5.423, c.1905). From that experience we learn, e.g., that the world is orderly, in that things and events belong to kinds; that we are able to represent and explain things; that we have beliefs, wishes, fears, etc.; that we are capable of investigation but also self-deception; etc. (2017a: 58, 68). These familiar truths prompt difficult questions about, e.g., the difference between the real and the imaginary, the nature of laws and kinds, the ways that strong evidence differs from weak, the origin of human mindedness, etc. (ibid.: 68). The sciences have contributory relevance to some of these questions, but it is philosophy's job to do the hard work of answering them (ibid.: 65–66). And so philosophy is neither a priori nor scientistic (note the Deweyan distaste for untenable dualisms) but continuous with ordinary empirical investigation (note the Peircean emphasis on continuity), and its subject matter is not just our concepts or our language but reality, i.e., *the one real world* (Haack 2017b: 41).

Reality and truth

The world is *real* in Peirce's sense, i.e., independent of how any individual believes it to be. And it's diverse: it is—as Haack says, borrowing James's (1977) phrase—a "pluralistic universe" containing

> particulars *and* generals: natural objects, stuff, phenomena, kinds, and laws; a vast array of human (and some animal) artifacts; mental states and processes, including our thoughts, dreams, etc.; social institutions, roles, rules, and norms; human languages and other sign-systems; a plethora of scientific, mathematical, and philosophical theories (and, in at least some instances, their objects); works of history and art criticism, etc.; myths, legends, and works of fiction, and the characters and places that figure in them.
>
> *(2016: 41)*

DOI: 10.4324/9781315149592-16

Despite its diversity, the world is *unified*—it is *one* real world so that, e.g., our intentional states are "*realized in* physiological states of our nervous systems—though not ... *reducible to* such physiological states," and particulars, including artifacts, are governed by real laws so that, e.g., we "can't make a working typewriter out of butter ... or a comfortable pillow out of brass" (ibid.: 42).

Haack calls her account of reality *innocent realism* because it is not guilty of "the various philosophically over-optimistic aspirations of some ambitious forms of Metaphysical Realism, scientific realism, etc." (ibid.: 40). For instance, she does not maintain that the only real things are those posited by physics.

> Are there really pieces of furniture, books, trees, etc., or only molecules, atoms, and such? The question rests on a false dichotomy. The suite of furniture in the living room consists of a sofa and two armchairs; a deck consists of 52 cards; and pieces of furniture, books, trees, etc., consist of atoms.
>
> (ibid.: 49)

And rather than assuming that reality consists of a fixed number of mind-independent objects, she recognizes that how many objects there are depends on one's concept of *object* (1998: 158). But that dependence does not mean that the world somehow depends on how we conceptualize it. The question "How many objects are in this drawer?" might have no determinate answer, since the pack of chewing gum in the drawer can count as one object or as several. But there *are* determinate answers to questions posed with more precise concepts. How many sticks of chewing gum are in the drawer? Exactly, and objectively, nine.

Haack also rejects the idea that there is a single vocabulary, like that of physics, in which all of reality can be described. But she argues that there is nonetheless "one uniquely true description of the world" (ibid.: 160): descriptions in different vocabularies—e.g., of physical objects and of intentional states—can be logically consistent and so can be conjoined into a single, true description of reality (ibid.). On Haack's tersely stated view of truth, which she calls *laconism* (a name for which she credits Kiriake Xerohemona), while there are different kinds of true proposition—empirical, logical, mathematical, etc.—there is only one concept of truth: "Whatever kind of proposition is said to be true, what is said of it is the same: that it is the proposition that p, and p; or ... that things are as it says" (2013: 56).

Evidence, justification, and warrant

The concepts of truth, evidence, and epistemic justification are deeply connected: the more justified someone is in believing that p, i.e., the better their evidence that p, the more likely that belief is to be true. Haack's *foundherentism* (see 2009a)—so named because it combines the strengths of foundationalist and coherentist theories while avoiding their weaknesses—explains what it is for one's belief to be more or less justified at a given time. Foundationalist theories assume that every justified belief is either *basic*—justified but without the support of other beliefs—or *derived*—justified with the support of basic beliefs; a weakness of those theories is that they cannot acknowledge that every justified belief derives support from other beliefs. Coherentists don't have that problem—they explain the justification of a belief entirely in terms of its membership in a coherent system of beliefs in which no belief is basic—but neither can they account for the role of sensory experience in justifying empirical beliefs.

Foundherentism avoids that pitfall of coherentism by insisting that one's evidence for a given empirical belief includes not just their other beliefs but also experiential evidence: visual, auditory, etc. experiences and the memories of those experiences. So the theory has both an *evaluative* aspect (a belief's being justified depends in part on the logical relationships among the belief's propositional content and the propositional contents of other beliefs) and a *causal* aspect (a belief's being justified depends in part on that belief state being caused by other mental states). Rejecting the dichotomy of direct and indirect theories of sensory perception, Haack accepts a Peirce-inspired account on which what we directly perceive in normal circumstances are external things and events, but on which perception is also interpretive, since there is "pervasive interpenetration of background beliefs into our beliefs about what we see, hear, etc." (ibid.: 158). As that suggests, foundherentism acknowledges "pervasive mutual support among beliefs" (ibid.: 38), thus avoiding the mistaken foundationalist view that some justified beliefs receive no evidential support from other beliefs.

Haack articulates foundherentism with the help of the analogy of a crossword puzzle. Sensory experiences and memories are like a puzzle's clues, and background beliefs are like already-completed entries in the puzzle. Whether an entry is correct depends on whether it is consistent with intersecting entries and with the corresponding clue; whether those other entries are correct, independent of whether they are consistent with the entry at hand; and how much of the puzzle has been filled in. Analogously, how justified someone is in believing that *p* is a function of how well their evidence *supports* that belief; how *secure* their reasons are for that belief, independent of any support given to those reasons by the belief in question; and how *comprehensive* their evidence is (ibid.: 126–127). The supportiveness of evidence is a matter of explanatory integration: one's evidence for a belief is more supportive the more the addition of that belief to that evidence improves the explanatory integration of the reasons that help constitute that evidence (ibid.: 128).

Haack distinguishes justification, which is always personal, from warrant, which can be either personal or social. Whether the claim that *p* is warranted for an individual at a given time depends on how good their evidence is for *p*, whether or not they actually believe *p* at that time. Someone can give a claim too little credence despite its being well warranted for them or too much credence despite its being poorly warranted for them (2003: 73). Unlike the concept of justification, the concept of warrant can be extended to groups, e.g., groups of scientists. A claim is warranted for a group to the degree that it would be warranted for a person

> whose evidence is the joint evidence of all the members of the group, ... construed as including ... the disjunctions of disputed reasons, and discounted by some measure of the degree to which each member is justified in believing that others are reliable and trustworthy observers, and of the efficiency or inefficiency of communication within the group.
>
> *(2003: 71)*

To say that a claim is warranted at a time, simpliciter, is just a concise way of saying that, at that time, it is warranted for the person or group whose evidence is the most secure and the most comprehensive (2003: 72). And since the security and the comprehensiveness of evidence are objective matters, so is the degree to which a claim is warranted at a time. This is true of both ordinary empirical claims and scientific claims, and the factors that determine the quality of the evidence for a scientific claim are "*worldly*, depending both on scientists' interactions with particular things and events in the world, and on the relation of scientific language to kinds and categories of things" (2003: 24).

The sciences

Like philosophy, science is continuous with ordinary empirical investigation. Haack's epistemology of science, which she calls *critical common-sensism* (a name borrowed from Peirce), steers between two objectionable extremes: undue cynicism and scientistic deference. No method of inquiry or form of inference demarcates science from non-science. Rather, physics, chemistry, etc. proceed, as does every other sort of empirical inquiry, by "making an informed conjecture, seeing how it stands up to the available evidence and any further evidence you can lay hands on, and then using your judgment whether to drop it, modify it, stick with it, or what" (2003: 167). The sciences are, "like all human enterprises, ... thoroughly fallible, imperfect, uneven in [their] achievements, often fumbling, sometimes corrupt, and of course incomplete" (ibid.: 19). Still, they are "epistemologically *distinguished*" (ibid.: 23). The natural sciences have been enormously successful in discovering truths, in part because of "helps" such as their communal and intergenerational nature; mathematical and statistical modeling; technologies that enable otherwise impossible observations; and technical concepts and terms, the meanings of which evolve over time, thus enabling more accurate representation of things and events belonging to real kinds (ibid.: 25).

The social sciences are "the same" as the natural sciences, "only different." Psychology, sociology, etc. study the real world, and some of the "helps" of the natural sciences are appropriate for investigation of physical aspects of the social world. But the intentional social sciences, which attend to humans' intentional states, rely on different "helps," e.g., interviews and questionnaires, not electron microscopes and cloud chambers (ibid.: 168). The claims of intentional social science cannot be "reduced" to claims of the natural sciences but can be integrated with them to form a more comprehensive account of the one real world, as a map of towns and roads can be superimposed on a contour map to show that "the roads go around the lake and through the pass in the mountains, that the town is on, not in, the river, and so on" (ibid.: 161).

Intellectual integrity

Expanding on Peirce's idea of *genuine inquiry*, which is motivated by a sincere desire to find the truth, Haack describes the genuine inquirer as "impartial, i.e., not motivated by the desire to arrive at a certain conclusion" (1998: 10). However, one who is engaged in *pseudo-inquiry* is defending a claim on which they settled in advance of the evidence. For example, someone engaged in *sham reasoning* defends a belief that they sincerely hold but that is immune to evidence. Another kind of pseudo-inquiry is *fake reasoning*, defending a claim not because you have a sincere commitment to it but because you think doing so will be to your benefit.

These ideas, as well as her thoughtful examination of Samuel Butler's *The Way of All Flesh* (Haack 2013: 211–213), enable Haack's explanation of *intellectual integrity* as a disposition to engage in genuine inquiry rather than pseudo-inquiry; one who lacks such integrity is prone to self-deception "about where evidence points, [and] temperamentally disposed to wishful and fearful thinking" (1998: 10). Far from promoting intellectual integrity, the extension to the humanities of the natural sciences' "culture of grants-and-research-projects ... has fostered an environment hospitable to sham and fake inquiry, inhospitable to the fragile intellectual integrity demanded by the genuine desire to find out" (ibid.: 195). This is part of the "research ethic" (2013: 252) of contemporary academia, as are the demand that everyone produce written research and the preposterous pretense that most of what gets published constitutes a significant contribution to knowledge.

The law

Haack's account of the corrosive incentives of today's academic culture illustrates her concern with, as Dewey put it, "the problems of men" (1980: 46; quoted at Haack 2020: 103), and so do her contributions to the philosophy of law, which focus "not exclusively on the *concept* of law but on the *phenomenon* of law—law as embodied in real legal systems" (2017c: 1050). Extending her work in epistemology and the philosophy of science to questions about evidence in the law, she sorts out the confusions involved in Justice Blackmun's majority opinion in the U.S. Supreme Court's *Daubert v. Merrell Dow Pharmaceuticals* decision, including his reliance on Popper's falsificationism in formulating criteria for the reliability, and thus the admissibility, of scientific evidence (2014: ch.5). Haack also argues that U.S. evidence law supports a kind of atomism that fails to appreciate the importance of combined evidence and thereby makes it more difficult for fact finders, like juries, to reach the conclusion best supported by the evidence (ibid.: ch.9). And she argues that, *contra* subjective Bayesianism, legal standards of proof—proof "beyond a reasonable doubt," "preponderance of the evidence," etc.—should be understood not in terms of statistical probability but instead as "epistemological likelihoods" (ibid.: 17), "degree[s] to which the evidence presented must warrant the conclusion (of the defendant's guilt or liability) for a case to be made" (ibid.: 47).

Beginning with her metaphysical view of social institutions, including legal institutions, as real, constantly evolving creations of human beings, Haack arrives at a "neo-classical legal pragmatism" (ibid.: 323), similar in some ways to the views of Oliver Wendell Holmes, Jr. and Benjamin Cardozo. Siding with Holmes, she argues that legal formalism—which understands legal reasoning as the analysis of legal concepts and the deductive inference of rulings from principles, precedents, facts—cannot recognize, or recognize the value in, the regular evolution of the meanings of legal terms and concepts (2007: 8). Extending her innocent realism, Haack characterizes the law as composing (she again borrows James's phrase) a "pluralistic universe"; she recognizes "the richness and variability of the legal systems of the world, past and present, their complicated interrelations, and their roots in commonalities of human nature and society" (2008: 457).

Conclusion

There is no single doctrine shared by Peirce, James, Dewey, and others that qualifies them as pragmatists, let alone one that they all share with contemporary philosophers who self-identify as pragmatists. But Haack advises us not to worry about who counts as authentically pragmatist; "probably it is better—potentially more fruitful, and appropriately forward-looking—to ask, rather, what we can borrow from the riches of the classical pragmatist tradition" (2006: 58). While she borrows from that tradition, her many interlocking applications and extensions of those ideas contribute to a unique, rich, and multifaceted body of work. This brief introduction has omitted several pragmatist themes: her recognition, shared with Peirce, that the meanings of our words and concepts—not just in the law but in science as well—shift and grow (2009b); elements of her philosophy of law inspired by Cardozo and Dewey (2017c); her moral fallibilism, inspired by Dewey (1984: ch.10) and by James (1979); her philosophy of mind, influenced by Mead (e.g., 2010, 2018); and her demolition of Rorty's "vulgar" pragmatism (e.g., 2009a: ch.9). It also neglects her scholarship in the history of pragmatism, her philosophy of logic (1978, 1996), and most of her contributions to social and cultural philosophy (see especially 1998 and 2013). But it nevertheless demonstrates that Haack's realist, worldly pragmatism, more than the work of any other living philosopher, represents the spirit of the classical pragmatists.

Acknowledgment

I am grateful to Susan Haack for her extremely helpful comments on an early draft of this chapter. Any errors in its descriptions of her work are, of course, mine alone.

References

Dewey, J. (1980) "The need for a recovery of philosophy" (1917), in Boydston, J. (ed.) *The middle works: 1899–1924, volume 10: 1916–1917*. Carbondale, IL: Southern Illinois University Press, pp. 3–48.

Dewey, J. (1984). *The quest for certainty. The later works: 1925–1953, volume 4: 1929*, ed. Boydston, J. Carbondale, IL: Southern Illinois University Press.

Haack, S. (1978) *Philosophy of logics*. New York: Cambridge University Press.
Haack, S. (1996) *Deviant logic, fuzzy logic: beyond the formalism*. Chicago, IL: University of Chicago Press.
Haack, S. (1998) *Manifesto of a passionate moderate*. Chicago, IL: University of Chicago Press.
Haack, S. (2003) *Defending science—within reason: between scientism and cynicism*. Amherst, NY: Prometheus Books.
Haack, S. (2005) "Not cynicism, but synechism: lessons from classical pragmatism." *Transactions of the Charles S. Peirce Society* 41(2), 239–253. Also in Margolis, J., and Shook, J. (eds.) (2006) *A companion to pragmatism*. Oxford: Blackwell, pp. 141–153.
Haack, S. (2006) "Introduction: pragmatism, old and new," in Haack, S. (ed.) *Pragmatism, old and new*. Amherst, NY: Prometheus Books, 15–67.
Haack, S. (2007) "On logic in the law: 'something, but not all.'" *Ratio Juris* 20(1), 1–31.
Haack, S. (2008) "The pluralistic universe of law: towards a neo-classical legal pragmatism." *Ratio Juris* 21(4), 453–480.
Haack, S. (2009a) *Evidence and inquiry: a pragmatist reconstruction of epistemology*, 2nd expanded ed. Amherst, NY: Prometheus Books.
Haack, S. (2009b) "The growth of meaning and the limits of formalism: in science, in law." *Análisis Filosófico* 29(1), 5–29.
Haack, S. (2010) "Belief in naturalism: an epistemologist's philosophy of mind." *Logos & Episteme* 1(1), 67–83.
Haack, S. (2013) *Putting philosophy to work: inquiry and its place in culture: essays on science, religion, law, literature, and life*, expanded ed. Amherst, NY: Prometheus Books.
Haack, S. (2014) *Evidence matters: science, proof, and truth in the law*. New York: Cambridge University Press.
Haack, S. (2016) "The world according to innocent realism: the one and the many, the real and the imaginary, the natural and the social," in Göhner, J., and Jung, E.-M. (eds.) *Susan Haack: reintegrating philosophy*. Cham: Springer, pp. 33–55.
Haack, S. (2017a) *Scientism and its discontents*. Rounded Globe [online]. Available at: https://roundedglobe.com/books/9935bace-bf5e-459c-80ae-ce42bfa2b589/Scientism%20and%20its%20Discontents/ (Accessed: 15 May 2021).
Haack, S. (2017b) "The real question: can philosophy be saved?" *Free Inquiry* 37(6), 40–43.
Haack, S. (2017c) "The pragmatist tradition: lessons for legal theorists." *Washington University Law Review* 95, 1049–1082.
Haack, S. (2018) "Brave new world: nature, culture, and the limits of reductionism," in Bartosz, B., Stelmach, J., and Kwiatek, Ł. (eds.) *Explaining the mind*. Kraków: Copernicus Center Press, pp. 37–68.
Haack, S. (2020) "Not one of the boys: memoir of an academic misfit." *Cosmos + taxis* 8(6 and 7), 92–106.
James, W. (1977) *A pluralistic universe*. Cambridge, MA: Harvard University Press.
James, W. (1979) "The moral philosopher and the moral life," in Burkhardt, F., Bowers, F., and Skrupskelis, I. (eds.) *The will to believe and other essays in popular philosophy*. Cambridge, MA: Harvard University Press, pp. 141–162.
Peirce, C. S. (1931–1960) *Collected papers of Charles Sanders Peirce*, 8 vols., Hartshorne, C., Weiss, P., and Burks, A. (eds.) Cambridge, MA: Belknap Press of Harvard University. Cited as "CP"; citations in decimal format by volume and paragraph number.
Wittgenstein, L. (1922) *Tractatus Logico-Philosophicus*, trans. C. K. Ogden. London: Routledge & Kegan Paul.
Wittgenstein, L. (1953) *Philosophical Investigations*, trans. G. E. M. Anscombe. Oxford: B. Blackwell.

15
NICHOLAS RESCHER'S METHODOLOGICAL PRAGMATISM

Michele Marsonet

Nicholas Rescher was born in Germany in 1928, and ten years later, he moved to the United States with his family. He studied first at Queens College and subsequently at Princeton. At Queens College (1946–1949), he met, among others, Herbert G. Bohnert and Carl Gustav Hempel, himself a German immigrant. Rescher studied mathematics and philosophy. In the period 1949–1952, he was a graduate at Princeton's Philosophy Department, where he met Alonzo Church, and received his PhD in 1952.

His academic career began in 1957 at Lehigh University, where he taught philosophy undergraduate courses. In that environment he met Adolf Grünbaum and wrote his first well-known works, concerning the history of Arabic logic. Grünbaum, who was Andrew Mellon Professor in philosophy at the University of Pittsburgh, helped him to become, in 1961, Professor of Philosophy at the same university, where he has worked ever since. Grünbaum and Rescher formed the nucleus of a Philosophy Department which was soon to became famous, including such figures as Kurt Baier, Alan R. Anderson and Wilfrid Sellars.

In 1964, Rescher founded the *American Philosophical Quarterly* and later *History of Philosophy Quarterly*. Subsequently, he served as Chairman of the Philosophy Department and as Director of the Center for Philosophy of Science, established in 1960. Outside Pittsburgh, Rescher was awarded honorary degrees from Lehigh University, from the Loyola University of Chicago and from the Argentina National Autonomous University of Cordoba, while in 1977, he was elected an honorary member of Corpus Christi College in Oxford.

For understanding Rescher's thought, it should be noted that Leibniz has always been his favorite philosopher because of Leibniz's many-sidedness and his ability in utilizing logic and mathematics toward philosophical ends. This differentiates Rescher from Dewey and other pragmatist thinkers, who deem symbolic tools less important.

Our author received a typical analytic training, with teachers like Hempel and Church leaning toward the logical empiricist brand of the analytic tradition. However, already at the beginning of his career, he tended to see formal logic not as an objective per se but rather as an instrument for pursuing larger, philosophical purposes.

In the beginning of his career, Rescher laid the foundations of his first well-known published works, all of them concerning the history of Arabic logic and philosophy. With the passing of time, however, his interests shifted from logic and analytic philosophy of science to epistemology, metaphysics, ethics and a philosophy of science conceived in broader terms. Ultimately his work has been mainly devoted to issues pertaining to social philosophy, political philosophy and metaphilosophy. We should not forget, however, that Rescher always maintained a constant interest in the history of philosophy.

Today pragmatism has gained new strength in the American philosophical circles. As a matter of fact, however, the contemporary neopragmatism actually thriving in the United States has a Rortyan flavor, while Putnam's rediscovery of William James's philosophy and of pragmatism in general is rather recent. Rescher's pragmatist stance is less well known than Rorty's, even though it is older, the reason being that Rescher's thought is essentially perceived as a form of idealism. His American colleagues seem to believe that Rescher's conceptual idealism is more important than his methodological pragmatism, while neither may they be distinguished by a neat border line nor can any of the two deemed to be more important than the other. Rescher's philosophy is a holistic system.

Rescher notes that there seem to be as many pragmatisms as pragmatists. Usually, however, those who are interested in pragmatism from a historical point of view tend to forget that a substantial polarity is present in this tradition of thought. It is a dichotomy between what Rescher calls "pragmatism of the left," which endorses a greatly enhanced cognitive relativism, and a "pragmatism of the right," namely a different position that sees the pragmatist stance as a source of cognitive security. Both positions are eager to assure pluralism in the cognitive enterprise and in the concrete conduct of human affairs, but the meaning they attribute to the term "pluralism" is not the same. Rescher sees Charles S. Peirce, Clarence I. Lewis and himself as adherents to the pragmatism of the right, or "objective" pragmatism, and William James, F.S.C. Schiller and Richard Rorty as representatives of the pragmatism of the left, or "subjective" pragmatism. Objective pragmatism is based on what works impersonally for the realization of some objective purpose, in an impersonal way. In Rescher's view, pragmatism, thus, is essentially a venture in validating objective standards.

But what does the "pragmatism of the right" really come to? Parochial diversity is something that a post-modern pragmatist like Richard Rorty gladly accepts in order to achieve results which are, at the same time, subjectivistic and relativistic. However, even Rescher sees practical efficacy as the cornerstone of human endeavors, but at the same time, he takes efficacy to be the best instrument we have at our disposal for achieving objectivity.

The social world that men themselves create requests that we constantly live having some purposes in mind, and objective pragmatism is just concerned with the effective and efficient achievement of purpose. However, the purposes that Rescher talks about are not mine or yours: they are not, in a word, correlated to the particular tastes of individuals or particular social groups. They can be rather taken to be all collective human endeavors whose rational roots are ultimately reducible to the nature of human condition as such. This means that all men qua men happen to share a natural environment to which they give order resorting to their rational-intellectual capacities. Of course, the largely autonomous social world assumes different shapes according to the different cultural traditions, but, still, we are somewhat compelled to assume a "principle of correspondence," according to which human purposes match the inputs that are set by the conditions of *homo sapiens*, as biological evolution on this planet and social evolution in our cultural environment have shaped us.

So Rescher's kind of pragmatism leads to objectivity, in the sense that objective constraint, and not personal preference, is the fundamental premise of our cognitive goals. What we mean to achieve in starting the process of empirical knowledge is control over the natural environment of which we are ourselves an essential part, and this control, in turn, may be both active (interactionistic) and passive (predictive). Although he openly declares his idealistic stance, Rescher recognizes the presence of a "reality principle" that is practically forced upon us just in view of our belonging in the natural world, and despite the fact that we play, in that same world, a very special role. Our control over nature, in turn, can never be total. We create the social-linguistic world but not natural reality. It must be admitted that we have access to natural reality only through social and linguistic tools, but it is wrong to draw, from this premise, the conclusion that men create the whole of reality, both social and natural.

If we accept this line of reasoning, any clear border line between the social and the natural world is illusory. We can, in other words, claim that nature imposes inescapable constraints upon us. But, at the same time, we are allowed to stress the fact that men always see nature from their point of view, this being their condition of "accessibility" to nature itself. However, there is no need to conceive of this condition in purely individualistic and solipsistic terms: it rather pertains to the human species at large.

We all know that different human groups categorize reality in different ways, even though these differences are never so great as to prevent a reasonably good communication among them. So we are bound to ask: How are these communal projects set up, given the inevitable difference among the many groups that actually form humankind? Can we really find a common basis which is shared by all human beings as such, so preventing the risk that talk about communal projects is just wishful thinking?

According to Rescher, we certainly can, and the basic reason relies on the view of human social life as a rational reaction of self-adaptation to the natural environment from which social groups themselves evolved. Thus, objective pragmatism claims that (1) our social-linguistic world evolved out of natural reality; (2) this social-linguistic world acquires an increasing autonomy; (3) between the social and the natural worlds, there is no ontological line of separation, but just a functional one; (4) however, the accessibility to natural reality is only granted by the tools that the social-linguistic world provides us with; (5) this means that our knowledge of natural reality is always tentative and mediated by our conceptual capacities; and (6) there is no need to draw relativistic conclusions from this situation, because the presence of "an objective reality that underlies the data at hand" puts upon personal desires objective constraints that we are able to overcome at the verbal level but not in the sphere of rational deliberations implementing actions.

Rescher always notes that the conceptual apparatus we employ itself makes a creative contribution to our view of the world, and his holistic (or systemic) stance is clearly influenced by Hegel and Bradley. It should be noted that Rescher immediately tied these idealistic insights to the philosophy of science, a sector that has always been at the core of his interests. The aforementioned statements, in fact, led him to the conclusion that scientific discovery is not a matter of simply "reading" what is written in the book of nature but is rather the outcome of a process we have already mentioned several times: the interaction between nature on the one side and human mind on the other. The contribution that mind gives to the construction of "our science" is at least as important as that provided by nature: no science – as we know it – would be possible without the specific contribution of the mind.

Rescher in the early 1970s launched his project of rehabilitating idealism. In particular, a "conceptual idealism" maintaining that we understand the real in somehow mind-invoking terms of reference is perfectly compatible with an ontological materialism, which holds that the human mind and its operations ultimately root in the machinations of physical process. On the one hand, Rescher accepts the Kantian view that our knowledge is strongly determined by the a priori elements present in our conceptual schemes and that they indeed have an essential function as long as our interpretation of reality is at stake. On the other hand, however, he tends to see these aprioristic elements as resting on a contingent basis and validated on pragmatic considerations.

The mind certainly makes a great contribution toward shaping reality-as-we-see-it, but the very presence of the mind itself can be explained by adopting an evolutionary point of view. There is no neat distinction between ontology and epistemology. Rescher refuses to draw a clear border line between the two, and one of the reasons for that lies in the aforementioned holistic character of his philosophical system. The separation between factual and conceptual is not sharp and clean but rather fuzzy. Yet there is another reason, which is connected to the ontological opacity of the real world. We can have access to the unconceptualized world only through conceptualization. And conceptualization is, in turn, the key feature that characterizes our cultural evolution.

The early analytic philosophy put forward various realistic theses that were directly opposite to the idealistic ones. For instance, the world, which was clearly mind-dependent for Bradley and the other British neoidealists, was instead seen as mind-independent by Russell and Moore; furthermore, reality, which was essentially one for the neoidealists, was conceived as formed by a multitude of separate objects by the new analytic thinkers. Rescher's sympathy for idealism grew constantly, following the attendance of a graduate seminar on Bradley's *Appearance and Reality* taught by Walter T. Stace in Princeton. This explains why Rescher endorses a coherentist approach to truth. He cites Bradley's contention that "Truth to be true must be true of something, and this something itself is not truth." Accepting the suggestions of an idealist like Ewing, we might even say that correspondence and coherence theories are not such traditional enemies as the philosophical tradition usually depicts them to be. Rescher's coherence theory implements Bradley's dictum that *system* (i.e. systematicity) provides a test criterion most appropriately fitted to serve as an arbiter of truth.

The idealistic component of Rescher's thought should not be overestimated but just given its proper weight within his philosophical system taken as a whole. No one can seriously doubt that there are strong idealistic features in his philosophy. For example, he never tires of stressing that the conceptual apparatus we employ itself makes a creative contribution to our view of the world, and his holistic stance is clearly influenced by Hegel and Bradley, i.e. thinkers who have long been quite unpopular within American philosophy. But idealism is just one element in a broader framework where pragmatism plays the key role, and other important components are detectable as well in his thought.

Rescher never diminishes the importance of biological evolution, which is specifically geared to the natural world and, after all, is supposed to precede our cultural development from the chronological point of view. The fact is, however, that it is cultural evolution that distinguishes us from all other living beings that share our planet with us. Just for this reason, Rescher claims that idealism, broadly speaking, is the doctrine that reality is somehow mind-correlative or mind-coordinated. However, his specifically conceptual idealism stands in contrast to an ontological doctrine to the effect that mind somehow constitutes or produces the world's material. Scientists, thus, will not find in his philosophy the basic anti-scientific attitude endorsed by the classical idealists and some contemporary neoidealist thinkers, who do not deem natural science important because it deals with a second-level reality.

Now, if we admit that the real (i.e. mind-independent) world exists, a distinction may be drawn between: (A) nature-as-we-understand-it and (B) nature itself. Is this distinction ontological or epistemological? To answer this question, we should be able to trace a line of separation between ontology and epistemology neater than the one Rescher is inclined to accept. Rescher's suggestion, however, is that our conceptual machinery is at work even when we try to gain access to (B) because our access to the world is always mind-involving. Going back to the aforementioned distinction between (A) and (B), it might even be said that it is a real distinction, because we know that our history is both biological and cultural, although our cultural life needs a preexistent biological basis in order to develop.

By using our scientific instruments and theories, we are able to shed some light on the natural history of the earth (and of the universe at large), but this natural history is always ours, because it is conducted by following our mental patterns and categories and by using the scientific instruments and theories that we build. We can push our sight so far as to imagine an era when no categorization of the world took place because no human beings were around. Still, even in this case, we must have recourse to categorization just to imagine a situation of this kind (which can be presumed to have been real because, after all, evolutionary epistemology gives good reasons for assuming that humankind was not present on our planet since the beginning). Any "absolutely objective" ontology is then left in the background, because precious little can be known about it, and it represents a *via negativa* that does not take us very far.

What can we possibly think about this reality, and how can we say what it is like? Even for imagining a world totally devoid of human presence, we must use human concepts. As it was said previously, conceptualization is not an option we can get rid of but a built-in component of our nature of human beings. According to our author, then, we must distinguish between the *that* and the *what* of this purported mind/thought independent reality. In this case, we are sensibly entitled to claim *that* it exists while simply rejecting the challenge to specify *what* it is like. Going back to the example of science, we know for sure *that* there are errors in present-day science but cannot say *what* they are. Rescher's conception of scientific realism is thus strictly tied to his distinction between reality-as-such and reality-as-we-think-of-it. He argues that there is indeed little justification for believing that our present-day natural science describes the world as it really is, and this fact does not allow us to endorse an absolute and unconditioned scientific realism. In other words, if we claim that the theoretical entities of current science correctly pick up the "furniture of the world," we run into the inevitable risk of hypostatizing something – i.e. our present science – which is only a historically contingent product of humankind, valid in this particular period of its cultural evolution.

As for political and ethical issues, if we want to be pluralists in the true spirit of Western democratic thought, we must abandon the quest for a monolithic and rational order, together with the purpose of maximizing the number of people who approve what the government, say, does. On the contrary, we should have in mind an acquiescence-seeking society where the goal is that of minimizing the number of people who strongly disapprove of what is being done. We should never forget, in fact, that the idea that all should think alike is both dangerous and anti-democratic, as history shows with plenty of examples. Since consensus is an absolute unlikely to be achieved in concrete life, a difference must be drawn between "being desirable" and "being essential." It can be said that it qualifies at most for the former status. The general conclusion is that consensus is no more than one positive factor that has to be weighed on the scale along with many others.

It is worth stressing the similarity between Rescher's epistemology and political/ethical philosophy: they both rest on his skepticism about idealization. In neither case we can get perfect solutions to our problems, short of supposing an unattainable idealization. We have to be fallibilists in epistemology because we are emplaced in suboptimal conditions, where our knowledge is not (and cannot be either) perfected. In other words, we have to be realistic and settle for imperfect estimates (i.e. the best we can obtain).

Since we cannot realize a Habermas-style idealized consensus, we must settle for what people will go along with, i.e. "acquiesce in." This may not be exactly what most or many of us would ideally like, but, in any case, if we insist on "perfection or nothing," we shall get either nothing or a situation very far away from our ideal standards. In the social and political contexts, "realism" means settling for the least of the evils because, as history teaches, disaster will follow if we take the line that only perfection is good enough.

According to Rescher, the overcoming of any ill-based foundationalism means neither the end of philosophy itself nor the refusal to recognize its cognitive value. He agrees with Rorty's assertion that philosophers cannot detach themselves from history or forsake the everyday and scientific conceptions that provide the stage setting of their discipline. But, nevertheless, he claims that the dissolution of philosophy is a deeply wrong answer. Skeptics of all sorts would like to liberate humankind from the need of doing philosophy, pointing out that it has thus far been unable to answer our questions in a proper way. Rescher, to the contrary, invites us to take sides because abandoning philosophical subjects is a leap into nothingness.

Rescher's Selected Bibliography

A System of Pragmatic Idealism (3 vol., Princeton: Princeton University Press, 1992–1994).
Conceptual Idealism (Oxford: Blackwell, 1973).
Fairness (New Brunswick and London: Transaction Publishers, 2002).
Many-Valued Logic (New York: McGraw-Hill, 1969).
Methodological Pragmatism (Oxford: Blackwell, 1977).
Philosophical Standardism (Pittsburgh: University of Pittsburgh Press, 1994).
Pluralism (Oxford: Clarendon Press, 1993).
Realistic Pragmatism (Albany: State University of New York Press, 2000).
Scientific Progress (Oxford: Blackwell, 1978).
Scientific Realism (Dordrecht: Reidel, 1987).
The Coherence Theory of Truth (Oxford: Clarendon Press, 1973).
The Development of Arabic Logic (Pittsburgh: University of Pittsburgh Press, 1964).
The Limits of Science (Berkeley: University of California Press, 1984).
The Strife of Systems (Pittsburgh: University of Pittsburgh Press, 1985).

16
ROBERT BRANDOM

Chauncey Maher

Many Interests, and One

A professor of philosophy at the University of Pittsburgh for more than 40 years, Robert Brandom has written extensively and insightfully on a wide range of topics, including language, mind, logic, norms, freedom, philosophical method, pragmatism (Brandom, 2011), Immanuel Kant (Brandom, 2009), G.W.F. Hegel (Brandom, 2019), and Wilfrid Sellars (Brandom, 2015). For some purposes it is useful to see him as having many different interests, but to introduce you to his work, here I will treat him as having at least one big interest that subsumes many others. He seeks to understand what is distinctive about human thought, the sort of thought we characteristically engage in. After briefly clarifying the question that guides him, I will present a few of his most intriguing and controversial claims, hoping to spur you to inquire further.

What Is Distinctive about Human Thought?

Consider an example of nonhuman thinking from my ordinary life. Butter, our dog, lies on the floor upstairs. Erin, my spouse, approaches our home in her car. Butter lifts her head, ears pricked up. She walks down the stairs and sits at the front door, just as Erin puts her key in the lock.

Historically, whether nonhuman animals can think has intermittently been a contentious topic. Over the past century, thanks to diligent and ingenious investigations into the behaviors of a wide variety of animals, denying that they do so is no longer very credible (Andrews & Beck, 2017; Andrews & Monsó, 2021).

Chimpanzees devise novel ways to get things they want, such as stacking boxes to reach a bunch of bananas. In response to different types of predators, Vervet monkeys make different types of calls, to which their fellows respond in fitting ways, such as climbing a tree to avoid a snake. Crows use tools to get insects that are otherwise difficult to get. An African gray parrot reliably distinguishes arbitrary items as being the same or different, two rather abstract categories. Returning from sources of food, honeybees dance on the walls of their hives, prompting other honeybees to head for and arrive at that same source. After visually spotting their prey, jumping spiders break visual contact, crafting effective detours to capture their prey (Bekoff et al., 2002; Shettleworth, 2009).

Now compare these examples with an example of human thinking.

DUSTY: Is Erin here?
SEAN: That's her car.

DUSTY: How can you tell? So many people have that same model and color.
SEAN: Yea, but that is her license plate.
DUSTY: Ah, so probably she is here.

How does the situation with Sean and Dusty differ from the situation with Butter, chimpanzees, or honeybees? What is distinctive about human thought?

Various answers look reasonable. One might say human thought is distinctive because of its topics or contents, what can be thought about. Humans think about things that matter to humans, like racism or famine. Humans think about abstract things, such as numbers and time and justice and the nature of thinking. Humans think about the distant past and future and purely hypothetical situations. Humans think about a wide variety of things that (one might reasonably hold) nonhuman animals don't or can't think about.

Alternatively, one might contend that human thinking is distinctive because we have a sort of control over it that other animals don't have. Although we often just find ourselves thinking something or other, as though it were foisted on us, we are also able to initiate and guide our thinking. Perhaps no other animals can do that or do it to the same extent.

A third answer: one might say that human thinking has a distinctive format. Sean and Dusty speak in sentences, which suggests their thoughts are also in a sentence-like format, having subjects and predicates. According to our current best evidence, although many animals communicate with their conspecifics, no other animals appear to use sentences. Perhaps their thoughts are restricted to a different format. Rather than being sentence-like, their thoughts might be image-like, or instead they might be diagrammatic.

Brandom contends that human thinking is distinctive because we are capable of propositional judgment. For example, Sean judges that *That is Erin's car*. He judges something; he takes a stand. And what he judges, what he takes a stand on, is a proposition.

Brandom's answer is controversial. Some theorists think human thinking is, at bottom, imagistic, not propositional (Prinz, 2002). Others think that thinking as such is propositional, because it must occur in a "language of thought" (Rescorla, 2019). Given that and given that other animals think, humans wouldn't be distinctive for thinking propositionally.

Brandom aims to clarify what is involved in propositional judgment, committing to several further claims that are intriguing and controversial. He contends that propositional judgment is normatively articulated, socially articulated, and inferentially articulated. I'll briefly discuss each of these claims in turn.

Human Thought Is Normatively Articulated

Brandom holds that making a propositional judgment—as when Sean judges *That is Erin's car*—necessarily has normative dimensions. Doing so involves taking on a type of responsibility and exerting a type of authority. Sean takes on the responsibility to provide reasons or justifications for this proposition. He also thereby acts as though he is authorized or has license to take on this responsibility. Being obliged and permitted to do things are necessarily normative, for they intrinsically involve being subject to norms.

With this initial step, Brandom distinguishes himself from some historically influential thinkers, such as Rene Descartes. He held that the human mind is an immaterial substance that occupies neither space nor time. He held also that a distinguishing feature of our minds is that each of us knows our own first and better than we know anything else. Brandom rejects both of these ways of characterizing what's distinctive about human thought. He maintains that it need not involve a special sort of substance, and it need not involve a special sort of knowing. Rather, according to him, it involves a distinctive sort of normative status.

Brandom's claim is controversial. In contemporary cognitive science—which studies human thought—researchers and teachers acknowledge that humans can think propositionally, but many, if not most, do not regard doing so as necessarily having any normative dimensions (Goldstein, 2019; Sternberg & Sternberg, 2017). This is not so much because they have arguments against Brandom's claim, but more because they do not regard normative matters as within their jurisdiction. They take themselves to be focused primarily on the causes, effects, and physiological implementation of thinking. They leave questions about what we are obliged or permitted to do to other inquirers, perhaps those in moral or political philosophy. To these cognitive scientists, it might seem that Brandom has confused two inquiries that should be treated as distinct.

Why does Brandom think that propositional judgment is inherently normative?[1]

Reacting reliably to a situation doesn't suffice for judging that this situation obtains. Imagine that on the road in front of our home, there is a collection of dirt. Erin's tires also happen to have a very distinctive tread. When she arrives in her car, the tires impress a track in that dirt. And that track occurs only when Erin's car arrives. The collection of dirt thus reacts reliably to the presence of Erin's car, but it does not judge that *That is Erin's car*. Dirt doesn't judge. This does not imply that propositional judgment is inherently normative. It raises the question of what more than reliable reaction is needed.

Any judging that a situation obtains can serve as and stand in need of reasons. When Sean judges that *That is Erin's car*, from this he may legitimately infer that *Something belongs to Erin*. And he may legitimately support his initial judgment by judging that *That is Erin's license plate* (on the car). From his initial judgment, he may not legitimately infer that *That is a sandwich*. And he is precluded from judging also that *That is not a car*. In these ways, judging that a situation obtains is governed by norms of rationality or good inference. These are norms concerning what may and may not be legitimately inferred from what.

Here is an argument for this controversial claim. Suppose that Sean's judging that *That is Erin's car* is not subject to any norms of inference. Then, there is nothing that he is permitted to infer from it. There is also nothing he is *not* permitted to infer from it. Furthermore, there is also nothing that he is permitted to infer it from. And there is nothing he is *not* permitted to infer it from. Then, since there is nothing whatsoever that this episode rationally permits or forbids, Sean's judging could just as well be a judging that *That is not Erin's car* rather than a judging that *That is Erin's car*. What Sean has allegedly judged has vanished. Nothing has been judged. Then, he has not judged anything at all. So, we must reject the supposition that Sean's judgment is not subject to any norms of inference.

Human Thought Is Socially Articulated[2]

Brandom contends further that propositional judgment is socially articulated. It necessarily involves relations between at least two entities, fellow propositional judgers. In this respect, being a judger is analogous to other types of thing that are socially articulated, such as being a parent, student, neighbor, doctor, rabbi, banker, citizen, or goalie. Each of these types of thing is defined in terms of relations to others. Parents necessarily have children. Students necessarily have teachers. Doctors necessarily are qualified to doctor to someone. And so on. These types of thing would not exist without standing in these relationships to others. Nothing could be one of these sorts of thing without being related in various ways to others.

Like the claim that propositional judgment is normatively articulated, this claim too is controversial. I suspect that if you were to ask a random sample of people whether making a propositional judgment requires more than one person, the majority would say, "No." I speculate further that they would take this to be obvious. For them, it would be perfectly clear that humans can and usually do make such judgments all on their own, without the aid or presence of anyone else. For them, being a judger would starkly contrast with being a parent or student or doctor, precisely because one doesn't need others to do it. Many, if not most, contemporary cognitive scientists would agree. They would grant that human relationships can deeply affect and be affected by propositional judgments. But that claim is different and weaker than Brandom's claim.

Why, then, does Brandom believe that propositional judgment is necessarily socially articulated?

His main reason is that normative statuses, such as being permitted to make certain inferences, being obliged to have reasons, and being prohibited from making other judgments, are necessarily socially articulated.

What's the argument? At this stage, Brandom takes for granted that propositional judgments are subject to norms of reasoning or inference. This means that they are assessable or evaluable in terms of such norms. It follows that someone can assess or evaluate whether the judger is in compliance with the norms. The "can" here is the "can" of license or permission. This someone has authority to assess, say, Sean's judging that *That is Erin's car*. A crucial further premise is that this someone with authority cannot be simply the judger herself or himself. Otherwise, whatever each judger did would automatically count as right; for whatever each judger took as right would be right. Sean, for instance, could do no wrong. For example, if he also judged that *That is not a car*, and thereby took doing so as right, it would be right, despite being incompatible with his first judgment. Without the possibility of going wrong, no norms would be in place. Thus, the someone with authority to assess a judging must differ from the judger. In addition, for someone to qualify as having this authority, the judger must take or treat this someone as having it. For if the judger does not acknowledge or recognize this person's authority, then any of their alleged assessments do not differ from mere causal compulsion, like being physically pushed by them; they would not be genuine assessments. Moreover, by treating the judger as someone whose conduct is an appropriate target of assessment, the assessor thereby treats the judger—in our case, Sean—as having the authority to take on the responsibilities of complying with the norms. So, ultimately, we have a symmetrical, mutual, or reciprocal arrangement between the judger and the assessor. Each treats and thereby confers authority on the other for certain things. This is why we should, according to Brandom, think normative statuses are socially articulated; they require inter-personal interaction and treatment.

Brandom contends further that a striking implication of the social articulation of propositional judgment is that there must be concrete or embodied interactions between propositional judgers. There must be ways for them to treat each other as authoritative over and responsible for complying with norms. Brandom calls this "the game of giving and asking for reasons." More plainly, it is speaking a language. Thus, Brandom maintains that distinctively human thought requires the capacity to speak a language.

Human Thought Is Inferentially Articulated

Brandom holds that human thought is inferentially articulated. So far, when talking of human thought or propositional judgment, we have focused mostly on the activity of think*ing* or judg*ing*. We turn now to the object of those activities, what is thought or judged in those activities, their contents. Brandom's contention is that the contents of propositional judgments are inferentially articulated.

When Sean judges that *That is Erin's car*, he judges something; there is a "what" that he judges; this is the content of the judging; it is the proposition *That is Erin's car*. This proposition is inferentially articulated in that it is defined by its role in good or proper or rationally legitimate inferences.

As we have seen, this proposition can serve as a premise or conclusion in various good inferences, such as:

PREMISE: That is Erin's car.
CONCLUSION: That is a car.

PREMISE: That is the vehicle with Erin's license plate.
CONCLUSION: That is Erin's car.

But there are many more:

PREMISE: That is Erin's car.
CONCLUSION: That is Erin's.

PREMISE: That is Erin's car.
CONCLUSION: I should start cooking dinner.

PREMISE: The toll booth's license plate program says that is Erin's car.
CONCLUSION: That is Erin's car.

Together with collateral premises, any proposition figures in a very large network of inferences, pieces of reasoning, or argument. This includes some bad ones, such as:

PREMISE: That is Erin's car.
CONCLUSION: Erin is in that car.

PREMISE: Erin is in that car.
CONCLUSION: That is Erin's car.

Brandom contends that a proposition—its content, what it means—is articulated and determined by its role in *good* inferences but not bad ones. It is defined by what it legitimately supports and is supported by; what can be properly inferred from it and what it can be properly inferred from. Commonly, this idea is called "inferentialism" and is regarded as a variety of conceptual-role semantics. The name "inferentialism" is sometimes used more broadly to include the controversial claims about normative and social articulation that we canvassed in preceding sections. (Peregrin, 2014).

Once again, Brandom's contention is controversial.

Most significantly, in standard logic and linguistics textbooks and courses, we are taught that the content or meaning of a proposition is its truth-condition, the condition under which it would be true. The meaning of the proposition expressed by the English sentence "That is Erin's car" is whatever condition would make that proposition true. We are told further that this is determined by the meaning of the sentence's parts and the way they are arranged. It is determined by what "That" refers to and by the meaning of "is Erin's car." The sentence is true if, and only if, what "That" refers to meets whatever the conditions are for being Erin's car. These words and phrases can be re-combined in infinitely many ways with other words and phrases, yielding utterly novel sentences, with new meanings, new conditions under which they would be true. From typical logic books, we learn also that while this sentence can and does enter into various good inferences, those inferences do not constitutively define the meaning of the sentence. Rather, inferences involving the sentence count as good or bad, given the prior, independent meaning of the sentence.

The meanings of sentences explain the goodness of inferences, not vice versa. From this point of view, by starting with inferential relationships, Brandom appears to be doing things backward.

Why does Brandom think the content of propositions must be inferentially articulated? He takes this to be a consequence of the prior idea that the act of judging is normatively and socially articulated in a certain way. These acts are exercises of *inferential* authority and undertakings of *inferential* responsibility. So, what is judged in a judging should be understood in terms of inference. In judging, what one has judged—what one exercises authority over and what one undertakes responsibility for—must be constitutively defined by its place in inferring. Because a propositional judgment is the fundamental unit of thought, Brandom holds that the whole content of that unit is the fundamental bearer of content. We can and should understand the content or meaning of a proposition's components or constituent elements (that which corresponds to words and phrases) in terms of their place in that larger whole. Thus, in Brandom's view, concepts should be understood in terms of their role in whole judgments (compare Carey, 2009).

Dig Further

With this short introduction to Robert Brandom's thinking about what is distinctive about human thought, I hope I have spurred you to dig further into his own work, as well as that of critics (Bouché, 2020; Weiss & Wanderer, 2010) and commentators (Koreň, 2021; Loeffler, 2018; Peregrin, 2014; Wanderer, 2008).

Notes

1 For some further details, see Brandom (1994: 5, 8–9, 11–12, 87–89, 214–215; 2000: 48, 80, 105–109; 2009: 32–35).
2 For some further details, see Brandom (2009: 59–64, 66–70, 78–79).

References

Andrews, K. & Beck, J. eds., 2017. *The Routledge Handbook of Philosophy of Animal Minds*. New York: Routledge.
Andrews, K. & Monsó, S., 2021. *Animal Cognition*. [Online] Available at: https://plato.stanford.edu/entries/cognition-animal/.
Bekoff, M., Allen, C. & Burghardt, G. M. eds., 2002. *The Cognitive Animal*. Cambridge, MA: MIT Press.
Bouché, G. ed., 2020. *Reading Brandom: On a Spirit of Trust*. New York: Routledge.
Brandom, R., 1994. *Making It Explicit*. Cambridge, MA: Harvard University Press.
Brandom, R., 2000. *Articulating Reasons*. Cambridge, MA: Harvard University Press.
Brandom, R., 2009. *Reason in Philosophy*. Cambridge, MA: Harvard University Press.
Brandom, R., 2011. *Perspectives on Pragmatism*. Cambridge, MA: Harvard University Press.
Brandom, R., 2015. *From Empiricism to Expressivism*. Cambridge, MA: Harvard University Press.
Brandom, R., 2019. *A Spirit of Trust*. Cambridge, MA: Harvard University Press.
Carey, S., 2009. *The Origin of Concepts*. New York: Oxford University Press.
Goldstein, E. B., 2019. *Cognitive Psychology*. 5th ed. Boston, MA: Cengage Learning.
Koreň, L., 2021. *Practices of Reason*. New York: Routledge.
Loeffler, R., 2018. *Brandom*. New York: Polity.
Peregrin, J., 2014. *Inferentialism: Why Rules Matter*. New York: Palgrave.
Prinz, J., 2002. *Furnishing the Mind*. Cambridge, MA: MIT Press.
Rescorla, M., 2019. *The Language of Thought Hypothesis*. [Online] Available at: https://plato.stanford.edu/entries/language-thought/.
Shettleworth, S., 2009. *Cognition, Evolution, and Behavior*. 2nd ed. New York: Oxford University Press.
Sternberg, R. & Sternberg, K., 2017. *Cognitive Psychology*. 7th ed. Belmont, CA: Wadsworth.
Wanderer, J., 2008. *Robert Brandom*. Montreal: McGill-Queen's University Press.
Weiss, B. & Wanderer, J. eds., 2010. *Reading Brandom*. New York: Routledge.

PART II

Pragmatism and Plural Traditions

17
PRAGMATISM'S FAMILY FEUD

Henry Jackman

Introduction

While William James and Charles Sanders Peirce are considered the two fathers of American Pragmatism, their overall philosophical outlooks were often remarkably different, with Peirce eventually labeling his position "Pragmaticism" to distinguish his views from those increasingly being associated with James.[1] A "two pragmatisms" narrative has remained with us ever since,[2] typically with the Peircean version being presented as the comparatively "objective" alternative to metaphysical realism and the Jamesian strand being castigated as an overly "subjective" departure from Peirce's position, a departure resulting from James's soft spot for religious belief and his "almost unexampled incapacity for mathematical thought" (Peirce CP 6. 182).

However, while James clearly does put more of an emphasis on "subjective" factors than does Peirce, his doing so was often the result of his simply drawing out consequences of Peirce's original framework. That framework was presented in an 1872 meeting of their Metaphysical Club where James and Peirce (along with, among others, Oliver Wendel Holmes and Chauncy Wright) famously discussed a number of the core ideas that have been associated with pragmatism ever since.[3] No official records were kept of those meetings, but it is believed that the two seminal papers that Peirce published five years later as "The Fixation of Belief" (1877) and "How to Make Our Ideas Clear" (1878) were revisions of a draft that he originally delivered to the club.[4] At roughly the time Peirce published these two papers, James published three of his first philosophical essays, essays that can be understood as responses to, and extensions of, their discussions of 1872. These papers, "Remarks on Spencer's Definition of Mind as Correspondence" (1878),[5] "The Sentiment of Rationality" (1879) and "Rationality, Activity and Faith" (1882),[6] laid the ground for a distinctly Jamesian strand of pragmatism.

In particular, while Peirce was still flirting with idealism at the time,[7] James's papers took those 1872 discussions and teased out some of the consequences that followed once they were placed more firmly in a naturalistic, particularly Darwinian, framework. Peirce was never comfortable with these consequences, thinking that James carried pragmatism "too far" (CP 8, 258, 1904), and in later work tried to distance himself from a number of positions defended in his earlier papers. James, by contrast, never rejected that early framework, which resulted in the increasing differences between the versions of pragmatism developed by the two. These differences show up most clearly in their conflicting conceptions of both when our beliefs are rationally justified and what it would take for those beliefs to be true.

Peirce and James on the Justification of Belief

The first major difference between James and Peirce revolves around the question of when we are justified in adopting, or holding on to, particular beliefs. According to the standard "two pragmatisms" narrative, Peirce defends the "moderate" pragmatist position that combines fallibilism (the view that none of our beliefs can be established with absolute certainty) with a type of anti-skepticism ("critical commonsensism") that holds that such certainty isn't required for our beliefs to be justified. James, by contrast, pushed this to a type of "extreme" pragmatism, where a belief's justification not only didn't require certainty but also could be grounded entirely in the belief's ability to make us successful or even just happy.[8] This characterization of James is certainly unfair, but it does reflect the fact that James did extend Peirce's position in ways that Peirce clearly wasn't happy with.

Peirce on the Fixation of Belief

Peirce and James both follow Alexander Bain's definition of belief as "that upon which a man is prepared to act" (a definition from which Peirce considered pragmatism to be "scarce more than a corollary") (C.P 5.12, 1907), and in "The Fixation of Belief," Peirce distinguishes belief and doubt in terms of the fact that "beliefs guide our desires and shape our actions" (1877: 114) while doubt "is an uneasy and dissatisfied state from which we struggle to free ourselves and pass into the state of belief" (1877: 114). A consequence of Peirce's view is that states that don't produce this sort of dissatisfaction aren't *real* doubts at all, since "the mere putting of a proposition into the interrogative form does not stimulate the mind to any struggle after belief" (1877: 115).

For Peirce, real doubt produces a type of "anxiety" (1968: 114) in us, and this "irritation of doubt causes a struggle to attain a state of belief" (1868: 24). Peirce refers to this struggle as "inquiry," and the rest of "The Fixation of Belief" evaluates various methods of inquiry in terms of their ability to "fix" our beliefs and produce a lasting end to doubt. These methods are meant to be evaluated *solely* on their ability to alleviate doubt, and crucially, not in terms of their ability to lead us to the truth. As Peirce puts it:

> the sole object of inquiry is the settlement of opinion. We may fancy that this is not enough for us, and that we seek, not merely an opinion, but a true opinion. But put this fancy to the test, and it proves groundless; for as soon as a firm belief is reached we are entirely satisfied, whether the belief be true or false.
>
> *(1877: 114–115)*

The justification of our beliefs, then, is understood not in terms of truth but in terms of what produces a sustained feeling of justification (or, perhaps better, a sustained absence of doubt). That said, the conclusion that Peirce reaches will not be that far from one tied to the traditional search for truth, as he goes on to argue that the methods of "tenacity," "authority" and the "a priori" method all fail to "fix" belief adequately, and that it is only the "method of science" that can really do the job.

James and the Sentiment of Rationality

Some of these basic ideas from Peirce's "Fixation of Belief" run through James's own papers "The Sentiment of Rationality" and "Rationality, Activity, and Faith." In particular, when James describes the sentiment of rationality as "This feeling of the sufficiency of the present moment, of its absoluteness,—this absence of all need to explain it, account for it, or justify it" (James 1897:

58), he is essentially describing the feeling that we have when our beliefs are entirely free from the "irritation" of doubt. James's investigation into what produces the sentiment of rationality can also be seen as in line with Peirce's project,[9] but what is importantly novel to James is his contention that the irritation of doubt is produced not only by generally epistemic reasons (say, our discovering that our beliefs are inconsistent, or seem to contradict experience) but by "passional" considerations as well. For instance, beliefs that frustrate our practical interests by suggesting that life is meaningless, and thus give us nothing to "press against" (James 1897: 70), produce a similar irritation and thus naturally come to be doubted.

It's important to note that for James this is primarily a *negative* claim. It is not that a belief's making us happy justifies us in believing it. Rather, it is that a belief's leaving us unsatisfied can cause us to doubt it, so beliefs that go against our practical interests will be more difficult to "fix" in precisely Peirce's sense.

This won't be the case with *every* "unhappy" belief. I'm not happy about the fact that I'm losing my hair, but that belief is confirmed every time that I look in the mirror, so the unhappiness doesn't produce doubt. However, for the beliefs that lack such constant evidential support, doubts can arise. For James, the belief in materialism is a paradigm case of this. The fact that there couldn't be anything "more" than matter out there isn't confirmed by everyday experience in the way that my hair loss is, and so doubts have the freedom to creep back in. Ockham's razor might favor the materialistic theory, but James sees that as just a reflection of our "passion for parsimony" (James 1897: 58), leaving it a question of which passions carry the most weight for each believer. For those with the temperament that James characterizes as the "sick soul,"[10] the hypothesis that there is nothing more than matter will, in spite of its parsimony, always produce feelings that it "just can't be true" or "doesn't make any sense," and doubts will come in their trail.

We see another application of this approach in James's discussion of nominalism, where he argues that we could never "fix" on a nominalistic system even if it were consistent and "fit" all of our experience, since doubts about the possibility of a more robust alternative would always arise.

> a consistently nominalistic account of things could never be generally accepted as the truth—the craving for a plus ultra the instant phenomenon, shut off today, would reassert itself tomorrow in some new mode of formulation and breed an everlastingly self-renovating protest against the reduction of all reality to actuality.
>
> *(James 1978: 367)*

It is just such cases that go beyond the available empirical evidence that are of the most interest to James, and "The Sentiment of Rationality" was originally meant to be part of a work on the psychology of philosophy that would analyze what makes us find a particular *philosophical* system rational.[11]

James is often portrayed here as endorsing something like Peirce's method of tenacity,[12] and one prominent example that Peirce gives of this method is frequently read as directed at James:

> Thus, if it be true that death is annihilation, then the man who believes that he will certainly go straight to heaven when he dies, provided he has fulfilled certain simple observances in this life, has a cheap pleasure which will not be followed by the least disappointment. A similar consideration seems to have weight with many versions in religious topics, for we frequently hear it said, "Oh, I could not believe so-and-so, because I should be wretched if I did."
>
> *(Peirce 1877: 116)*

However, tenacity involves holding on to a belief in the face of *contrary* evidence, and this is very different from refusing to adopt a belief that you aren't evidentially compelled to hold. James is defending the more modest claim that a view that truly makes someone "wretched" will "afflict the mind with a ceaseless uneasiness" (James 1897: 100) and thus lead it to be doubted if any compelling reason isn't given for it.

Religious belief was the highest profile example of such an evidence-transcendent case,[13] but the range of such beliefs was considerably larger, and more general methodological assumptions such as that the future would be like the past, that our investigations would lead to the truth or that our beliefs were capable of being true at all arguably all fell into this camp. James rejected transcendental arguments guaranteeing the truth of such "regulative" principles, but he insisted that we still had the right to believe in them, since such beliefs all fall into the class where, in Peirce's own terms, we "begin with all the prejudices which we actually have," and for which we don't have any "positive reason to doubt" (Peirce 1868: 28, 29). But while Peirce relied on such "regulative ideals" as well, he eventually argued that James's epistemology was too permissive and that while we were entitled to *hope* that such regulative principles were true, we weren't entitled to *believe* in them.[14]

Unfortunately, Peirce's appeal to hope in these cases runs into problems with the earlier account of belief and its connection to action. If belief really is "that upon which a man is prepared to act," and hopes are capable of producing the same habits of action as beliefs, then it would seem that these hopes should also count as beliefs themselves.[15] One needs to find some behavioral difference between belief and hope in these cases, and since the obvious suggestion that we often don't act on our hopes isn't going to work here, Peirce seems to find his difference between his special action-guiding version of hope and belief in the way that the two states are responsive to evidence. While James characterized faith "belief in something concerning which doubt is still theoretically possible" (James 1893: 76), Peirce imagined faith to be something stronger; in particular, he took it to produce a type of belief that was actively *resistant* to contrary evidence. According to Peirce, while faith was "highly necessary in affairs," it was "ruinous in practice" because "you are not going to be alert for indications that the moment has come to change your tactics" (CP 8.251, 1897, CWJ 8:244). However, faith (especially in James's thin sense) needn't be viewed as having this consequence, and there is no reason for James to think that we couldn't remain fallibilists for beliefs based on faith.[16] It may be that Peirce was misled by James's later talk about belief in terms of a "willingness to act irrevocably" (James 1893: 14), but that willingness should be understood as willingness to take an irrevocable action (such as the mountain climber's leaping over the canyon),[17] not as an irrevocable willingness to act in a certain way (such as the aforementioned mountain climber continuing to believe that he can make the jump as he plummets into the abyss).[18]

Peirce famously said that the motto that "deserves to be inscribed upon every wall in the city of philosophy" was "Do not block the path of inquiry" (CP 1.135), and while James would certainly agree with this, the two early pragmatists had very different ideas about what would block inquiry. James understood inquiry in evolutionary terms, and so he was in favor of a comparatively promiscuous set of starting points combined with a confidence that experience would weed out candidates that conflicted with it.[19] For James, proliferation and selection was seen as the best way for inquiry to succeed. It was the *responsiveness* of our attitudes to experience that was of primary importance, and a belief that we give up in the light of contrary experience will be more responsive than a "hope" that we cling to come what may.

Peirce and James on the Nature of Truth

The difference between James's and Peirce's views on the fixation of belief had immediate consequences for their conceptions of truth when these views of justification are used to fill out the Peircean idea that "what we mean by the truth" is the "opinion which is fated to be ultimately agreed to by all who investigate."[20]

Peirce's tying truth to our practices of inquiry represents a decisive break from the sort of metaphysical realism that makes the truth about the world radically independent (at least in principle) from what we might come to know about it. Nevertheless, Peirce's comparatively conservative conception of inquiry made the break less radical than it became for James.

For James, if truth is tied to inquiry in the way that Peirce suggests, then *any* factors that contribute to inquiry have the potential to contribute to truth, and so it seemed to James that metaphysical systems that frustrate our practical needs could never be true, because doubts are invariably bound to arise about them. As he put it in his notes on "The Sentiment of Rationality":

> If universal acceptance be, as it surely is, the only mark of truth which we possess, then any system certain not to get it, may be deemed false without further ceremony, false at any rate for us, which is as far as we can inquire.[21]

It needn't follow from this that every *pleasant* ideas would thereby be *true* (though James is often accused of thinking this). Such ideas may tempt us, but if they bump up against recalcitrant experience, doubts will arise, and so they will also fail to be true. James's conception of truth is not, then, more forgiving than Peirce's. On the contrary, it is significantly more demanding. To be "absolutely"[22] true, a belief must *not only* fit with current and future experience in the way that Peirce requires *but also* be in line with our "spontaneous powers" in the way that James describes.[23]

While Peirce may have identified truth with the "opinion which is fated to be ultimately agreed to by all who investigate," he had little enthusiasm for James's idea that such ultimate agreement must account for our passional nature.[24] Indeed, Peirce drew an increasingly bright line between his "all who investigate" and James's "universal acceptance" and placed a number of implicit restrictions on his community of inquiry that would have been quite alien to James.

If James was right about the contributions of our practical interests to the sowing of doubt, then it's quite possible that no system of the world could ever be "fixed" in the sense that Peirce's account of truth requires. Peirce was aware of this, and even in his original paper, he admits that truth might not be understandable in terms of *human* inquiry:

> Our perversity and that of others may indefinitely postpone the settlement of opinion; it might even conceivably cause an arbitrary proposition to be universally accepted as long as the human race should last. Yet even that would not change the nature of the belief, which alone could be the result of investigation carried sufficiently far; and if, after the extinction of our race, another should arise with faculties and disposition for investigation, that true opinion must be the one which they would ultimately come to.
>
> (Peirce 1878a: 139)

Of course, *simply* appealing to a possible future race isn't really going to help, since exactly the same "perversity" might arise with them, and even if there eventually were a future race that got things right, after *their* extinction, there might be yet another race whose investigation drifted toward a contrary opinion.

It is for these reasons that for Peirce, the "all" in "all who investigate" is implicitly restricted to a community of scientists.[25] Indeed, it is restricted to not only a community of scientists but also to a community of idealized Peircean scientists who lack any sort of practical interest that might affect inquiry in any of the ways that James highlights. The Peircean scientist is, after all, supposed to have no "vital" interests,[26] and while this idealized inquirer can seem noticeably inhuman, it may be the only thing that can be plugged into Peirce's definition that would give him the results he wants. What Peirce needs is a set of "ideal" successors, where this idealization includes a lack of interest in practical matters (or perhaps just a restriction of these practical interests to the pursuit of the "development of concrete reasonableness") (see Peirce 1902).

The superiority of the scientific method was argued for in "The Fixation of Belief" in terms of that being the best method of fixing belief and eliminating doubt. However, Peirce's argument can seem a little disingenuous if it turns out that this method only works if you restrict yourself to a sub-community of disinterested scientists. If we only achieve community consensus by whittling down the community, defenders of the other methods could help themselves to this strategy as well. After all, Peirce himself says that the method of authority might be the best for "the mass of mankind," with only the doubts of "a few individuals" persisting in the "most priest-ridden states" (Peirce 1877: 118). If the method of science can only permanently fix belief when we restrict ourselves to a possible future race of passionless scientists, the defender of authority might rightly insist that they could successfully fix beliefs as well by restricting *their* community to an excessively deferential set of possible successors.

If all of the methods can fix beliefs equally well by restricting their communities in this way, Peirce's preference for the scientific method may just be a reflection of the fact that it is the method that works best for those with his particular temperament,[27] but while Peirce was happy to understand truth in terms of something like these passionless inquirers, James sees little reason to think that *we* should judge ourselves by the standards of this possible future race. Indeed, James addresses just this issue in a paper published in the same year as "How to Make Our Ideas Clear," "Remarks on Spencer's Definition of Mind as Correspondence." While that paper presents itself as a critique of Herbert Spencer's work, one of the main issues that he took Spencer to task for (the prospect of a purely "disinterested" intellect) was precisely the issue that here divides James and Peirce.[28]

James's inclusion of fully human inquirers, passions and all, into the community which could determine the truth also made his version of pragmatism more open to the possibility of their being *normative* truths,[29] and his remarks on Spencer end with a defense of the potential objectivity of such normative truths that is cashed out in explicitly Peircean terms.

> Mental interests, hypotheses, postulates, so far as they are bases for human action—action which to a great extent transforms the world—help to *make* the truth which they declare. In other words, there belongs to mind, from its birth upward, a spontaneity, a vote. It is in the game, and not a mere looker-on; and its judgments of the *should-be*, its ideals, cannot be peeled off from the body of the *cogitandum* as if they were excrescences ... The only objective criterion of reality is coerciveness, in the long run, over thought. Objective facts, Spencer's outward relations, are real only because they coerce sensation. Any interest which should be coercive on the same massive scale would be *eodem jure* real. ... If judgments of the *should-be* are fated to grasp us in this way, they are what "correspond."
>
> *(James 1878: 21–22)*

Conclusion

James's first biographer, Ralph Barton Perry, notoriously claimed that "the modern movement known as pragmatism is largely the result of James's misunderstanding of Peirce" (1935, Vol. 2: 409), but in many respects James understood Peirce all too well. He adopted Peirce's central ideas of understanding inquiry in terms of what could "fix" belief and truth in terms of what inquiry would converge upon, but did so while holding on to the idea that we were looking at specifically *human* inquiry and rejecting Peirce's frankly Procrustean conception of who the pragmatist's inquirer must be. The resulting view was often a more radical departure from the traditional conception of truth and inquiry than what Peirce presented, but it remained a natural extension of the central tenets of their original 1872 discussions.

Notes

1 See Peirce (CP 5.414; 1905). Though there are some questions of whether Peirce was concerned to distance himself from *James's* position, rather than just the views found in the "literary journals" that were "inspired" by James (for a discussion of this, see Pihlström 2004: 28).
2 See, for instance, Apel (1981), Mounce (1997), Haack (1998), Rescher (2000) and Misak (2000, 2013)
3 For an extensive discussion of the club and its members, see Menand (2002).
4 Kiryushchenko (2016: 147).
5 Published in *Mind* in 1879 but written mostly in 1877 (Perry 1935, Vol. 1: 782).
6 These last two were combined into the version of "The Sentiment of Rationality" that appeared in James (1897).
7 See Meyers (2005: 326).
8 See, for instance, Russell (1946).
9 See Lamberth (2014: 136).
10 See James (1902), lectures 6 and 7.
11 He later suggested that "The Psychology of Philosophizing" would have been a better title for the essay (James 1978: 359.).
12 Many (starting at least with Dewey [1916] and running up through Misak [2013: 64]) thought that Peirce himself read James this way.
13 And I should note that there is thus some fairness in Aikin's complaint (Aikin 2014: 85, 175) that James's suggestion in "The Will to Believe" that he is providing a justification of the religious beliefs of his undergraduate audience amounts to something of a "bait and switch," since while his own "religious hypothesis" ("the best things are the more eternal things," and we are better off for believing that [James 1897: 29–30]) is arguably an evidence-transcendent one, the religious beliefs of some of his audience (such as that the earth was created in seven days about 6,000 years ago) arguably *do* go against the available evidence.
14 See Misak (2013: 50–52).
15 See Jackman (2020) and Pihlström (2004: 40–41).
16 Indeed, in James (1897: 79) he seems to explicitly contrast his more fallibilist version of faith with the uncritical version that Peirce considers (with the more uncritical conception of faith being associated, as was sadly typical of James, with Catholicism).
17 Indeed, it isn't far from Peirce's own characterization of "Full belief" as a "willingness to act upon the proposition in vital crises" (RLT 112).
18 Russell (1946: 815), Aikin (2014: 90) and Atkins (2016: 26–30) seem to follow Peirce in misreading James in this way.
19 See Klein (2013). It often isn't appreciated how this evolutionary model puts an underlying *social* foundation to James's epistemology. The epistemic norms that he recommends are the ones that he thinks be most successful for a *population* of inquirers.
20 Peirce (1878a: 138–139). See also Peirce (1868: 52, 54–55).
21 Notes on "The Sentiment of Rationality," in James (1978: 360).
22 To the extent that James is very forgiving in his talk of truth, it is for the more "temporary truths" that represent the temporary resting points of inquiry, not the "absolute" truth that is cashed out in these Peircean terms. (See James 1907: 106–107.)
23 For a discussion of how this leads James to a type of pessimism about the prospects of our attaining absolute truth, see Jackman (2019, 2021).
24 Indeed, over the years he drifted away from understanding truth in terms of "agreement" at all, replacing it with something closer to a belief's "indefeasibility" (see Misak 2013).
25 For further discussion of how James and Peirce differ in the size of the community that they are willing to tie truth to, see Klein (2013).
26 See his claim that "pure science has nothing at all to do with *action*" and so "what is properly and usually called *belief* ... has no place in science at all" (Peirce 1898: 112).
27 Indeed, he suggests as much himself (Peirce 1877: 119–120).

28 While James was free with his attacks on Spencer, he was always reluctant to explicitly criticize Peirce (especially in print).
29 Though some Peirceans, particularly Misak (2000) and Heney (2016), argue that this can be done within a strictly Peircean framework as well.

Works Cited

Aikin, S. F. 2014. *Evidentialism and the Will to Believe*. New York: Bloomsbury.
Apel, K. O. 1981. *Charles S. Peirce: From Pragmatism to Pragmaticism*. Translated by J. M. Krois. Amherst: University of Massachusetts Press.
Atkins, K. 2016. *Peirce and the Conduct of Life*. New York: Cambridge University Press.
Dewey, J. 1916. "The Pragmatism of Peirce." *The Journal of Philosophy*, Vol. 13, No. 26, pp. 709–715.
Haack, S. 1998. *Manifesto of a Passionate Moderate: Unfashionable Essays*. Chicago and London: University of Chicago Press.
Heney, D. 2016. *Towards a Pragmatist Metaethics*. New York: Routledge.
Jackman, H. 2019. "William James on Moral Philosophy and Its Regulative Ideals." *William James Studies*, Vol. 15, No. 2 (Fall 2019), pp. 1–25.
Jackman, H. 2020. "No Hope for the Evidentialist: On Zimmerman's Belief: A Pragmatic Picture." *William James Studies*, Vol. 16, No. 1 (Fall 2020), pp. 66–81.
Jackman, H. (2021). "Putnam, James, and 'Absolute' Truth." *European Journal of Pragmatism and American Philosophy*, Vol. XIII, No. 2, pp. 1–13.
James, W. 1878. "Remarks on Spencer's Definition of Mind as Correspondence." *The Journal of Speculative Philosophy*, Vol. 12, No. 1 (January 1878), pp. 1–18. Reprinted in James's *Essays in Philosophy* (Cambridge: Harvard University Press, 1978).
James, W. 1897, 1979. *The Will to Believe and Other Essays in Popular Philosophy*. Cambridge: Harvard University Press.
James, W. 1902, 1985. *The Varieties of Religious Experience*. Cambridge: Harvard University Press.
James, W. 1907, 1979. *Pragmatism*. Cambridge: Harvard University Press.
James, W. 1978. *Essays in Philosophy*. Cambridge: Harvard University Press.
Klein, A. 2013. "Who Is in the Community of Inquiry?" *Transactions of the Charles S. Peirce Society*, Vol. 49, No. 3, pp. 413–423.
Klein, A. 2021. "William James: Where All Consciousness is Motor," interview on 3–16 am. https://www.3-16am.co.uk/articles/william-james-where-all-consciousness-is-motor.
Kiryushchenko, V. 2016. *The Social and the Real: The Idea of Objectivity in Peirce, Brandom, McDowell*. Dissertation, York University, June 2016.
Lamberth, D. 2014. "*A Pluralistic Universe* a Century Later: Rationality, Pluralism and Religion." In Halliwell, M. & Rasmussen, J. D. S. (Eds.) *William James and the Transatlantic Conversation*. New York: Oxford University Press, pp. 133–150.
Menand, L. 2002. *The Metaphysical Club: A Story of Ideas in America*. New York: Farrar, Straus and Giroux.
Meyers, R. G. 2005. "Peirce's 'Cheerful Hope' and the Varieties of Realism." *Transactions of the Charles S. Peirce Society*, Vol. 41, No. 2 (Spring 2005), pp. 321–341.
Misak, C. J. 2000. *Truth, Politics and Morality: Pragmatism and Deliberation*. London and New York: Routledge.
Misak, C. J. 2004. *The Cambridge Companion to Peirce*. New York: Cambridge University Press.
Misak, C. J. 2013. *The American Pragmatists*. New York: Oxford University Press.
Mounce, H. O. 1997. *The Two Pragmatisms: From Peirce to Rorty*. London and New York: Routledge.
Peirce, C. 1867–1868. "Critique of Positivism, 1867–1868." In Eward C. Moore, Max H. Fisch, Christian Kloesel, Don D. Roberts, Lynn A. Ziegler (Eds.), *Writings of Charles S. Peirce: A Chronological Edition, Volume 2: 1867–1871*. Bloomington: Indiana University Press (1982), pp. 122–130.
Peirce, C. 1868. "Some Consequences of Four Incapacities." *Journal of Speculative Philosophy*, Vol. 2, pp. 140–157. Reprinted in Houser and Kloesel (Eds.), *The Essential Peirce*, vol. 1 (Bloomington: Indiana University Press, 1992).
Peirce, C. 1877. "The Fixation of Belief." *Popular Science Monthly*, Vol. 12 (November 1877), pp. 1–15. reprinted in Houser and Kloesel (eds.), *The Essential Peirce*, vol. 1 (Bloomington: Indiana University Press, 1992).
Peirce, C. 1878a. "How to Make Our Ideas Clear." *Popular Science Monthly*, Vol. 12 (January 1878), pp. 286–302. reprinted in Houser and Kloesel (eds.), *The Essential Peirce*, vol. 1 (Bloomington: Indiana University Press, 1992).

Peirce, C. 1878b. "The Doctrine of Chances." *Popular Science Monthly*, Vol. 12 (March 1878), pp. 604–613. reprinted in Houser and Kloesel (eds.), *The Essential Peirce*, vol. 1 (Bloomington: Indiana University Press, 1992).

Peirce. C. 1898, 1993. *Reasoning and the Logic of Things*. Cambridge: Harvard University Press.

Peirce, C. 1902. "Pragmatic and Pragmatism." In Baldwin, J. M. (Ed.) *Dictionary of Philosophy and Psychology*. New York: The Macmillan Co., vol. 2, pp. 321–322.

Perry, R. B. 1935. *The Thought and Character of William James* (2 vols). Boston: Little, Brown, and Company.

Pihilström, S. 2004. "Peirce's Place in the Pragmatist Tradition." In Misak, *The Cambridge Companion to Peirce*. New York: Cambridge University Press (2004), pp. 27–54.

Rescher, N. 2000. *Realistic Pragmatism*. Albany: State University of New York Press.

Russell, B. 1946. *History of Western Philosophy*. London: George Allen and Unwin.

18
ONE HUNDRED YEARS OF PRAGMATISM AT HARVARD

Douglas McDermid

The pragmatism or pluralism which I defend has to fall back on a certain ultimate hardihood, a certain willingness to live without assurances or guarantees.

– *William James*[1]

There is no God's Eye point of view that we can know or usefully imagine; there are only the various points of view of actual persons reflecting various interests and purposes that their descriptions and theories subserve.

– *Hilary Putnam*[2]

Introduction

Harvard's most famous professors of pragmatism – William James (1842–1910), Clarence Irving Lewis (1883–1964), Willard Van Orman Quine (1908–2000), and Hilary Putnam (1926–2016) – defended three closely connected theses about knowledge and reality. The first thesis is that traditional empiricism (understood to include positivism) has failed to appreciate the extent to which our worldview is mind-made or shaped by subjectivity. The second thesis is that this first thesis – that our minds do not 'copy' reality or relate to things disinterestedly – need not lead to skepticism about knowledge or relativism about truth. The third thesis is that philosophy can avoid skepticism and relativism only by overcoming certain conventional oppositions: between creation and discovery, making and finding, freedom and constraint.[3]

How did each of our Harvard pragmatists develop these ideas? It is time to consider some concrete examples in chronological order.

William James

William James began teaching physiology part-time at Harvard in 1872, having graduated from Harvard Medical School in 1869. Within a few years, he had turned his hand to psychology and philosophy, quickly distinguishing himself in both fields. By the time James retired in 1907, his writings were widely admired and immensely influential. A popular and inspiring teacher, his students at Harvard included many future philosophers of note: George Santayana (1863–1952), W.E.B. Dubois (1868–1963), Dickinson S. Miller (1868–1963), John Elof Boodin (1869–1950), Arthur O. Lovejoy (1873–1962), William Ernest Hocking (1873–1966), Ralph

Barton Perry (1876–1957), Horace Kallen (1882–1974), C.I. Lewis (1883–1964), and Alain Locke (1885–1954).

The center of James's philosophical vision is his conviction that the human mind, far from being a receptive and malleable tabula rasa, is an active principle which creatively shapes our experience. If he is right, even the most mundane experiences are unthinkable without the manifold contributions of mind. Here are three examples.

Selective Attention

Suppose you are looking at a Japanese maple tree in your neighbor's large, lush garden. In a case like this, it may seem self-evident that you are letting the world imprint itself on the surface of your wide-open mind, much like a signet ring leaves an impression on pliant wax. But is that really what is happening? Take a moment to consider how things actually appear or present themselves to you in experience. For instance, do you experience the Japanese maple on its own, in splendid isolation from the rest of the garden – flowers, moss, stones, and so forth? Of course not; you see the tree against a more or less definite background, which is experienced *as a background*. Yet this distinction between foreground and background does not exist in Nature independently of us; such structures of salience are plainly brought into being by consciousness. And this, in effect, is James's first point: that there is no observation without attention, no attention without selection, no selection without interests, and no interests without subjectivity or a definite point of view. Far from being a species of disinterested contemplation, your ordinary awareness of trees and rocks is conditioned by your practical projects and purposes, your values and needs. Note what follows: if our minds thus have the power to shape the Protean stuff of sense, objective factors do not fully determine the What and the How of human experience. Locke's tabula rasa metaphor, James concludes, is apt to mislead the unwary.

Observation as Theory-Laden

In the second place, you cannot see a tree *as a tree* (or, for that matter, as an enduring thing of any kind) unless you already have certain ideas and principles in your repertoire. Since sensation in the absence of a conceptual filter is sheer unmeaning chaos – "one great blooming, buzzing confusion" (James 1890 I: 488) – observation is not prior to all theory. James makes this broadly Kantian point in several ways, one of which is to remind us how our experience is shaped by expectation. As every fledgling prestidigitator knows, what an audience sees depends on what they are prepared to see, and what we are prepared to see depends on our interpretation of past experience – that is, on a set of background assumptions and classifications. Since "[t]he trail of the human serpent is thus over everything" (James 1907: 33), observation must lean on theory for guidance and support.

What are the consequences if James is right, and there is no hard-and-fast distinction between observation and theory? For one thing, the demise of this pre-Kantian dualism reinforces fallibilism: if observation inherits the uncertainty inherent in all theory, even garden-variety perceptual judgments – "This is a maple tree" – must be treated as hypotheses which may need to be revised at some point down the road. Furthermore, empiricist foundationalism is undermined if the distinction between theory and data collapses, because there is absolutely no point looking for theory-free apprehensions of fact if there are no perceptions without preconceptions. The bottom line, then, is this: no scientific theory can appeal to indubitable data obtained without theory's aid, because there are no such data. Here James partly anticipates the celebrated attack on the Myth of the Given mounted by Wilfrid Sellars (1912–1989).

Anti-representationalism

In the third place, we have no right to think that our concepts copy or mirror the nature of things. This anti-representationalist thesis can be derived from James's previous point – viz., that observation is theory-laden. For if we can only see the world through the lens of a theory or scheme, we have no access to an unconceptualized reality, and if we have no access to an unconceptualized reality, we cannot compare its nature with our concepts to see whether the two sides match. Consequently, it is impossible in principle for us to determine whether any framework corresponds to something beyond all frameworks: "When we talk of a reality 'independent' of human thinking ... it seems a thing very hard to find" (James 1907: 112). Substance and property, identity and change, cause and effect, mind and body: these forms of thought may be expedient or even indispensable in practice, says James, but we must not allow ourselves to think that they conform to the structure of some absolute reality.

Some philosophers think this line of reasoning leads to the skeptical-sounding conclusion that we can never know things-in-themselves but only appearances or mere phenomena. James disagrees; like Berkeley, whom he regards as a proto-pragmatist, he thinks the only meaningful distinction between appearance and reality is one which can be drawn within experience. As far as James is concerned, then, a Kantian world of noumena with which we have no dealings is as good as nothing; a pallid and empty abstraction, it is as confused in theory as it is useless in practice.

Truth and Correspondence

Since we have no use for things-in-themselves, truths cannot be said to correspond or agree with them. What then becomes of the idea that truth means correspondence with reality? Like his fellow pragmatist John Dewey (1859–1952) at Columbia, James thinks philosophical talk about 'correspondence' has no meaning for us unless it picks out a relation which falls wholly within experience.[4] Theories and ideas thus 'correspond' with reality, James maintains, to the degree that they respond satisfactorily to the sensory flux, leading us smoothly from one point in experience to another until we reach a destination which is agreeable, albeit temporary. To discharge this correspondence function properly, however, our thinking must get out in front of the facts, anticipating concrete changes and preparing us for them. Truth, in short, is a matter of transitions. Accordingly, all our assertions must look to the future for their vindication; ideas must be judged by their fruits or practical consequences. This forward-looking view supports fallibilism, because nothing can be taken as final or certain if our thoughts about things are underdetermined by available data. The moral, James thinks, is that we must always leave room in life for the radically unexpected; that is, for possibilities so novel that our current worldview cannot accommodate them. This maxim is meant to apply not only to science and philosophy but also to ethics and religion.

William James died in 1910, the same year that Dewey published *The Influence of Darwin on Philosophy*. By 1916, all James's most distinguished departmental colleagues – Josiah Royce (1855–1916), George Santayana (1863–1952), Hugo Munsterberg (1863–1916), and George Herbert Palmer (1842–1933) – had all either retired or died. To be sure, there were still talented philosophers in Cambridge during the First World War; for instance, Ralph Barton Perry (1876–1957) and William Ernest Hocking (1873–1966), both of whom had studied under James and Royce.[5] Nevertheless, Harvard could not boast of a philosopher comparable to James or Dewey in stature and influence until 1924, when Alfred North Whitehead (1861–1947) was imported from England. Unfortunately, Whitehead does not play an important part in our story; as much as he admired "that adorable genius, William James" (Whitehead 1926: 2), his vision and sensibility were very much his own. To understand the next major stage in the development of Harvard pragmatism, we must look elsewhere.

C.I. Lewis

At first blush, Clarence Irving Lewis may seem like an unlikely successor to William James – and not only because his dissertation was supervised by James's dialectical sparring partner, the absolute idealist Josiah Royce. Hired by Harvard in 1920, a full decade after James's death, the young Lewis represented the new breed of American academic: the PhD-toting puzzle-solver who was more fox than hedgehog, more logician than lay moralist or *litterateur*. A philosopher's philosopher with a Peircean passion for inquiry, Lewis wrote almost exclusively for an audience of academic specialists, and his impersonal professorial prose has little in common with James's impassioned, impressionistic style. Furthermore, Lewis was always a keen student of philosophy's past – an interest he shared with his mentor, the preternaturally erudite Royce. James, in contrast, showed little interest in parsing the dicta of the mighty dead. Be that as it may, no one did more than C.I. Lewis to keep the tradition of pragmatism alive at Harvard during the interwar period.

According to Lewis, philosophy's aim is not the discovery of brave new worlds or the multiplication of paper doubts. Since "[t]here can be no Archimedean point for the philosopher" (Lewis 1929: 23), we must be content to reflect patiently on our practices, extracting the normative criteria implicit in them. Among other things, this means that epistemology's primary task is not to refute skepticism but to analyze the nature of knowledge. Lewis himself attempted to provide such an analysis in *Mind and the World Order* (1929), a treatise whose title hints at the constructivist sympathies of its author.

The Analysis of Knowledge

Echoing Kant's dictum that "intuitions without concepts are blind" (Kant 1781: A 51 / B 76), Lewis maintains that unfiltered sensory input – the so-called Given – can be converted into knowledge only if our minds can impose order or structure upon it via the application of categories. In other words, our knowledge of Nature is made possible by the union or integration of two elements: the immediate data of sense (courteously furnished by the world) and a scheme of concepts or principles (kindly supplied by the mind). Without the first element, there is nothing for thought to grasp, because knowledge has no content unless some datum is presented to the mind. Without the second element, error becomes inconceivable, because the distinction between truth and falsity vanishes if there is nothing but James's 'blooming buzzing confusion.' Idealists tend to minimize the givenness of the Given, Lewis observes, while realists tend to underestimate the significance of the mind's contribution. His own position, like the one favored by James, seeks to avoid such one-sidedness. On the one hand, human knowledge is not a mere copy or duplicate of sensory data; on the other, it plainly depends on something which our thought did not create and cannot alter.

Conceptualistic Pragmatism

As we can see, Lewis thinks Kant was right about three things: (a) there can be no knowledge of the world unless the Given is interpreted, (b) the mind cannot interpret the Given without an a priori framework of categories, and (c) the critical study of the a priori is the true task of philosophy. However, an important point of disagreement must not be overlooked: Kant thinks there is one fixed and permanent framework of a priori categories, whereas Lewis thinks many internally consistent frameworks are available. This may seem puzzling; how can such pluralism about frameworks be reconciled with Lewis's firm belief in necessary truths?

The answer lies in Lewis's distinction between two senses of necessity: necessity as the opposite of what is contingent and necessity as the opposite of what is voluntary. Although Kant treats a priori truths as necessary in both senses, Lewis maintains that the a priori element in our knowledge is necessary only in the first sense. According to his "conceptualistic pragmatism" (Lewis 1929: 5), a priori categories and principles do not impose themselves *on us*; rather, these mind-forged forms are imposed *by us* on the Given. Since nothing compels us to adopt any interpretive framework in particular, the a priori element in knowledge may be regarded as the product of the mind's free and creative activity.

The Primacy of the Practical

To be sure, some of these mind-made frameworks seem better than others. But what can "better" or "worse" mean when applied to schemes? Such judgments can only be justified by appealing to pragmatic criteria, Lewis maintains, because human knowledge cannot be understood apart from human agency. Since we are beings with practical interests, we need a world which is orderly and legible; for action is blind and barren in the face of the absolutely incomprehensible. And there would be nothing for us but indecipherable chaos – the aforementioned 'blooming buzzing confusion' – unless the Given were interpreted with a scheme of concepts. It follows that the primary function of a conceptual scheme is to organize sensory data for us in ways we find useful or practically satisfactory. Yet if our schemes are akin to tools or instruments, we are free to repair or replace them when they let us down. Hence our interpretive frameworks, Lewis concludes, are open to revision.

With Lewis, Harvard pragmatism entered the era of self-conscious academic professionalism. Yet there was a price to be paid for such respectability – namely, the loss of cultural impact and relevance. In contrast to James and Dewey, who were public philosophers decades before the term had been coined, virtually everything Lewis published – about symbolic logic, epistemology, ethics, and social philosophy – was *research*. It is hardly surprising that his subtle oeuvre had little influence on the laypersons who readily lent their ears to James and Dewey.[6]

W.V.O. Quine

Lewis's course on Kant became a rite of passage for Harvard graduate students, one of whom was a young logician from Ohio named Willard Van Orman Quine. After finishing his PhD under the supervision of Alfred North Whitehead in 1932, Quine spent the better part of a year in Europe, having been awarded a prestigious Sheldon Travelling Fellowship. While overseas, he met with Rudolf Carnap (1891–1970) and other members of the Vienna Circle, an anti-metaphysical club dedicated to the proposition that science should set the agenda for philosophy.[7] Quine agreed with the Viennese, and his mature philosophy put science first in more ways than one.

Empiricism and Holism

Like Lewis, Quine thought hard about the nature of human knowledge. Yet unlike his old teacher, Quine was primarily interested in understanding *scientific* knowledge. Indeed, the central question of his epistemology is a question about scientific inquiry: How does experience function as a constraint on theory? Empiricist foundationalism, though acknowledging theory's underdetermination by observational data, nevertheless held that certain statements – 'basic' or 'foundational' ones – could be verified individually in virtue of their allegedly direct relation to experience.

According to Quine, this atomistic view of verification should be permanently shelved, because no empirical statement – not even "This is a maple tree" – can be confirmed in isolation from the network in which it is embedded. Nor can any statement be conclusively falsified, since theory-defying data or "recalcitrant experience" (Quine 1953: 43) can only prove that error lurks *somewhere* in our system. Experience alone, therefore, cannot determine the status or fate of any individual statement; a statement's connections with other statements, including its distance from our system's periphery, must also be taken into account.

Empiricism and Contingency

Quine's holistic view of verification, implying as it does that no statement is unrevisable in principle, lends aid and comfort to fallibilism as well as to anti-foundationalism. Since recalcitrant experience is powerless to dictate how truth values will be redistributed, we must undertake this reconstructive work ourselves, revising our scheme in the light of certain familiar standards: coherence, conservatism, predictive power, and so forth. Yet there is no objectively correct way to rank or weigh these vague desiderata, and they may easily come into conflict with one another. When this happens, how are we to proceed?

For Quine, the answer is clear: we must look to our subjective preferences for guidance, because there is nowhere else to look. We are therefore free to discard any statement we choose, including laws of logic, provided we mend our web of belief accordingly. By the same token, we are free to retain any statement we please in the face of recalcitrant experience – an observation which helps us to undo the hoary dualism between analytic truths (which remain true come what may, because they are independent of experience) and their synthetic brethren (which are open in principle to revision, because they are not independent of experience). Necessary or a priori truths, we must conclude, have no place in an empiricism which remains true to itself. Stripped of the last vestige of rationalism, philosophy must confront the radical contingency of things.

Naturalism

Quine's attack on the a priori is much more radical than anything dreamt of in Lewis's philosophy. True, there is a sense in which Lewis relativized or humanized the a priori, but he never attacked the distinction between necessary and contingent truths. Indeed, he could not attack that distinction, because he subscribed to the Kantian view that philosophy is either the study of the a priori or it is nothing. Quine, who sees things differently, describes his position as "a more thorough pragmatism" (Quine 1953: 46). Once we drop the dogmatic belief in two kinds of truth, empiricism becomes practically indistinguishable from naturalism – the Deweyan thesis that philosophy is continuous with science, broadly construed. Among other things, this means that metaphysics can no longer treat physics with lofty condescension, because there is no super-sensible Reality beyond Nature to which metaphysicians have privileged access. It also means that epistemology must become the study of science's methods from science's point of view, because naturalism leaves it no room to be anything else. Science certainly cannot be scrutinized from an external point of view, Quine observes, because there is no external point of view or "cosmic exile" (Quine 1960: 275) if philosophy shares science's conceptual scheme. And if there is no external point of view, philosophical skepticism about science per se requires no rebuttal from the naturalist camp.

Conceptual Schemes

Science thus has no foundations and needs none. Does this imply that our current scheme is somehow sacrosanct or beyond criticism? By no means; Quine has not forgotten his fallibilism.

Admittedly, there is nothing in our worldview which cannot be questioned, but not everything can be questioned at the same time. Our predicament thus parallels that described by the positivist Otto Neurath (1882–1945). How can sailors repair their ship, asks Neurath, when drydock is not an option? The answer is simple: they must be content to take their stand on planks which do not currently need to be fixed, because there is nothing else they can do. Similarly, a conceptual framework can only be changed and improved from within; no external standpoint or Archimedean point is available. Hence, the true test of our posits and concepts is not whether they copy an unconceptualized world but whether (as Lewis and James had stressed) they organize sensory input effectively. As Quine puts it, "[o]ur standard for appraising basic changes of a conceptual scheme must be, not a realistic standard of correspondence to reality, but a pragmatic standard" (Quine 1953: 79; cf. 44).

Quine was an exemplary professional, as was Lewis. Yet their ways of doing philosophy differed significantly, as did their interests. Not to put too fine a point on it, Quine had absolutely nothing to say about value theory (including social and political philosophy), and his approach to the technical problems he cared about was ruthlessly ahistorical.[8] For better or worse, this was a sign of the times; the ideals of professionalism in American philosophy were undergoing a sea change during the 1940s and 1950s, at least at institutions like Harvard. Old-fashioned breadth and edifying visions were on their way out; sub-specialization and the linguistic turn were on their way in, as was the related idea that philosophy should be inspired by science. No doubt the debonair ghost of Moritz Schlick (1882–1936), the leader of the Vienna Circle, smiled approvingly at these developments. The shade of William James, to whom scientism was anathema, almost certainly did not.

Hilary Putnam

Before becoming Quine's colleague at Harvard in the mid-1960s, Hilary Putnam wrote his PhD dissertation under the supervision of one eminent logical positivist – Hans Reichenbach (1891–1953) – at UCLA, then worked with another – Quine's old friend, Rudolf Carnap (1891–1970) – at Princeton's Institute for Advanced Study. Although Putnam agreed with them about the need for philosophy to follow science's lead, he soon found himself frustrated with the narrowness and one-sidedness of their approach.

By the late 1970s, Putnam had become a fierce critic of scientism in general and of scientific realism in particular. However, he was also critical of the irrationalist alternative to scientism espoused by Thomas Kuhn (1922–1996) and Paul Feyerabend (1924–1994), neither of whom had any use for the notion of objective truth. Putnam was similarly unimpressed by the Protagorean neo-pragmatism of Richard Rorty (1931–2007), who had the temerity to suggest that truth was "what our peers will, *ceteris paribus*, let us get away with saying" (Rorty 1979: 176).

Pragmatic Realism

Enter Putnam's "pragmatic realism" (Putnam 1987: 17), which attempts to find a Jamesian via media between the extremes of tough-minded metaphysical realism and tender-minded cultural relativism. According to the philosophers Putnam labels metaphysical realists, truth is essentially a matter of conformity with Nature, and our statements must be judged by whether they correspond to the way the world is in itself. In contrast, Putnam's cultural relativists think of truth as conformity with Convention, acknowledging no normative authority external to a community's discursive practices, paradigms, or forms of life. Although Putnam is convinced that both of these doctrines are flawed, he also believes they contain complementary insights which are ripe for synthesis. On one level, our discourse is plainly answerable to objects, so there is an important sense in which relativism is wrong and realism is right. On a deeper level, however, what we call objects actually depend on our discourse, so there is an important sense in which realism is wrong and relativism is right. Let us see how Putnam proposes to bring about this bold synthesis.

A False Dichotomy about Truth

Metaphysical realism asserts that objects exist in themselves. What does this mean? To be more precise, what do the words 'in themselves' betoken? There is no way to make philosophical sense of this kind of old-fashioned realist rhetoric, Putnam claims, because there is no way to say what constitutes an object apart from the choice of a conceptual scheme which stipulates the criteria and conditions of objecthood. Since there is more than one valid or acceptable way to understand what an object is, objects can only be said to exist as such relative to a given conceptual scheme, and since schemes cannot be judged true or false, talk about objects 'in themselves' involves a category mistake. Rorty and his fellow relativists, Putnam concludes, are right at least about this much: there is no domain of objects independent of discourse, no ready-made reality for language or mind to mirror. In the words of Putnam's colleague Nelson Goodman (1906–1998), "there is no world independent of description" (Goodman 1988: 154).[9]

Nevertheless, Putnam assures us that the consequences of conceptual relativity are not as dramatic as the *maitre penseurs* of postmodernism suppose. To be more precise, the fact that there are no scheme-independent objects does not imply that "anything goes" (to quote Feyerabend, quoting Cole Porter). Obviously, if there are no such objects, no descriptive statements can correspond with them. However, postmodern Porterism cannot be derived from conceptual relativity unless we add the following assumption: "*Either* truth means correspondence with scheme-independent objects, *or* truth is a name for nothing objective." Yet this assumption, implicitly endorsed by Rorty, is false: it is perfectly possible to reject radical relativism about truth without thinking of truth as correspondence with things-in-themselves. The first step toward such a chastened realism – the kind Putnam calls "pragmatic" or "internal" – is to recognize that scheme-constituted objects are the only kind of objects we can refer to or think about. Like Kantian phenomena or appearances, such entities do not exist absolutely or 'in themselves' (whatever that means), but this does not prevent our descriptions of them from being simply true or simply false within the conceptual scheme we have adopted.

Truth and Assertibility

In one sense, then, the objects we think and talk about are created by our discourse; in another sense, they constrain it. The problem with Rorty's neo-pragmatism, Putnam thinks, is not that he endorses the first claim; he is surely right in thinking that truth cannot be equated with correspondence to things-in-themselves. The problem is that Rorty is unable to reconcile the first claim with the second. This failure can be seen in his logic-defying leap from "Nature does not tell us how she is to be described" to "Truth is whatever your cultural peers will let you get away with saying." Apart from the fact that this is a non sequitur, Putnam observes, Rorty's conclusion is patently indefensible. For one thing, the truth of statements cannot be identified with assertibility; for mere assertibility, unlike truth, is a property which can be lost. Moreover, what are we to make of the curious fact that Rorty's cultural peers will not let him get away with saying that "truth is what your cultural peers will let you get away with saying"? The view, it seems, effectively subverts or refutes itself.

According to Putnam, we would do better to think of truth as assertibility *under epistemically ideal conditions* – a view which brings to mind the Peircean understanding of truth as the opinion that would be accepted at the end of inquiry. Cartesian-style skeptical scenarios – "How do you know you aren't really a brain in a vat?" – can thus be swiftly dismissed, because such unfalsifiable fantasies make sense only if truth is a radically non-epistemic notion, having no essential connection with rational acceptability or warrant. Putnam thus reaches the same conclusion as Dewey: there is no philosophical problem of knowledge in general.

Beyond Kant and Back to James

As we have seen, Putnam wants to retain the notion of objective truth at the level of individual statements (e.g. "This is a Japanese maple"). But why, it may be asked, does he think that the conceptual schemes used to formulate such statements cannot themselves be judged true or false? According to Putnam, the idea of a correspondence between schemes and a world beyond them is meaningless unless there is "a God's Eye point of view" (Putnam 1981: 49) from which we can compare a world with our ways of representing it. But the very idea of a God's Eye point of view is absurd: there is no perception without interpretation, no experience apart from classification, no observation untouched by theory and interests. Since we cannot know the world without concepts, "the notion of comparing our system of beliefs with unconceptualized reality to see if they match makes no sense" (Putnam 1981: 130; cf. 54). If it is manifestly impossible for us to measure schemes against an unconceptualized reality, however, what is the point of saying there must be a way the world is in itself? The good news is that no such world is needed to make sense of our practices, let alone to improve them. As Putnam sees it, James, Lewis, and Quine were fundamentally right: conceptual schemes must be judged not by whether they mirror a ready-made Reality but by how well they satisfy human needs and promote human flourishing.[10]

Putnam and Rorty showed analytic philosophers that the rumors of pragmatism's demise had been greatly exaggerated. Unlike the free-wheeling Rorty, however, Putnam always made it crystal-clear that pragmatism was not a new name for relativism, skepticism, quietism, or the end of philosophy. In addition, Putnam distinguished pragmatism from scientism, declared war on fashionable forms of reductionism, and denounced the old positivist dichotomy between facts (as objective or cognitive) and values (as subjective or non-cognitive). Finally, he insisted that a commitment to professionalism does not give academic philosophers a license to ignore the concerns of their fellow citizens, or what Dewey called "the problems of men" (Dewey 1946: 4). For these and other reasons, Putnam's version of pragmatism can be seen as a return to the liberal and humane vision of William James.

Notes

1 James (1909: 229).
2 Putnam (1981: 50).
3 There are, of course, other Harvard pragmatist philosophers of note: Nelson Goodman (1906–1998), Morton White (1917–2016), Israel Scheffler (1923–2014), and Cornel West (b. 1953). However, I cannot do justice to their work in this chapter.
4 Dewey's statement of this view is as clear as can be: "[T]ruth is an experienced relation, instead of a relation between experience and what transcends it" (Dewey 1910: 156; cf. 95, 109).
5 For a detailed history of the Harvard philosophy department from 1860 until 1930, see Kuklick (1977).
6 Lewis's influence as a teacher, I should add, was considerable. For instance, no fewer than three of his PhD students became celebrated epistemologists: Yale's Brand Blanshard (1892–1987), Brown's Roderick Chisholm (1916–1999), and Harvard's Roderick Firth (1917–1987).
7 For a lively history of the Vienna Circle, see Sigismund (2017).
8 Unlike Quine, Dewey at Columbia and his student Sidney Hook (1902–1989) at New York University developed versions of pragmatic naturalism which were neither ahistorical nor apolitical.
9 Goodman's PhD advisor, incidentally, was C.I. Lewis.
10 Compare Putnam's treatment of conceptual relativity with Carnap's (1950) distinction between 'internal questions' and 'external questions.'

Works Cited

Carnap. R. (1950) "Empiricism, Semantics, and Ontology." *Revue International de Philosophie* 4: 20–40.
Dewey. J. (1910) *The Influence of Darwin on Philosophy and Other Essays.* New York: Henry Holt.
Dewey, J. (1946) *Problems of Men.* New York: Philosophical Library.
Goodman, N. (1988) *Reconceptions.* Indianapolis, IN: Hackett.
James, W. (1890) *The Principles of Psychology.* 2 volumes. New York: Dover, 1950.
James, W. (1907) *Pragmatism.* Indianapolis, IN: Hackett, 1981
James, W. (1909) *The Meaning of Truth.* London: Longmans, Green, and Co.
Kant, I. (1781) *Critique of Pure Reason.* Trans. N. Kemp Smith. London: Macmillan, 1964.
Kuklick, B. (1977) *The Rise of American Philosophy: Cambridge, Massachusetts, 1860–1930.* New Haven, CT: Yale University Press.
Lewis, C.S. (1929) *Mind and the World Order.* New York: Scribners.
Putnam, H. (1981) *Reason, Truth and History.* Cambridge: Cambridge University Press.
Putnam, H. (1987) *The Many Faces of Realism.* LaSalle, IL: Open Court.
Quine, W.V.O. (1953) *From a Logical Point of View.* Cambridge, MA: Harvard University Press.
Quine, W.V.O. (1960) *Word and Object.* Cambridge, MA: MIT Press.
Rorty, R. (1979) *Philosophy and the Mirror of Nature.* Princeton, NJ: Princeton University Press.
Sigismund, K. (2017) *Exact Thinking in Demented Times: The Vienna Circle and the Epic Quest for the Foundations of Science.* New York: Basic Books.
Whitehead, A. N. (1926) *Science and the Modern World.* New York: The Free Press, 1967.

19
PRAGMATISM IN BRITAIN AND ITALY IN THE EARLY 20TH CENTURY

Gabriele Gava and Tullio Viola

Introduction

When we think of pragmatism in its early stages, we consider it an exclusively North American affair. In a sense, this is correct. After all, the United States is the birthplace of pragmatism. We can locate the origins of the movement in the discussions among the members of the "Metaphysical Club" in Cambridge, Massachusetts, around 1872. However, it is equally important to keep in mind that pragmatist ideas soon began to spread beyond North America. Moreover, the development of pragmatism in new geographical areas was not a question of passive reception. In some instances, thinkers outside of North America placed their own stamp on this originally American school of philosophy.

In this chapter, we focus on the early reception of pragmatism in Britain and Italy. We will start by presenting the position of the British philosopher Ferdinand Canning Scott Schiller and then direct our attention to the pragmatist school that flourished in Italy at the beginning of the 20th century with the work of Giovanni Vailati, Mario Calderoni, Giovanni Papini, and Giuseppe Prezzolini. There are at least three factors that make the British and Italian reception particularly noteworthy. First, these thinkers explicitly called themselves pragmatists, and their North American colleagues likewise considered them to be full-blown members of the pragmatist school. Second, they were able to enrich the original pragmatist doctrine with new ideas. And third, these ideas traveled back to America and exerted a significant impact on the thought of Peirce, James, and Dewey.

The combination of these factors is essential to our argument, given that the reception of pragmatism in Europe is by no means limited to the episodes considered here. Pragmatism was discussed and creatively developed in many other areas of Europe, most notably France (Pudal, 2011) and German-speaking regions (Ferrari, 2017; Sölch, 2018). Furthermore, the British reception of pragmatism is not limited to the work of Schiller (see, e.g., Misak, 2016). What does make Schiller and the Italian group stand out, however, is their explicit claim to membership in the pragmatist tradition combined with their ability to engage in close dialogue with their American interlocutors.[1]

Moreover, as we will demonstrate, this dialogue raises questions about the conventional schema used to trace the development of pragmatism. According to the most widely accepted understanding, one ought to distinguish between "two" pragmatisms (see, e.g., Mounce, 1997). The first allegedly safeguards a strong notion of truth, is more "objective," and has its root in Peirce's position, while the second is "subjective" and goes back to James. What we will learn by considering the dialogue between European and American authors, however, is that this schema is, to say the least, oversimplified.

Ferdinand Canning Scott Schiller

Ferdinand Canning Scott Schiller was born to German parents on August 16, 1864, in Ottensen, near Altona, a city which at that time belonged to the Danish Realm. His university education was at Balliol College, Oxford. After obtaining a master's degree, he moved to Cornell University, where he was an instructor in logic and metaphysics from 1893 to 1897. It is during this time at Cornell that Schiller met William James, who had a lasting influence on him. In 1897, Schiller went back to Oxford, where he spent most of his career. During the last years of his life, he regularly taught at the University of Southern California. He died in Los Angeles on August 9, 1937.

Even though Schiller is hardly well known today, he was a prominent figure during his lifetime and was considered one of the most influential defenders of pragmatism. Bertrand Russell, for instance, called James, Schiller, and Dewey the three founders of pragmatism (1928: 61). Schiller's pragmatism was markedly influenced by William James, whom he followed in thinking that a distinctive approach to truth was an integral part of pragmatism (Schiller, 1903: 58–59). Furthermore, he believed that reality is not to be considered static and independent of human agency; rather, reality is "plastic" and is molded in response to the way in which we interact with it in an attempt to realize our aims (Schiller, 1902: 60–61).

Some aspects of Schiller's approach, however, make his understanding of pragmatism even broader than James. For example, in the first essay in *Studies in Humanism*, titled "The Definition of Pragmatism and Humanism," he provides seven definitions of pragmatism, some of which go beyond the classical descriptions. This applies especially to the last definition, according to which pragmatism is "*a conscious application to epistemology ... of a teleological psychology, which implies, ultimately, a voluntaristic metaphysics*" (Schiller, 1907: 12, italics in the original). In this way, Schiller gives center stage to the notion of purpose in order to describe, first, our psychology, and second, reality at large. While the notion of purpose is certainly important for all pragmatists, few would include voluntaristic metaphysics in a definition of pragmatism.

Axioms as postulates. A good point of departure to single out various distinctive features of Schiller's pragmatism is the essay "Axioms as Postulates," which appeared in 1902 in a collection of essays entitled *Personal Idealism*. In this article, Schiller provides an account of the "axioms" that have a central role in our knowledge. These are principles that at least appear to have a fundamental and unquestionable validity. Principles that are regarded as "axioms" in this sense are the principle of identity (Schiller, 1902: 94–104), the principle of contradiction (106) and the principle of causation (108).

Schiller argues that, traditionally, there are two strategies for explaining how these "axioms" came to be. The first strategy is pursued by empiricism. According to this view, the possession of these axioms is the result of our mind being in contact with features of reality through experience (Schiller, 1902: 65–66). Empiricist accounts of axioms have two fundamental flaws. First, they provide an incorrect picture of the mind as altogether passive (65, 84). Second, they do not realize that these principles cannot be passively derived from experience "because they must already be possessed before experience can confirm them" (66). The second strategy to explain our possession of axioms contends that axioms are pieces of a priori knowledge. The case for apriorism is built on the failure of empiricism. Insofar as axioms are "presupposed by all experience," they must be a priori truths. However, Schiller complains that this inference is fallacious: "It does not follow that because the necessary truths are presupposed in all experience they are, in the technical sense, *a priori*" (68).

Schiller provides an alternative explanation of the "constitutive" status of axioms. In order to be "presuppositions" of experience, axioms do not need to be a priori truths. Rather, they can also be the result of a "postulation" that we carry out in our dealing with reality in an attempt to realize our practical ends.

> It suffices that we should hold them *experimentally*, as principles which we need *practically* and would like to be true, to which therefore we propose to give a trial, without our adoring them as ultimate and underivable facts of our mental structure. In other words they may be prior to experience as *postulates*.
>
> (Schiller, 1902: 69)

That they are called "axioms" and not "postulates" is dependent on the fact that they have proved extremely successful in our practical dealings with reality (92).

In the context of his account of axioms, Schiller introduces some aspects of his approach that will be developed further in later texts. The first point is the subordination of logical and epistemic principles to practical and ethical principles. In "Axioms as Postulates," Schiller already claims that principles of knowledge cannot be treated separately from our volitional nature and that, in turn, our practical ends fundamentally determine knowledge (Schiller, 1902: 85–86). Second, in Schiller's view, pragmatism implies a different but related doctrine, which he calls "anthropomorphism." The doctrine asserts that since "every attempt to know rests on the fundamental methodological postulate that the world is knowable, we must also postulate that it can be interpreted *ex analogia hominis* and anthropomorphically" (118). A third aspect of Schiller's approach that emerges in "Axioms as Postulates" is his critique of formal logic. Since logical principles are ultimately tools for dealing with reality, it makes no sense to analyze them from the perspective of "pure thought" and independently of our psychology (128).

The subordination of logic and metaphysics to ethics. In the first chapter of *Humanism* (Schiller, 1903), "The Ethical Basis of Metaphysics," Schiller argues that a subordination of logic (or epistemology) and metaphysics to ethics directly follows from the principle of pragmatism. He begins with the observation that one characteristic of the "absolute idealism" championed by philosophers like F. H. Bradley is that it makes ethical considerations completely irrelevant to metaphysical investigations. Pragmatism opposes this approach:

> Instead of regarding practical results as irrelevant, it makes Practical Value an essential ingredient and determinant of theoretic truth. ... [I]t regards knowledge as derivative from conduct and as involving distinctively moral qualities and responsibilities in a perfectly definite and traceable way.
>
> (Schiller, 1903: 4)

Ethics provides standards for the evaluation of our practical pursuits, and such standards also play a fundamental role in logic and metaphysics:

> Now inasmuch as such teleological valuation is also the special sphere of ethical inquiry, Pragmatism may be said to assign metaphysical validity to the typical method of ethics. At a blow it awards to the ethical conception of *Good* supreme authority over the logical conception of *True* and the metaphysical conception of *Real*.
>
> (Schiller, 1903: 8–9)

On the one hand, these claims clearly recall James's contention that truth is a form of the good, like wealth, health, etc. (James, 1975: 104). On the other hand, they point toward a conception of logic as a normative science (Schiller, 1903: 54–55; 1907: 78). Truth is conceived as a particular kind of "good" or "value" for which logic provides the standards of validity (Schiller, 1903: 10). By considering logic a normative science, Schiller comes close to Peirce. Accordingly, it is no surprise that Peirce, in a draft for a letter to Schiller from May 12, 1905, highlights precisely this commonality in their views:

> As to valuation, in 1903 I delivered a course of lectures in Boston ... in which I explained at length how reasoning was analogous [to] – and in fact, *a particular case of*, – *moral self-control*, how Logic ought to be founded on Ethics and Ethics on a transfigured *Esthetics* which would be the *Science of values*, although now wrongly treated as a part of Ethics.
>
> (Scott, 1973: 371–372)

Yet, even though Schiller's description of logic as a normative science brings him closer to Peirce, Schiller thought that this conception of logic implied a continuity between logic and psychology, which Peirce opposed.

Anthropomorphism and humanism. In "Axiom as Postulates," Schiller refers back to *Riddles of the Sphinx*, where he provides a more extensive discussion of anthropomorphism. In a preliminary sense, anthropomorphism expresses a simple truism: since all of our conceptions are *by definition* ours, it follows that our thinking is inescapably anthropomorphic (Schiller, 1891: 145). However, the notion of anthropomorphism can be made more informative when we distinguish between good and bad anthropomorphism. Bad anthropomorphism can be either false or confused. The former is "the ascription to beings other than ourselves of qualities or attributes which we know they cannot possess because of their difference from ourselves" (145). The latter is a kind of anthropomorphism that dresses itself as its negation and for this reason leads to contradictions and difficulties (146; for an example, see 60–62). By contrast, the ideal of good or true anthropomorphism is to devise explanations that make "use of no principles which [are] not self-evident to human minds, self-explanatory to human feelings" (147).

In later writings, Schiller prefers to use the term "humanism" instead of anthropomorphism. However, humanism is not just a new name for what was previously called anthropomorphism. While in "Axioms as Postulates" anthropomorphism was a corollary to pragmatism, Schiller describes humanism as a wider doctrine, of which pragmatism is just a particular application to the problem of knowledge (Schiller, 1903: xxi).

> But Humanism will seem more universal. It will seem to be possessed of a method which is applicable universally, to ethics, to aesthetics, to metaphysics, to theology, to every concern of man, as well as to the theory of knowledge.
>
> (Schiller, 1907: 16)

Its motto is Protagoras's saying that "man is the measure of all things" (Schiller, 1903: xvii; Schiller, 1907: 33). If we then ask how we should understand the "man" to which the motto refers, Schiller clearly stresses that it refers to the *individual* person. Accordingly, humanism *points toward* a specific metaphysical view that Schiller calls "personalism" (Schiller, 1907: 19). The latter contends that the metaphysics we adopt is ultimately an expression of the aims, experiences, expectations, and idiosyncrasies that we have as unique individuals.

The critique of formal logic. Central to Schiller's attacks on Bradley and the British idealists is his critique of a conception of logic that depicts it as concerned with the laws of "pure thought," for whose identification psychological investigations would be irrelevant. In Schiller's view, this account makes logic completely useless and, to a certain extent, paradoxical. Since the purpose of logic is to provide a theory of how we attain knowledge, the attempt to devise a logic that does not take into consideration processes of actual knowing is flawed from the start (Schiller, 1907: 73–74; see Abel, 1955: chaps. 3 and 7; Schiller, 1912).

Even though Schiller insists on the importance of psychological investigations for logic, he does not entirely blur the distinction between the two disciplines. Psychology is characterized as a "descriptive science, whose aim is the description of mental process as such" (Schiller, 1907: 75). By contrast, as we have already seen, logic is a "normative" science, whose task is to evaluate the processes of thinking that psychology describes. The need for such evaluations arises from the fact that the actual processes of thinking often lead us to error (78). It would be wrong, however, to think that logic can identify standards of correctness independently of psychological investigations regarding the actual processes of thinking: "its normative function arises quite naturally out of our actual procedures, from the observation that some cognitive processes are in fact more valuable" (78).

The impact of Schiller's pragmatism on Peirce and Dewey. It is customary to see Schiller as James's close ally, with the latter strongly influencing the former while, in turn, taking up some of Schiller's ideas. Here, we wish to point out that some aspects of continuity between Schiller, Peirce, and Dewey can also be highlighted. One effect of focusing on these aspects is that the rigid story of "two pragmatisms" falls apart. According to this story, Schiller clearly belongs to the "subjectivists." However, Peirce, the supposed champion of "objectivism" within the tradition, claimed that his own views were much closer to Schiller's than to James's (Scott, 1973: 372).

Peirce's encounter with Schiller's anthropomorphism and humanism was for him the occasion to reflect again on the characteristics and implications of his pragmatism. While he disliked the term "humanism" because it implied an overly broad conception of philosophy that conflicted with his scientific approach (Peirce, 1931–1952: 5.537; see also 8.262), he appropriated the term "anthropomorphism," which, in his view, captured one of the fundamental tenets of the position that he had been advocating for a long time. In his correspondence with Schiller, Peirce expresses a central aspect of his anthropomorphism by stressing that

> Man's faculties ... are pretty nicely adjusted to the needs of his life; and he is so immersed and submerged in conceptions of the *pragmatisch* (I don't say the *praktisch*) in such entirety that no conception, direct or indirect, can be had of an exterior standpoint.
>
> *(Scott, 1973: 376)*[2]

In a way similar to Schiller, Peirce stresses that it makes no sense to speak of a conception that is not anthropomorphic, as if we were able to take a God's eye point of view. He stresses that this anthropomorphism renders it meaningless to speak of the "limits of cognition."

> The strict consequence of this is, that it is all nonsense to tell [man] that he must not think in this or that way because to do so would be to transcend the limits of a possible experience. For let him try ever so hard to think anything about what is beyond that limit, it simply cannot be done.
>
> *(Peirce, 1931–1952: 5.536)*

This agrees with Peirce's rejection of the Kantian "thing-in-itself" as an object that is in principle unknowable and supports his idea that we should not "block the road of inquiry." Peirce's appropriation of anthropomorphism is not an incorporation of Schiller's ideas within his system but prompts a reflection on the anthropomorphistic implications of his own position.

As for Dewey, Schiller's contentions on the relationship between logic and psychology display many points of convergence with Dewey's naturalistic account of logic. Given the similarities between their projects, it is difficult to determine who influenced whom. However, on one point at least, it might have been Schiller who influenced Dewey. In reviewing Schiller's *Humanism*, Dewey insists on the importance of the book for its ability to derive logical standards starting from psychological facts: "For the contention of the humanist is precisely the unreality of the separation from each other of describable facts and normative values" (Dewey, 1904: 337). Moreover, in *Logic: The Theory of Inquiry*, he proposes a picture of the relationship between psychological facts and logical norms that closely resembles Schiller's. He criticizes the usual interpretation of this relationship according to which the "logical" consists "of 'norms' provided from some source wholly outside of and independent of experience" (Dewey, 1986: 107). In opposition to this view on the origin of logical norms, Dewey proposes an analogy with legal norms, which, while arguably originating from social interactions and transactions, are nonetheless able to acquire a genuinely normative character (106). This account of logical norms as having a properly normative character appears to deviate from Dewey's position in *Studies in Logical Theory*. There, Dewey characterizes logic as a "natural history of thinking" (Dewey, 1976: 309), where this seems to imply a purely "descriptive" account of logic. If we keep in mind that Schiller did not read *Studies in Logical Theory* before *Humanism* appeared (Schiller, 1903: ix) and that Dewey liked *Humanism* precisely because it provided a "naturalist" way to account for the normativity of logic, it is at least possible that Schiller partly influenced Dewey's later position in *Logic: The Theory of Inquiry*.

The Italian Pragmatists

The story of Italian pragmatism parallels the story of the cultural and philosophical journal *Leonardo*, published in Florence from 1903 until 1907. With its engaged and provocative style of cultural criticism, the journal sought to express a philosophical sensitivity that deviated from both the positivism of the Italian academy and the most prominent Italian anti-positivist philosophy of the time, namely, the idealism of Benedetto Croce (Casini, 2003). The editors of the journal and leaders of the pragmatist movement were two very young intellectuals: Giovanni Papini (1881–1956) and Giuseppe Prezzolini (1882–1982). In the course of the first years of their editorial venture, the two young men articulated a variety of pragmatism based on a voluntaristic and quasi-Nietzschean reading of James and Schiller. In the case of Prezzolini, the influence of French philosophy, from Sorel to Bergson (see Prezzolini, 1904a), and Catholic modernism was equally significant.

The two other main representatives of Italian pragmatism were Giovanni Vailati (1863–1909) and Mario Calderoni (1879–1914). Vailati, the oldest member of the group, had discovered pragmatism prior to the foundation of *Leonardo* and was, in a sense, the pivotal figure of the movement. Like Peirce, he was primarily a logician, as well as a mathematician and a historian of science. His intellectual style was rather distinct from that of Papini and Prezzolini, and he focused mostly on the significance of pragmatism for mathematical logic and epistemology (see Thayer, 1981: 335–336; Vailati, 1911). His parallel interests in economics and the social sciences motivated his student and friend Calderoni to work at the crossroads between economics, jurisprudence, and moral philosophy. Calderoni entered the scene with a thesis that used the pragmatist doctrine of meaning to address the debate of determinism versus individual responsibility in the philosophy of law. In subsequent work, he developed an anti-normative approach to morals that bears some resemblances to the almost contemporaneous work of G. E. Moore while also making use of concepts derived from economic thought, such as the concept of marginal value (see Calderoni, 1924).

In 1907, *Leonardo* ceased publication, as its original impetus succumbed, in part, to the looming disagreement between the more anti-positivistic strand of pragmatism represented by Papini and Prezzolini and the more logically oriented style of Vailati and Calderoni. With Vailati's death and the philosophical reorientation of its two young leaders, pragmatism was bound to disappear from the

Italian intellectual scene for decades to come. The apex of its fortune can perhaps be located around 1905, a year in which the pragmatist tone of the texts published in *Leonardo* reached their full maturity and the small group of its contributors garnered international recognition, thanks to lively exchanges with James, Peirce, and Schiller. These exchanges have one common trait: they pivot on the problem of how to define the core message of pragmatism while respecting the need for pluralism.

James and the Corridor Theory. The exchange with William James was perhaps the most momentous. James had personally met the Italian pragmatists in Rome, at the Fifth International Congress of Psychology, in April 1905. He was enthusiastic about this small group of thinkers, who appeared to him "a very serious philosophic movement, apparently *really* inspired by Schiller and myself."[3] In a short article on "G. Papini and the Pragmatist Movement in Italy" (James, 1906), James placed special emphasis on Papini's brilliant literary style, a welcome alternative, he thought, to the academic literature of North America. James also highlighted Papini's pluralist approach by taking up a simile that was to become famous:

> Pragmatism, according to Papini, is thus only a collection of attitudes and methods, and its chief characteristic is its armed neutrality in the midst of doctrines. *It is like a corridor in a hotel, from which a hundred doors open into a hundred chambers.*
>
> (James, 1906: 339, emphasis added)

However, this profession of pluralism did not prevent Papini from offering a minimal definition of the core idea of pragmatism, which James summarized as "an unstiffening of all our theories and beliefs by attending to their instrumental value" (338).

Papini's pluralistic definition of pragmatism, to which James felt so attracted, had originally developed as a reaction to a lively debate between Calderoni and Prezzolini, which animated the Florentine group during the two years prior to the Rome meeting (Colella, 2021). Leaning toward a "logical" and Peircean understanding of pragmatism, Calderoni (1904) had criticized the voluntarist and anti-intellectualist attitude of Prezzolini. He had done so by distinguishing three versions of pragmatism. The first version, Calderoni's favorite, was of Peircean inspiration and was reducible to a method of verification of the meaning of propositions on the basis of the experience we can conceivably make of their practical consequences. The second version centered on the "will to believe." A third and intermediate version construed the concept of experience less statically than the first, i.e., by pointing out that what we take to be an experience has always already been voluntarily constructed or selected. The mediating character of this third version, however, was, in Calderoni's opinion, not sufficient to avoid an irreconcilable clash between the first two. Prezzolini (1904b) replied to these arguments by accusing Calderoni of falling victim to the same positivistic spirit that pragmatism was meant to dispel.

The April 1905 issue of *Leonardo* sought to wrap up the dispute by publishing two additional contributions: Ferdinand Schiller's "Definitions of Pragmatism" (1905) and, more importantly, an essay signed by the "Florentine Pragmatist Club," but ascribable to Papini, titled "il pragmatismo messo in ordine" ("pragmatism put in order," Papini, 1905). This text essentially reiterates the distinction between three different versions of pragmatisms: a Peircean version, centered on the logic of induction; a Jamesian version, based on the will to believe; and an intermediate version that establishes a deeper connection between the will and the logical procedures of induction. But, unlike Calderoni, Papini's text did not declare these three versions to be incompatible. Indeed, it is at the end of this article that the conciliatory simile of the corridor first makes its appearance.

A Common Ground? Ironically, it was Papini himself, the creator of the "corridor-theory" and an advocate of pluralism, who also initiated an extremely successful interpretation of Italian pragmatism as divided into two virtually incompatible camps: the "magical" and the "logical" (Papini, 1977: 7; cf. Garin, 1963). The "magicians" were Papini himself and Prezzolini. Their primary goal was to use pragmatism as a tool for the transformation of reality – to conjure up the magical forces that bestow a creative power on words. The logicians, however, were Vailati and Calderoni: these philosophers were only preoccupied with action to the extent that it could shed light on the meaning of words.

This distinction was very successful, partly because it immediately resonated with the classical scheme of the "two pragmatisms" already mentioned in the Introduction. However, it risks concealing more than it reveals. At least for a certain period of time, the four main representatives of Italian pragmatism were carrying out a common philosophical project that can hardly be appreciated if we keep dividing them into two opposed camps. We can locate the philosophical center of these common projects around the possibility, already foreshadowed in the course of the 1904–1905 debate, of an intermediate definition of pragmatism that stresses the role of the will both in logic and in action (Maddalena, 2019).[4] More generally, all Italian pragmatists betray an interest in applying the scope of philosophical inquiry not merely to scientific rationality but to culture at large. In this sense, the question is perhaps not so much one of choosing between logic and magic but of exploring the continuous spectrum that extends between these two poles.

In 1906, Vailati left Florence, and the points of convergence between the Italian pragmatists became increasingly difficult to discern (Maddalena, 2019). Papini, however, continued his attempt to clarify pragmatism in other writings, most prominently in the essay "What Pragmatism Is Like," published in the *Popular Science Monthly* (Papini, 1907). Although Papini eschewed, once again, an explicit definition, his paper does seek to give an answer to the question it poses in the title. Pragmatism, argues Papini, is, first of all, a love for the concrete and a distaste for metaphysical systems. It is also a fascination with the "power" of humanity and its ability to control reality by means of action rather than contemplation. Although its distaste of contemplative attitudes may indicate an overlapping with positivism, Papini argues, the overlap is only superficial. Pragmatists are not "agnostic" like the positivists. They do not reject traditional metaphysical ideas because they are "too lofty for our intelligence" but because they are "too devoid of sense, too stupid."

Peirce's Intervention. Both Papini's repeated attempts to define pragmatism and the debate in *Leonardo* from which they originally stemmed reached Peirce's ears. The Italian pragmatists made contact with Peirce shortly after their meeting with James. Probably at the suggestion of James himself, Vailati began sending him some of his publications. Calderoni, for his part, had sent him the three issues of *Leonardo* that contained his debate with Prezzolini as well as Papini's final contribution (Fisch and Kloesel, 1982: 70). Peirce replied to Calderoni with a long letter, which provides an extremely detailed presentation of his major philosophical ideas, from semiotics to logic and the theory of the categories.[5] In that letter, he also expressed his sympathy with Calderoni's position in the debate and upheld his first definition of pragmatism.

This exchange with the Italian pragmatists contributed substantially to Peirce's decision that he too would take on the project of articulating a definition of pragmatist philosophy. In a series of texts written from 1905 on, he sought not only to offer such a definition but also to lay out a "proof" of pragmatism's validity. One of these texts, an unpublished manuscript, was written in the form of a dialogue between a pragmatist philosopher and an opponent of pragmatism, with the latter called "Jules" (Peirce, 1931–1952: 5.497–501). This was, by Peirce's own admission, the anglicized version of "Giuliano il Sofista," Prezzolini's nickname in *Leonardo* (5.497n; see Colapietro, 2021).

More importantly, in the article "What Pragmatism Is," published in *The Monist* in 1905, Peirce first announced his decision to change the name of his philosophy from "pragmatism" to "pragmaticism," as a reaction to the increasingly ambiguous meaning that the original concept had acquired (5.414). Contrary to what one might perhaps expect, however, Peirce's primary reference was neither to James nor to Schiller when he denounced that ambiguity. He wrote that James's radical empiricism "substantially answer[ed]" to his definition of pragmatism, "albeit with a certain difference in the point of view." Also, he deemed Schiller's "Axioms as Postulates" "most remarkable" and his use of the concept of pragmatism relatively unproblematic. Peirce was instead suggesting that the confusion ensued when the concept of pragmatism began to travel from philosophical prose to "literary journals."

In retrospect, Peirce tended to identify this "literary" usage of pragmatism precisely with the *Leonardo* group, and in particular with Papini's claim that pragmatism cannot be defined (see Peirce, 1931–1952, 5.495; 6.482; 6.490). Again, the fact that Peirce did not see an immediate continuity between Papini and James clashes with the received picture of the "two pragmatisms." However, we should note that Peirce's original article was published in the same eventful month of April 1905 in which *Leonardo* published "il pragmatismo messo in ordine," and the Florentine group met with James in Rome. It seems, therefore, more reasonable to read his polemic against literary journals as the expression of a general unease about the ways his original philosophical insight was deployed on both shores of the Atlantic, rather than as a direct reference to the *Leonardo* group.[6]

Notes

1 Another, more troubling commonality between Schiller and Italian pragmatism is the presence of anti-democratic tendencies that would later lead some of these thinkers to flirt with fascism. See Vogt (2002).
2 This is only one of the three propositions that Peirce connects to his anthropomorphism. See Scott (1973: 376). For a more complete account of Peirce's anthropomorphism, see Bergman (2014).
3 From a letter to James's wife, cited in Fisch and Kloesel (1982: 69).
4 "From induction to the Will to Believe there is a continuity which is provided by the common goal: the aspiration to be able to act" (Papini 1905: 46, cited by Maddalena, 2019).
5 The letter is published partly in 8.205–213, partly in Fisch and Kloesel (1982).
6 Peirce's letter to Calderoni gives us some more hints of whom Peirce had in mind while coining the term "pragmaticism." See Peirce (1931–1952): 8.205):

> In the April number of the Monist ["What Pragmatism Is," 1905] I proposed that the word "pragmatism" should hereafter be used somewhat loosely to signify affiliation with Schiller, James, Dewey, Royce, and the rest of us, while the particular doctrine which I invented the word to denote, which is your first kind of pragmatism, should be called "pragmaticism."

It is equally worth keeping in mind that Peirce by no means had an exclusively negative view of Papini's work. See, e.g., 5.494. See also Peirce's long letter to Papini, conserved at the archive of the Fondazione Primo Conti in Fiesole (Italy).

References

Abel, R. (1955) *The Pragmatic Humanism of F. C. S. Schiller*. New York: King's Crown Press.
Bergman, M. (2014) 'The Curious Case of Peirce's Anthropomorphism,' in Thellefsen, T. and Sorensen, B. (eds.) *Charles Sanders Peirce in His Own Words*. Berlin: De Gruyter, pp. 315–323.
Calderoni, M. (1904) 'Le varietà del pragmatismo.' *Leonardo*, II (November), pp. 3–7.
Calderoni, M. (1924) *Scritti*. Campa, O. and Papini, G. (eds.). Florence: La Voce.
Casini, P. (2003) *Alle origini del novecento. 'Leonardo', 1903–1907*. Bologna: Il Mulino.
Colapietro, V. (2021) '"Tell Your Friend Giuliano…": Jamesian Enthusiasms and Peircean Reservations,' in Maddalena, G. and Tuzet, G. (eds.) *The Italian Pragmatists: Between Allies and Enemies*. Leiden and Boston: Brill-Rodopi, pp. 142–162.
Colella, E. P. (2021) 'Two Faces of Italian Pragmatism: The Prezzolini-Calderoni Debate, 1904–1905,' in Maddalena, G. and Tuzet, G. (eds.) *The Italian Pragmatists: Between Allies and Enemies*. Leiden and Boston: Brill-Rodopi, pp. 115–141.
Dewey, J. (1904) 'Review of Humanism.' *The Psychological Bulletin*, 1 (10), pp. 335–340.

Dewey, J. (ed.) (1976) *Studies in Logical Theory*. Part of vol. 2 of *The Middle Works of John Dewey, 1899–1924*. Boydston, J. A. (ed.). Carbondale: Southern Illinois University Press, 1976–1983.

Dewey, J. (1986) *Logic: The Theory of Inquiry*. Vol. 12 of *The Later Works of John Dewey, 1925–1953*. Boydston, J. A. (ed.). Carbondale: Southern Illinois University Press, 1981–1990.

Ferrari, M. (2017) 'William James and the Vienna Circle,' in Pihlström, S., Stadler, F. and Weidtmann, N. (eds.) *Logical Empiricism and Pragmatism* (Vienna Circle Institute Yearbook, 19). Cham: Springer, pp. 15–42.

Fisch, M. H. and Kloesel, C. J. W. (1982) 'Peirce and the Florentine Pragmatists: His Letter to Calderoni and a New Edition of His Writings.' *Topoi*, I, pp. 68–73.

Garin, E. (1963) 'G. Vailati e la cultura italiana del suo tempo.' *Rivista critica di storia della filosofia*, XVIII, pp. 287ff.

James, W. (1906) 'G. Papini and the Pragmatist Movement in Italy.' *The Journal of Philosophy, Psychology and Scientific Methods*, III (13), pp. 337–341.

James, W. (1975) *Pragmatism*. Bowers, F. and Skrupskelis, I. K. (eds.). Cambridge: Harvard University Press.

Maddalena, G. (2019) 'Vailati, Papini, and the Synthetic Drive of Italian Pragmatism.' *European Journal of Pragmatism and American Philosophy*, XI (1). DOI: 10.4000/ejpap.1533.

Misak, C. (2016) *Cambridge Pragmatism. From Peirce and James to Ramsey and Wittgenstein*. Oxford: Oxford University Press.

Mounce, H. O. (1997) *The Two Pragmatisms: From Peirce to Rorty*. London: Routledge.

Papini, G. (1905) 'Il pragmatismo messo in ordine.' *Leonardo*, III (April), pp. 45–48.

Papini, G. (1907) 'What Pragmatism Is Like,' translated by K. Royce. *Popular Science Monthly*, 71 (October), pp. 351–358.

Papini, G. (1977) *Opere. Dal 'Leonardo' al futurismo*. Milano: Mondadori.

Peirce, C. S. (1931–1952) *Collected Papers of Charles Sanders Peirce*. Hartshorne, C. and Weiss, P. (volumes 1–6), and Burks, A. (volumes 7–8) (eds.). Cambridge: Harvard University Press (followed by volume and paragraph number).

Prezzolini, G. (1904a) *Il linguaggio come causa d'errore: Henri Bergson*. Firenze: Spinelli.

Prezzolini, G. (1904b) 'Le varietà sul pragmatismo. Risposta a Mario Calderoni.' *Leonardo*, II (November), pp. 7–9.

Pudal, R. (2011) 'Enjeux et usages du pragmatisme en France (1880–1920): Approche sociologique et historique d'une acculturation philosophique.' *Revue Française de sociologie*, 52 (4), pp. 747–775.

Russell, B. (1928) *Sceptical Essays*. London: George Allen and Unwin.

Schiller, F. C. S. (1891) *Riddles of the Sphinx: A Study in the Philosophy of Humanism*. London: S. Sonnenschein.

Schiller, F. C. S. (1902) 'Axioms as postulates,' in Sturt, H. (ed.) *Personal Idealism*. New York and London: Macmillan, pp. 47–133.

Schiller, F. C. S. (1903) *Humanism: Philosophical Essays*. New York and London: Macmillan.

Schiller, F. C. S. (1905) 'The Definitions of Pragmatism.' *Leonardo*, III (April), pp. 44–45.

Schiller, F. C. S. (1907) *Studies in Humanism*. New York and London: Macmillan.

Schiller, F. C. S. (1912) *Formal Logic: A Scientific and Social Problem*. New York and London: Macmillan.

Scott, F. J. D. (1973) 'Peirce and Schiller and Their Correspondence.' *Journal of the History of Philosophy*, 11, pp. 363–386.

Sölch, D. (2018) 'Deutschsprachiger Raum,' in Festl, M. G. (ed.) *Handbuch Pragmatismus*. Stuttgart: Metzler, pp. 297–303.

Thayer, H. S. (1981) *Meaning and Action: A Critical History of Pragmatism*. 2nd edition. Indianapolis: Hackett.

Vailati, G. (1911) *Scritti*. 3 vols. Calderoni, M., Ricci, U. and Vacca, G. (eds.). Leipzig, Barth and Florence: Seeber.

Vogt, P. (2002) *Pragmatismus und Faschismus. Kreativität und Kontingenz in der Moderne*. Weilerswist: Velbrück Wissenschaft.

20
PRAGMATISM AND ANALYTIC PHILOSOPHY

Henrik Rydenfelt

Introduction and background

Analytic philosophy is a philosophical tradition that – like pragmatism – resists a clear-cut definition or encapsulation in a limited number of philosophical theses. The work of analytic philosophers is typically characterized by conceptual rigor, the use of formal logical tools and argumentation with explicit premises and conclusions. The beginnings of this tradition can be located in Gottlob Frege's development of new logical devices for the formal representation of thoughts and inferences, including his first-order predicate logic and his accounts of proof and definition, as well as his quest to uncover the logical foundations of mathematics. In Great Britain, Frege's views were received by a generation of philosophers dismayed by (British) idealism, led by G. E. Moore and Bertrand Russell. Initially concentrated on logic, philosophy of mathematics and philosophy of language, the new logical tools were soon applied in metaphysics and ethics. Russell's logical atomism and Ludwig Wittgenstein's early work inspired a number of thinkers on the continent, especially in the Vienna Circle of Moritz Schlick, Hans Hahn, Rudolf Carnap and Otto Neurath, as well as the Society for Empirical Philosophy led by Hans Reichenbach in Berlin, resulting in the heyday of logical positivism and logical empiricism. During and after the Second World War, analytic philosophy gradually became dominant in philosophy departments in the United States with the immigration of notable European thinkers and established itself as the philosophical mainstream in the United States, United Kingdom, and many Nordic countries, eventually spreading to universities in South America and Asia. In addition to philosophical logic and philosophy of mathematics, the contemporary fields of philosophy of science, meta-ethics and philosophy of mind can be viewed as the extension of the projects and tools developed in this tradition to philosophical issues concerning science, ethics and consciousness.

The history of interactions between analytic philosophy and philosophical pragmatism is complex. Pragmatists prefigured many "analytic" developments; in particular, Charles S. Peirce's pragmatist account of meaning has been seen as foreshadowing logical positivist or empiricist views. Early pragmatists engaged in discussions with prominent analytic philosophers, with William James and Russell debating the concept of truth, and Russell criticizing John Dewey's work on logic. However, the influx of European philosophers to the United States in the 1930s and 1940s was also among the reasons for the decreasing attention to pragmatism in the following decades. Already beginning in the early 1950s, American philosophers influenced by both pragmatism and the analytic tradition – W. V. O. Quine and Wilfrid Sellars in particular – turned their critical attention to some of the key presuppositions of logical atomism, logical positivism and logical empiricism. Although Wittgenstein's later contention that meaning is use – for "a *large* class of the employment of the word 'meaning'" (1953: § 43) – has later been celebrated as fundamentally pragmatist view, it was initially hardly connected with pragmatism; the same could be said of the accounts of the use of "ordinary language" that were developed by Gilbert Ryle, J. L. Austin and P. F. Strawson, among others, mostly in Oxford around the same time. Beginning in the 1970s, Peirce's logical advances and his semiotics have informed the developments of formal logic, in particular through the works of Jaakko Hintikka (e.g. Hintikka 1980). By the 1980s, philosophers trained in the analytic tradition – such as Hilary Putnam and Richard Rorty – began to argue for views that they traced back to James and Dewey. During the 21st century, many leading philosophers of the analytic tradition, including Robert Brandom and Huw Price, have advanced pragmatist accounts of central philosophical questions. At present, the scope of pragmatism appears to be extending within analytic philosophy (e.g. Misak 2016).

Analysis, logical atomism and language use

"Analytic" in "analytic philosophy" is customarily taken to refer to logical analysis; however, beginning already with Russell and Moore, logical analysis was used to refer to a number of philosophical endeavors. In the first place, analysis was used to refer to the interpretation and clarification of ordinary language statements by means of the logical tools provided by Frege and further developed by Russell and Whitehead (1910–1913/1925–1927). The second sense of analysis is that of splitting larger units of language (statements, propositions, terms) to their constituent parts. In Moore's (1903) account, complex terms can be defined in terms of their components; it is because moral terms such as "good" resist such splitting into component parts that they admit of no definition. This view of analysis is embedded in Russell's (1914) *logical atomism*, the contention that our language can be decomposed to basic concepts and vocabulary, and the world itself is similarly composed of atomic "facts" concerning the qualities of and relations between individual things. Russell's view informed Wittgenstein's (1922) position, in the *Tractatus*, that analyses of language take us beyond the constituents of propositions to the constitution of reality itself.

Peirce's (1885) independent development of first-order logic (in part together with his student, O. H. Mitchell; see Mitchell 1883) entailed both of the two senses of analysis already discussed: roughly, accounting for the composition of ordinary language statements in terms of complexes of quantifiers, subjects and predicates. However, beginning with Peirce, pragmatists have decidedly resisted atomistic views both in logic and in metaphysics. Peirce's pragmatist account of meaning is holistic (a point to which we will presently return); by contrast, metaphysical atomism of the Russellian sort is closely connected to nominalism that denies the reality of generals and of which Peirce tended to accuse most of his contemporaries. Although many later pragmatists have not advanced (modal) realism of a Peircean type, pragmatist philosophers have consistently emphasized continuity and holistic views in metaphysics, semantics and logic.

The concept of analysis can also be extended to encompass the aspirations of many analytic philosophers to provide accounts of the "use" of words and of the ambiguities within ordinary language, producing such refinements as J. L. Austin's (1962) distinctions between locutionary, illocutionary and perlocutionary (speech) acts. Again, the most natural point of comparison is Peirce and his semiotics – the study of signs and their interpretation – that influenced these developments, despite not being limited to narrowly linguistic phenomena. During the past decades, many accounts in analytically minded exploration of language that focus on "use" rather than "content" – or pragmatics rather than semantics – have been referred to as "pragmatist" positions; for example, the pragmatist label has been extended to the moral expressivism and quasi-realism developed by Simon Blackburn (1984). While perfectly pragmatist in spirit, there is a difference in emphasis. In Peirce's semiotic account, the focus was largely on the purposes of the interpretation of signs rather than on the function or use of language by an utterer or speaker.

Finally, the tradition of analytic philosophy also includes revisionary projects of (linguistic) analysis. Following Wittgenstein, Carnap thought that philosophical problems were largely due to the inadequacies of our linguistic tools. Instead of suggesting that philosophical problems are meaningless, or "analyzing" them away, however, Carnap preferred to provide what he called "explications" – revisions or replacements of terms and concepts within our linguistic frameworks. While working on his *Logical Syntax of Language* (1934), Carnap arrived at his principle of tolerance that maintains that choice of formal language is free, later suggesting that linguistic forms are to be tested by their success and failure in practical use (1950: 221). Carnap's tolerant pluralism bears resemblance to some long-standing pragmatist ideas, such as William James's (ontological) pluralism; moreover, it has been argued that, in Carnap's view, the choice of linguistic framework is a "pragmatic" question (Kraut 2020).

Indeed, Carnap's position has close affinities to contemporaneous proposals by C. I. Lewis, whose views were heavily influenced by James and who was a doctoral student of Josiah Royce's (nowadays often included in the number of the classics of pragmatism) as well as a teacher of Nelson Goodman, Roderick Chisholm and Quine, among others. Lewis famously developed the notion of strict implication; for Lewis, this development had the implication that the number of possible systems of logic is unlimited. Moreover, Lewis (1929) argued that while our interpretations of experience (such as beliefs) are open to empirical testing and revision, such testing does not extend to the validity of our concepts and their logical relationships. Rather, conceptual systems face the "pragmatic" test of utility or convenience. Quine's famous criticism of the analytic-synthetic distinction has been taken to imply the shipwreck of Lewis's and Carnap's project: Quine showed that a strict dichotomy between questions of definition and (empirical) questions of fact is untenable. The resulting view is closer to the position of the early pragmatists, who (put in terms closer to their views) did not distinguish between empirical and "pragmatic" testing of theories but, contrariwise, attempted to explicate the former by the latter.

Meaning and meaningfulness

A central aspiration of analytic philosophy has been to rid philosophy of meaningless jargon, unclear concepts and pointless debates. A similar ambition was central to Peirce's early presentations of his pragmatism in his articles "The Fixation of Belief" (1877) and "How to Make Our Ideas Clear" (1878) and was taken up by both James and Dewey. Beginning with his first philosophical papers, Peirce repudiated the Cartesian picture of "ideas" as representing objects to the mind and replaced it with his triad of signs, interpretants and objects. A part of this account is his maxim of pragmatism that pertains to those signs that have logical interpretants – such interpretants that are thoughts as opposed to feelings and actions. As a thought can be interpreted by another thought (or a sentence translated by another sentence) indefinitely, to trace the ultimate meaning of such signs, Peirce thought we must step outside of thinking and language and anchor meaning in conduct. The ultimate interpretants of these signs, Peirce argued, are habits of action. Any meaningful sentence, if accepted by a speaker, would result in action under some conceivable circumstances. If the conceivable conduct resulting from the acceptance of two sentences in no way differs, their meaning is the same. In this way, pragmatism enables us to detect meaningless statements as well as weed out pointless debates in which superficially distinct views ultimately have the same meaning.

Peirce's views have affinities with the stringent account of meaningfulness produced by early logical positivism: verificationism, the view that only empirically verifiable statements are meaningful (cf. Schlick 1936), perhaps most evident in Wittgenstein's (1922) early position that mathematical and logical – and in many cases, philosophical – vocabularies lack meaning. In Peirce's early formulations, habits of conduct were connected with expectations concerning sensations. Action based on habits is spurred by stimuli which are "derived from perception", and the purpose of action is to "produce some sensible result" (Peirce 1878: 131). Thus, "our action has exclusive reference to what affects the senses, our habit has the same bearing as our action, our belief the same as our habit, our conception the same as our belief; [...]" (ibid.). This is suggestive of a test of meaning or meaningfulness in terms of empirical consequences. Moreover, the "sensible results" referred to may be identified with the conditions of verification of the belief or statement in question.

However, a number of salient differences are to be noted. First, Peirce's account is able to avoid the issues revolving around theoretical and dispositional predicates, which produced the logical empiricists so much concern. In brief, the logical empiricists initially identified "meaning" with the actual conditions of verification, while in Peirce's account, the practical consequences were always counterfactual in nature, pertaining to how the acceptance of a sentence *would* make one act under *conceivable* circumstances. Second, in many of Peirce's formulations of pragmatism, the connection between conduct and sensation is far less rigid. In his later writings, Peirce emphasizes that the relevant practical consequences are the effects on deliberate conduct and are not limited to "sensible effects".

A third salient difference between pragmatism and early verificationism (or more broadly, logical empiricism) is holism, or the idea that sentences have experiential implications only in the context of larger theories (cf. White 1956). Some logical empiricists connected the verification of a sentence with a limited number of observations (or observation sentences). Beginning in the 1950s, this "reductionism" about meaning came under heavy criticism by Quine (1951) and others. Holism was already central to Peirce's view: a single meaningful sentence or judgment may enforce innumerable habits of action depending on the circumstances and aims at issue. The depth of pragmatist holism can be elucidated by the two puzzles later made popular by Quine (1960, 1975). The first, the underdetermination of theories by evidence, is the issue of whether there could be two different theories that have all the same observational consequences. The second, the indeterminacy of translation, is the question whether there could be two equally correct but different translations of a language. While Quine's stance concerning these puzzles is somewhat ambivalent, both issues, from Peirce's point of view, are motivated by too little holism concerning the conceivable circumstances where these theories and translations would differ (provided that they indeed differ in meaning).

A fourth difference also pertains to the scope of holism. Among contemporary pragmatists, Robert B. Brandom (1994) has advocated the inferentialist view that mastering the meaning of an assertion is to grasp its inferential proprieties: what else one would be committing oneself to by making the assertion, what would entitle one to the assertion and what would preclude such entitlement. Brandom has rightly noted that Peirce's pragmatism is concentrated on the practical differences "downstream" – the commitments that incur from the acceptance of a sentence rather than the "upstream" of other sentences that would justify its acceptance. However, Peirce's pragmatism may better fit our ordinary conceptions of "sameness" in meaning, for example when two speakers differ over what would justify commitment to, or the assertion of, a sentence while agreeing on its import to conduct.

Fifth and finally, Peirce's maxim of pragmatism was not intended to provide an account of the "meaning" of all signs. For example, in his semiotics, Peirce developed a refined account of indices and icons. Peirce's account of indexical terms, in particular, anticipated Saul Kripke's (1980) and Putnam's (1975) "new theory of reference", the account of reference and meaning that owes to an actual (or causal) connection between words and objects (Short 2007, ch. 10). Relatedly, Peirce's accounts of modal logic and his modal metaphysics anticipated the contemporary possible world semantics that also reinvigorated the issue of the ontological status of possible worlds, in particular in the works of David Lewis (1973) and David Armstrong (1978a, 1978b), who interpreted the new logical devices as implying new metaphysical puzzles.

Truth, logic and science

Pragmatism has traditionally been identified with a theory of truth that maintains that truth is "what works", or that true sentences are those that, if accepted, would result in success in attaining the aims of action. While the building blocks of this account are in Peirce's pragmatist view of meaning, already explored above, it was popularized by William James (1907), and further developed by Dewey (1938) in his accounts of truth, knowledge and "warranted assertibility". However, the central aspiration of these pragmatists was to explicate the notion of truth in terms of practical consequences rather than to offer a competing "theory" of what truth means. It was Russell who arguably invented the whole idea of *competing* "theories" of truth. A part of the reason for some confusion on this score is likely due to James's readiness to accept some of Russell's characterizations of the situation, even titling a collection of papers that included many of his responses to Russell and other critics *The Meaning of Truth* (1909).

The actual views of the classical pragmatists are rather more complex and multifaceted. Peirce (1877) proposed what may be the first instance of a deflationary or minimalist account of truth (cf. Rydenfelt 2021; Short 2000) that later became so central to analytic philosophy due to the appropriation of Alfred Tarski's (1935) truth definitions for fully interpreted formal languages for the purposes of semantical projects (cf. Davidson 1990; Horwich 1990). In addition, Peirce, James and Dewey all proposed versions of what (after Russell) has been called the "correspondence" account. Peirce is also commonly associated with the view that truth is the belief that would stand at the ideal "end of inquiry". An "epistemic" view of truth, along these lines, as justified belief under ideal conditions was later proposed by Hilary Putnam (1981), who then identified it with pragmatism (in papers printed in Putnam 1990) only to later recant and criticize the pragmatists for (Putnam 1995). It should be noted that in Peirce's extensive corpus, this view makes its main appearance in a single central passage that is difficult to interpret. Peirce's consistent position was to approach truth in terms of the practice of inquiry as its aim (rather than as its "end", in the sense of culmination).

Indeed, rather than providing a theory – let alone a definition – of truth, early pragmatists were occupied by developing both formal and informal tools of inquiry, the enterprise that they called *logic*. Much of what currently is called formal and mathematical logic would, in Peirce's division of sciences, belong to mathematics, the science that draws necessary conclusions. Logic, as a part of normative science (itself part of philosophy), is already informed by and intended to inform concrete inquiry. Similarly, Dewey's grand work, *Logic: The Theory of Inquiry* (1938), was broadly concerned with what most philosophers would now call the philosophy of science. As indicated by Peirce's placement of logic next to esthetics and ethics, logic is the study of how we *ought* to act in our inquiries (or how we *should* infer). This pragmatist view of logic is antithetical to the version of "naturalized" epistemology as a branch of (descriptive) psychology that Quine (1990) later proposed. Nevertheless, the pragmatists maintained that their position is fundamentally naturalist (cf. Rydenfelt 2011a). Dewey, in particular, sometimes preferred the label "cultural naturalism" for his pragmatism.

Issues of normativity have motivated some contemporary reactions against classical pragmatism. In the account that we owe to Richard Rorty (1979), it was two American thinkers, W. V. O. Quine and Wilfrid Sellars, whose central insights spelled the defeat of the logical empiricist project, paving the way for a new era of pragmatism. Quine's criticisms of the analytic-synthetic distinction and of "reductionism" about meaning were already discussed above. Sellars's (1956) criticism of the myth of the Given – often considered the "third" dogma of logical empiricism – was the argument that there cannot be any cognitive states that are both basic in the sense that they have a positive epistemic status (such as being true or justified) without reference to other such states *and* epistemically efficacious in the sense that they could serve to provide an epistemic status to other, non-basic states.

Rorty's take from these criticisms of the empiricist project was to argue that, instead of the search for truth, or placing the foundation for such a search in an "outside" reality, our human projects can at best aim at justification among an expanding number of peers. This resulted in a criticism of the early pragmatists, especially Peirce, for their scientific realism. However, the classics of the pragmatist tradition were not enchanted by the myth of the Given: in their view, sensations (or perceptions) enter consciousness in the form of beliefs (judgments or perceptual judgments) and only then may inferentially justify other beliefs (judgments). Even though inquiry is viewed as a social, human practice, in the scientific project, it is reality that is vested with a crucial normative status (Rydenfelt 2021).

Ethics and political philosophy

Peirce's writings on logic and semiotics are both a natural point of comparison and an important source for the projects central to analytic philosophy concerning meaning, logic and science. It was John Dewey's work, however, that brought pragmatism to bear strongly on ethics and social and political philosophy. Although Peirce provides a perspective on ethics as a form of scientific inquiry, his brief discussions of such inquiry are notably individualistic, typically involving only the reflective comparison of actual and potential practices of conduct with an ideal and omitting the role of the community of inquiry otherwise so central to Peirce's view of science. Dewey (1927, 1935) developed these notions further, proposing that ethical and – in many ways, by extension – political issues are to be settled by a community in a practice that is (or at least resembles) scientific inquiry (cf., Bernstein 2010; Putnam & Putnam 1990; Rydenfelt 2013, 2019b).

John Rawls's work, especially his *A Theory of Justice* (1971), elevated social and political philosophy to the main agenda of analytic philosophy. Rawls proposed both substantial theses concerning justice in a liberal democracy – in particular, his two principles of justice as fairness – as well as a meta-ethical vision of how such theses are justified and legitimized as what the representatives of citizens would agree upon in an "original position" cleansed of particular interests and knowledge of their position in the society behind a "veil of ignorance". After Rawls, much of the analytic debate in political philosophy has centered around possible defenses of democracy as preferable to alternative forms of political decision-making and as acceptable to (all) citizens, often with reference to its epistemic or cognitive merits. Proponents of deliberative democracy, such as Jürgen Habermas (1990) and Rawls (1996), have proposed that the very validity or justification of political decisions depends on their acceptability to all citizens, or at least all reasonable citizens. Others have argued that democracy is epistemically preferable among the societal arrangements that are acceptable to all citizens (Estlund 2008).

Although many contemporary philosophers have hailed deliberative views of democracy as pragmatist and Deweyan in spirit, Dewey's work does not appear to include an attempt to produce an argument for democracy and against anti-democratic views on epistemic grounds. Accordingly, many contemporary pragmatists have turned to Peirce for the building blocks of such as defense, arguing that Peirce's view of belief and the scientific method imply the necessity of a consideration of the views and experiences of others, in a deliberative vein, in the pursuit of truth (Misak 2000; Talisse 2007). However, doubts may be raised on whether such arguments based on pragmatist ideas can succeed without begging the question against the anti-democratic opponents who do not appreciate the views and experiences of (at least some) others (Rydenfelt 2011b, 2019a). Indeed, from the point of view of Dewey – as well as Peirce – the scientific method is a social achievement rather than a presupposition of (full-fledged) belief. Pragmatist views of democracy are perhaps best viewed, along Dewey's lines, as attempts at explicating the consequences of the scientific method to our social arrangements. As such, they also invite complex but timely questions concerning the role of scientific experts and scientific experimentation in democratic processes.

Conclusion

Pragmatists have typically – but in different ways – resisted dichotomies such as those between mind and matter, the definitional and the empirical, logic and empirical inquiry and science and ethics, emphasizing continuity and holism against nominalism and atomism. In this, they seem to have been on the right side of the philosophical argument of the 20th and early 21st century. However, at times, this has also come at the expense of some clarity that analytic philosophers tend to appreciate, such as clear-cut views concerning where pragmatism stands in current philosophical debates and indeed concerning what pragmatism is or may be. All the while, pragmatism seems to have gained a more prominent role. In some ways, analytic philosophy appears to be slowly incorporating some lessons from the history of philosophy of the past centuries, including the teachings of the pragmatist tradition. Sellars famously described his project as that of steering analytic philosophy out of its "Humean" and into its "Kantian" phase, in particular by the realization of the blindness of a "Given" that is inferentially inert in the normative space of reasons. Almost as famously, Rorty (1997) described Brandom's inferentialist project as that of ushering analytic philosophy from its Kantian to its "Hegelian" phase with the realization that normative statuses are social achievements based on recognition. It remains to be seen if analytic philosophy can finally be brought to a "Peircean" stage with the realization that within our human, social practices, such a normative status can be conferred to reality.

References

Armstrong, D. M. (1978a) *Universals and Scientific Realism, Volume I: Nominalism and Realism*, Cambridge: Cambridge University Press.
Armstrong, D. M. (1978b) *Universals and Scientific Realism, Volume II: A Theory of Universals*, Cambridge: Cambridge University Press.
Austin, J. L. (1962) *How to Do Things with Words*, 2nd edition, M. Sbisà and J. O. Urmson (eds.), Oxford: Oxford University Press, 1975.
Bernstein, R. (2010) *The Pragmatic Turn*, Cambridge: Polity Press.
Blackburn, S. (1984) *Spreading the Word*, Oxford: Oxford University Press.
Brandom, R. B. (1994) *Making It Explicit*, Cambridge, MA: Harvard University Press.
Carnap, R. (1934) *Logische Syntax der Sprache*, Vienna: Springer. English translation by Amethe Smeaton, *The Logical Syntax of Language*, London: Routledge, 1937.
Carnap, R. (1950) "Empiricism, Semantics, and Ontology", *Revue Internationale de Philosophie* 4(11): 20–40.
Davidson, D. (1990) "The Structure and Content of Truth" (The Dewey Lectures 1989), *Journal of Philosophy* 87: 279–328.
Dewey, J. (1927) "The Public and Its Problems", in J. A. Boydston (ed.), *The Later Works of John Dewey*, Volume 2. Carbondale: Southern Illinois University Press, pp. 235–370.
Dewey, J. (1935) "Liberalism and Social Action", in J. A. Boydston (ed.), *The Later Works of John Dewey*, Volume 11. Carbondale: Southern Illinois University Press, pp. 1–65.
Dewey, J. (1938). "Logic: The Theory of Inquiry", in J. A. Boydston (ed.), *The Later Works of John Dewey*, Volume 12. Carbondale: Southern Illinois University Press, pp. 1–529.
Estlund, D. (2007) *Democratic Authority*, Princeton: Princeton University Press.
Habermas, J. (1990) *Moral Consciousness and Communicative Action*, trans. C. Lenhardt and S. W. Nicholsen, Cambridge: Polity Press.
Hintikka, J. (1980) "C. S. Peirce's 'First Real Discovery' and Its Contemporary Relevance", *The Monist* 63(3): 304–315.

Horwich, P. (1990) *Truth*, Oxford: Basil Blackwell.
James, W. (1907) *Pragmatism: A New Name for Some Old Ways of Thinking*, Cambridge, MA: Harvard University Press, 1975.
James, W. (1909) *The Meaning of Truth: A Sequel to "Pragmatism"*, Cambridge, MA: Harvard University Press, 1975.
Kraut, R. (2020) "Rudolf Carnap. Pragmatist and Expressivist about Ontology", in R. Bliss and J. T. M. Miller (eds.), *The Routledge Handbook of Metametaphysics*. London: Routledge, pp. 32–47.
Kripke, S. (1980) *Naming and Necessity*, Cambridge, MA: Harvard University Press.
Lewis, C. I. (1929) *Mind and the World Order: Outline of a Theory of Knowledge*, New York: Charles Scribners.
Lewis, D. (1973) *Counterfactuals*, Oxford: Blackwell Publishers and Cambridge, MA: Harvard University Press (reprinted with revisions, 1986).
Misak, C. (2000) *Truth, Politics, Morality, Pragmatism and Deliberation*, London: Routledge.
Misak, C. (2016) *Cambridge Pragmatism: From Peirce and James to Ramsey and Wittgenstein*, Oxford: Oxford University Press.
Mitchell, O. H. (1883) "On a New Algebra of Logic", in C. S. Peirce (ed.), *Studies in Logic by Members of the Johns Hopkins University*, Boston: Little, Brown, and Company, pp. 72–106.
Moore, G. E. (1903) *Principia Ethica*, Cambridge: Cambridge University Press.
Peirce, C. S. (1877) "The Fixation of Belief", in The Peirce Edition Project (ed.), *Writings of Charles S. Peirce: A Chronological Edition*, Volume 3. Bloomington: Indiana University Press, pp. 242–256.
Peirce, C. S. (1878) "How to Make Our Ideas Clear", in The Peirce Edition Project (ed.), *Writings of Charles S. Peirce: A Chronological Edition*, Volume 3. Bloomington: Indiana University Press, pp. 257–275.
Peirce, C. S. (1885) "On the Algebra of Logic: A Contribution to the Philosophy of Notation", in The Peirce Edition Project (ed.), *Writings of Charles S. Peirce: A Chronological Edition*, Volume 5. Bloomington: Indiana University Press, pp. 162–190.
Putnam, H. (1975) "The Meaning of 'Meaning'", in *Mind, Language and Reality: Philosophical Papers*, Volume 2, Cambridge: Cambridge University Press, pp. 215–271.
Putnam, H. (1981) *Reason, Truth and History*, Cambridge: Cambridge University Press.
Putnam, H. (1990) *Realism with a Human Face*, Cambridge, MA: Harvard University Press.
Putnam, H. (1995) *Pragmatism: An Open Question*, Oxford: Blackwell.
Putnam, H. and Putnam, R. A. (1990). "Epistemology as Hypothesis", *Transactions of the Charles S. Peirce Society* 26(4): 407–433.
Quine, W. V. O. (1951) "Two Dogmas of Empiricism", *Philosophical Review* 60: 20–43.
Quine, W. V. O. (1960) *Word and Object*, Cambridge, MA: M.I.T. Press.
Quine, W. V. O. (1975) "Empirically Equivalent Systems of the World", *Erkenntnis* 9: 313–328.
Quine, W. V. O. (1990) *Pursuit of Truth*, Cambridge, MA: Harvard University Press; revised edition, 1992.
Rawls, J. (1971) *A Theory of Justice*, Cambridge, MA: Harvard University Press.
Rawls, J. (1996) *Political Liberalism*, New York: Columbia University Press.
Rorty, R. (1979) *Philosophy and the Mirror of Nature*, Princeton: Princeton University Press.
Rorty, R. (1997) "Introduction", in W. Sellars (ed.), *Empiricism and the Philosophy of Mind*. Cambridge, MA: Harvard University Press, pp. 1–12.
Russell, B. (1914) *Our Knowledge of the External World*, London: Allen & Unwin.
Russell, B. and Whitehead, A. N. (1910–1913/1925–1927) *Principia Mathematica*, 3 vols, Cambridge: Cambridge University Press (second edition 1925–1927).
Rydenfelt, H. (2011a) "Naturalism and Normative Science", in J. Knowles and H. Rydenfelt (eds.), *Pragmatism, Science and Naturalism*, Berlin and New York: Peter Lang, pp. 115–138.
Rydenfelt, H. (2011b) "Epistemic Norms and Democracy: A Response to Talisse", *Metaphilosophy* 42(5): 572–588.
Rydenfelt, H. (2013) "Sensitive Truths and Sceptical Doubt", in H. Rydenfelt and S. Pihlström (eds.), *William James on Religion*. New York: Palgrave Macmillan, pp. 128–144.
Rydenfelt, H. (2019a) "Democracy and Moral Inquiry: Problems of the Methodological Argument", *Transactions of the Charles S. Peirce Society* 55(2): 254–272.
Rydenfelt, H. (2019b) "Pragmatism, Social Inquiry and the Method of Democracy", in K. Holma and T. Kontinen (eds.), *Practices of Citizenship in East Africa. Perspectives from Philosophical Pragmatism*, London: Routledge, pp. 29–43.
Rydenfelt, H. (2021) "Realism without Representationalism", *Synthese* 198(4): 2901–2918.
Schlick, M. (1936) "Meaning and Verification", *Philosophical Review* 45(4): 339–369.
Sellars, W. (1956) "Empiricism and the Philosophy of Mind", in H. Feigl and M. Scriven (eds.), *Minnesota Studies in the Philosophy of Science*, volume I, Minneapolis: University of Minnesota Press, pp. 253–329.

Short, T. L. (2000) "Peirce on the Aim of Inquiry: Another Reading of 'Fixation'", *Transactions of the Charles S. Peirce Society* 36(1): 1–23.
Short, T. L. (2007) *Peirce's Theory of Signs*, Cambridge: Cambridge University Press.
Talisse, R. B. (2007) *A Pragmatist Philosophy of Democracy*, London: Routledge.
Tarski, A. (1935) "Der Wahrheitsbegriff in den formalisierten Sprachen", *Studia Philosophica* 1: 261–405. Originally published in Polish in 1933; trans. by L. Blaustein.
White, M. (1956) *Toward Reunion in Philosophy*, Cambridge, MA: Harvard University Press.
Wittgenstein, L. (1922) *Tractatus Logico-Philosophicus*, C. K. Ogden (trans.), London: Routledge & Kegan Paul. Originally published as "Logisch-Philosophische Abhandlung", in Annalen der Naturphilosophische, XIV (3/4), 1921.
Wittgenstein, L. (1953) *Philosophical Investigations*, G. E. M. Anscombe and R. Rhees (eds.), G. E. M. Anscombe (trans.), Oxford: Blackwell.

21
PRAGMATISM AND CONTINENTAL PHILOSOPHY

Paul Giladi

I think it is safe to contend that the relationship between pragmatism, whose leading thinkers in its various iterations and the constellation of positions brought under the umbrella term 'continental post-Kantian European philosophy', is one of the most complex and polydimensional in the Western canon.

This is, in part, due to how (a) there are clear and substantive genuine philosophic disagreements between pragmatists and continental thinkers such as Michel Foucault and Jacques Derrida, (b) there are clear and substantive philosophic disagreements between pragmatists, (c) there are clear and substantive philosophic disagreements between continental thinkers, (d) there are unfortunate and regrettable philosophic mischaracterizations of both pragmatist and continental thoughts borne out of either innocent ignorance or deliberate misrepresentation, (e) there are very unlikely bedfellows, and (f) there are bedfellows whose at-homeness is strengthened by acts of discursive revivification.

My aspiration in this chapter is to shed some light on at least three major philosophical themes that directly bring pragmatist and continental philosophers into conversation, where doing so can start the arduous-but-enjoyable process of experimenting with the complex and polydimensional ways in which pragmatism and continental philosophy may hang together. As one may reasonably expect, Richard Rorty will occupy a central position in the dialectic here. This may provoke a range of reactive attitudes: from exasperation to joy.

Horkheimer and Gramsci: Pragmatism as Ideology

I would like to start with the way in which two key exponents of critical social theory, Max Horkheimer and Antonio Gramsci, viewed and responded to pragmatism. For Horkheimer and Gramsci, in their respective but not purely commensurable Western Marxist conceptual schemes, the position of pragmatism was – at best – unsophisticated positivism, the philosophy of corporatism, and, to use a helpful expression from Alison Kadlec, "a crude apology for the *status quo*" (Kadlec 2006: 525). At worst, pragmatism was in direct league with fascist ideology.

While the Frankfurt Schoolers' task of rebuking Hegel principally fell to Theodor Adorno,[1] Horkheimer took up the task of bringing pragmatist philosophy before the tribunal of critique. *The Eclipse of Reason* contains a plethora of excoriations of pragmatism (see Horkheimer 2004: 29–42). For the most part, Horkheimer identified pragmatism, represented by C. S. Peirce and John Dewey's respective commitments to experimentalism and valorization of scientific inquiry, with a corrosive, scientistic variety of philosophical naturalism.[2] In this respect, Horkheimer's argument is that pragmatism instantiates the social pathology of industrial homogenization operating through totalizing patterns of formal rationality. Such a type of rationality focuses exclusively on ensuring that natural phenomena conform to laws of fundamental physics. The practices of formal reason not only disenchant nature because their operational drive is the logic of domination but also – in proto-Foucauldian ways – serve as tools of disciplinarity and the coercive regulation of society. However, by the time pragmatists had finished reeling from Horkheimer's vitriolic (and regrettable) characterization of them, additional rage against Peirce and Dewey is levelled by Gramsci:

> Hegel can be considered the theoretical precursor of the democratic revolutions of the nineteenth century, [while] the pragmatists, at the most, have contributed to the creation of the Rotary Club movement and to the justification of conservative and reactionary movements.
> *(Gramsci 1971: 373)*

For Gramsci, pragmatism's staunch commitment to *common sense* (regardless of Peirce's articulation of 'critical commonsensism) props up an oppressive *status quo* by espousing 'utilitarian attitudes that are emblematic of institutional environments built on and sustained by exploitation and instrumentalization. In other words, pragmatism is judged to be conceptually bound up with the logic of capitalism itself, which is inherently geared towards social fragmentation and reified subjectivity. For that matter, the alleged valorization of 'traditional' theory (i.e. formal scientific inquiry exclusively geared towards description and explanation) by Dewey and Peirce means that pragmatism is "unable to identify hegemonic structures at work in the generation of common sense" (Kadlec 2006: 526). To put this another way, because pragmatism is, so the Gramscian argument goes, ideologically wedded to traditional theory, pragmatism thereby fails to identify the urgent need to transform current social reality and articulate a vision of human emancipation.

This is the crucial point I wish to challenge, because there is compelling reason to suppose that the following passage in Dewey's *Reconstruction in Philosophy* illustrates *critical* dimensions to Dewey's social theory:

> the process of growth, of improvement and progress, rather than the static outcome and result, becomes the significant thing. Not health as an end fixed once and for all, but the needed improvement in health – a continual process – is the end and good. The end is no longer a terminus or limit to be reached. *It is the active process of transforming the existent situation.* Not perfection as a final goal, but the ever-enduring process of perfecting, maturing, refining is the aim in living. Honesty, industry, temperance, justice, like health, wealth and learning, are not goods to be possessed as they would be if they expressed fixed ends to be attained. They are directions of change in the quality of experience. Growth itself is the only moral "end".
> *(Dewey 1920: 177, emphasis added)*

What Dewey writes about "the active process of transforming the existent situation" would hardly be out of place in *any* critical social theory – whether feminist, critical race theory, queer theory, Adorno and Horkheimer's neo-Marxism, Gramsci's philosophy of *praxis*, Jürgen Habermas's colonization thesis, or Axel Honneth's worries about systemic patterns of individual and social misrecognition and nonrecognition. This is not least because Dewey's position helps diagnose the normative deficiencies of contemporary social reality: the plurality of contemporary social pathologies and misdevelopments brought about by the iterations of capitalism. For Dewey, ameliorative normative content is weaved into the very structure of progressive social processes and institutions, insofar as the proper function of these processes and institutions is the purpose of fostering growth and nurturing "the critical, inquiring spirit" (Dewey 1920: 16). To put this another way, progressive social processes and institutions are organized in a way that realizes intersubjectively constituted agency.

Crucially, a democratic way of life is the life of inquiry, where inquiry is open, non-dogmatic, inclusive, fallibilist, ceaseless, critical, problem-solving experimentation. To this extent, the democratic life and the inquiring life are *mutually supportive*, insofar as democratic environments promote and sustain inquiry, and inquiry promotes and sustains democracy. This dimension of Deweyan democracy is, crucially, of *critical* use, in that it helps disclose how *current* social institutions fail to be relational institutions, since *current* social institutions fail to promote those relational institutional practices of inquiry, symmetrical recognition, and deliberation necessary for growth (cf. Honneth 2014: 176).

Of course, Dewey is not as critical of modernity as Gramsci and Adorno and Horkheimer respectively are. However, importantly, this does not *eo ipso* disqualify Dewey's pragmatist resources from either involving commitments to progressive social transformation or being legitimately deployed in such a way to articulate some kind of emancipatory narrative. Deweyan democracy's intimate relationship with relational institutions reveals concern about currently deficient modern social reality, namely that antidemocratic trends gradually undermine the realization of, what Horkheimer calls, an 'expressive totality'.

Unlike false totalities, expressive totalities involve a conception of a social whole in which heterogeneous needs and interests of members of society are expressed and also fully developed at no cost to social stability. For Horkheimer, as well as Gramsci, the consequence of a situation in which there is no expressive whole, but only a crystallization into well-ordered homogeneous complexes under the steering mechanism of instrumental practices and unfettered market forces, is that the plight of individuality is almost hopeless. This is because the subjective and objective conditions for exercising freedom and achieving solidarity are eroded by increasing patterns of reification, social homogenization, and cultural hegemonization.

In Deweyan terms, these intersecting social pathologies and misdevelopments stunt growth and stultify self-development. Indeed, there is good reason to think that the vocabulary of reconstruction, transformation, direction, control, education, experimentation, learning, intelligence, and democracy, which is central to Dewey's variety of pragmatism, not only finds a home in critical social theories *tout court* but also is part of their emancipatory grammar. Though the first part of this chapter has drawn attention to a highly problematic reception of pragmatism by two major continental figures, which, in turn, gave pragmatists little reason to happily engage with their continental cousins, I would now like to turn attention to a 'rosier' relationship.

Rorty and Lyotard (and Adorno)

Richard Rorty construes metaphysics as a specific kind of realist *Weltanschauung*:

> [f]or our notion of the world – it will be said – is … a hard, unyielding, rigid *être-en-soi* which stands aloof, sublimely indifferent to the attentions we lavish upon it.
>
> *(Rorty 1982: 13)*

Such a way of portraying metaphysics as a genus of inquiry principally concerned with establishing a 'God's-eye view' is summed up by Rorty in a later work, in which he writes that he uses "'metaphysics' as the name of the belief in something non-human which justifies our deep attachments" (Rorty 2001: 89). By presenting metaphysics as comprising 'non-human' dimensions, where what is 'non-human' appears to refer to something which transcends the locus of *social and cultural practice*, Rorty regards metaphysics as the great nemesis of pragmatism, so much so that "[t]he pragmatist … does not think of himself as *any* kind of metaphysician" (Rorty 1982: xxviii).

According to Rorty, pragmatism is the apotheosis of the secular age that runs through Feuerbach, Marx, Nietzsche, and Dewey, where the vocabulary of foundationalism and essentialism had been decidedly abandoned in favour of an altogether different mode of discourse and vocabulary: the exigencies underpinning agents' socio-political and cultural practices are now the proper sites of meaning and normativity. Crucially, for Rorty, the central aspect of the paradigm shift to "post-Philosophical culture" (Ibid.: xlii) is the gradual withering away of representationalism, whose conceptual scheme is wedded to traditional categorial/ontotheological dualisms, such as essence/accident, appearance/reality, freedom/nature, mind/body, etc. The problem with these binary categories of thought, indispensable to the representationalist logic and practice of metaphysics itself, so Rorty's argument goes, is that they exhibit a pathological sadomasochistic cognitive propensity for regarding normative constraints and the site for the justification of our beliefs as beyond *our* practices.[3]

According to Rorty, in order to get over the effects of this pathological sadomasochistic propensity, the concept of value undergoes radical critical redescription through the development of anti-representationalism: one shifts from making sense of norms as extra-human dictates to making sense of norms as "social achievements" (Brandom 2002: 216). Norms are the pragmatic results of intersubjective doxastic practices between agents, norms get their normative purchase by virtue of being assented to and acknowledged by a community of agents, and fallibilistic normative content itself is the product of communal warranted assertibility.

The ubiquity of human practice in constituting the norms of inquiry signifies, for Rorty, that we have not only broken free from the mythopoetic conception of human mindedness as 'the mirror of nature' but that we have also – as Nietzsche famously puts it – emerged from the "shadows of God" (Nietzsche 2001: 109). In writing that "[t]he pragmatists' anti-representationalist account of belief is, among other things, a protest against the idea that human beings must humble themselves before something non-human, whether the Will of God or the Intrinsic Nature of Reality" (Rorty 2006: 257), Rorty sees the function of pragmatism as the dismantling of the two pillars of representationalism: a *metaphysical* conception of truth and a *metaphysical* conception of objectivity. Pragmatism, in other words, for Rorty, signifies the cultural "substitution of fraternity for authority" (Ibid.: 262).

On this substantive point, Jean-François Lyotard's articulation of the postmodern condition as deeply "incredulous towards metanarratives" (Lyotard 1984: xxiv) comes into the story. By 'metanarratives', Lyotard refers to concepts deemed to have totalizing and universalizing logics, such as law, inquiry, truth, science, meaning, reason, capital, God, the state, etc. For Lyotard, unlike Rorty, however, the issue with metanarratives is not so much how some of them have particular fondness for submission and non-anthrocentred notions of authority. The issue, rather, is the prefigurative role all metanarratives – *whether they are humanistically constituted* – play in the development of left-wing and right-wing totalitarianism and deformations of subjectivities. In other words, while Rorty laments sadomasochistic vassalage to *être-en-soi*, Lyotard is concerned with structural violence committed by metaphysics.

Metanarratives play a role in subjection, because the driving force of metanarratives *qua* metanarratives is coercive legitimation, domination, and systematic ordering of its subject matter *via* 'grand narrative'. Understood in this way, metanarratives – which are central to all modes of material and cultural production in modernity – are unmasked by the Lyotardian postmodernist as ideological, insofar as metanarratives are often held to be emancipatory, but they in fact function to *nullify* emancipation.

Lyotard's position on the normative problems of modern metanarratives, therefore, ups the critical ante, so to speak, and hits harder than Rorty's contention that "the vocabulary of Enlightenment rationalism, although it was essential to the beginnings of liberal democracy, has become an impediment to the preservation and progress of democratic societies" (Rorty 1989: 44). As Fredric Jameson writes, "Lyotard [is an] explicitly political figure with an overt commitment to the values of an older revolutionary tradition" (Jameson 1991: 61). Indeed, the postmodern way of making sense of things aims to counter the force of metanarratives, in order to institute "the crisis of metaphysical philosophy … [and] wage war on totality" (Lyotard 1984: xxiv; 82).

Indeed, Lyotard's own notion of the *différend*, which involves agonistic social relations and processes, prioritizes heterogeneity, alterity, and difference over commonality, commensurability, and identity. For, "consensus obtained through discussion … does violence to the heterogeneity of language games" (Ibid.: xxv). As Fred Dallmayr puts it,

> Lyotard debunks [the primacy of commonality, commensurability, and identity] as logically untenable (and politically obnoxious). Seen from the vantage of Nietzsche and the theory of language games, he insists, resort to a metadiscourse or metaphysics is no longer feasible.
> *(Dallmayr 1997: 40)*

Under the postmodern condition, then, the space of reasons is an arena invariably comprising rupture, opposition, contestation, challenge, disturbance, and discontinuity rather than a public sphere exemplified by practices of communication, uptake, and solidarity (viz. Lyotard 1984: 81). Given what I have written about Lyotard, I think it is worth bringing Adorno into the discourse here: contestation, challenge, disruption, disturbance, and discontinuity, for Adorno, are the trace-effects of the ineliminable presence of 'non-identity' (i.e. difference).

The logical structure of modern social organization is typified by drives towards the domination (and even obliteration) of difference. As Adorno notes, "we are dealing with the principle of mastery" (LND: 9). These steering mechanisms geared towards universal reification produce a false totality and – at the material-psychological level – result in a damaged subjectivity, damaged life. The function of Adorno's negative dialectics and "logic of disintegration" (Ibid., p. 6), therefore, is not to offer *resistance* to these totalizing dispositions and ideological forms of modern social organization.

Rather, the function of negative dialectics is radical, construed as a *reversal* of logico-metaphysical power, according to which the category of difference (i.e. non-identity) has *priority* over totalizing categorial frameworks such as those exemplified by their affiliation with Kant's account of determinate judgement. To quote Adorno, "chang[ing] this direction of conceptuality, to give it a turn toward non-identity, is the hinge of negative dialectics" (ND: 12).

To achieve success in philosophy, under Lyotard's *différend* and Adorno's logic of disintegration, would be to 'know one's way around' with respect to internal tension rather than with respect to welding into one unified, coherent image. Our discursive forms of life require multiple images, multiple pictures, which are in conflict with one another, because conflict, rather than a transcending *Aufhebung*, is emblematic of cognitive life itself.

Rorty's writings indicate more sympathy for Lyotard than for Adorno. This is, in part, because Adorno's negative dialectics is unashamedly metaphysical with remnants of mythopoetic yearning for transcendence (see Adorno 1981, 2008), whereas Lyotard's contempt for *all* varieties of metaphysics chimes with Rorty's idiosyncratic conceptualization of pragmatism as 'liberation from the Primal Father'. However, at the same time, Rorty's fondness for Lyotard is clearly tempered by what he sees as a failing endemic to *all* critical social theories:

> Lyotard unfortunately retains one of the left's silliest ideas – that escaping from institutions is automatically a good thing, because it insures that one will not be "used" by the evil forces which have "co-opted" these institutions. Leftism of this sort necessarily devalues consensus and communication, for insofar as the intellectual remains able to talk to people outside the *avant-garde* he "compromises" himself.
> *(Rorty 1991: 175)*

For Rorty, a constitutive feature of radical social critique is the contention that existing social institutions, the products of modernity, are irredeemably pathological. To put this another way, critical social theories are individuated by their commitment that modern social democracies are too far gone and that incremental piecemeal reform as a mechanism for amelioration is normatively impotent.

For example, the welfare state that is an essential institution of social democracy *principally* structures the provision of welfare under the framework of reifying capitalist practices: since the structure of social democracy is constituted by the systems of money (market capitalism) and power (the state), the provision of welfare will invariably fail to fulfil the function of mitigating conflict. Under the social-welfare state, there is little or no way to avoid ideological encroachment and 'colonization by systems', since what is the *base* of the societal *superstructure* is the capitalist mode and relations of production. If the base is constituted by systems, then the entire whole is vulnerable to encroachment by systems. Securing and protecting the lifeworld, therefore, is effectively impossible under the welfare state.

This kind of discourse, which is common in Western Marxist critical theories of society is antidemocratic, for Rorty, since (i) he deems ideology-disclosure as displaying elitist contempt for 'co-opted' and 'pathological' everyday consciousness and ordinary practices, and (ii) he views the idea of liberal reformism as normatively impotent as revealing disdain for consensus and reconciliation. Indeed, the kind of animosity towards ordinary vocabulary and quotidian discursive formations, which is best exemplified by Herbert Marcuse's Hegelian-Marxism, deeply troubles Rorty.

This is because Rorty is part of that cultural tradition, comprising figures such as Ralph Waldo Emerson, Anton Chekov, Henry David Thoreau, and Stanley Cavell, which construes the banal as beautiful, and the quotidian and habitual as marvellous and redemptive. In this respect, critical social theories are judged by Rorty to not only loathe liberal values but also to ironically reproduce a representationalist and essentialist metaphysics: "social institutions can be viewed as experiments in cooperation rather than as attempts to embody a universal and ahistorical order" (Rorty 2006: 270).

Rorty's vision of a post-Philosophical culture is explicitly anti-authoritarian. However, as one can tell, its anti-authoritarianism is of a romantic liberal ironist variety and deeply hostile to ideology-disclosure and the romantic, radical critique of modernity. This is despite Rorty's own Nietzsche-inspired articulation of pragmatism as secular humanist 'liberation' from Platonizing drives towards purification.

For Rorty, *contra* the respective critical social theories of Rousseau, Hegel, Nietzsche, Marx, Gramsci, Lukács, Adorno and Horkheimer, Foucault, Pierre Bourdieu, and Lyotard, the normative deficiency of social reality is not any type of structural violence and systemic oppression. As Rorty writes, responding to left-wing critiques of liberalism, "I just can't think of anything I learned from post-Mill writings that added much" (Rorty 1998: 64). Rather, according to Rorty, the problem with contemporary political economy and contemporary political communities is the amount of cruelty on show.

The central moral-political aspiration of Rorty's liberal ironism is that "suffering will be diminished, that the humiliation of human beings by other human beings may cease" (Rorty 1989: xv). In this respect, then, Rorty's liberal, who disparages the revisionary impulse for convulsive radical overhaul of existing social institutions and forms of production in favour of empathetic coping with existing social institutions, contends that "the only way to avoid perpetuating cruelty within social institutions is by maximising the quality of education, freedom of the press, educational opportunity, opportunities to exert political influence, and the like" (Ibid., p. 67). This is where Habermas's notion of 'postmetaphysical thinking' joins the fray.

Rorty and Habermas

Habermas places significant philosophical as well as socio-political emphasis on the intrinsically social character of language: meaning, normativity, and knowledge are mediated by practices that are rooted in communicative action. For Habermas, communicative action is the type of action aimed at establishing consensus (i.e. mutual understanding) through the agonistic establishment of legitimate and valid norms for persons (i.e. language-using individuals). As Habermas frames it:

> The concept of communicative action presupposes language as the medium for a kind of reaching understanding, in the course of which the participants, through relating to a world, reciprocally raise validity claims that can be accepted or contested.
> *(Habermas 1984: 99)*

Crucially, for Habermas, successful communication between agents involves the hearer being able to transparently grasp the reasons motivating the propositions put forward by the speaker:

> *We understand a speech act when we know what makes it acceptable.* From the standpoint of the speaker, the conditions of acceptability are identical to the conditions for his illocutionary success. Acceptability is not defined here in an objectivistic sense, from the perspective of an observer, but in *the performative attitude of a participant in communication*.
> *(Habermas 1984: 297–298; emphasis added)*

On the corresponding socio-political front, Habermas contends that all social processes are assessed with respect to how well (or invariably not) they foster communicability and the development of 'discourse', namely *non-coercive* arenas for the agonistic, public use of reason. As he writes, "[an] ego-identity can only stabilise itself in the anticipation of symmetrical relations of unforced reciprocal recognition" (Habermas 1992: 188). Democracy and communication are necessarily tied together and *mutually supporting*: the failure to develop communicative action is a barrier to democracy as promised in the public sphere, and the failure to develop democratic values is a barrier to communicative action. This shares much in common with Rorty's own contention that "[a] liberal society is one which is content to call 'true' (or 'right' or 'just') whatever the outcome of undistorted communication happens to be, whatever view wins in a free and open encounter" (Rorty 1989: 67).

Communication and discourse, for Habermas, are products of postmetaphysical thinking, typified by Habermas's fallibilist methodology of rational reconstruction and move to the *pragmatics of communication*, rather than resource to a representationalist mirror of nature: "[The former] concedes primacy to world-disclosing language – as the medium for the possibility of reaching understanding, for social cooperation, and for self-controlled learning processes – over world-generating subjectivity" (Habermas 1992: 153).

For that matter, G. H. Mead's pragmatist social psychology is a significant influence on Habermas's 'Intersubjectivist Turn': the practice of navigating one's way in a team/group by understanding the various roles and behavioural habits/associations of 'the generalized other' enables a person to develop self-consciousness, which involves the internalization of socializing practices (viz. Mead 2015: 154). Mead's well-known claim that individuation occurs through socialization and his focus on both gestural and linguistic forms of interaction become central to Habermas's own theory. As Habermas writes:

> I see the more far-reaching contribution of Mead in his having taken up themes that ... individuation is pictured not as the self-realization of an independently acting subject carried

out in isolation and freedom but as a linguistically mediated process of socialization ... Individuality forms itself in relations of intersubjective acknowledgement and of intersubjectively mediated self-understanding ... Mead will shift all fundamental philosophical concepts from the basis of consciousness to that of language.

(Habermas 1992: 152–153, 162)

In other words, Habermas credits Mead with the postmetaphysical insight that language-use involves norms requiring discursive exchange, a variety of an I-thou relation, rather than the I-they and/or I-it relation, and that the most basic linguistic unit is "the relationship between ego's speech-act and alter's taking a position" (Habermas 1992: 163).

On this subject, Habermas identifies and lays out what he sees as Hegel's *better* conception of intersubjectivity than Fichte's. From Habermas's perspective, given that Hegel – much like Fichte – articulates the communicative normative content of modern ethical life in *metaphysical* ways, neglecting the *pragmatic* dimensions of language-use and communication, Hegel, at best, multiplies beyond necessity his development of a proto-form of communicative rationality and action. Habermas construes his own postmetaphysical model as Hegelian without any "metaphysical mortgages" (Habermas 1987: 316). From Habermas's postmetaphysical perspective, given that Hegel articulates the communicative normative content of modern ethical life in *metaphysical* ways, Hegel – at best – multiplies beyond necessity in his development of a proto-form of communicative rationality and action.

For Rorty and Habermas, the principal problem with metaphysical thinking is that it is "a permanent neutral matrix for inquiry" (Rorty 1982: 80), and – especially in Hegel's case – its elaborate, speculative, conceptual toolkit neglects the pragmatic dimensions of everyday language-use, which, in turn, suppresses "the desire for as much intersubjective agreement as possible, the desire to extend the reference of 'us' as far as we can" (Rorty 1991: 23). Such an orientation also reveals how Habermas and Rorty share a Shklarian-liberal concern for vulnerability, specifically a concern for the kind of vulnerability unique to agents *qua* agents.[4]

Indeed, Habermas connects vulnerability to a mode of communicative *praxis*:

[a]rgumentation insures that all concerned in principle take part, freely and equally, in a cooperative search for truth, where nothing coerces anyone except the force of the better argument ... Moral intuitions are intuitions that instruct us on how best to behave in situations where is it in our power to counteract the extreme vulnerability of others by being thoughtful and considerate. In anthropological terms, morality is a safety device compensating for a vulnerability built into the sociocultural form of life.

(Habermas 1990: 198, 199)

For example, in cases of perversions of communicative dynamics, such as testimonial injustice, these alienate a speaker from both their own communicative rationality and from the speech-based practices which necessarily constitute discourse between peers. This experience "carries with it the danger of an injury that can bring the identity of the person as a whole to the point of collapse" (Honneth 1995: 132–133), where the identity under threat here is a person's self-interpretation as agential, since speech involves vulnerability with respect to revealing oneself in the transparent, trusting communicative act of sharing propositional content for uptake by the listener. To use Andrea Lobb's expression, the kind of 'epistemic injury' (Lobb 2018: 1) endured here can made sense of in relation to what Rorty calls 'mute despair' and 'intense mental pain'. For Rorty, this notion of agential pain – the type of pain unique to agents

– reminds us that human beings who have been socialised ... can all be given a special kind of pain: they can all be humiliated by the forcible tearing down of the particular structures of language and belief in which they were socialised (or which they pride themselves on having formed for themselves).

(Rorty 1989: 177)

The failure to properly recognize and accord somebody the epistemic acknowledgement they merit is an act of abuse in the sense of forcibly depriving individuals of a progressive communicative environment in which the epistemic recognition accorded to them plays a significant role in enabling and fostering their self-confidence as a communicative agent.

From a critical theoretic perspective, Rorty's liberal ironist is more Last Man than Übermensch, because the political *praxis* of the liberal ironist is "dull, banal, bourgeois – values Rorty tried to rehabilitate" (Allen 2015: 174). Liberal ironism's commitment to incremental, piecemeal reforms – as opposed to system-change – serves to preserve the *status quo* rather than make way for the principal engine of progressive social change: subaltern discursive formations, which actively resist assimilation or integration into the ever-expanding and 'legitimating' 'we' of the romantic, utopian liberal. As Foucault and Gilles Deleuze & Felix Guattari respectively put it,

> I do not appeal to any "we" - to any of those "we"'s whose consensus, whose values, whose traditions constitute the framework for a thought and define the conditions in which it can be validated. But the problem is, precisely, to decide if it is actually suitable to place oneself within a "we" in order to assert the principles one recognizes and the values one accepts; or if it is not, rather, necessary to make the future formation of a "we" possible, by elaborating the question.
>
> *(Foucault 1984: 385)*

> Western democratic conversation between friends ... [only produces] pleasant or aggressive dinner conversations at Mr. Rorty's.
>
> *(Deleuze & Guattari 1994: 6, 144)*

Prima facie, though Deleuze's aggression towards Rorty suggests little commonality between the two thinkers, there is every reason to think Deleuze and Rorty still share much substance in common – not least because of their respective anti-representationalism and staunch preference for historicism and contingency.[5] In what follows, I would like to focus on one such commonality, by sketching a metaphilosophical affinity between Rorty's Harold Bloom-inspired 'postmodern' liberal ironist hero – the "strong poet" – and Deleuze and Guattari's revisionary metaphysician, *le personnage conceptuel*.

Rorty and Deleuze

Rorty and Deleuze rarely engage with one another's work, which is frustrating and strange considering Rorty's Francophilia and intellectual attraction to the bevvy of 'postmodernists', of whom only Deleuze openly and warmly engaged with pragmatism (particularly Dewey's process philosophy). Rorty's most substantive discussion of Deleuze is his 1983 *Times Literary Supplement* review of *Nietzsche and Philosophy*, in which "Deleuze is confidently dismissed" (Allen 2015: 166). However, one prophetic positive remark by Rorty about Deleuze is the staging ground for what I argue in this final section:

> On my view, James and Dewey were not only waiting at the end of the dialectical road which analytic philosophy travelled, but are waiting at the end of the road which, for example, Foucault and Deleuze are currently traveling.
>
> *(Rorty 1982: vxiii)*

'The dialectical road' in question is the pathway from abandoning representationalist discourse and justification *qua* grounding to playfulness with categorial redescription. What philosophically matters, for Rorty, is not the ability to successful unravel and decipher the structure of reality but the development of novel, interesting, and important concepts. In this way, Rorty's utopia is not centred on the practices and achievements of the earnest Peircean, nor on those of the Justin Trudeau-esque politician, nor on those of the unbowed and proud Deweyan secularist. Rather, "an ideally liberal polity would be one whose culture hero is Bloom's 'strong poet' rather than the warrior, the priest, the sage, or the truth-seeking, 'logical,' 'objective' scientist" (Rorty 1989: 53).

If one brackets away the liberal utopia, a political ideal that Deleuze and Guattari would abrasively abjure, what Rorty envisions here is strikingly similar to the following passages from *What Is Philosophy?*:

> To think is to experiment, but experimentation is always [concerned with] that which is in the process of coming about – the new, remarkable and interesting – which replace the appearance of truth and which are more demanding.
>
> *(Deleuze & Guattari 1994: 111)*

> Philosophy does not consist in knowing and is not inspired by truth. Rather, it is categories like Interesting, Remarkable, or Important that determine success or failure ... Thought as such produces something interesting when it accedes to the infinite movement that frees it from truth as supposed paradigm and reconquers an immanent power of creation.
>
> *(Ibid., pp. 82, 140)*

The kind of revisionary project Rorty and Deleuze & Guattari respectively produce is a type of creative "subversion" in Deleuze's technical sense (Deleuze 1983: 53). From this perspective, the sense of achievement in categorial redescription and perennial discursive experimentation is not the pride that one feels having solved a philosophical problem or made a scientific discovery. It is in saying "Thus I willed it" (Rorty 1989: 37):

> Ironists specialise in redescribing ranges of objects or events in partially neologistic jargon, in the hope of inciting people to adopt and extend that jargon. An ironist hopes that by the time she has finished using old words in new senses, not to mention introducing brand-new words, people will no longer ask questions phrased in the old words.
>
> *(Ibid., p. 78)*

As Deleuze & Guattari express a similar point:

> some concepts must be indicated by an extraordinary and sometimes even barbarous or shocking word, whereas others make do with an ordinary, everyday word that is filled with harmonics so distant that it risks being imperceptible to a nonphilosophical ear. Some concepts call for archaisms, and others for neologisms, shot through with almost crazy etymological exercises.
>
> *(Deleuze & Guattari 1994: 7–8)*

Rather than provide a topography of our ordinary conceptual framework and offer tweaks, Rorty and Deleuze & Guattari respectively redesign the way in which we make sense of things through the development of new discursive architectures. This Nietzsche-inspired expressive practice of creating new and interesting concepts means that, for *les personnages conceptuels*, "metaphoric redescriptions are the mark of genius and of revolutionary leaps forward" (Rorty 1989: 28).

For these thinkers, as Barry Allen rightly puts it, "[p]hilosophy is, then, not a matter of finding the truth, not even the true questions. It is creating concepts. Problems are productive not for the truth of their answers but the creativity of their concepts" (Allen 2015: 165). Interestingly, it is ironic that Rorty's continental valorisation of massive redescription may be said to appear in tension with Rorty's analytic pioneering of the descriptive metaphysical 'linguistic turn'. This is because P. M. S Hacker's descriptive metaphysical 'linguistic turn' commitment to "concepts and categories that we could not abandon without ceasing to be human" (Hacker 2001: 368) runs against the traffic of the revisionary projects of Rorty and Deleuze and Guattari[6]: there are concepts, categories, and frameworks we must abandon to not simply retain, but more saliently, even *perfect* our humanity.

Notes

1 See Finlayson (2017) and Giladi (2020).
2 According to Hans Joas,

> [Horkheimer] relies by and large, however, on the pertinent book by Max Scheler [*Erkenntnis und Arbeit: Eine Studie über Wert und Grenzen des pragmatischen Motivs in der Erkenntnis der Welt*], in which pragmatism appears as a philosophy that reduces human life to labour and is therefore not adequate for a portrayal of what is authentically spiritual or personal. In these works Horkheimer therefore continues the tradition of decades of arrogant and superficial German snubbing of the most ingenious stream of American thought. Scheler's interpretation suits Horkheimer's attempt to treat pragmatism throughout as the inconsistent brother of logical positivism.
>
> (Joas 1993: 81)

3 See Rorty (2002: 391). Cf. "we can hear the mutterings of the desire for a return of terror" (Lyotard 1984: 82).
4 For further on the Rorty-Habermas relation, see Bacon & Rutherford (2021).
5 See Allen (2015) and Patton (2015) for outstanding discussions of how Deleuze and Rorty hang together.
6 Deleuze's notion of problematic ideas involves the redescriptions of existing social kinds (particularly sexuality) that, as Paul Patton writes, "disturb habitual ways of thinking and talking" (Patton 2015: 157).

References

Adorno, T. W. (1981) *Negative Dialectics*, E. B. Ashton (trans.) and (ed.), London: Continuum.
———. (2008) *Lectures on Negative Dialectics: Fragments of a Lecture Course 1965/1966*, R. Tiedemann (ed.), R. Livingstone (trans.), Cambridge: Polity Press.
Allen, B. (2015) "The Rorty-Deleuze *Pas de Deux*," in S. Bowden, S. Bignalli and P. Patton (eds.), *Deleuze and Pragmatism*, New York: Routledge, pp. 163–177.
Bacon, M. & Rutherford, N. 2021. 'Rorty, Habermas, and Radical Social Criticism', in G. Marchetti (ed.), *The Ethics, Epistemology, and Politics of Richard Rorty*. New York: Routledge, pp. 191-208.
Brandom, R. B. (2002) *Tales of the Mighty Dead: Historical Essays in the Metaphysics of Intentionality*, Cambridge, MA: Harvard University Press.
Dallmayr, F. (1997) "The Politics of Nonidentity: Adorno, Postmodernism – And Edward Said," *Political Theory* 25: 33–56.
Deleuze, G. (1983) "Plato and the Simulacrum," R. Krauss (trans.), *October* 27: 45–56.
Deleuze, G. & Guattari, F. (1994) *What Is Philosophy?* H. Tomlinson and G. Burchill (trans.), London: Verso.
Dewey, J. (1920/1976) "Reconstruction in Philosophy", in J. A. Boydston (ed.), *The Middle Works*: 1899–1924, Volume 12, Carbondale, IL: Southern Illinois University Press, pp.77–201.
Gramsci, A. (1971) *Selections from the Prison Notebooks*, Q. Hoare and G. Nowell (trans.), London: Lawrence and Wishart.
Finlayson, J. G. (2017) "Hegel and the Frankfurt School," in D. Moyar (ed.), *The Oxford Handbook of Hegel*, Oxford: Oxford University Press, pp. 718–740.
Foucault, M. (1984) *The Foucault Reader*, P. Rabinow (ed.), New York: Pantheon.
Giladi, P. (2020) "The Dragon Seed Project: Dismantling the Master's House with the Master's Tools?" in P. Giladi (ed.), *Hegel and the Frankfurt School*, New York: Routledge, pp. 195–240.

Habermas, J. (1984) *The Theory of Communicative Action: Volume One. Reason and the Rationalisation of Society*, T. McCarthy (trans.), Boston, MA: Beacon Press.
———. (1987) *The Philosophical Discourse of Modernity: Twelve Lectures*, F. Lawrence (trans.), Cambridge: Polity Press.
———. (1990) *Moral Consciousness and Communicative Action*, C. Lenhardt and S. Weber Nicholsen (trans.), Cambridge, MA: MIT Press.
———. (1992) *Postmetaphysical Thinking*, W. M. Hohengarten (trans.), Cambridge, MA: MIT Press.
Hacker, P. M. S. (2001) *Wittgenstein: Connections and Controversies*, Oxford: Oxford University Press.
Honneth, A. (1995) *The Struggle for Recognition: The Moral Grammar of Social Conflicts*, J. Anderson (trans.), Cambridge, MA: MIT Press.
———. (2014) *Freedom's Right: The Social Foundations of Democratic Life*, J. Ganahl (trans.), Cambridge: Polity Press.
Horkheimer, M. (2004) *Eclipse of Reason*, London: Continuum Press.
Jameson, F. (1991) *Postmodernism, or, The Cultural Logic of Late Capitalism*, London and New York: Verso.
Joas, H. (1993) *Pragmatism and Social Theory*, Chicago, IL: University of Chicago Press.
Kadlec, A. (2006) "Reconstructing Dewey: The Philosophy of Critical Pragmatism," *Polity* 38: 519–542.
Lobb, A. (2018) "'Prediscursive Epistemic Injury': Recognising Another Form of Epistemic Injustice?," *Feminist Philosophy Quarterly* 4 (Article 3): 1–23.
Lyotard, J.-F. (1984) *The Postmodern Condition: A Report on Knowledge*, G. Bennington and B. Massumi (trans.), Manchester: Manchester University Press.
Mead, G. H. (2015) *Mind, Self, and Society: The Definitive Edition*, C. W. Morris, D. R. Huebner and H. Joas (eds.), Chicago, IL: University of Chicago Press.
Nietzsche, F. (2001) *The Gay Science*, J. Nauckhoff (trans.), New York: Cambridge University Press.
Patton, P. (2015) "Redescriptive Philosophy: Deleuze and Rorty," in S. Bowden, S. Bignalli and P. Patton (eds.), *Deleuze and Pragmatism*, New York: Routledge, pp. 145–160.
Rorty, R. (1982) *Consequences of Pragmatism*, Minneapolis: University of Minnesota Press.
———. (1989) *Contingency, Irony, and Solidarity*, New York: Cambridge University Press.
———. (1991) *Objectivity, Relativism, and Truth: Philosophical Papers*, 4 vols, Volume 1, Cambridge: Cambridge University Press.
———. (1998). *Against Bosses, Against Oligarchies: A Conversation with Richard Rorty*, D. Nystrom and K. Puckett (eds.), Charlottesville, VA: Prickly Pear Pamphlets.
———. (2001) "Justice as a Larger Loyalty," in M. Festenstein and S. Thompson (eds.), *Richard Rorty: Critical Dialogues*, Malden, MA: Polity, pp. 223–235.
———. (2002) "Words or Worlds Apart? The Consequences of Philosophy for Literary Studies," *Philosophy and Literature* 26: 369–396.
———. (2006) "Pragmatism as Anti-authoritarianism," in J. R. Shook and J. Margolis (eds.), *A Companion to Pragmatism*. Oxford: Blackwell Publishing.

22
PROSPECTS FOR "BIG-TENT" PRAGMATIC PHENOMENOLOGY

J. Aaron Simmons

Introduction

In this essay, I argue that phenomenology and pragmatism should be read as important resources for each other. After arguing that technical definitions of each discourse are not only hard to come by but also likely to prevent serious engagement (due to the overriding disagreements in method and core commitments), I contend for a "big-tent" approach such that the two discourses help each other to enact humility in light of lived realities. Then, I conclude with suggestions of philosophical areas where a pragmatic phenomenology is likely to be especially fruitful.

When I sat down to write the introduction to my co-authored book, *The New Phenomenology: A Philosophical Introduction* (Simmons and Benson 2013), I remember struggling with how to frame the group of thinkers I was considering. I wanted to present new phenomenology as a distinctive philosophical trajectory despite the fact that its representative thinkers all disagreed substantively with each other on a variety of topics. Aware of the risk of overstating their coherence, I looked for an example of some other group of thinkers who, despite profound disagreements, were rightly viewed as all generally headed in a broadly unified philosophical direction and, thus, warranted being labeled as doing something in common. The example to which I turned was *pragmatism*.

What compelled me in that example was that, as Robert Talisse and Scott Aikin note, pragmatism "is a *living philosophy* rather than a historical relic" (Talisse and Aikin 2008: 3). And, as they go on to explain,

> the lack of a precise definition of pragmatism does not derive from some excess of imprecision on the part of pragmatist philosophers; rather, it is a product of the fact that, relatively speaking, pragmatism is a new phenomenon on the philosophical scene. It has been just over 100 years since the term was first used in print in a philosophical context.
>
> *(Talisse and Aikin 2008: 3)*

What struck me then, and continues to strike me now, about their characterization was that you could replace "pragmatism" with "phenomenology" and it was all just as true. Indeed, William James first introduced the term "pragmatism" in 1898 (see Malachowski 2013: 1), just two years before Edmund Husserl offered the first expansive defense of "phenomenology" as a distinct option for philosophical inquiry in *Logical Investigations* (two volumes published in 1900 and 1901). Moreover, both pragmatism and phenomenology quickly transitioned from the strict views of the founder(s) to a wide range of variants. If phenomenology is a tradition that consists of *Husserlian heresies* (see Simmons and Benson 2013: 43–72), then we could say that pragmatism is a tradition that consists of *Peircean heretics* (see Hookway 2013). Consider some other remarkable similarities: Although both pragmatism and phenomenology are often described as "methodologies," neither is reducible to singular methods. Also, rather than being merely a rigid set of doctrines, we could say of phenomenology what Talisse and Aikin say of pragmatism—namely, that both emerge "as a collection of more or less loosely connected philosophical themes, arguments, and commitments" (Talisse and Aikin 2008: 3–4). Moreover, despite their relatively short histories, there are "classical" and "new" alternatives for both movements (see Malachowski 2010).

The analogy that I drew between phenomenology and pragmatism in *The New Phenomenology* was offered as simply a way of trying to stabilize the inherent dynamism of the philosophical view that I was considering. Therein I suggested that new phenomenology should be understood more as a group of thinkers who all formed a kind of philosophical *family* than as a strict philosophical school (Simmons and Benson 2013: 4). This metaphor, I believe, applies equally well to pragmatism. Both C.S. Peirce and also Edmund Husserl established lines of philosophical inheritance that then were received, and deeply transformed, by those who would be formed by their influence. Perhaps it was a benefit to the history of philosophy that neither Husserl nor Peirce left a clear "executor" to their philosophical wills, as it were. Instead, their own lives and works are marked by repeated rethinkings, rephrasings, revisions, and rearticulations of the views that they had originally presented. So, I believe, the best way to inherit their legacy is not to allow that legacy to become rigid but to keep it flexible enough to allow more "children" to be birthed by it. Even admitting that phenomenology and pragmatism are decidedly not the same thing, there remain good reasons to think that they need to draw upon each other more often. That is, even if they might be families that live in the same neighborhood, they do not seem to visit each other's houses with enough frequency.

In this chapter, then, I will begin by offering a literature review in order to achieve some definitional ground clearing in order to set the stage for how we are best able to set a table at which pragmatists and phenomenologists would be inclined to pull up a chair. I will suggest that due to the contested identities of each discourse, it is easy to find both sites of critical resonance and also seemingly central points of opposition between them (which many other scholars have considered in detail).[1] However, if we start with such technical definitions, we are unlikely to get much constructive engagement that is not fraught with problems. Accordingly, I will offer a "big-tent" approach to understanding each discourse as offering an important corrective to the potential excesses of the other. I will then conclude by suggesting a few areas where I think phenomenology and pragmatism ought to lean on each other in order to be better at what they each hope to do: be faithful to human experience while making a difference in people's lives. Such a big-tent approach makes the idea of pragmatic phenomenology, or phenomenological pragmatism, much more promising.[2]

Contested Concepts and Why Definitions Matter

It might seem that defining pragmatism and phenomenology would be a straightforward manner. Due to the histories of contested identity that defines them both, however, easy definitions are hard to come by. As Donald Cerbone notes, "introducing phenomenology, is no easy matter, in part because there are so many ways to begin and no one way is ideal" (Cerbone 2006: 1). Similarly, Dermot Moran rightly admits that

> It is important not to exaggerate … the extent to which phenomenology coheres into an agreed method, or accepts one theoretical outlook, or one set of philosophical theses about consciousness, knowledge, and the world.

(Moran 2000: 3)

Shaun Gallagher echoes this point and suggests that the difficulties of definition extend all the way to the very beginning of phenomenology itself: "what phenomenology is, is itself open to question. Indeed, it might be better to talk of *phenomenologies* (in the plural)" (Gallagher 2012: 10). Moreover, when Dan Zahavi asks "is there really something like a phenomenological tradition, let alone a phenomenological method?" he quickly responds: "Opinions are divided" (Zahavi 2012b: 1).

The same definitional messiness also applies to pragmatism. As Alan Malachowski notes, "Even the trio of founding figures, Peirce, James and Dewey, ... did not generally conceive or speak of pragmatism in unison. Indeed, both Peirce and Dewey were wary of the very name 'pragmatism'" (Malachowski 2013: 2–3). This diversity of perspective leads Bruce Wilshire to conclude that "We should not think that because the single name 'pragmatist' has been applied to James and the others that there is one body of thought to which all subscribe" (Wilshire 1977: 51). Patrick Baert nicely summarizes the situation when he claims that "pragmatism was, and still is, a heterogeneous entity. From the beginning, pragmatism entailed competing branches and antithetical positions" (Baert 2011: 26).

In full acknowledgment of the complicated and plural histories of both movements, it is important to realize that an engagement between phenomenology and pragmatism depends heavily on which definition we use to get things started. For example, if we follow Merleau-Ponty and define phenomenology as "from the start a rejection of science" (Merleau-Ponty 1967: 358), then we are bound to run into problems trying to bring this perspective together with that of C.S. Peirce's account of pragmatism in "The Fixation of Belief," which Christopher Hookway summarizes as having "defended the 'method of science' as the only defensible method for the conduct of inquiry" (Hookway 2013: 18). Due to such seemingly striking oppositions between phenomenology and pragmatism, Scott Aikin concludes that there is potentially no compatibility to be found between them: "If the reasons behind the phenomenological tradition's anti-naturalism are right, then pragmatism is an incoherent philosophical program" (Aikin 2006: 317; see also Hills 2013). It is due to oppositional tendencies, such as those noted by Aikin, that Sandra Rosenthal, an early proponent of bringing phenomenology and pragmatism together, readily acknowledges that "The philosophic methodology of phenomenology is not the philosophic methodology of pragmatism, and of this methodological assimilation one must beware" (Rosenthal 1974a: 77).

In light of Aikin's challenge and Rosenthal's judgment, we should be honest about the fact that there might be dim prospects indeed for pragmatic phenomenology. One could even conclude, as James Edie does, that when it comes to arguments for the overlaps between James and Husserl, for example, "that one instinctively feels that some of the evidence, at least, has been rigged" (Edie 1965: 14; as qtd. in Rosenthal and Bourgeois 1980: 4).

Despite such legitimate worries, however, there remain many who think that engagement remains not only possible but promising. For example, Wilshire contends that phenomenology and pragmatism "should be seen as mutually assisting philosophical efforts" (Wilshire 1977: 54; see also Wilshire 2000: 180). Alternatively, Shaun Gallagher claims of his own work that he takes "a pragmatic approach to phenomenology" (Gallagher 2012: 2). Additionally, Victor Kestenbaum refers to the "phenomenological sense of Dewey's philosophy" (Kestenbaum 1977: Preface), and Jason Hills claims that "classical pragmatism already has phenomenological methods" (Hills 2013: 318). Drawing on James and Bergson, in particular, Megan Craig defends the "possibility of a newly pragmatic phenomenology" (Craig 2016: 272). Finally, and perhaps most interestingly, even despite her warning about the potential incompatibility between pragmatic and phenomenological methodologies, Rosenthal also suggests that "the two philosophies manifest striking similarities in their basic stances toward certain key issues" (Bourgeois and Rosenthal 1983: 1).

When confronted with such varied accounts of the potential relationships of these two perspectives, it is not immediately clear how to proceed. Often the oppositional narratives begin with overly narrow conceptions, but the compatibilist approaches struggle to acknowledge fully enough the potential incongruities. Traditional definitions of both phenomenology and pragmatism actually prove quite unhelpful for navigating through this morass. Consider, for example, the following recent attempts at definitions of phenomenology:

- Phenomenology is the study of human experience and of the ways things present themselves to us in and through such experience" (Sokolowski 2000: 2).
- "Phenomenology can be characterized as the study of intentionality" (Cerbone 2000: 4).
- "Phenomenology ... emphasizes the attempt to get to the truth of matters, to describe phenomena, in the broadest sense as whatever appears in the manner in which it appears, that is as it manifests itself to consciousness, to the experiencer" (Moran 2000: 4).
- "Phenomenology should ...be understood as a philosophical analysis of the different types of world-disclosure" (Zahavi 2012b: 3).

All four definitions place their emphasis slightly differently, but they all indicate the centrality of some key ideas: appearance, experience, consciousness, intentionality. With the exception of experience, these are not ideas that are frequently deployed as fundamental in pragmatism (whether classical or new). But, more troubling is the way that these definitions actually raise the oppositional worries from the outset. In particular, the focus on conscious experience seems to sidestep naturalistic/causal narratives in favor of subjectivity as the relevant domain for inquiry. Accordingly, phenomenology seems to eschew the pragmatic approach to empirical, causal inquiry and replace it with a concern for the conditions of lived experience. Similarly, whereas pragmatism operates with a general commitment to ontological realism (in some sense), phenomenology seems ready to disregard all such debates about the existence of objects as belonging to the "natural attitude." Engaging in a series of "reductions," phenomenology attempts to bracket the ordinary conception of, and concern for, existence by turning its attention to experiential modes of appearance rather than the objective status of that which appears. Further, even if we acknowledge that existential phenomenology is concerned with existence, we should note that it is as a subjective dimension of lived experience, not a pragmatically situated view about the social world. Additionally, Heidegger's phenomenology locates ontology as the central phenomenological issue, but his approach to the topic considers "ontology" as detached from most contemporary debates about epistemic and metaphysical realism/anti-realism that occupy many pragmatic-inspired debates. The point is that as soon as we get very technical and precise with our definitions, the points of contact become much harder to sustain because they seem to come at the cost of downplaying points of resistance. And yet, if we throw our hands up in light of the oppositions, we miss the possible areas of engagement.

In order to motivate an alternative way forward then, it is worth considering an example from each tradition that allows us to get a sense of the stakes of engagement itself. I will, thus, look at the classic Husserlian account of looking at a cup on the table and James's account of the squirrel going around the tree.

Cups and Squirrels

Phenomenological Cups: Consider the cup on the table in front of me. Depending on my specific identity, I might see the cup on the table in a variety of ways related to the assumptions that operate in my life. For example, if I am a craftsperson, I might see the cup as a work of artistic creativity. If I am a mathematician, I might see it as an example of a particular sort of volume equation. If I am someone dying of thirst, I might see it as a lifesaver. If I am a store clerk, I might see it as an item that needs cataloged in inventory. And so on. The point is that in the natural attitude, I assume the existence of the cup as located in relation to a variety of contexts of meaning that are merely "there."

Husserl rightly worries that all of these situational frameworks yield theoretical baggage that gets in the way of giving an account of the actual experience of seeing the object itself. Now, by "itself," here, Husserl would not mean some noumenal actuality that lies behind or beyond the appearance. That is the sort of representational distancing that he considered to result in an unavoidable skepticism. Instead, by developing the idea of intentionality—that consciousness is always conscious of something—phenomenology starts from a denial of representationalism such that the world is "out there" and my consciousness is "in here." Instead, phenomenology sets aside questions of empirical reality and begins with the essential connection between consciousness and its object. Phenomenology then continues on to consider the mode in which the object is presented. For example, I can currently *perceive* the cup as immediately present in front of me, but I can also *remember* the cup that was present in front of me yesterday, and I can *desire* to have that cup in front of me tomorrow. Importantly, though, in all of these modes, the intentional object is the same: the cup. This is why Husserl says that the motto of phenomenology is "back to the things themselves."

Pragmatic Squirrels: James's famous example of the squirrel running around the tree does not start by trying to describe the experience of seeing a squirrel but instead is inaugurated by a specific metaphysical question asked about the squirrel (see James 1975: 27–28). James's characteristically pragmatic response is that everything depends on "what you *practically* mean" (James 1975: 27). As James goes on to explain, "the pragmatic method in such cases is to try to interpret each notion by tracing its respective practical consequences. What difference would it practically make to anyone if this notion rather than that notion were true?" (James 1975: 28). James's approach is reflected in Malachowski's definition of pragmatism more broadly: "Pragmatism's key feature is the primacy it gives to practice" (Malachowski 2013: 11).

When presented this way, as a matter of describing the experience of cups and the practical consequences of squirrels, it is easy to see why technical definitions of the two movements are unlikely to open onto much space for agreement. Precisely what are viewed as key strengths by one are viewed as core weaknesses by the other and vice versa. It is not just that they disagree about something, but that they are simply doing entirely different things from the start. It seem that if James were to tell Husserl to look at the squirrel, that Husserl would respond by saying that he was not thirsty, and that if Husserl were to ask James about the experience of seeing the cup, that James would respond by climbing a tree.

And yet, both traditions reject representationalism; both stress the importance of experience to philosophical inquiry (though they understand experience in different ways); both value conceptual clarity; both are skeptical of abstract metaphysical speculation; and both are entirely committed to fallibilism due to the realities attending embodiment and cultural contexts. How, then, can we build upon these sites of resonance without attempting to cover over the core differences? I think the best option is to start with broader definitions than usually found in the scholarly literature.

Open Questions and Big Tents

When it comes to fostering constructive dialogue between pragmatism and phenomenology, I suggest that we understand them both as more of big-tent conceptions that invite new directions in philosophy rather than simply historical positions that force narrow constraints on what should and can be done within philosophy. The goal should not be to remain loyal to Husserl or Heidegger, or to Peirce or James, but instead to figure out what they get right and argue for its truth and value for where we find ourselves now. In all cases, heresy might be the best form of fidelity. Here I am drawing on Hilary Putnam's encouragement not to view pragmatism

> as a movement that had its day at the end of the nineteenth and the beginning of the twentieth century, but as a way of thinking ... of lasting importance, and an option (or at least an "open question") that should figure in present-day philosophical thought.
>
> *(Putnam 1995: xi)*

What Putnam says is right not only when applied to pragmatism but also to phenomenology.

Drawing on Putnam, then, I want to offer the following big-tent (even if slightly heretical) definitions:

- *Phenomenology* focuses on achieving accurate descriptions of human experience from a first-personal perspective. Its guiding question is "Is this account faithful to experience?"
- *Pragmatism* focuses on whether, how, and to what extent our descriptions make a difference in our lives. Its guiding question is "Does this account matter?"

When presented this way, we can begin to see how phenomenology needs pragmatism and pragmatism needs phenomenology. The goal, I believe, of bringing them together is not to achieve some sort of philosophical agreement—indeed, in light of debates about naturalism, methodology, and evidence, say, widespread agreement is unlikely—but instead to achieve appropriate humility about one's preferred approach as not being adequate to life itself. Life is neither a singular matter of accurate description nor of practical utility. It takes something like a pragmatic phenomenology to appreciate the philosophical implications of such a realization.

Returning to cups and squirrels, unless we take seriously the contours of human experience, James's practical concerns can't get off the ground, but unless we attend to the practical consequences of our beliefs, Husserl's descriptive rigor is irrelevant. Although phenomenology's emphasis on intentionality highlights that we are already situated in a world of meaning, pragmatism's emphasis on practice illustrates that meaning-making is an ongoing task that depends on what we choose to think and to do. When we take a big-tent approach, we better understand scholarly debates as occurring within a shared world in which conscious experience is of utmost practical importance, and in which practical concerns impact our conscious experience. Pragmatic phenomenology helps to protect against the reductive temptations that threaten to derail both pragmatism and phenomenology, when considered individually. Specifically, pragmatism without phenomenology risks becoming scientistic power play, and phenomenology without pragmatism risks becoming quasi-mystical self-indulgence. Or as William Luijpen pragmatically writes, we should "ask ourselves whether phenomenology can yield anything but interesting diaries or entertaining novels with philosophical implications" (Luijpen 1966: 6).

In light of pragmatic phenomenology, we can see how accurate accounts of experience are what help to guide our practical concerns by anchoring them in intersubjective life.[3] Similarly, accounts of human experience must open onto a concern for transformed living if it is not to instantiate solipsism as an empirical fact. Phenomenology thus begins to be understood as much more practically minded than it might initially seem. As Sokolowski notes, "phenomenology helps us reclaim a public sense of thinking, reasoning, and perception. It helps us reassume our human condition as agents of truth" (Sokolowski 2000: 12). And, as Gallagher explains, channeling James's situationalist awareness, phenomenology "begins with the fact that we, as agents who must act, and as thinkers who try to get a grasp on what we are doing, are always *already situated* in the world" (Gallagher 2012: 1). That gets it exactly right. We must act, but in order to act well, we must appreciate who we are and where we find ourselves. When phenomenology and pragmatism are isolated from each other, they both threaten to slide into dangerous arrogance, but when they serve to humble each other, they invite reflective progress for conscious subjective individuals who are always also engaged social beings.

Concluding Case Studies

In conclusion, I want to offer suggestions for ways in which pragmatic phenomenology might prove especially fruitful in contemporary areas of philosophical debate and social life.

- Dewey's (1938) pragmatic notion of experiential education should be viewed as a practical instantiation of Michel Henry's (2012) phenomenological call for education to be a site of culture-creation. And yet, Henry's account helps to illustrate why experiential education must be anchored in a robust conception of subjective life.
- Rorty's (1989) pragmatic account of liberalism as opposed to cruelty, and Jacques Derrida's (2005) phenomenological notion of justice as related to the "democracy to come" can be read as two sides of the same coin. Derrida provides the account of how justice is a call forward into the sorts of beings that we hope to become, while Rorty shows how this call must be accompanied by practical steps toward more inclusive communities in historical contexts (see also Mouffe 1996).
- Pragmatic considerations of embodiment and social identity, like that offered by Cornel West (2002), should be viewed as important resources for phenomenological discussions of embodied affectivity, as on display in Merleau-Ponty (2014). Phenomenological accounts of experience should, thus, also appreciate the importance of making it matter in the lives of the oppressed and marginalized (see Ahmed 2014).
- Although I noted that phenomenology often tends to sidestep debates regarding realism and anti-realism, the neo-pragmatism of Hilary Putnam (1992) offers an especially rich resource for thinking about the ways in which some versions of realism might be wedded to phenomenological awareness regarding embodied investment in a world of meaning. Aligned with Putnam in many ways, Hubert Dreyfus and Charles Taylor (2015) explicitly illustrate just such a pragmatically minded phenomenological realism might involve.

More examples could easily be offered, but hopefully these suggestions serve as a general indication of what is possible for pragmatic phenomenology. The point in each case is that the truth of human experience and subjectivity matters for our continued living only when we allow it to have a practical impact on how we navigate the world that we share with others, but practical suggestions for that world are misguided unless grounded in the truth of such experience.

The big-tent approach that I have proposed is not meant to be a replacement for the careful work of technical scholarship. Instead, it is offered as an encouragement for such scholarship to be undertaken in light of the humility that lived reality should bring to bear upon all philosophical programs when they think that they are sufficient unto themselves. Ultimately, pragmatic phenomenology invites us to declare Robert Sokolowski's assessment of phenomenology to be true of pragmatism as well: each discourse "can continue to make important contribution[s] to current philosophy. [Because their] intellectual capital is far from spent, and [their] philosophical energy is still largely unexploited" (Sokolowski 2000: 2). By reading pragmatism and phenomenology together, without reducing them to each other, we are better able to exploit the philosophical energies of each. In the end, phenomenology is benefited when it is pragmatic and pragmatism is well served when it is phenomenological—a big-tent, definitional approach allows us more easily to see exactly that.

Notes

1 For some general discussions of how phenomenology and pragmatism might be brought together, see Kestenbaum (1977); Wilshire (1968, 2000); Rosenthal and Bourgeois (1980); Edie (1965, 1967); and Bourgeois and Rosenthal (1983). See also the discussion that spans across Rosenthal (1974a, 1974b); Wilshire (1977); and Rosenthal (1977). For more focused discussions of particular pragmatists and specific phenomenologists, see Craig (2010); Kessler (1978); and Simmons and Hackett (2016): Part V. As concerns the possible connections of neo-pragmatism and phenomenology, see Baert (2011).
2 Henceforth my references to "pragmatic phenomenology" will be shorthand for both of these alternatives.
3 Also, approaching phenomenology in this way allows for a variety of interdisciplinary possibilities to emerge—see Bruzina and Wilshire (1982); Ihde and Zaner (1977); and Petitot, Varela, Pachoud, and Roy (1999). For the range of phenomenological directions in contemporary philosophy, see Zahavi (2012a,b).

References

Ahmed, Sara. (2014) *The Cultural Politics of Emotion*, New York, NY and London: Routledge.
Aikin, Scott F. (2006) "Pragmatism, Naturalism, and Phenomenology," *Human Studies* 29: 317–340.
Baert, Patrick. (2011) "Neo-Pragmatism and Phenomenology: A Proposal," *European Journal of Pragmatism and American Philosophy* 3, no. 2: 24–40.
Bourgeois, Patrick L. and Sandra B. Rosenthal. (1983) *Thematic Studies in Phenomenology and Pragmatism*, Amsterdam: B.R. Grüner Publishing Co.
Bruzina, Ronald and Bruce Wilshire, eds. (1982) *Phenomenology: Dialogues and Bridges*, Albany, NY: State University of New York Press.
Cerbone, David R. (2006) *Understanding Phenomenology*, New York, NY and London: Routledge.
Craig, Megan. (2016) "Vitalism, Pragmatism, and the Future of Phenomenology," in J. Aaron Simmons and J. Edward Hackett (eds.), *Phenomenology for the Twenty-First Century*, London: Palgrave MacMillan, pp. 271–296.
———. (2010) *Levinas and James: Toward a Pragmatic Phenomenology*, Bloomington, IN and Indianapolis, IN: Indiana University Press.
Derrida, Jacques. (2005) *Rogues: Two Essays on Reason*, trans. Pascale-Anne Brault and Michael Naas, Stanford, CA: Stanford University Press.
Dewey, John. (1938) *Experience and Education*, New York, NY: Touchstone.
Dryfus, Hubert and Charles Taylor. (2015) *Retrieving Realism*, Cambridge, MA and London: Harvard University Press.
Edie, James M., ed. (1967) *Phenomenology in America: Studies in the Philosophy of Experience*, Chicago, IL: Quadrangle Books.
Edie, James. (1965) "The Philosophical Anthropology of William James," in James Edie (ed.), *An Invitation to Phenomenology: Studies in the Philosophy of Experience*, Chicago, IL: Quadrangle Books, pp. 110–133.
Gallagher, Shaun. (2012) *Phenomenology*, New York, NY: Palgrave Macmillan.
Henry, Michel. (2012) *Barbarism*, London: Continuum.
Hills, Jason L. (2013) "Pragmatism and Phenomenology: A Reconciliation," *American Catholic Philosophical Quarterly* 87, no. 2: 311–320.
Hookway, Christopher. (2013) "'The Principle of Peirce' and the Origins of Pragmatism," in Alan Malachowski (ed.), *The Cambridge Companion to Pragmatism*, Cambridge: Cambridge University Press, pp. 17–35.
Ihde, Don and Richard M. Zaner, eds. (1977) *Interdisciplinary Phenomenology*, The Hague: Martinus Nijhoff.
James, William. (1975) *Pragmatism*, Cambridge, MA and London: Harvard University Press.
Kessler, Gary E. (1978) "Pragmatic Bodies Versus Transcendental Egos," *Transactions of the Charles S. Peirce Society* 14 (Spring): 101–119.
Kestenbaum, Victor. (1977) *The Phenomenological Sense of John Dewey: Habit and Meaning*, Atlantic Highlands, NJ: Humanities Press.
Luijpen, William A. (1966) *Phenomenology and Humanism: A Primer in Existential Phenomenology*, Pittsburgh, PA: Duquesne University Press.
Malachowski, Alan. (2013) "Introduction: The Pragmatism Orientation," in Alan Malachowski (ed.), *The Cambridge Companion to Pragmatism*, Cambridge: Cambridge University Press, pp. 1–13.
———. (2010) *The New Pragmatism*, Montreal and Kingston: McGill-Queen's University Press.

Merleau-Ponty, Maurice. (2014) *Phenomenology of Perception*, trans. Donald A. Landes, New York, NY and London: Routledge.
———. (1967) "What Is Phenomenology?" in Joseph J. Kockelmans (ed.), *Phenomenology: The Philosophy of Edmund Husserl and Its Interpretation*, Garden City, NY: Anchor Books, pp. 356–374.
Moran, Dermot. (2000) *Introduction to Phenomenology*, New York, NY and London: Routledge.
Mouffe, Chantal, ed. (1996) *Deconstruction and Pragmatism*, New York, NY and London: Routledge.
Petitot, Jean, Francisco J. Varela, Bernard Pachoud, and Jean-Michael Roy, eds. (1999) *Naturalizing Phenomenology: Issues in Contemporary Phenomenology and Cognitive Science*, Stanford, CA: Stanford University Press.
Putnam, Hilary. (1995) *Pragmatism: An Open Question*, Cambridge, MA and Oxford: Blackwell.
———. (1992) *Realism with a Human Face*, Cambridge, MA and London: Harvard University Press.
Rorty, Richard. (1989) *Contingency, Irony, and Solidarity*, Cambridge: Cambridge University Press.
Rosenthal, Sandra B. (1977) "Phenomenology and Pragmatism: The Significance of Wilshire's Reply," *Transactions of the Charles S. Peirce Society* 13 (Winter): 56–66.
———. (1974a) "Recent Perspectives on American Pragmatism Part One," *Transactions of the Charles S. Peirce Society* 10 (Spring): 76–93.
———. (1974b) "Recent Perspectives on American Pragmatism Part Two," *Transactions of the Charles S. Peirce Society* 10 (Summer): 166–184.
Rosenthal, Sandra B. and Patrick L. Bourgeois. (1980) *Pragmatism and Phenomenology: A Philosophic Encounter*, Amsterdam: B.R. Grüner Publishing Co.
Simmons, J. Aaron and Bruce Ellis Benson. (2013) *The New Phenomenology: A Philosophical Introduction*, London: Bloomsbury.
Simmons, J. Aaron and J. Edward Hackett, eds. (2016) *Phenomenology for the Twenty-First Century*, London: Palgrave MacMillan.
Sokolowski, Robert. (2000) *Introduction to Phenomenology*, Cambridge: Cambridge University Press.
Talisse, Robert B. and Scott F. Aikin. (2008) *Pragmatism: A Guide for the Perplexed*, London: Continuum.
West, Cornel. (2002) *Prophesy Deliverance!: An Afro-American Revolutionary Christianity*, Louisville, KY: Westminster John Knox.
Wilshire, Bruce W. (2000) *The Primal Roots of American Philosophy: Pragmatism, Phenomenology, and Native American Thought*, University Park, PA: Pennsylvania State University Press.
———. (1977) "William James, Phenomenology, and Pragmatism: A Reply to Rosenthal," *Transactions of the Charles S. Peirce Society* 13 (Winter): 45–55.
———. (1968) *William James and Phenomenology*, Bloomington, IN and Indianapolis, IN: Indiana University Press.
Zahavi, Dan, ed. (2012a) *The Oxford Handbook of Contemporary Phenomenology*, Oxford: Oxford University Press.
———. (2012b), "Introduction," in Dan Zahavi (ed.), *The Oxford Handbook of Contemporary Phenomenology*, Oxford: Oxford University Press, pp. 1–4.

23
PRAGMATISM AND ITS PROSPECTS

Michael Bacon

The chapters in this collection illustrate that Richard Bernstein is correct to say that there is no essence to pragmatism (Bernstein 1995: 61). Bernstein is, I think, also right when he argues that "what we call pragmatism is itself – to use a Kantian turn of phrase – *constituted* by the narratives we tell about pragmatism" (Bernstein 1995: 55, emphasis in original). Any narrative is necessarily selective, and will always be developed out of a particular and partial idea of what is of value today and likely to be useful going forward. In this chapter, I draw on the work of several pragmatists to sketch a narrative that takes the most interesting and important work in contemporary pragmatism to be that which, explicitly or implicitly, brings together its theoretical commitments in epistemology and political theory.

I'll begin by noting that while Bernstein is right that there is no essence to pragmatism, all pragmatists are naturalists, in the minimal sense at least of regarding human behaviour as being continuous with that of non-human animals. At the same time, pragmatists deny that human activity is reducible to that of the non-human animals. Accounting for the difference is a central part of Robert Brandom's rationalist formulation of pragmatism.

In Brandom's terms, all animals are *sentient*, in the sense that they respond to the world around them. Humans differ in that we are also *sapient*, for we and we alone are capable of using language and engaging in linguistic practice. In his favourite illustration of the difference between sentience and sapience, Brandom compares the ability of a competent English-speaker to report the colour red with a parrot trained to squawk whenever it encounters red objects. Drawing on Sellars, Brandom argues that knowledge is a matter of standing in the "space of reasons." To illustrate, in order to understand the concept "red," one must be able to place it in a web of *inferential* relations so that when couched in propositions, it can serve as both a reason for other beliefs and can itself, if questioned, stand in need of reasons. To be taken as making a claim when asserting "That's red," a person must also be taken to be committed to what that claim entails. In contrast, the parrot does not know that "That's red" commits it to (for example) "That's scarlet" and precludes it from uttering "That's green" – any more than a piece of iron knows that it is raining when it responds by rusting. Brandom concludes:

> What's the difference that makes the difference here? ... I think the answer is that *you*, but *not* they, can use your response as the premise of *inferences*. For *you*, but *not* for them, your response is situated in a network of connections to other sentences, connections that underwrite inferential *moves* to it and from it.

(Brandom 2009: 170, emphasis in original)

These relations of responsibility are captured in Brandom's account of what he dubs *deontic scorekeeping*. As he describes it, members of social linguistic practices keep score of each other's commitments and entitlements, acknowledging valid moves and issuing sanctions for invalid ones.

> We sapients are discursive scorekeepers. We keep track of our own and each other's propositionally contentful deontic statuses. Doing that requires being able to move back and forth across the different perspectives occupied by those who undertake commitments and those who attribute them.
>
> *(Brandom 1994: 591)*

Deontic scorekeeping instantiates a difference between that to which we may *think* ourselves committed and entitled, and that to which our fellow participants in a linguistic practice *take us* to be committed and entitled. Once one has committed oneself, one is committed to the inferential consequences that follow, whether or not one acknowledges or is even aware of them.

A long-standing debate in pragmatism is which, if any, theory of truth it should endorse. On Brandom's formation, truth is the property that is preserved in correct inferences. The key here is the notion of what he terms a *material inference*. A correct inference is governed by the norms of social practices, as these are characterised in the game of giving and asking for reasons. This distinction is structural – it is one which all communities rely on if they are to function successfully: "the distinction between claims or applications of concepts that are objectively correct and that are merely taken to be correct is a structural feature of each scorekeeping perspective" (Brandom 1994: 595).

Brandom is aware of the objection that his view amounts to a form of relativism. He replies by challenging the assumption that relativism can only be avoided by identifying a perspective outside of our social practices. He recognises that we respond to the physical world and are constrained by it, but the important point is that these constraints must be seen as standing *inside* our social practices. We recognise an independent objective reality, but that recognition results from processes of interaction and scorekeeping rather than being something that exists prior to that recognition.

Brandom distinguishes his version of rationalism from what he regards as two irrationalist alternatives. The first is the Nietzschean idea that the game of giving and asking for reasons is nothing more than the expression of power; in particular, the power to manipulate the language in which we understand ourselves and the world. The second form of irrationalism Brandom identifies is the Romantic view that the practice of exchanging reasons is merely one practice among others, such as artistic uses of language. For Brandom, both are equally mistaken. There is, he claims, more to reason than manipulation; denying this would be to say that there is no way to distinguish between the different forms manipulation can take, a consequence that Brandom notes even Foucault resisted. And the game of giving and asking for reasons cannot be seen simply as one practice among others, for the reason that it is the practice that *institutes* the meanings which are presupposed by all other practices. Brandom writes:

> I am here disagreeing with Wittgenstein, when he claims that language has no downtown. On my view, it does, and that downtown (the region around which all the rest of discourse is arrayed as dependent suburbs) is the practices of giving and asking for reasons.
>
> *(Brandom 2009: 120, emphasis in original)*

According to Brandom, the classical pragmatists understood sapience exclusively in *instrumental* terms, taking it to be no more than a matter of using means-end rationality to satisfy given and fixed ends. He takes pragmatism to contain some of the resources needed to avoid instrumentalism, but thinks them drawn out best not by the classical pragmatists but by Kant and (in particular) Hegel. Following his interpretation of Hegel, Brandom claims that by entering into relations of commitment and entitlement, our understanding of ourselves and our relations with each other and the world change. This, in turn, entails a new conception of freedom, which he calls *expressive* freedom:

> The ... most sophisticated conception of sapience is Hegel's account of us as creatures of our positive expressive freedom – beings whose essence it is to have no essence, no nature, but only a history structured and driven by the description of ourselves that we endorse at each stage in our development – hence self-creating beings, who can change what we are *in* ourselves by changing what we are *for* ourselves, by identifying with new descriptions of ourselves, by adopting new vocabularies.
>
> (Brandom 2009: 154–155, emphasis in original)

Brandom devotes little attention to the question of what institutional arrangements might be needed best to ensure the capacity for individuals to express themselves freely. From what he does say, it is clear that he has in mind broadly democratic institutions. He writes that:

> Our moral worth is our dignity as potential contributors to the Conversation. This is what our political institutions have a duty to recognize, secure, and promote. ... [O]ur overarching *public* purpose should be to ensure that a hundred private flowers blossom, and a hundred novel schools of thought contend.
>
> (Brandom 2011: 152, 153, emphasis in original)

I will return to the thoughts about political institutions and public purposes contained in this passage later in the chapter, but turn first to Huw Price, who can be seen to extend the role Brandom accords to expressivism and place it at the heart of pragmatism.

Price's interest in pragmatism arises principally from his interpretation of the work of philosophers such as Hume and Wittgenstein rather than the classical pragmatists. In his understanding,

> Pragmatism ... begins with linguistic explananda rather than material explananda; with phenomena concerning the use of certain terms and concepts, rather than with things or properties of a non-linguistic nature. It begins with linguistic behaviour, and asks broadly anthropological questions: How are we to understand the roles and functions of the behaviour in question, in the lives of the creatures concerned? What is its practical significance? Whence its genealogy?
>
> (Macarthur and Price 2007: 95, emphasis in original)

The priority Price grants to language is of course not exclusive to pragmatists, with some recent work in metaphysics also taking a linguistic approach to philosophical questions; for example, in examining semantic terms such as truth and reference. Typically, however, metaphysicians ask the question of what it is that *makes* sentences true. This leads to the task of trying to account for the relationship between the world and the languages which are taken to represent it. In doing so it raises the problems – scepticism, the external world and of other minds – that pragmatists from Peirce to Putnam have argued that we should set aside. Price shares their concern, thinking it unhelpful to view knowledge in terms of a relation between a sentence and an item or event which it purports to represent, for the reason that doing so encourages us to shift our attention from questions about the utility of different languages and vocabularies to unhelpful metaphysical questions. For him, the result of the combination of a linguistic starting point coupled with a rejection of representationalism leads to a form of expressivism, which he associates with pragmatism.

One influential version of expressivism is the quasi-realism of philosophers such as Simon Blackburn. Quasi-realism holds that we are entitled to regard our commitments as being true, but we cannot see them as true because they represent some item in the world. Taking this much from quasi-realism, Price goes beyond it in one respect. Quasi-realists are *local* expressivists – they think that some claims *are* genuinely descriptive in the sense assumed by representationalism. The local expressivist asks whether or not some particular area of language, for instance, moral or aesthetic language, should be thought of as genuinely descriptive. In contrast, Price claims that we should dispense with the idea that representation is *ever* a useful way of understanding the relation between language and the world. As he remarks,

> The right thing to do, as theorists, is not to say that it turns out that none of our statements is a genuine representation; it is to stop talking about representation altogether, to abandon the project of theorizing about word-world relations in these terms.
>
> *(Price 2011: 10)*

Rather than adopting a local expressivism in certain cases, Price proposes a *global* expressivism.

Price's expressivism goes global by drawing on Brandom's inferentialism. In answer to the question of what gives content to assertions, Brandom argues that it is not a matter of their representing (or failing to represent) some item in the world, but rather that content is given in the game of giving and asking for reasons. Price takes Brandom's inferentialism to be entailed by global expressivism, according to which the use of linguistic expressions is what gives rise to the content of those expressions.

Global expressivism, in Price's account, is committed to what he calls a *vertical* form of pluralism, which holds that there is an irreducible plurality of kinds of vocabulary. This form of pluralism further challenges representationalism:

> One of the reasons that the representationalist model is a bad theory, the pragmatist wants to say, is that it does not pay enough attention to these factors. It is blind to the 'located' character of various bits of language – to their dependence on various contingent features of the circumstances of the natural creatures who use them.
>
> *(Price 2011: 12)*

Different vocabularies are suited for different purposes, but none of them is metaphysically privileged. For the purposes of predicting and controlling the physical environment, the language of the natural sciences enjoys privilege over others, say that of theology. However, this is not a *metaphysical* but *pragmatic* claim, because it is to say no more than that the language of natural science is better suited to a particular set of tasks.

By arguing that there is a pluralism of different vocabularies, there might be thought to be a significant difference between Price and Brandom. As we have seen, Brandom thinks that language has a "downtown": the practice of undertaking commitments and of giving and asking for reasons is central to *all* vocabularies. Price, in proposing a vertical pluralism about kinds of vocabulary, might seem to be arguing for something very different. However, he thinks there is no difference to be made out here. Vertical pluralism is consistent with a unitary account of language according to which different vocabularies are all characterised by relations of inference; the different uses to which we can put languages are united in being genuine assertions. Price concludes:

> Thus I think we can follow Brandom here – agree that language has a downtown – without abandoning the pluralistic aspect of Blackburn's expressivism. … To preserve the pluralism,

> what we need is the idea that although assertion is indeed a fundamental language game, it is a game with multiple functionally distinct applications – a multifunction tool, in effect.
>
> (Price 2013: 33)

As is the case with Brandom, the concept of truth plays an important role in Price's pragmatism. He holds that if we inspect them, social and linguistic practices can be seen to be comprising three different norms, which he labels sincerity, justification and truth. The norm of sincerity is a matter of asserting what one believes to be the case. The norm of justification goes further: it is the norm of asserting only that which one is justified in asserting according to the reasons and evidence one possesses. The third norm is the norm of truth, which holds that when we make assertions, we are asserting what we believe to be correct, and thereby take those who think differently to be mistaken. What this norm adds to the other two is that it enables us to make sense of ourselves as participants in conversation. Price writes that "what the third norm provides ... is the automatic and quite unconscious sense of engagement in common purpose that distinguishes assertoric dialogue from a mere roll call of individual opinion" (Price 2011: 165).

Price does not discuss the political implications that might follow from his expressivist understanding of pragmatism. For his part, Brandom indicates that expressive freedom requires a form of democracy, but he does not explore in any detail what that form might take. Other pragmatists have done so. The title of Cheryl Misak's book *Truth, Politics, Morality: Pragmatism and Deliberation* captures the very close connection she sees between epistemological issues, on the one hand, and moral and political issues, on the other (Misak 2000). Among pragmatists, the figure most closely associated with political theory and democracy is Dewey, but Misak turns not to him but to Peirce in order to develop a substantive account of, and justification for, an original form of deliberative democracy.

Central to Misak's work is a redescription of Peirce's theory of truth. Peirce is often taken to have understood truth to be that which would be identified at what he called "the end of inquiry." Critics have argued that the notion of the end of inquiry is empty, for the simple but important reason that inquirers can never know that they have reached it. Misak agrees with this objection, but also notes that the idea of the end of inquiry is not the only understanding of Peirce's position and in her work offers an alternative. As she puts it, Peirce's

> considered and much better formulation is this: a true belief is such that it would withstand doubt, were we to inquire as far as we fruitfully could into the matter. A true belief is such that, no matter how much further we were to investigate and debate, that belief would not be overturned by recalcitrant experience and argument.
>
> (Misak 2013: 37)

Misak minimalises one formulation of Peirce's position (truth is what would be reached at the "end of inquiry") and advocates another (a true belief would be indefeasible). This has the advantage of avoiding the problems in Peirce's account. In her reformulation, truth as indefeasibility provides not a *goal* at which to aim but rather a *method* by which inquiry ought to be conducted. A commitment to truth as indefeasibility requires that inquirers not limit themselves to fellow members of a social practice but address reasons and arguments no matter where they come from, for as long as it is productive to do so. Misak proposes that

> a methodological requirement falls out of the idea that a true belief would be the best belief, were inquiry to be pursued as far as it could fruitfully go. That methodological principle is that the experience of others must be taken seriously.
>
> (Misak 2000: 6)

She concludes that what Peirce called the method of science, where no view is in principle above scrutiny and where beliefs are put to the test in communities of inquiry, applies to *all* communities: "deliberative democracy in political philosophy is the right view, because deliberative democracy in epistemology is the right view" (Misak 2004: 15).

Robert Talisse has joined Misak in formulating an account of democracy from Peirce's writing, which he developed through greater reference to what he takes to be the limitations of forms of deliberation set out by non-pragmatist philosophers and political theorists such as John Rawls, Amy Gutmann and Dennis Thompson. He takes them to be (to use Brandom's preferred term) instrumentalists, presenting a conception of the person as possessing a set of beliefs, needs and interests before they enter society. Deliberation is thus conceived of merely as a means to reach a settlement to accommodate them. Talisse objects that

> despite their ostensible turn from aggregation toward deliberation, liberal theorists of deliberative democracy have retained precisely the element which rendered the adversarial model unsatisfactory, namely, the view that citizens come into the political arena distinct, independent entities with competing and irreconcilable fixed interests.
>
> *(Talisse 2004: 89–90)*

In contrast, the Peircean account of deliberation that he sets out does not regard citizens and their interests in this way but as things to be reflected upon, challenged and transformed through deliberation.

A related issue that Talisse argues ought to be of concern for pragmatists is how far the institutions in which members of social practices come together enable them all to participate as equals (a concern raised against deliberative forms of democracy by, for example, Sanders 1997). Talisse and Scott Aikin note that the classical pragmatists were conspicuously quiet about the concept of justice, and seek to address it by turning not to Peirce or Dewey but to Rawls. Specifically, Talisse and Aikin argue that pragmatists should embrace what Rawls calls the Difference Principle, which holds that social and economic inequality is acceptable only if it is to the advantage of the worst-off members of society. They write:

> The Difference Principle instructs us to look first to the least well-off and to fashion the basic structure so as to maximize their social position. As the least well-off are also those most vulnerable to political marginalization, the Difference Principle seeks to secure the conditions necessary for democratic participation among equal members of a community of social inquiry. This is a prima facie reason in favor of the Difference Principle that should compel pragmatists.
>
> *(Talisse and Aikin 2018: 232)*

I think this right, and that a commitment to the Difference Principle can usefully be taken to set out the conditions in which we might all express ourselves freely in the game of giving and asking for reasons.

Bringing these thoughts together, the narrative I have briefly sketched centres on the idea that the most interesting and useful work in contemporary pragmatism sees interconnection between its positions in epistemology and in political theory. In saying this, I have though tried to keep in mind Bernstein's point that there is no "essence" of pragmatism. And accordingly, of Brandom's remark that, of every tradition of thought,

> at any one time there will be diverse interpretations, complete with rival canons, competing designations of heroes, and accounts of their heroic feats. Making canons and baking

traditions out of the rich ingredients bequeathed to us by our discursive predecessors is a game that all can play.

(Brandom 2009: 113)

It is, however, a game that can be played with different degrees of skill and success. In his *Preludes to Pragmatism*, Philip Kitcher argues that an understanding of pragmatism such as this one neglects the distinct contributions the classical pragmatist made to philosophy and the potential their work holds out for us. Referring to Brandom and Hilary Putnam, Kitcher writes that despite his admiration for their work, they

> strike me as too conservative. Pragmatism should not be domesticated and brought into the precincts of "normal philosophy," so that James and Dewey can join the pantheon of respectable philosophers. To paraphrase Marx, the point is not to continue philosophy-as-usual, but to change it.

(Kitcher 2012: xiv)

By connecting pragmatism to themes and debates in "normal philosophy," Kitcher regards Brandom, Putnam and others as failing to make good on the hopes of the classical figures. On my contrasting view, the prospects for pragmatism lie in the work of those who change "philosophy-as-usual" precisely by engaging with the ideas and debates that concern non-pragmatist philosophers and political theorists today.

I close by suggesting that it is a strength of the work of the writers I have discussed that, when they look to the history of pragmatism and to the classical figures, they are happy to read, and misread, their work. Brandom, as we have seen, develops his rationalist version of pragmatism with reference to both the classical pragmatists and (more importantly for him) German Idealism. In both cases, he has been taken to task by critics for what they regard as his misreading of figures in those traditions (see, for example, Levine 2013; Houlgate 2020). Misak presents her account of truth as indefeasibility as one to which Peirce was himself committed, but arguably, it amounts to a significant redescription of what he meant or intended (see, for example, Gascoigne and Bacon 2020). Misak acknowledges Dewey's historical importance for pragmatism's understanding of democracy, but his work plays no role in her account of deliberation (see, for example, Misak 2019: 628–629). Talisse joins Misak in working out a recognisably Peircean account of democracy, but recent statements of his position make little (Talisse 2009) or no (Talisse 2019) reference to Peirce. And, in what I take to be the largest difference between Talisse and Misak, Talisse argues that there is little of value and much that is unwelcome in Dewey's understanding of democracy as "a way of life," and invites us to join him in bidding it farewell (Talisse 2010). All of this has provoked often accusatory, sometimes acrimonious and occasionally anguished responses from those hoping to remain faithful to the letter of their predecessors' works. But the right attitude for any pragmatist to take, toward the classical figures and to every other heroic precursor, is best expressed by Richard Rorty when he recommends that the critic "asks neither the author nor the text about their intentions but simply beats the text into a shape which will serve his own purpose" (Rorty 1982: 151). So long as pragmatists continue to do so, pragmatism's prospects look good.[1]

Note

1 For their helpful comments, my thanks go to Neil Gascoigne, James Lewis and Nat Rutherford.

Work Cited

Bernstein, R. J. (1995) "American Pragmatism: The Conflict of Narratives," in H. J. Saatkamp Jr. (ed.) *Rorty and Pragmatism: The Philosopher Responds to His Critics.* London: Vanderbilt University Press, pp. 54–67.

Brandom, R. B. (1994) *Making It Explicit: Reasoning, Representing, and Discursive Commitment.* London: Harvard University Press.

———. (2009) *Reason in Philosophy: Animating Ideas.* London: The Belknap Press of Harvard University Press.

———. (2011) *Perspectives on Pragmatism: Classical, Recent, and Contemporary.* London: The Belknap Press of Harvard University Press.

Gascoigne, N. and M. Bacon (2020) "Taking Rorty Seriously: Pragmatism, Metaphilosophy, and Truth." *Inquiry: An Interdisciplinary Journal of Philosophy.* DOI: 10.1080/0020174X.2020.1820375.

Houlgate, S. (2020) Review of "A Spirit of Trust: A Reading of Hegel's Phenomenology," *Norte Dame Philosophical Reviews.* Available at: https://ndpr.nd.edu/reviews/a-spirit-of-trust-a-reading-of-hegels-phenomenology/.

Kitcher, P. (2012) *Preludes to Pragmatism: Toward a Reconstruction of Philosophy.* Oxford: Oxford University Press.

Levine, S. (2012) "Brandom's Pragmatism." *Transactions of the Charles S. Peirce Society* 48.2, 125–140.

Macarthur, D. and H. Price (2007) "Pragmatism and Quasi-Realism," in C. Misak (ed.) *New Pragmatists.* Oxford: Clarendon Press, pp. 91–120.

Misak, C. (2000) *Truth, Politics, Morality: Pragmatism and Deliberation.* London: Routledge.

———. (2004) "Making Disagreement Matter: Pragmatism and Deliberative Democracy." *Journal of Speculative Philosophy* 18.1, 1–8.

———. (2013) *The American Pragmatists.* Oxford: Oxford University Press.

———. (2019) "The Impact of Pragmatism," in K. Becker and I. D. Thomson (eds.) *The Cambridge History of Philosophy, 1945–2015.* Cambridge: Cambridge University Press, pp. 624–631.

Price, H. (2011) *Naturalism Without Mirrors.* Oxford: Oxford University Press.

———. (2013) *Expressivism, Pragmatism and Representationalism.* Cambridge: Cambridge University Press.

Rorty, R. (1982) *Consequences of Pragmatism.* Minneapolis: University of Minnesota Press.

Sanders, L. M. (1997) "Against Deliberation." *Political Theory* 25.3, 347–376.

Talisse, R. B. (2004) *Democracy after Liberalism: Pragmatism and Deliberative Politics.* London: Routledge.

———. (2009) *Democracy and Moral Conflict.* Cambridge: Cambridge University Press.

———. (2010) "A Farewell to Deweyan Democracy." *Political Studies* 59.3, 509–526.

———. (2019) *Overdoing Democracy: Why We Must Put Politics in its Place.* Oxford: Oxford University Press.

Talisse, R. B. and S. F. Aikin (2018) *Pragmatism, Pluralism, and the Nature of Philosophy.* London: Routledge.

PART III
Pragmatism's Reach

24
PRAGMATISM AND LOGIC

F. Thomas Burke

Pragmatism is an approach to semantics that is geared to agents' possible actions and the consequences of such actions—an approach to semantics in which contemporary extensional (set-theoretic) techniques play a useful but subordinate role. Pragmatist *logic*, then, is an approach to logic that accommodates a pragmatist approach to semantics.

Pragmatist Semantics

Peirce's original statement of the pragmatic maxim stipulates that the meanings of our words are best couched in terms of possible effects of our actions in the world—"effects that might conceivably have practical bearings" etc. (1878; WP3:266). Actions have practical bearings in that they tend to produce tangible results. There are at least four distinct notions of practical bearings that classical pragmatists have entertained in different contexts.

1   First, we may *operationalize* our terms. Many, if not most, of Peirce's illustrations of how to use the pragmatic maxim were of this sort. We may understand this semantic methodology in narrow terms historically associated with Bridgman (1927)—cf. Peirce's definitions of "lithium" (1903) and "diamond" (1905)—but we might also adopt a James-esque "radical operationalism" and allow broader conceptions of what we mean by "operations" and respective "outcomes"—again, cf. Peirce's definitions of the words 'reality' and 'truth' (1877–1878). Notably, Dewey (1929, LW4:88–92) was quick to endorse the "operational thinking" of Bridgman (1927) and Eddington (1928), though he distinguished it from his "instrumental" theory of conceptions in ways that acknowledged "ambiguities in the notion of pragmatism" (LW4:90n2).

2   Dewey's *instrumentalism* identifies a second notion of "practical bearings" that may contribute to clarifying the meanings of various "conceptions." Namely, we employ various concepts in more or less useful ways to solve problems. Conceptions become instrumentally *meaningful* in the context of a given inquiry to the extent that they may usefully serve as *means* to furthering successful completion of that inquiry. For instance, the various words and laws of Newtonian mechanics are meaningful as elements of an abstract mathematical framework, but they take on new kinds of meaning when one uses that framework to send humans into orbit around the Earth or land them on the Moon. Such mathematical frameworks thus have "practical bearings" over and above the intrinsic practicalities of the mathematics as such. The relevant actions in this second sense are actions of choosing and applying one or more conceptions in

DOI: 10.4324/9781315149592-28

3   Practicalities of mathematics constitute a third notion of "practical bearings" that may contribute to clarifying the meanings of various "conceptions." This is the essence of *inferential-role* semantics. Brandom's (2000, 2007) and Sellars's (1958, 1975) inferentialism purports to be "a functional theory of concepts which would make their role in reasoning, rather than their supposed origin in experience, their primary feature" (Sellars 1975, 285).

For instance, consider a formal first-order linguistic framework such that for $i, n = 1, 2, 3, \ldots$, we allow infinitely many names c_i, infinitely many n-ary function symbols f_i^n, infinitely many n-ary predicate symbols P_i^n, along with the two standard first-order quantifier symbols $\forall$ and $\exists$ and infinitely many first-order variables x_i. Also include the five standard truth-functional connectives $\wedge, \vee, \neg, \rightarrow,$ and $\leftrightarrow$, and include the standard identity symbol $=$. With this first-order toolkit, we can then build a particular first-order language L using just one name c_1, one unary function symbol f_1^1, and one binary function symbol f_1^2. We may also introduce some axioms like the following (assuming to begin with some appropriate but mysteriously nondescript first-order domain of quantification, D):

- $\forall x_1 \exists x_2 (x_2 = f_1^1(x_1))$
- $\forall x_1 \forall x_2 (f_1^1(x_1) = f_1^1(x_2) \rightarrow x_1 = x_2)$
- $\neg \exists x_1 (c_1 = f_1^1(x_1))$
- $\forall x_1 (f_1^2(x_1, c_1) = x_1)$
- $\forall x_1 \forall x_2 (f_1^2(x_1, f_1^1(x_2)) = f_1^1(f_1^2(x_1, x_2)))$
- $(\forall x_1 (\varphi(x_1) \rightarrow \varphi(f_1^1(x_1))) \wedge \varphi(c_1)) \rightarrow \forall x_1 \varphi(x_1)$ for any L-formula $\varphi(x_1)$ with free x_1

The aim here, in one regard, is to be as obscure as possible, respecting any requirements characteristic of a first-order language with identity but otherwise using meaningless symbols. Using standard first-order elimination and introduction rules, one can commence to prove various theorems, e.g.,

- $\forall x_1 \exists x_2 (f_1^2(x_2, x_2) = x_1 \vee f_1^1(f_1^2(x_2, x_2)) = x_1)$
- $f_1^2(f_1^1(f_1^1(c_1)), f_1^1(f_1^1(c_1))) = f_1^1(f_1^1(f_1^1(f_1^1(c_1))))$

It can also be meta-proven that the first-order L-theory containing all and only such deductive consequences of the given axioms is consistent, complete, and decidable.

What might we be talking about (e.g., what is domain D) if we were to use such a language with these axioms? Inferential role semantics says that if we look carefully at how, e.g., f_1^1 and f_1^2 are used in various proofs and what various L-theorems might "say" about them, we will soon discover that f_1^1 is working "inferentially" like a unary *successor* function while f_1^2 works like a binary *addition* function (and c_1 is working like the familiar name "zero"). What we have here is essentially the language of Presburger arithmetic! This will be even more obvious if we were to introduce a second name c_2 to L with the stipulation that

- $c_2 = f_1^1(c_1)$

We could then prove as a theorem that the successor function is just an "add one" function. Namely,

- $\forall x_1 (f_1^2(x_1, c_2) = f_1^1(x_1))$

If we include a third function symbol f_2^2 to generate an extended first-order language L^+ with a few more respective axioms (so that the new function works inferentially like binary multiplication, to give away the surprise), we will then essentially have Peano arithmetic.

Unlike Presburger arithmetic, Peano arithmetic, if consistent, is not complete and is in any case not decidable. All of this would in principle be discoverable (putting aside the existence of nonstandard models of arithmetic for the moment) simply by focusing on how the various pieces of L^+ work together inferentially, this being different from how like pieces of L work together inferentially.

None of this, of course, reflects how mathematics and other kinds of theory construction are actually done, or not typically. Mathematicians already know what they are wanting to talk about when they commence in some inquiry or other to design or modify; nomenclatures and build or rebuild respective theories and/or models. The point of presenting the arithmetic examples, as if we could reverse engineer a bit of disembodied language to figure out how and why it might have been used to talk about whatever, is only to highlight how much of this theory-building process takes place in purely inferential terms. With only inferential roles at our disposal, we actually might succeed in such reverse-engineering efforts. Much of what is meant by the various fixed names and function symbols is reflected solely in their respective inferential roles. It is in this sense that they have "practical bearings" of a third sort. The practical actions in this case are actions of reasoning. The reasoning in this case is decidedly deductive reasoning—perhaps in the form of demonstrating consequence or non-consequence. Theorems and other kinds of demonstrative conclusions will be the respective tangible results.

It must be acknowledged that Brandom's and Sellars's ambitions extended beyond the merely deductive inferentialism alluded to above. They readily admitted that the design and use of such languages do not take place solely within a materially vacuous space of purely syntactic reasons. Reasons are to be found also in the earthy circumstances of real-life inquiry. Brandom and Sellars were recommending a "broadly functional approach to the semantics of [modal] concepts that focuses on their *inferential roles*—as it were, looking *downstream* to their *use*, as well as *upstream* to the circumstances that elicit their application" (Brandom 2009). "Non-logical" concepts, then, will have "inferential connections" that link the circumstances and consequences of their application. Brandom's obligation to explain what is meant by "material, non-logical inferential connections" between "circumstances and consequences of application of non-logical concepts" might be met by first acknowledging that inferentialism, as something distinct in itself, is sandwiched between Peirce's operationalism and Dewey's instrumentalism. Operationalism in effect looks "upstream" toward clarifying, grounding, and otherwise coming to concrete terms with circumstances that elicit the formulation and use of this or that linguistic and/or conceptual apparatus. Instrumentalism tends rather to look "downstream" toward assessing and predicting consequences of using any such apparatus in such circumstances.

There are thus at least these three functionally different though related kinds of "practical bearings" to consider when clarifying the meanings of our words—operational, instrumental, inferential—and broadly speaking, all three have equal standing in pragmatist semantics.

4 A fourth kind of "practical bearings" to consider is what we might refer to as *moral* repercussions of our judgments. In Peirce's kind of logic, so-called practical bearings are gauged at least in part relative to particular inquiries. Various conceptions are used allegedly to affect the course of some respective inquiry. Among the three kinds of practical bearings outlined above (operational, instrumental, inferential), Dewey's instrumentalism perhaps most explicitly promotes an inquiry-based contextualization of semantics, given its focus on meaningfulness of specific conceptions in terms of their usefulness in advancing this or that inquiry. If we view inferentialism as being sandwiched between operationalism (upstream) and instrumentalism (downstream), then inferentialism and operationalism both inherit the same context-dependency that is explicit in instrumentalism. More to the point, operationalism, instrumentalism, and inferentialism are all addressed internally to successful completion of

respective local inquiries. They are answerable to particular circumstances. Proposed solutions to particular problems are conclusive to the extent that such results are stable if not legitimate responses to whatever instigates respective inquiries. It is required in principle that a proposed solution indeed solves a respective problem.

One may always ask, nevertheless, about longer-term, contextually independent consequences of such final solutions outside of whatever situations stimulated their establishment. A solution may seem fully warranted insofar as it solves the problem at hand and, moreover, may have been established in accordance with the best methods of inquiry; but will it continue to work in the long run beyond what was carefully considered in the course of that inquiry? These are *moral* concerns. This gets at what the word "moral" means. Have all of the bases really been covered? What was not foreseen? What surprises may be just over the horizon? These are practical bearings that are gauged globally if they may be gauged at all. Often, because of their contextual independence, such practical bearings will be unfathomable but not therefore insubstantial. People live or die on such grounds.

No one of these four conceptions of practical bearings holds sway over the others as being *the* one correct or primary reading of what Peirce was talking about in his various formulations of the pragmatic maxim. Rather, arguably, they are *all* part of what he meant even if the differences may be or have been only vaguely conceived. Having distinguished them, one question would be: how do they work together as the basis of a study of meaning and reasoning?

Pragmatist Logic

So far, we have addressed pragmatist semantics insofar as that is essentially what pragmatism is—a school of thought addressing how best to specify what we mean when we converse and use various words or sentences or theories or when we build various models in various cooperative efforts to make sense of things. Pragmatist logic may then be viewed as what logic must be to accommodate a pragmatist semantics.

This can work given that a pragmatist approach to semantics is couched within a doctrinal framework centered around a particular conception of beliefs as fallible products of inquiry. That is, logic—in line with Peirce (1877–1878) and Dewey (1938)—can be cast as a study of inquiry. It would in that case be a normative study seeking to distinguish better and worse kinds of inquiry. We now outline some basic features of this way of conceiving logic to show how it may accommodate pragmatist semantics (including the four notions of practical bearings introduced earlier).

Any particular inquiry is essentially a problem-solving process. An inquirer (1) is presented with some problem or question or challenge; (2) surveys possible solutions with various degrees of clarity and plausibility; (3) analyzes, organizes, and systematizes these options; (4) maps out research trajectories to explore and compare these options, to formulate various empirical tests of what is expected versus what actually happens when adopting certain solutions; and (5) repeats and refines all of this as needed until it is possible to determine which solutions are best for the problem at hand.

The process of "an inquiry" as just depicted can be viewed as having "practical bearings" in all four ways outlined earlier. Figuring out how such matters contribute to an account of what it is to inquire well is what is meant by "pragmatist logic," particularly in terms of how such practicalities help to establish standards of clarity of thought and discourse (see Burke 2013, 2018). This is wholly compatible with contemporary conceptions of logic and scientific inquiry.

First, one of the more sophisticated versions of deductive *inferentialism* so far articulated is the so-called semantic view of scientific theories that emerged out of some of Suppes's work in the 1960s. We will start there. This "semantic view" is better referred to as a *multiple-modeling view* of scientific theories insofar as scientific theories are cast as collections of models rather than as singular syntactic systems with carefully chosen axioms. In effect, it is alleged that actual scientific practice proceeds by building models rather than just syntactic axiom systems (hence the label "semantic view"). Suppes (1960, 1962, 1967) emphasized formal set-theoretic modeling where a given theory might be composed of a hierarchy of different types of models, extending from sketchy but flexibly applicable models carrying suggestive ontological import to empirically concrete "models of data" presenting potential results of hands-on measurement activities. This clearly affords richer "spaces of reasons" determined not just by internal structural features of single models but also by structural "morphisms" of various kinds reflecting hierarchical relationships among the respective models for a given theory.

We should note that Suppes's kind of inferentialism qua a multiple-modeling view of scientific theories focuses on set-theoretic structures as media for deductive inference. Brandom's and Sellars's inferentialism respectively casts Gentzen-like deduction as the primary measure of what constitutes "the space of reasons" by which the legitimacy of rational discourse relative to any such model is gauged and/or to which it is otherwise answerable. However, Peirce (1877–1878, 1903) distinguished three complementary kinds of inference that determine characteristics of such spaces of reasons. That is, there are three kinds of "inference"—abduction, deduction, induction—where "reasoning" synthesizes all three kinds of inference. This coordination proceeds in the course of inquiry presumably with the intention of warranting final conclusions of respective inquiries. It follows that any such spaces of reasons are in principle to be surveyed by logical means that recognize and synthesize all three kinds of inference, not just one of the three. The three generic aspects of reasoning may be distinguished as follows, including a fourth cumulative synthesis:

1 *Abductive inference*:

   A Describe unexplained data that are already on hand.
   B State possible hypotheses that would explain these data.

2 *Deductive inference*:

   C Explicate background assumptions, theoretical frameworks, common sense, etc. that might be used along with each hypothesis in phase B to *predict* new data.
   D State deducible predictions (under such and so circumstances) for each hypothesis.

3 *Inductive inference*:

   E Describe experimental methods, observational procedures, etc. that may be used to yield new data relevant to testing these hypotheses.
   F Do such experimental/observational work, and state what new data are actually obtained (in respective cases) along with any relevant statistical/probabilistic analyses of the methods and results.
   G Induce initial assessments of the truth-values of respective hypotheses.

4 *Cumulative (synthetic) reasoning*:

   H Evaluate the overall argumentation in light of any (mis)matches between predicted data and data actually obtained, considering the soundness of the predictions and the adequacy of empirical procedures to obtain new data.

These aspects of reasoning are presented here in a linear stepwise fashion, but inquiry is assumed to proceed (especially when it is a collective enterprise) on all of these fronts at once until the problem is solved. Suppes's multiple-modeling view of scientific theories focuses exclusively on structural features of set-theoretic models and static relations among such models. Pragmatism focuses more fundamentally on the process of constructing and refining such sets of models in the course of given inquiries wherein spaces of abductive, deductive, and inductive reasons tend to evolve, hopefully guiding respective inquiries toward successful conclusions.

A second point to notice regarding the multiple-modeling view of scientific theories is that the pluralistic nature of this view as it has developed since the 1980s, say, is not limited to the structural hierarchies of set-theoretic models that Suppes championed. Consider things like wave models versus particle models of light. At any given point in the history of science, when confronted with some problematic conceptual scheme or some troublesome class of phenomena, there will be such competing accounts, each of which is at least partially (in its own way) successful as a tentative explanation of the targeted problem but where none of the options is decisively better than the others. At one point, say, before Young's double-slit experiment seemed to tip the scale toward a wave model, the theory of light was simply dualistic. The two distinct models taken together determined what was meant by the word "light" at that time. Science abhors inconsistencies and explanatory lacunae, and efforts to better understand the nature of light led to a kind of synthesis of the two types of models that preserved neither in its original form—unlike, say, the competition between Ptolemaic versus Copernican models of celestial motions where efforts to better understand planetary motions led to a rejection of the one kind of modeling in favor of the other.

It must be stressed, though, that multiple-modeling pluralism need not provoke an eliminative search for singular unification. By virtue of different kinds and degrees of simplification, idealization, abstraction, and the like, some pluralities of models may be openly embraced as such. For example, simpler Copernican models of planetary motions might be more practical as rough-and-ready calculating devices even though it is known that more complex Copernican models are more accurate in predicting such things. Or consider how Newtonian physics is quite up to the task of sending a rover to Mars even when we have relativistic and quantum physics at our disposal. Despite known flaws, Newtonian models maintain a first-class status in state-of-the-art theories of gravitation (etc.).

This illustrates many of the main features of *instrumentalism*, namely, different models that constitute a given theory may be useful in different ways, under different circumstances. While inferentialism locates meaningfulness in internal inferential structure, instrumentalism locates meaningfulness in a different functional sense, namely, in the *usefulness* of various models as *means* to achieving some particular objective. Such usefulness will hinge on the structural features of respective models, of course, but their usefulness is something over and above the *deductive-inferential* space of reasons internal to such models. By pragmatist lights, such instrumental usefulness of one or more models in respective circumstances falls within the scope of "semantics" not because we are talking about models as such but because we are talking about their respective potential *practical bearings* in an instrumental sense. Talking about how a given model can be used to achieve these-or-those respective ends is but one of the four ways to better clarify the semantic content of that model.

A third point to highlight regarding the multiple-modeling view of scientific theories has been mentioned in passing but is worth bringing into sharper focus, namely, Suppes's original version of this view focused on models as set-theoretic structures. As Brandom and Sellars might put it, most of the interesting formal work in this original view addressed how such models sanction various natural-deductive inferences internally, how various models might be structurally related to constitute a structurally coherent set of models addressing some common scientific subject matter, and how as a result such a coherent set of models thus determines a meaningful "space of reasons" concerning that scientific subject matter. The significant point here is that, while it was only in the 1980s that Suppes's 1960s work began to take hold as an influential view of scientific theories (Giere 1988; Suppe 1977; van Fraassen 1980), various proponents of this view in the 1990s were looking for more, particularly regarding how the structural features of such sets of models connect with actual (real) scientific subject matters. This reflected a move toward scientific realism in contrast with Suppes's set-theoretic coherentism (see Suppe 1989). Models in a multiple-modeling view are supposed to be *about* something. Models of planetary systems are about actual planets and such, and their legitimacy hinges on how well they represent what they are about. This gave rise to what may be termed the *representational multiple-modeling view* of scientific theories (Hughes 1996, 1997; Suárez and Pero 2019; Suppe 2000; van Fraassen 2000, 2008).

These later sentiments require more than just the mutual coherence of multiple set-theoretic structures. Proponents of the representational multiple-modeling view may nevertheless too easily succumb to what Dewey called the "spectator theory of knowledge"—a fallacious notion, namely, that we scientists might presume to work with a God's-eye-view of how our models are structured as well as what those models are about. Without succumbing to this fallacy, we nevertheless must do more than legitimize "models of data" as conduits for connecting theories qua mutually coherent sets of models to concrete subject matters. This is where Peirce-the-laboratory-man's *operationalism* enters the semantic picture, namely, there is no ground to stand on that would support a God's-eye-view of a scientific theory's targeted subject matter, but we might *operationalize* our models to connect them concretely with actual targeted subject matters. This applies to any kind of model, not just models of data.

For instance, Suppes (2002) updates many of his earlier examples that illustrate a multiple-modeling view of scientific theories. Suppose we build a set-theoretic structure to model our own solar system using a descriptive language referring to planets (as particles), times, masses, positions, forces, etc. The Sun along with its orbiting planets are such that elliptical paths of motion of the respective planets could be derived (with the Sun's fixed position as one of the foci of such paths), and positions of respective planets at different times relative to the Sun's position could be predicted. Call this *Model One*. We might also construct "models of data" in other set-theoretic terms including records of apparent positions of the various planets relative to the fixed stars and from the perspective of some particular telescope on Earth's rotating surface, where such records include readings on a given clock in the immediate vicinity of the telescope in the respective instances in which apparent positions are recorded. Call this *Model Two*. Procedures specifying how the rotating-telescope and clock readings are produced are admittedly not part of Model Two as such, but the readings might consist of ordered tuples of rational numbers presenting one-dimensional clock readings along with angular coordinates relative to Polaris as a reference point and a ray from Polaris through Dubhe as a reference direction. Then, a *theory* about our own solar system might consist of just these two models, properly related so that certain sequences of data in Model Two would correspond to certain predicted paths of motion of respective particles in Model One. Alternatively, and more likely, the theory might include one or more other intermediate models that help to establish this correspondence between Models One and Two. In either case it is not entirely implausible to allege that the *theory* will have been at least implicitly "operationalized" by virtue of the role of Model Two as the product of using clocks and telescopes in prescribed ways.

Suffice it to say that having recourse to such models of data is not what Bridgman (1927) meant by operationalism (see Chang 2009/2019). Model One employs numerous concepts (regarding particles, times, mass, positions, motions, forces, etc.), any of which might be directly or indirectly subject to respective measurement operations. In this sense Model One might itself be operationalized, *not just externally* in terms of its relations to separate models of data. Generally speaking, any particular model might be operationalized without its being a model of data.

This was a core feature of Einstein's theory of special relativity (which strongly affected Bridgman's thinking), consisting essentially of a set of models, each of which employs concepts of spatial intervals, temporal intervals, relative motions, etc., regarded as measurable variables using clocks, meter sticks, light signals, etc., ignoring prior conceptions of an ether or absolute space and time that our Model One seems to presuppose. The idea for Einstein (and for Bridgman) was to allow real physical subject matters to manifest themselves only by way of possible interactive measurements of what we take to be some of their key features, otherwise assuming little else. Procedures specifying how the measurements are made become an integral part of the modeling.

Similar sentiments arose within the development of quantum mechanics, particularly considering things like Heisenberg's uncertainty relations and the principle that only quantities that are in principle observable should play a role in quantum-mechanical modeling. An explicitly operationalist approach to quantum physics was in fact explored in detail by David Foulis and Charles Randall in the 1970s and 1980s (see Foulis, Piron, and Randall 1983 for entry into that literature). The larger point is that virtually any well-designed modeling, regarding quantum mechanics or otherwise, can be linked to concrete subject matters that such modeling is allegedly "about" by explicitly associating data-generating operations with some or all of the key variables employed in such models. For example, a commitment to "times" as absolute entities in some extensional domain of discourse begs too many questions, but a suitably structured domain of possible time-interval data that might result from reading one or more clocks is pretty much as real as it gets. One can basically do all of quantum mechanics or any other kind of physics by working with such operationalized variables.

So here is the point regarding a post-Suppes *representational multiple-modeling view* of scientific theories as of the 1990s. In place of a *representational* conception, we should instead prefer an *operational* multiple-modeling conception of scientific theories. It is only by way of interactive connections with concrete subject matters that our theories have realist as opposed to merely coherentist ballast. Operationalizing some or all of the terms used in various modeling efforts thus constitutes a third way in which scientific concepts may be better clarified specifically by grounding those concepts in data-producing activities. This is a third way that logic qua study of inquiry can accommodate pragmatist semantics.

Finally, a fourth way that logic (as an inquiry into the nature of inquiry) can accommodate a pragmatist semantics becomes clearer when we consider how and why it is that logic is a *normative* discipline. Logic is not merely a descriptive study of inquiry. It is concerned rather with distinguishing *better and worse* kinds of inquiry. Why is *modus ponens* a good deductive rule while affirming the consequent is not? What makes an explanation a *better* if not *best* explanation? Why is a 5% margin of error *good enough* in some circumstances and unacceptable in others? These are technical questions pertaining to how we fathom and constrain multiple-modeling spaces of abductive, deductive, and inductive reasons *for the better*. Similar questions address inquiry more broadly. Why might it be the case that one manner of inquiring (e.g., using scientific methods) is better than another (e.g., holding tenaciously to prior beliefs) (see Peirce 1877)? How is it and why does it matter that some theories as sets of models are preferable to others? All else being equal, theory one may be simpler than theory two, but why does that lend support to thinking theory one is the better theory (see Kuhn 1977), as opposed to being just a different theory?

Such questions about what distinguishes *better and worse* kinds of inference or theory choice can be answered in ways that appeal at bottom to what Peirce calls "the first rule of reason" and one of its corollaries:

> 135. Upon this first, and in one sense this sole, rule of reason, that in order to learn you must desire to learn, and in so desiring not be satisfied with what you already incline to think, there follows one corollary which itself deserves to be inscribed upon every wall of the city of philosophy:
> Do not block the way of inquiry.
>
> *(Peirce CP1:135; also EP2:48)*

This is only a statement of Peirce's first rule of reason (*to learn, one must want to learn in open-minded ways*) and of the corollary (*do not block the way of inquiry*). What is the argument? Why accept the rule? How does the corollary follow from the rule?

By *learning*, Peirce means the growth and correction of one's beliefs. Peirce (EP2:42–47; CP5:574–584) discusses in detail how inquiry can be self-corrective and open-minded. Each kind of inference by itself can work in self-corrective ways, assuming in each case "a hearty and active desire to learn what is true" (EP2:47). Inquiry, and especially open-minded self-corrective inquiry, requires effort. One must work to perform experiments, and careful deductive analyses do not just happen without purposely employing a critical eye. If we do not want to do such work, we will almost certainly not do it. This establishes the first rule of reason: *to learn, we must want to learn in open-minded ways*.

In this regard, recall Peirce's familiar doubt/belief depiction of inquiry in 1877 (EP1:114–115). Inquiry is initiated and spurred on by the arousal of doubt. Doubt is unpleasant. In the presence of doubt, we want its removal. We instinctively want to inquire. This is just a general fact about inquiry, more or less of the same order as the natural fact that, at the surface of the Earth, water flows downhill. The question is: do we instinctively want to learn with an open mind? The latter requires some proactive effort, and thus, we have the first rule of reason as a general matter of fact about "rational" inquiry that is meant to promote learning without prejudice.

How does it follow then that we should not block the way of inquiry? One might object: in a do-or-die emergency, we may have to stop merely thinking and start doing something. But Peirce has in mind unforced blockages like making "absolute assertions" as if they were beyond any further critique, or "maintaining that this, that, and the other never can be known" (EP2:49–50; CP1:137–140). In Peirce's view, we can never be sure that the results of inquiry cannot be further corrected. We should therefore not without compelling and objective reasons stop ourselves from trying to do so. Such blocks on further inquiry negate the desire to learn and thus halt learning (with or without prejudice). This contrapositive version of the first rule of reason is just a fact about inquiry and learning in particular, and the corollary follows as an "ought" to the extent that we place a positive value on learning.

With an eye then on explaining the normativity of logic, we may clarify the first rule of reason and its corollary by noting that the word "inquiry" is used in two different ways: on one hand, as a *mass noun*, and on the other hand, as a *count noun*. There is never-ending inquiry, in the mass sense, which is manifested in myriads of particular inquiries, in the count sense. This is perhaps best understood as a distinction between inquiry as an infinite game versus particular inquiries as finite games (Carse 1986; Sinek 2019).

A particular inquiry has a beginning, a middle, and an end. It proceeds as if there are one or more "winning" solutions to the problem that has given rise to the inquiry; it proceeds according to certain pre-established, relatively fixed rules that sanction what counts as a solution; and it proceeds as if problems and potential solutions are real problems and solutions imposed on inquirers from without. In some cases, a particular inquiry might be cast as a two-player game between a group of inquirers as one player and a larger unpredictable world as the other player. Or a particular inquiry might sometimes be better cast as a multi-player game of different schools of inquirers vying against other schools of inquirers, each proposing and promoting competing solutions that are themselves answerable to that same larger world that imposes problems on these inquirers from without.

Inquiry in the mass sense, however, has no beginning or end. There are no winners or losers as such, and the aim is not to win but to continue the play. It does not presuppose any fixed rules but continually addresses what kinds of rules are suitable for its particular constituent inquiries so as to best promote the perpetuation of inquiry. Any rules that might govern inquiry in this mass sense will be like "the grammar of a living language … always evolving to guarantee the [continued] meaningfulness of discourse" (Carse 1986).

On this score, the corollary *do not block the way of inquiry* states an "ought" that in one sense just states a fact about inquiry in the mass sense, i.e., as an infinite game. The corollary is just a matter of fact about the nature of infinite games as such where the point of insisting on not blocking the way of the infinite game is equivalent to emphasizing that one is always to stay in the game.

All of the questions we asked earlier concerning how it is that logic is a normative and not just a factual discipline can now be answered in much the same way. For example, *modus ponens* does not just preserve truth as a matter of fact. Preserving truth in this way is also in fact a *good* thing. It is why that kind of deduction is *better* than things like affirming the consequent, namely, "truth preserving" rules like *modus ponens* do not destroy inquiry, whereas bad rules like affirming the consequent afford no such guarantee and thus tend to "block the way of inquiry." It is as simple as that.

The first rule of reason and its corollary are closely related to another key fact about games, finite or infinite:

> It is an invariable principle of all play, finite and infinite, that whoever plays, plays freely. Whoever *must* play, cannot play.
>
> (Carse 1986: 4)

Focusing specifically on learning as "play," this may be regarded as an addendum to Peirce's corollary of the first rule of reason. Learning without prejudice cannot be forced. One cannot be made to learn, with or without prejudice, if one does not want to learn. So do not try to impose a will to inquire where it does not already freely exist. Thus, neither block *nor force* the way of inquiry. This comports with the pedagogical methods practiced in Dewey's Laboratory School in late-1890s Chicago (Knoll 2014). Such compulsion must emerge naturally from the cultivated instincts and interests of the learner. This applies generally to inquiry in the mass sense, i.e., as an infinite game, and to inquiries in the count sense, i.e., as finite games. Particular inquiries as finite games have rules, depending on the nature of the inquiry so that when you freely choose to play, you are freely choosing to follow certain rules or else suffer penalties and possibly lose the game. Freely choosing to play by certain pre-established rules is not a case of being forced to play.

In any case, then, the fourth way that logic can accommodate a pragmatist semantics exhibits the fourth kind of "practical bearings" that we previously referred to as *moral* repercussions (possible moral reverberations) of particular inquiries. We have cast this in terms of inquiries (plural) as finite games and inquiry (singular) as an infinite game, the point being that over and above any of the inferential, operational, and instrumental details peculiar to particular inquiries as finite games, we may also consider independently the repercussions that various particular finite wins and losses and manners of play have on inquiry as an infinite game. This goes beyond merely explaining the normativity of deduction rules and the like to asserting that logic as a study of inquiry is an inherently moral discipline. Scientific accomplishments affect just about every aspect of our lives for better or worse. Such affects are included among the "practical bearings" of particular theories, or of particular models that comprise theories, or of particular words and concepts used to formulate such models, thus constituting a fourth dimension of pragmatist semantics besides the other three discussed previously. What a given "hard word" means includes not only its inferential, operational, and instrumental characterizations in particular inquiries but also a broader moral characterization of the repercussions of its various uses upon inquiry in general.

We have argued here that pragmatist logic is a study of inquiry that incorporates pragmatist semantics as characterized by the pragmatic maxim (along with a certain prior conception of the nature of inquiry and belief, etc.) where the latter maxim appeals to a notion of "practical bearings" that can be interpreted in at least four distinct ways. We have attempted to cash this out more concretely by casting and expanding a Suppes-like multiple-modeling conception of scientific theories in ways that exhibit all four kinds of "practical bearings": Brandom's and Sellars's inferentialism, Peirce's operationalism, Dewey's instrumentalism, and a Jamesian kind of moralism that focuses on experience and thus inquiry as an infinite game.

References

Brandom, Robert B. 2000. *Articulating Reasons: An Introduction to Inferentialism.* Cambridge, MA: Harvard University Press.

Brandom, Robert B. 2007. Inferentialism and Some of Its Challenges. *Philosophy and Phenomenological Research* 74(3):651–676.

Brandom, Robert. 2009. Pragmatism, Inferentialism, and Modality in Sellars's Arguments against Empiricism. In Cristina Amoretti, Carlo Penco, and Federico Pitto, eds., *Towards an Analytic Pragmatism.* Aachen, Germany: CEUR Workshop Proceedings, Vol-444. (ceur-ws.org/Vol-444/).

Bridgman, Percy Williams. 1927. *The Logic of Modern Physics.* New York: Macmillan.

Burke, F. Thomas. 2013. *What Pragmatism Was.* American Philosophy Series. Indianapolis: Indiana University Press.

Burke, F. Thomas. 2018. Dewey's Chicago-Functionalist Conception of Logic. In Steven A. Fesmire, ed., *The Oxford Handbook of Dewey*, 507–530. Oxford: Oxford University Press.

Carse, James P. 1986. *Finite and Infinite Games: A Vision of Life as Play and Possibility.* New York: Ballantine Books.

Chang, Hasok. 2009/2019. Operationalism. In *Stanford Encyclopedia of Philosophy*. First published 2009 with substantive revision 2019 (plato.stanford.edu/entries/operationalism/).

Dewey, John. 1929. *The Quest for Certainty: A Study of the Relation of Knowledge and Action.* New York: Minton, Balch. Reprinted in LW4.

Dewey, John. 1938. *Logic: The Theory of Inquiry.* New York: Henry Holt. Reprinted in LW12.

Eddington, Arthur Stanley. 1928. *The Nature of the Physical World.* New York: Macmillan.

Foulis, David, Constantin Piron, and Charles Randall. 1983. Realism, Operationalism, and Quantum Mechanics. *Foundations of Physics* 13(8):813–841.

Giere, Ronald. 1988. *Explaining Science.* Chicago: University of Chicago Press.

Hughes, R. I. G. 1996. Semantic View of Theories. *Encyclopedia of Applied Physics* 17:175–180.

Hughes, R. I. G. 1997. Models and Representation. Proceedings of the 1996 Biennial Meetings of the PSA, Part II: Symposia Papers. *Philosophy of Science*, 64(Supplement):S325–S336.

Knoll, Michael. 2014. Laboratory School, University of Chicago. In D. C. Philips, ed., *Encyclopedia of Educational Theory and Philosophy*, vol. 2, 455–458. Thousand Oaks, CA: Sage Publications.

Kuhn, Thomas S. 1977. *The Essential Tension: Selected Studies in Scientific Tradition and Change.* Chicago: Chicago University Press.

Peirce, Charles Sanders. 1877–1878. Illustrations of the Logic of Science. In six parts. *Popular Science Monthly* 12–13. Reprinted together in EP1, chaps. 7–12, and in WP3, chaps. 60–65.

Peirce, Charles Sanders. 1903. Sundry Logical Conceptions. Third section of *A Syllabus of Certain Topics of Logic* (unpublished). In EP2, chap. 20.

Peirce, Charles Sanders. 1905. Issues of Pragmaticism. *The Monist* 15:481–499. Reprinted in EP2, chap. 25.

Peirce, Charles Sanders. 1998. *The Essential Peirce: Selected Philosophical Writings*, vol. 2 (1894–1914). Ed. Peirce Edition Project. Bloomington: Indiana University Press. Items in this collection are indicated by EP2 followed by page or chapter numbers.

Sellars, Wilfrid. 1958. Counterfactuals, Dispositions, and Causal Modalities. In Herbert Feigl, Michael Scriven, and Grover Maxwell, eds., *Concepts, Theories, and the Mind-Body Problem*, Minnesota Studies in the Philosophy of Science, vol. II, 225–308. Minneapolis: University of Minnesota Press.

Sellars, Wilfrid. 1975. Autobiographical Reflections. In Hector-Neri Castañeda, ed., *Action, Knowledge, and Reality.* The Matchette Foundation Lectures for 1971, 277–293. Indianapolis: Bobbs-Merrill.

Sinek, Simon. 2019. *The Infinite Game.* New York: Portfolio Penguin.

Suárez, Mauricio, and Francesca Pero. 2019. The Representational Semantic Conception. *Philosophy of Science* 86:344–365.

Suppe, Frederick. 1977. The Search for Philosophic Understanding of Scientific Theories. In Frederick Suppe, ed., *The Structure of Scientific Theories*, second edition, 3–241. Urbana: University of Illinois Press.

Suppe, Frederick. 1989. *The Semantic Conception of Theories and Scientific Realism.* Urbana: University of Illinois Press.

Suppe, Frederick. 2000. Understanding Scientific Theories: An Assessment of Developments, 1969–1998. Proceedings of the 1998 Biennial Meetings of the PSA, Part II: Symposia Papers. *Philosophy of Science*, Supplement 67:S102–S115.

Suppes, Patrick. 1960. A Comparison of the Meaning and Use of Models in the Mathematical and Empirical Sciences. *Synthese* 12(2/3):287–301. Reprinted in Suppes 1969, 10–23.

Suppes, Patrick. 1962. Models of Data. In E. Nagel, P. Suppes, and A. Tarski, eds., *Logic, Methodology, and the Philosophy of Science: Proceedings of the 1960 International Congress*, 252–261. Stanford: Stanford University Press. Reprinted in Suppes 1969, 24–35.

Suppes, Patrick. 1967. What Is a Scientific Theory? In Sidney Morgenbesser, ed., *Philosophy of Science Today*, 55–67. New York: Basic Books. Revised and updated in Suppes 2002, 2–10.

Suppes, Patrick. 1969. *Studies in the Methodology and Foundations of Science: Selected Papers from 1951 to 1969*. Dordrecht: D. Reidel Publishing Company.

Suppes, Patrick. 2002. *Representation and Invariance of Scientific Structures*. Stanford: CSLI Publications.

van Fraassen, Bas C. 1980. *The Scientific Image*. Oxford: Oxford University Press.

van Fraassen, Bas C. 2000. The Semantic Approach to Scientific Theories. In Lawrence Sklar, ed., *The Nature of Scientific Theory*, vol. 2, 175–194. New York: Garland.

van Fraassen, Bas C. 2008. *Scientific Representation: Paradoxes of Perspective*. Oxford: Oxford University Press.

25
PRAGMATISM AND METAPHYSICS

Claudine Tiercelin

Pragmatism and metaphysics suffer both from a bad reputation: just as there are as many pragmatisms as pragmatists, there are as many metaphysics as metaphysicians. What *is* pragmatism? A philosophical doctrine, a historical trend? What is *common* to its various representatives? Shouldn't we rather stress the *oppositions* between "vulgar" or "revolutionary" pragmatists – some scientistic, some literary – and "reformist" ones, the latter being more concerned with objectivity, truth and *knowledge* than by "human" practice and action, the primacy of *life*, of vital, psychological *and* religious matters? Shouldn't we stop wondering who has the right version and adopt a new name ("pragmaticism", as Peirce did)? Or even consider that, since "the reformist aspirations of the classical pragmatist tradition have been transformed by contemporary neo-pragmatists into one or another form of revolutionary anti-intellectualism", an evolution which can be "traced from Peirce, James, Dewey, and Mead, through Schiller, Lewis, Hook, and Quine, to Rorty's literary-political neo-pragmatism", we should "begin to worry that Russell was right to predict that pragmatism would lead to 'cosmic impiety,' or at any rate to fascism" (Haack, 2004: 5)?[1] As far as metaphysics is concerned, the situation is even worse. Throughout its history, the definition, nature and object of the queen of the sciences have been problematic, its links with science and theology a matter of constant debate from the Middle Ages on, its importance questioned, and at the turn of the 20th century, we have been taught by many (from logical positivists to Heideggerians) that it was just dead.

However, despite their various misfortunes, both pragmatism and metaphysics are still alive, and even among the most flourishing areas of contemporary philosophy. And this may be partly due to their mutual reinforcement. Most (classical and recent) pragmatists agreed on attacking metaphysics, but it did not prevent them from having their own "metaphysical" agenda. Most of all, provided pragmatism *and* metaphysics are understood in keeping with logic, semantics, inquiry and science, along certain *scientific* (though not scientistic) but also *realistic* lines, there is a genuine *pragmatist* way to deal with metaphysics and to face the "Integration Challenge" "binding metaphysics and epistemology together"(Peacocke, 1999: 1–2), so as to explain *why* and *how metaphysical knowledge* is possible, which is a key "to the nature of ourselves as rational thinkers" (Peacocke, 2004: 267). In its turn, *metaphysics*, so conceived, enriches the *pragmatist* agenda the founders may have had in mind when they gathered in Harvard in the 1870s and chose, "half-ironically, half-defiantly", the name "Metaphysical Club" for what became the birthplace of pragmatism.

DOI: 10.4324/9781315149592-29

Claudine Tiercelin

Pragmatism and the fight against metaphysics

However opposed they may be, most pragmatists concur in fighting *against* metaphysics. It is striking to see Rorty, in one of his earliest inspiring papers (1961), praise Peirce – shown to be in total opposition with him (Haack, 1997) – for his "sound and important" identification of the major vices of any metaphysical undertaking. In a nutshell:

> (1) What Peirce called 'nominalism' and what present-day philosophers call 'reductionism' are forms of a single error. (2) The error in both cases goes back to 'the Protean metaphysical urge to transcend language.'[2] (3) Peirce's attempt to give sense to the notion of *universalia ante rem* is not a result of succumbing to this urge, but is rather his device for repudiating it as strongly as possible. (4) When Peirce says that 'vagueness is real', and when Wittgenstein points to the difference between causal and logical determination, the only differences between what they are saying are verbal (or, to give the cash value of this overworked word, uninteresting). (5) The similarity of their insights about language reflects that fact that the slogans 'Don't look for the meaning, look for the use' and 'The meaning of a concept is the sum of its possible effects upon conduct' reciprocally support each other.
>
> *(1961: 197–198; Tiercelin, 2019: 13–16)*

Rorty is right in locating some crucial metaphysical vices, as I have done myself for years (Tiercelin, 1986, 2019) in the various ways in which "nominalism" – or "reductionism" – involves a dramatic lack of attention, on the level of language, meaning, knowledge, inquiry and plain ontology, to such an important phenomenon as "vagueness"; he is right, too, in his recognition of major links, on that count, between Wittgenstein and Peirce, and of the insights common to the numerous branches pragmatism has grown into, including Peirce, James and Dewey, indeed, but so many others, Wittgenstein, Ramsey,[3] Putnam[4] and, more widely, all those who objected to a "Platonistic", "Great Mirror" like (Rorty, 1980) or representationalist (Price, 2011) approach to metaphysics.

Of course, such attacks against metaphysics are not new. Metaphysics has been accused of endless scholastic verbal disputes; of being unable (Aristotle, Aquinas) to separate itself from theology (and, for some, from ontology); of having excessive ambitions about entities outside phenomenal experience; of falling into antinomies about God, the Soul, Immortality, Substance or Free-Will; of trespassing the limits of knowledge concerning a supposed "essence" or "nature" of *reality* and of its *properties*. We can understand why Berkeley or Kant could be taken as forerunners of pragmatism, why parallels are drawn between Peirce and Wittgenstein, and, more and more, with Ramsey, to say nothing of the Vienna Circle,[5] and their heirs (Carnap, Quine). But we could also add that, today, rather than praising metaphysical "boldness", many contemporary metaphysicians *themselves* (Langton, 1998; Lewis, 2009; Jackson, 1998) still favor "Humean", "Kantian" or "Ramseyan" types of "humility".[6]

No doubt, Peirce was the one who launched the battle against metaphysics and who provided the most detailed reasons for it. He denounced the "deplorably backward" and "immature" condition of metaphysics, "injurious to the physical sciences" and "greatly hampered the progress of such special sciences as the moral or psychical sciences" (6.2). For him, who was a deeply pious man, the "puny, rickety and scrofulous" state of current metaphysics stemmed from its remaining too long, not so much in the hands of "parish priests" or "Christian ministers" – who although they were "mainly practical men", hence not able to work in the "scientific spirit", were trying at least to "bring home to men's hearts the truth of the Gospel of Love" – as in those of "professional theologians" or "seminarian philosophers", a "much more tragic fate", because "the principal business of theologians" – "who must be judged as scientific men, since theology pretends to be a science"– is to make men feel the enormity of the slightest departure from the metaphysics they assume to be connected with the standard faith"(6.3). Now

> nothing can be more unscientific than the attitude of minds who are trying to confirm themselves in early beliefs. The struggle of the scientific man is to try to see the errors of his beliefs – if he can be said to have any beliefs.
>
> *(6.3)*

Hence,

> historically we are astonished to find that it has been a mere arena of ceaseless and trivial disputation. But we also find that it has been pursued in a spirit the very contrary of that of wishing to learn the truth, which is the most essential requirement of the logic of science.
>
> *(6.5)*

As a result, pragmatism was to get rid of the pseudo-problems by which metaphysics was encumbered. For in blocking the way of inquiry, "the one unpardonable offence in reasoning, metaphysicians have in all ages shown themselves the most addicted to" (1.136). Hence, pragmatism was to be "no doctrine of metaphysics, no attempt to determine any truth of things", no "*Weltanschauung*" (5.131), but "a mere maxim of logic instead of a sublime principle of speculative philosophy" (5.18), "merely a method of ascertaining the meanings of hard words and of abstract concepts" (5.464), so as to separate mere grammatical from real distinctions, in order for us neither "to mistake the sensation produced by our own unclearness of thought for a character of the object we are thinking" (5.398) nor "to mistake a mere difference in the grammatical construction of two words for a distinction between the ideas they express" (5.399). In a pre-Wittgensteinian way, pragmatism had first to tidy up the room: "It solves no real problem. It only shows that supposed problems are not real problems" (8.259). "What is wanted, therefore, is a method for ascertaining the real meaning of any concept, doctrine, proposition, word, or other sign" (5.6). And

> pragmatism will serve to show that almost every proposition of ontological metaphysics is either meaningless gibberish – one word being defined by other words, and they by still others, without any real conception ever being reached – or else is downright absurd.
>
> *(5.423)*

Hence, a program of eradication of pseudo-problems, among which the "salad of cartesianism", and of such foundationalist metaphysicians who, from Aristotle to Descartes, Locke or Hume, thought they could get out of the maze of words and reach a proper foundation in relying on intuitions, *sense data* or ultimate first principles, as the 1868 papers of the *Journal of Speculative Philosophy* were to show, whereas the pragmatic maxim formulated a few years later in "How to Make Our Ideas Clear" was to provide the criteria for a correct analysis of meaning and the right method to follow so as to fix true beliefs.

For sure, Peirce was not alone among the early pragmatists to criticize metaphysics. Praising the "pragmatic maxim", "the principle of Peirce" (P, 1975: 29), James too was allergic to "quibbling", "scholastic hairsplitting" and "ferocious metaphysical disputes" (e.g. "Does the man go round the squirrel or not?" [*ibid*.: 27–28]). He insisted on the value of "*the pragmatic method*, which is primarily a method of settling metaphysical disputes that otherwise might be interminable" (e.g. "Is the world one or many – fated or free? – material or spiritual?"), "a method only", then, which "does not stand for any special results" (*P*: 31), whose "astonishing" efficiency at making "many philosophical disputes collapse into insignificance" consists in "trying" "to interpret each notion by tracing its respective practical consequences" (*P:* 28). For

> there can *be* no difference anywhere that doesn't *make* a difference elsewhere – no difference in abstract truth that doesn't express itself in a difference in concrete fact and in conduct consequent upon that fact, imposed on somebody somehow, somewhere and somewhen.
>
> *(P: 30)*

Therein lies the virtue of the *empiricist* attitude, represented in pragmatism, "both in a more radical and in a less objectionable form than it has ever yet assumed", since

> a pragmatist turns his back resolutely and once for all upon a lot of inveterate habits dear to professional philosophers. He turns away from abstraction and insufficiency, from verbal solutions, from bad *a priori* reasons, from fixed principles, closed systems, and pretended absolutes and origins. He turns towards concreteness and adequacy, towards facts, towards action, and towards power. That means the empiricist temper regnant, and the rationalist temper sincerely given up. It means the open air and possibilities of nature, as against dogma, artificiality and the pretence of finality in truth.
>
> *(P: 30–31)*

Pragmatism, then,

> has no dogmas, and no doctrines save its method. As the young Italian pragmatist Papini has well said, it lies in the midst of our theories, like a corridor in a hotel. Innumerable chambers open out of it. In one you may find a man writing an atheistic volume; in the next someone on his knees praying for faith and strength; in a third a chemist investigating a body's properties. In a fourth a system of idealistic metaphysics is being excogitated; in a fifth the impossibility of metaphysics is being shown. But they all own the corridor, and all must pass through it if they want a practicable way of getting into or out of their respective rooms. No particular results then, so far, but only an attitude of orientation, is what the pragmatic method means. *The attitude of looking away from first things, principles, "categories", supposed necessities; and of looking towards last things, fruits, consequences, facts.*
>
> *(ibid.: 32)*

Examining "The Subject-Matter of Metaphysical Inquiry" (1915: 337–345), Dewey too claimed that he was "not concerned to develop a metaphysics; but simply to indicate one way of conceiving the problem of metaphysical inquiry as distinct from that of the special sciences", which he defined as "a way which settles upon the more ultimate traits of the world as defining its subject-matter, but which frees these traits from confusion with ultimate origins and ultimate ends – that is, from questions of creation and eschatology" (1915: 345). And, he added, "such traits are found in any material which is the subject-matter of inquiry in the natural science", a "kind of inquiry, not far from what Aristotle had in mind – provided one does not confuse it with the 'superior' study of 'the divine', or theology – to which the name metaphysical may be given" (*ibid.:* 340). Like Peirce, Dewey took pragmatism primarily as a *critical* tool (MW4: 107), and for him, "(contrary to James), pragmatists should not attempt to 'settle metaphysical disputes that might otherwise be interminable', they should *dispose* of them; we 'do not solve' the standard philosophical problems, 'we get over them' (MW4: 14)" (Talisse and Aïkin, 2008: 15–16). Dewey's "deconstruction" of traditional metaphysics, so much praised by Rorty, is well known (Boisvert, 1988; Gale, 2002). In a nutshell, metaphysics is guilty of having four major undesirable consequences. Because it locates true being in some timeless supernatural realm, it saps our incentive to take our workaday world seriously and fosters an undemocratic, hierarchical society in which one class exercises authority over other classes. Its theses are unverifiable and thus without cognitive meaning, and, as a result, it is completely aloof from the concerns and activities of ordinary people.[7]

Such an attitude against metaphysics could of course be extended to many contemporary pragmatists or neo-pragmatists from Rorty to Putnam. Very early, Putnam saw that, in order to reach a correct realistic position, the main enemy was Metaphysical Realism (a variant of Platonism), i.e. the "parochial" illusion of a readymade world (or "dough") out there, fixed once for all, dictating one single "true" description, and independent of our thoughts about it. However, the more Putnam moved from Internal realism to Commonsense (or "natural" or "pragmatist") realism, the more it seemed to him that we should free ourselves from any kind of metaphysical temptation, i.e. the illusion that we should "explain" what takes place, for example, in mathematics or in ethics, by introducing extraneous reasons to those areas, instead of simply looking at what is going on there, metaphysics being, almost by definition, contrary to commonsense. So, in *Ethics without Ontology*, still in line with the pragmatists, Putnam called for the burial of Ontology, after a diagnosis of the "disastrous consequences" of metaphysics of all sorts, either inflationary (Platonist or Metaphysical Realist) or deflationary (whether they be eliminationist or simply reductionist) (2004: 18–20). The therapy is so sweeping that the book might not only be titled, as Peter van Inwagen suggested, "Everything without Ontology" (2005: 11) but "Everything without Metaphysics" (Tiercelin, 2006: 55; and Putnam's reply to Tiercelin, 2006: 95–97). And, to a certain extent, we could today extend similar criticisms against major metaphysical vices to Price (2011)[8] or Kitcher (2018).

A variety of metaphysical agendas among the pragmatists

However, far from being *radical* foes of metaphysics, most pragmatists had a *metaphysical* agenda. For Peirce, things are crystal-clear: there is no repudiation of metaphysics. Pragmatism is not an *end* but a *means*: its aim, once metaphysics had been "purified" (5.423), and "all rubbish being swept away", is to erect a "scientific realist metaphysics", i.e. basically, a way of handling "a series of problems capable of investigation by the observational methods of the true sciences" (*ibid*.) although distinguishable from other "species" of prope-positivism, *first*, by "its retention of a purified philosophy"; *second*, by "its full acceptance of the main body of our instinctive beliefs"; and *third*, by "its strenuous insistence upon the truth of scholastic realism (or a close approximation to that)" (*ibid*.). In so doing,

> instead of merely jeering at metaphysics, like other prope-positivists, whether by long drawn-out parodies or otherwise, the pragmaticist extracts from it a precious essence, which will serve to give life and light to cosmology and physics. At the same time, the moral applications of the doctrine are positive and potent; and there are many other uses of it not easily classed.
> (*ibid*.)

Neither is James radically anti-metaphysical. *First*, "metaphysics" is but the other name for "philosophy", in its "narrow and original sense", "meaning the completest knowledge of the universe", thus "including the results of all the sciences, rather than to be contrasted with the latter" (*Some Problems of Philosophy*, 1979: 20); and in this "more worthy sense",

> as the results of the sciences get more and more available for coordination, and the conditions for finding truth in different kinds of question get more methodically defined, we may hope

that the term will revert to its original meaning. Science, metaphysics and religion may then again form a single body of wisdom, and lend each other mutual support.

(ibid.: 9–20)

Second, what James objects to is not the *empiricist* but the *rationalistic* way of looking at it (*ibid.*: 21–25). Indeed, James sympathizes with "nominalism, for instance, in always appealing to particulars; with utilitarianism in emphasizing practical aspects; with positivism in its disdain for verbal solutions, useless questions and metaphysical abstractions", which could be summed up as "*anti-intellectualist tendencies*" (P: 32). But, pragmatism *realized* also means devoting one's time to "Some Metaphysical Problems" now "Pragmatically considered", thus ensuring the "general triumph" of the pragmatic method, "an enormous change" in "the 'temperament' of philosophy".

> Teachers of the ultra-rationalistic type would be frozen out, much as the courtier type is frozen out in republics, as the ultramontane type of priest is frozen out in protestant lands. Science and metaphysics would come much nearer together, would in fact work absolutely hand in hand.

Hence, following the pragmatic method simply means that

> you must bring out of each word its practical cash-value, set it at work within the stream of your experience. It appears less as a solution, then, than as a program for more work, and more particularly as an indication of the ways in which existing realities may be *changed*. *Theories thus become instruments, not answers to enigmas, in which we can rest.* We don't lie back upon them, we move forward, and, on occasion, make nature over again by their aid. Pragmatism unstiffens all our theories, limbers them up and sets each one at work.

(P: 31–32)

In brief, "Pragmatism is uncomfortable away from facts. Rationalism is comfortable only in presence of abstractions" (*ibid.*:38), and there *can* be "genuine metaphysical debates", provided "some practical issue, however conjectural and remote, is involved" (P: 52). So, clearly enough, when James claims, that "at the outset, at least, [pragmatism] stands for no particular results" (P: 31–32), we cannot take him seriously. As Talisse and Aikin rightly noted, not only did James "broaden" Peirce's maxim "in no longer limiting the notion and meaning of 'practical consequence' to 'what is tangible and conceivably practical'", viewed mostly in Kantian terms by Peirce (Tiercelin, 1993a, 1993b), but in designing his pragmatism to include within a given proposition's pragmatic meaning the psychological effects of *believing* it (2008: 11), as shown in James's analysis of the materialism *versus* spiritualism dispute (P: 54–55), according to what suits best our "temperament" (rather than evidence): hence, even if spiritualism is inferior to materialism in terms of "clearness", it has "practical superiority" over it, since contrary to materialism, which does not provide "a permanent warrant for our more ideal interests", is not a "fulfiller of our remotest hopes" but results in "utter final wreck and tragedy", spiritualism guarantees an ideal order that shall be permanently preserved, obeys the "need of an eternal moral order", which is "one of the deepest needs of our breast". Now, it is far better to live with a world of *promise* ("spiritualist faith") than with the "materialism's sun", which "sets in a sea of disappointment. Remember what I said of the Absolute: it grants us moral holidays. Any religious view does this" (*ibid.*: 55–56). So, clearly enough, James's pragmatism involves endorsing "specific results", even a strong metaphysical agenda which is detailed not only in his sophisticated theory of truth (*ibid.*: 33)[9] but also *via* his radical empiricism, his pluralism and the promotion of meliorism, "a twist on existentialism that is deeply American" (Putnam, 1990: 229; Tiercelin, 2002a: 95; 2005, 2014).

Like James, although he viewed pragmatism primarily as a *critical* tool for assessing the value of metaphysical issues, Dewey had his own metaphysical agenda: a glimpse at *Experience and Nature* whose aim is "to discover some of these general features of experienced things and to interpret their significance for a philosophical theory of the world in which we live" (LW1:14) would convince anyone that for the ardent "Hegelian absolute idealist", "attempting to exercise his own personal devil, and believing and practicing Christian" Dewey (Gale, 2002: 477), who uses "ontology" and "metaphysics" interchangeably (Boisvert, 1988: 17–18), metaphysics does count: not so much as the mere "cognizance of the generic traits of existence" (LW1: 50) but in its capacity "to relate them to goods and values. Metaphysical claims may be unverifiable and thus devoid of cognitive content, but they *express* human ideals and aspirations" (MW3: 73; cf. LW15: 16, quoted by Gale, 2008: 482–483). Such a metaphysical program, in which Hegel's and Darwin's influences play such an important role, within a naturalistic, "immediate", then "deliberate" empiricist and basically instrumentalist framework, could indeed be read along deflationist lines: after all, "[instrumentalism] involves the doctrine that the origin, structure, and purpose of knowing are such as to render nugatory any wholesale inquiries into the nature of Being" (MW6: 89). And "the chief characteristic trait of the pragmatic notion of reality is precisely that no theory of Reality in general, *überhaupt*, is possible or needed" (MW10: 39). However, far from being modest, Dewey's program has a very wide scope, not only because of the focus put on some concepts which metaphysics – and altogether any sound theory of *inquiry* (LW12: 108–109) – should pay attention to, such as "traits", "generic traits", but also because of the overwhelming part played by experimentation, "experience *as*", or live (rather than passive) "experiencing" within a biological, social and *existential* environment, "situations" (preferred to objects or events) being viewed in terms of a "construction" or "creation" of "problems", first, from a "psychological" standpoint (EW1: 123), then in terms of an "immediate empiricism"(1905) and "experiences *as*" (MW3: 158) or, as in *Experience and Nature*, *via* a "denotative method". It is hard to see, Rorty claims, how Dewey's displaying of generic traits "could either avoid banality or dissolve" (*CP*: 74) and easier to think of *Experience and Nature*, Dewey's fundamental statement of metaphysics, "as an explanation of why nobody needs a metaphysics, rather than as itself a metaphysical system" (*CP*: 72). For another admirer of Dewey, "James ... succumbs less than Dewey to the temptation to offer a metaphysics of value" (Putnam, 1992: 196–197), and Dewey is "too quick in concluding from the overlap between scientific values and ethical values to their complete identity" (1994: 174); in that respect, Peirce was more lucid when he denounced the relativistic and skeptical risk there was in not maintaining a distinction between the two, and in completely giving up the idea of "pure knowledge" (Putnam, 1994: 204–205; Tiercelin, 2002b: 318). At all events, it makes sense to claim that "in Dewey's version, pragmatism is expanded to include in addition an aesthetic theory, a social philosophy, a philosophy of religion, a philosophy of science, and a philosophy of education" (Talisse and Aikin, 2008: 23).

As to Putnam himself, even if he was ready to bury Ontology, after a diagnosis of the "disastrous consequences" of metaphysics of all sorts, either inflationary (Platonist or Metaphysical Realist) or deflationary (whether they be eliminationist or simply reductionist) (2004: 18–20), his strategy was not primarily of destruction (to cure a "disease") but of positive "replacement". As a "strategic optimist," a man of the Enlightenment, his aim was to give up the *vehicle* while retaining the philosophical *insights* (2004: 18, 85),[10] so as to replace a dead and stinking corpse by something alive, i.e. *pragmatic pluralism*. And Putnam's goal was even wider: in keeping with his ongoing emphasis on the impossible demarcation between epistemology and ethics (1978), between scientific and "non-scientific" knowledge, he stressed the relatedness of the issues in the philosophy of mathematics and in ethics, clearly aspiring to an integrated (*architectonic*?) vision of philosophy, eager to "take the ways of thinking that are indispensable in everyday life much more seriously than the onto-theological tradition has been willing to do", hence, in so far as ethics is defined as "being concerned with the solution of practical problems", having ethics at its core (2004: 32).

Pragmatism and metaphysics: a mutual reinforcement

It is time to say how and why pragmatism and metaphysics can, each in its way, be a source of mutual enhancement. Rorty was right, I claimed, in his diagnosis of major metaphysical vices, well identified by pragmatists, among which, to try to transcend language, to sink into nominalism, another name for reductionism, to neglect vagueness. However, it is often alleged that most pragmatists favored various forms of instrumentalism (Dewey), verificationism, operationalism and nominalism ("pragmatism is always appealing to particulars", James, *P*: 32) instead of realism, in terms of their accounts of truth, meaning, inquiry, experience or reality (Haack, 1977: 378). I would like to qualify such an analysis and, most of all, explain briefly why, however paradoxical the alliance may seem, pragmatism *implies* realism, properly conceived, namely, a realism involving some *metaphysical* commitment, and why, provided this is the case, both pragmatism *and* metaphysics have the best chance to flourish.

First, even if in their most contemporary versions, many pragmatists or neo-pragmatists (Price, Kitcher and Putnam) are reluctant to call themselves "realists" and prefer such names as "natural realists" (Putnam), "quasi-realists" (Price; see Tiercelin, 2013), and they do not seem keen on describing their position as straightforwardly anti-realistic or nominalistic. Even the standard supposed – though partly justified[11] opposition between the "nominalist" pragmatism of James *versus* Peirce's "realist" version seems far-fetched (Haack, 1977: 392–393): there is a "Peircean strain", in James's account of truth: ideas must, if not *correspond*, at least *"agree with* reality" (*P*: 96; Hookway, 2000: 292), and the difference is not that Peirce accepted and James denied the reality of universals but that Peirce denied that real universals could be reduced to particulars, while James thought they could. James was never a straightforward "verificationist" in logic (Klein, 2016: 17ff), and he even approached the realistic position in his mature writings, especially in the *Pluralistic Universe* (Rosenthal, 2000: 94; Seigfried, 1990: 267; 399n5).

So, the real opposition between most pragmatists does not lie in the attitude toward metaphysics – despite the differences in the metaphysical agendas – not even that much on realism itself but rather on a certain understanding of what *realism* means and implies. Peirce saw what the genuine issues were: he admitted that "everybody ought to be a nominalist at first", "because it is simpler than realism and it is true that the economy of research prescribes to try the simpler one first". He said he would continue in that opinion until he is driven out of it "by the force majeure of irreconcilable facts" (4.1). He also recognized that "no realist is so foolish as to maintain that *no* universal is a fiction" (NEM. IV: 343). But he took the reasons for realism to be "pretty compelling" (6.605); although he had been "blinded by nominalistic preconceptions" (6.103), he had "never been able to think differently on that question of realism" (1.20). He even went so far as to call himself a "scholastic realist of a somewhat extreme stripe" (5.470), adapting the virtues of Duns Scotus's theory to the requirements of modern science. So, for him, at least, there was "a logical affinity" (2.58) between pragmatism *and* realism, and we may safely add a logical affinity too between pragmatism, realism *and* a realistic scientific metaphysics: "Pragmaticism could hardly have entered a head, he wrote, that was not already convinced that there are real generals" (5.503).

Space prevents me from detailing Peirce's sophisticated scientific and scholastic realistic metaphysics, but let me sketch what I take to be fruitful insights in his views, from the common standpoint of a pragmatist *and* a metaphysician, today. Pragmatism was introduced first as a principle of logic, "a method of ascertaining the meaning of hard words and abstract conceptions" (5.464). Why is such a therapeutic value of pragmatism priceless as applied to metaphysics? Not because it is "abstract": it is not more "abstract" than mathematics, nor "beyond the reach of human cognition" nor "inscrutable", but because, paradoxically, "metaphysics, even bad metaphysics really rests on observations", and "upon kinds of phenomena with which every man's experience is so saturated that he usually pays no particular attention to them" (6.3). As Emile Meyerson said: "L'homme fait de la métaphysique comme il respire" (Tiercelin, 2011b). Metaphysics is unavoidable, because an implicit metaphysics accompanies all our judgments (1.129; 6.2), including, of course, those of the scientists. Hence, "there is no escape from the need of a critical examination of 'first [*metaphysical*] principles'" (1.129; 6.4), which, incidentally, should be distinguished from *theological* principles (Tiercelin, 2020), in order for metaphysics to gain its autonomy and the conditions of its possibility *as a science* (as Peirce's inspirer, Duns Scotus, contrary to Aquinas, insisted on). This is why, too, metaphysics can be viewed as "a science" and a "positive" science, whose "business is merely to study the most general features of reality and real objects", for doing metaphysics does not require any special technique; one makes hypotheses, which are then tested in light of experiences which anybody can have, such "critical commonsensist" experiences benefiting, then, from the logician's conceptual analyses, the epistemologist's *inquiries* led in a *fallibilist* spirit and the scientist's experimentations in the laboratory.

"Realism" is indeed one of the concepts requiring clarification, as classical and recent pragmatists and metaphysicians have realized. Peirce's lesson is huge: as Rorty noted, he was both convinced of the necessity to give a far-reaching solution to the classical and medieval problem of universals and aware that "the most deep-rooted and crude misunderstanding [lay] in believing that the problem of universals [had] anything to do with 'Platonic ideas'" (8.17),[12] in opposing Metaphysical Realism (or Platonism) and nominalism, or in wondering whether there *existed* universals apart from our ideas or words, while it lay in the "*fundamentum universalitatis*" (6.37; 6.361), since "the real is that which is not whatever we may happen to think it, but is unaffected by what we may think of it" (8.12). For *scholastic realism*, the sound realistic option, no universal or singular entities should be seen as *utterly* independent of thought and signification: "The real is that which *signifies* something real" (5.320). Thus, we cannot get out of language and representation and dream of reaching simple, ultimate elements (Haack, 1992: 22–23; Tiercelin, 1986, 1992). A decisive consequence of this, drawn from pragmatism itself, is that any genuine realism should be viewed, first and foremost, as some form of *semantic realism*. We have a great lesson here, in terms of metaphysical method: as metaphysicians keep saying today, we should always start by securing the *semantic level* and the meaning of our property attributions, in particular, if we look for real *properties*, which are not given simply by the *meaning* of our predicates, and defend, as I think we should, the reality of *dispositional* properties, which was also part of Peirce's aim in formulating the pragmatic maxim.[13]

But we have a second great lesson for contemporary metaphysics. Indeed, we have to prove that the logical principles are not only "regulatively valid" but may be taken as "truths of being" (1.487). No "linguistic turn" without an "experientialist turn", so to speak. The discoveries of science, only, afford the means of showing that things that have the same name are really similar and that there are *real laws* and not mere accidental uniformities. Pragmatism yields the acceptable connection: "The question of realism and nominalism … means the question how far real facts are analogous to logical relations, and why" (4.68). Thus, its task will be to hold to the Scotistic inheritance while adapting it to the requirements of modern science, so true it is (as shown by F.H. Abbot's *Scientific Theism*) that "science has always been at heart realistic and always must be" (1.20). Hence, scholastic realism turns nicely into a piece of *scientific metaphysics* (Haack, 1992: 24): it works as a kind of high-level *abductive* hypothesis, as scientific realists often claim today: it is not a pure a priori choice or a matter of taste, but "required to explain how science is possible" (1.20; 5. 423), for science seeks to *explain* and not merely *describe* natural phenomena, which it can do only – obeying the "no miracles argument" – if there are *real laws* to be discovered, which, in turn, requires that there be real kinds of things (or universals) in the world (4.1): in pragmatist terms, it involves the evolution of the pragmatic maxim meant in 1878 to give the grounds of the method to be followed in *science* in order to determine the truth of a given proposition, from an indicative (and material or Philonian) interpretation of the conditional (with its operational and verificationist accents) to a subjunctive reading, expressing a law of nature "as real and as having a kind of *esse in futuro*" (5.48), thus forbidding the reduction of any statement of the maxim to singular or discrete events (nominalism): universals are not mere creations of the mind but real *active operative principles in nature*, tendencies, *dispositions* to behave, habits or "*would be*", describable in general terms (Tiercelin, 1986, 1992). Peirce's scientific realism is assumed by *dispositional realism*.[14]

Now we can see why a further step toward intelligibility is needed; *scientific realism* cannot but take a *metaphysical* turn and assume real *metaphysical possibilia*. Thus, realism well-conceived involves an irreducible indeterminacy, at both ends, of the knowing *as inquiry* process, under both forms of vagueness and generality: our tendency to generalize and to take habits is real, but vagueness too is real, i.e. the element of possibility, spontaneity or randomness which lies at the bottom of habit and cannot be reified (Peirce's name for tychism, i.e. the view that "absolute chance is a factor of the universe" (6.201). As a consequence, any sound realism must be defined as an anti-reductionist realism of the indeterminate on the three logical, epistemological *and* ontological levels (5.312; 6.348; 5.425). I think we have here all the ingredients for developing, within a pragmatist framework, a very nice program, along dispositional realistic lines (Tiercelin, 2011a: 247–359).

Conclusions

How and why does such a mutual relation between pragmatism and metaphysics allow us to face the Integration Challenge; to resist deflationist positivistic or Heideggerian tendencies but also Humean, Kantian or Lewisian forms of skepticism; and to fill the gap between our epistemology and our metaphysics?

From the *epistemological* side of the challenge, the various tools provided by the rich analyses of the pragmatists (Peirce, Dewey, Ramsey, Putnam) of such crucial concepts as "science", "knowledge" (Tiercelin, 2005: conclusion, 2008, 2011b), inquiry, the distinction between "scientific" and "scientistic", the focus on fallibilism, the various modes of logic, inference, probability, truth, belief, and, more generally, the types of beliefs, "truths" and justifications we may be dealing with in metaphysics are a great contribution. Are they beliefs whose truth stems from common sense? Scientifically established truths and therefore *necessarily* contrary to the "manifest image" reflected to us by the world? Or beliefs that are really of a very different nature? Either way, what reasons, what justifications do we have for maintaining these beliefs, for favoring such and such conception of truth, for judging the knowledge of the things that it reveals as merely conceivable, or possible or even necessary? Are our metaphysical *doubts* mere paper doubts, nonsensical or justified?

From the *metaphysical* side, we can also follow the pragmatist inspiration: metaphysical knowledge is possible, provided we are no longer obsessed by the search for eternal, universal, overarching truths but seek to understand our relationship with the *real* – which we can do only by starting off from where we are and not from "nowhere" through the following steps (Tiercelin, 2001b): first, in the "therapeutic" phase, adopt a formal logical and aprioristic framework of conceptual analysis, examine the conditions of possibility – in terms of truth and meaning – of the concepts we use, so as to make the necessary modal distinctions and proceed to the continuous critical massaging of our folk (or "commonsensical") intuitions (Jackson, 1998); second, metaphysics being in continuity with science, reflect on the way in which our categories and judgments constitute a genuine *knowledge*, within the framework of a basically pragmatist, realistic and fallibilist strategy, in which knowledge is not so much viewed *as inquiry* as integrating some of its aspects and as aiming at the fixation of true beliefs (Tiercelin, 2005, conclusion): hence, trust the a posteriori results of science, so as to avoid any highfalutin "armchair" metaphysics, without indulging into a straightforward naturalized or scientistic metaphysics; determine also whether and why, for example, scientific theories are true or can be viewed as *justified true beliefs*. But never forget that

> empirical science at most tells us what *is* the case, not what *must* or *may* be (but happens not to be) the case. Metaphysics deals in possibilities. And only if we can delimit the scope of the possible, can we hope to determine empirically what is actual. This is why empirical science is dependent upon metaphysics and cannot usurp the latter's proper role.
>
> *(Lowe, 1998: 5, 9 and 16)*

Third, determine whether, and in what sense, the categories of our thinking and our language are not mere "functions of judgment", as Kant believed, but the mirror, well and truly, of the categories of *reality*. And finally, examine the type of *reality* such theories are talking about, the nature of the actual (i.e. categorical or dispositional) properties constituting it, the causal and nomic relations between them and what is necessary to guarantee the unity, in short, "the cement" of things (Tiercelin, 2011a, 2011b). Now, we start to understand why such a metaphysical and realistic commitment is not only possible but desirable: indeed, if the founder of pragmatism thought the problem of universals was "as pressing today as ever it was" (something Russell praised him to have seen), and constituted a real alternative (see W2: 486–487), it was because, finally, for a pragmatist, ethics is primary, and if ethics, from a pragmatist standpoint, is also mostly viewed as an *education* of our moral capacities and dispositions which have more to do with moral *perceptions* than with either some a priori faculty, moral sense or normative prescriptions we should follow (Tiercelin, 2005a: 204ff), then, there is an obligation for the pragmatist, namely, to find a solid basis for a correct account of how our *perceptions* relate us to the *real* world. And this may explain why, from the beginning, Peirce thought the only way to have one's ideas clear about this was to try and find out a correct account of *reality* itself (which incidentally was the motivation for his search into truth, and not the other way round) and to opt for a *realist* position in what he thought was the major point that had to be settled, if one wanted to make any progress at all (and not a "pseudo-problem", as Putnam, 2004, came to think).

Thanks to the pragmatists, we know better too why metaphysics should remain experimental, in touch with the laboratory, which implies the exact opposite both of a literary (or theological) attitude and of a scientistic or dogmatic stance. We should less try and reach a *systematic* or *dogmatic* metaphysics than solve special metaphysical issues and adopt a critical commonsensist and fallibilist position (Tiercelin, 2016), even if it may bring us close, at times, to skepticism (NEM.IV: 343–344; 8. 43). For "it is certainly conceivable that this world which we call the real world is not perfectly real but that there are things similarly indeterminate. We cannot be sure that it is not so" (4. 61). However, "Every attempt to understand anything – every research – supposes, or at least hopes, that the very objects of study themselves are subject to a logic more or less identical with that which we employ" (6.189). Hence, even if the path is rather narrow, it is possible to overcome metaphysical humility and to promote a reasoned form of metaphysical boldness. Provided such an undertaking is done in the spirit beautifully described by the Logician of Milford:

> My philosophy may be described as the attempt of a physicist to make such conjecture as to the constitution of the universe as the methods of science may permit, with the aid of all that has been done by previous philosophers. I shall support my propositions by such arguments as I can. Demonstrative proof is not to be thought of. The demonstrations of the metaphysicians are all moonshine. The best that can be done is to supply a hypothesis, not devoid of all likelihood, in the general line of growth of scientific ideas, and capable of being verified or refuted by future observers.
>
> *(1.7)*

Who could dream of a better way for a pragmatist metaphysician to block moral skepticism and find "a genuine course between the soggy sands of relativism and the cold rocks of dogmatism" (Blackburn, 2001: 29)?

Notes

1 For a recent evaluation of the Italian Pragmatists, see Maddalena and Tuzet (2021).
2 The phrase is from D. F. Pears's "Universals" (1951).
3 Ramsey's pragmatism has received more and more attention (Dokic and Engel, 2002; Hookway, 2000; Misak, 2016; Sahlin, 1990; Tiercelin, 1993a, 2004, 2005, 2014, 2015). See his rejection of "scholasticism" and of the linguistic and grammatical ambiguities surrounding "universals" (Tiercelin, 2004). For example, compare Peirce (5.453; 6.348; 4.344) with Ramsey:

> The chief danger to our philosophy, apart from laziness and woolliness is *scholasticism*, the essence of which is treating what is vague as if it were precise and trying to fit it into an exact logical category. A typical piece of scholasticism is Wittgenstein' view that all our everyday propositions are completely in order and that it is impossible to think illogically.
>
> (*PP*: 7)

4 See his rejection of "metaphysical realism" and his long search for an "internal" then "natural" realism, which drove him to view pragmatism as "a way of thinking ... of lasting importance and an option (or at least an 'open question') that should figure in present-day philosophical thought" (1994; 1995: xi; Tiercelin, 2002a).
5 For recent work, Pihlström, Stadler and Weidtmann (2017).
6 For a criticism of them and of various followers of the "Canberra plan", see Tiercelin (2011a: 361–380).
7 Gale (2002: 479–480). See MW3: 127; MW3:76; LW1: 17; LW1, 194; LW1: 332–333; MW6: 52; LW1: 17.
8 Price joins the pragmatists (James, Dewey, Rorty) in denying that language is made to mirror the world – hence the representationalist picture of the "Great Mirror". He joins Wittgenstein in holding that language has a variety of distinct uses: one must attend to the plurality of ways in which it functions ("linguistic pluralism"). He also expresses some sympathy not only for semantic anti-realism and for inferentialism but also for semantic deflationism. On the metaphysical branch, Price rejects ontological realism and reductionism and opts for a version of expressivism, but one which is global and not only local. He also favors, alongside deflationism, Carnapian neutralism, resuscitating Carnap's distinction between external (metaphysical) questions and internal (linguistic) questions. Price denies that the resulting combination, his "pragmatist", "expressivist" and "pluralist" metaphysics and semantics, is incompatible with naturalism. It is naturalism "without mirrors", a "subject naturalism", centered on the genealogy of our practices (a piece of anthropology), in opposition to "object-naturalism", which focuses on the way in which we represent the external world. His own version of pragmatism is meant to respect Wittgenstein's *motto* "I'll teach you differences". I have raised some doubts about Price's version of naturalistic anti-representationalism (Tiercelin, 2013: 661).
9 For more connections between his views and those of Peirce and Ramsey, in particular, but also with the main contemporary theories of truth, see Haack (1976, 1977); Misak (1991); Tiercelin (2005, 2014 and 2018).
10 "Being both fallibilists *and* anti-sceptics" is "*the* unique insight of American pragmatism," Putnam noted (1994: 152). But how can one maintain the truth there is in fallibilism without giving everything to the skeptic or to the irrationalist relativist? Both James and Peirce were eager to "value tolerance and pluralism" and to avoid "the epistemological scepticism that came with that tolerance and pluralism" "without tumbling back into moral authoritarianism" (Putnam, 1995: 2).
11 Concerning, in particular, the psychologization of pragmatism because of the fierceful individualism it induces, exposing us to the arbitrariness and fantasies of the "literary circles" and to "sham reasoning", Which, indeed, makes real differences among the pragmatists on such (still very alive and important) questions as: should one view philosophy as closer to literature rather than to science? On this, however, Peirce finds Schiller guiltier than James (Tiercelin, 2017, 2018).
12 "It must not be imagined that any notable realist of the thirteenth or fourteenth century took the ground that any "universal" was what we in English should call a "thing", as it seems, that, at an earlier age, some realists and some nominalists too, had done ... their very definition of "universal" admits that it is of the same generic nature as a word, namely, is: "*Quod natum aptum est praedicari de pluribus*". Neither was

it their doctrine that any "universal" *itself* is real. They might, indeed, some of them, think so; but their realism did not consist in *that* opinion, but in holding that what the word *signifies*, in contradistinction to what it can be truly said of, is real" (1.27n1).

13 "Anybody may happen to opine that "the" is a real English word; but that will not constitute him a realist. But if he thinks that, whether the word "hard" itself be real or not, the property, the character, the predicate, *hardness* is not invented by men, as the word is, but is really and truly in the hard things and is one in them all, as a description of habit, disposition, or behavior, *then,* he is a realist" (1.27n1).

14 "I myself went too far in direction of nominalism when I said that it was a mere question of the convenience of speech whether we say that a diamond is hard when it is not pressed upon, or whether we say that it is soft until it is pressed upon. I *now* say that experiment will prove that the diamond is hard, as a positive fact. That is, it is a real fact that it *would* resist pressure, which amounts to extreme scholastic realism" (8. 208).

References

Blackburn, S. (2001) *Being Good*, Oxford: Oxford University Press.
Boisvert, R. D. (1988) *Dewey's Metaphysics*, New York: Fordham University Press.
Dewey, J. (1915) "The Subject-Matter of Metaphysical Inquiry", *Journal of Philosophy*, vol. 12, no. 13, 337–345.
———. (1967–1972) *The Early Works (EW), 1882–1898*, 5 vols., Carbondale: Southern Illinois University Press.
———. (1976–1983) —*The Middle Works (MW), 1899–1924*, 15 vols., Carbondale: Southern Illinois University Press.
———. (1981–1990) *The Later Works (LW), 1925–1953*, 17 vols., Carbondale: Southern Illinois University Press.
D. F. Pears's "Universals" (1951) *The Philosophical Quarterly*, vol. 1, no. 3, 1951, 218–227, repr. in *Logic and Language*, A. N. Flew (ed.), Oxford, 1955.
Dokic, J., and Engel, P. (2002) *Ramsey, Truth and Success*, London: Routledge.
Gale, R. M. (2002) "The Metaphysics of John Dewey", *Transactions of the Charles S. Peirce Society*, vol. 38, no. 4, 477–519.
Haack, S. (1976) "The Pragmatist Theory of Truth," *British Journal for the Philosophy of Science*, vol. 27, 231–249.
———. (1977) "Pragmatism and Ontology: Peirce and James," *Revue Internationale de Philosophie*, vol. 31, 377–400.
———. (1997) "We, Pragmatists…Peirce and Rorty in Conversation," *The Partisan Review*, vol. LXIV, no. 1, 91–107.
———. (2004) "Pragmatism: Old and New", *Contemporary Pragmatism*, vol. 1, no. 1, 3–41.
Hookway, C. (2000) *Truth, Rationality and Pragmatism: Themes from Peirce*, Oxford: Clarendon Press.
Jackson, F. (1998) *From Metaphysics to Ethics. A Defense of Conceptual Analysis*, Oxford: Clarendon Press.
James, W. (1975–1988) *The Works of William James (WWJ)*, F. H. Burkhardt, F. Bowers and I. K. Skrupskelis (eds.), Cambridge: MA: Harvard University Press.
———. (1975) *Pragmatism (P)* in *WWJ.*, vol. 1.
———. (1979) *Some Problems of Philosophy (SPP)*, in *WWJ.*, vol. 7.
Kitcher, P. (2018) "James and Pragmatism: The Road Not Taken", *The Oxford Handbook of William James*, A. Klein (ed.), available online at https://www.oxfordhand James books.com/view/10.1093/oxfordhb/9780199395699.001.0001/oxfordhb-9780199395699-e-35.
Klein, A. (ed.) (2018) *The Oxford Handbook of William James*, Oxford: Oxford University Press.
Langton, R. (1998) *Kantian Humility. Our Ignorance of Things in Themselves*, Oxford: Oxford University Press.
Lewis, D. (2009) "Ramseyan Humility," in D. Braddon-Mitchell, R. Nola and D. Lewis (eds.) *The Canberra Programme*, Oxford: Oxford University Press, 203–222.
Lowe, E. J. (1998) *The Possibility of Metaphysics*, Oxford: Clarendon Press.
Maddalena, G., and Tuzet, G. (eds.) (2021) *The Italian Pragmatists: Between Allies and Enemies*, Leiden: Brill.
Misak, C. (1991) *Truth and the End of Inquiry a Peircean Account of Truth*, Oxford: Oxford University Press.
———. (2016) *Cambridge Pragmatism: From Peirce and James to Ramsey and Wittgenstein*, Oxford: Oxford University Press.
Peacocke, C. (1999) *Being Known*, Oxford: Clarendon Press.
———. (2004) *The Realm of Reason*, Oxford: Clarendon Press.
Pears, D. F. (1951) "Universals", *The Philosophical Quarterly*, vol. 1, no. 3, 218–227.

Peirce, C. S. (1931–1958) *Collected Papers of Charles Sanders Peirce* (1931–1958), vols. 1–6, C. Hartshorne and P. Weiss (eds.) (1931–1935), vols. 7–8, A. Burks (ed.) (1958), Cambridge: The Belknap Press of Harvard University Press. Abbrev: Volume number, paragraph number, e.g.: (5.348).

———. (1976) *The New Elements of Mathematics* (NEM), C. Eisele (ed.), 4 vols., Mouton: The Hague.

———. (1982–2000) *Writings of Charles S. Peirce: A Chronological Edition* (1982–2000), 6 vols, Peirce Edition Project (ed.), Bloomington: Indiana University Press. (W+ Volume number + page number (e.g.: W2: 234).

Pihlström, S., Stadler, F., Weidtmann, N. (eds.), (2017) *Logical Empiricism and Pragmatism*, Dordrecht: Springer, Vienne Circle Institute Yearbook.

Price, H. (2011) *Naturalism without Mirrors*, Oxford: Oxford University Press.

Putnam, H. (1978) *Meaning and the Moral Sciences*, London: Routledge & Kegan Paul.

———. (1987) *The Many Faces of Realism*, Open Court: La Salle, Illinois.

———. (1990) *Realism with a Human Face*, J. Conant (ed.), Cambridge, MA: Harvard University Press.

———. (1992) *Renewing Philosophy*, Cambridge, MA: Harvard University Press.

———. (1994) *Words and Life*, J. Conant (ed.), Cambridge, MA: Harvard University Press.

———. (1995) *Pragmatism, an Open Question*, Oxford: Blackwell.

———. (2004) *Ethics without Ontology*, Cambridge, MA: Harvard University Press.

Ramsey, F. P. (1990) *Philosophical Papers* (PP), D. H. Mellor (ed.), Cambridge: Cambridge University Press.

Rorty, R. (1961) "Pragmatism, Categories and Language", *The Philosophical Review*, vol. 70, 197–223.

———. (1979) *Philosophy and the Mirror of Nature*, Princeton: Princeton University Press.

———. (1982) *Consequences of Pragmatism*, Minneapolis: University of Minnesota Press.

Rosenthal, S. B. (2000) "William James on the One and the Many," in K. Oehler (ed.), *William James, Pragmatismus*, Berlin: Akademie Verlag, 93–108.

Sahlin, N.-E. (1990) *The Philosophy of F. P. Ramsey*, Cambridge: Cambridge University Press.

Seigfried, C. H. (1990) *William James's Radical Reconstruction of Philosophy*, Albany: SUNY Press.

Talisse, R. B., and Aikin, S.F. (2008) *Pragmatism, a Guide for the Perplexed*, London and New York: Continuum.

Tiercelin, C. (1986) "Le vague est-il réel? Sur le réalisme de C.S. Peirce", *Philosophie*, vol. 10, 69–96.

———. (1992) "Vagueness and the Unity of Peirce's Realism," *Transactions of the C.S. Peirce Society*, vol. XXVIII, no. 1, 51–82.

———. (1993a) *La pensée-signe: études sur Peirce*, Nîmes: Editions Jacqueline Chambon, available at: http://books.openedition.org/cdf/2209.

———. (1993b) *C. S. Peirce et le pragmatisme*, Paris: Presses Universitaires de France, available at: http://books.openedition.org/cdf/1985.

———. (2002a) *Hilary Putnam, l'héritage pragmatiste*, Paris: Presses Universitaires de France, available at: http://books.openedition.org/cdf/2010.

———. (2002b) "Philosophers and the Moral Life," *Transactions of the C.S. Peirce Society*, Essays in Honor of Richard S. Robin, vol. XXXVIII, no. 1, 307–326.

———. (2004) "Ramsey's Pragmatism," *Dialectica*, special issue on F. P. Ramsey, P. Engel and J. Dokic (eds.), vol. 58, no. 4, 529–547.

———. (2005) *Le doute en question: parades pragmatistes au défi sceptique*, Paris: Editions de l'éclat. Re-edit. 2016 with a new postface.

———. (2006) "Metaphysics Without Ontology?", On Putnam's *Ethics without Ontology*", *Contemporary Pragmatism*, 55–66; Putnam's reply, 92–94.

———. (2008) "The Fixation of Knowledge and the Question-Answer Process of Inquiry", in F. Lihoreau (ed.), *Knowledge and Questions, Grazer Philosophische Studien*, vol. 77, 23–44.

———. (2011a) *Le Ciment des Choses: petit traité de métaphysique scientifique réaliste*, Paris: éditions d'Ithaque.

———. (2011b) *La connaissance Métaphysique*, Paris: Fayard; English version *Metaphysical Knowledge*, available at: https://docs.google.com/viewer?a=v&pid=sites&srcid=ZGVmYXVsdGRvbWFpbnxjbGF1ZGlu-ZXRpZXJjZWxpbnxneDo0N2U4ODEzZDA5MWExY2E2.

———. (2013) "No pragmatism without realism," Review of Huw Price's *Naturalism without mirrors*. Oxford: Oxford University Press, 2011, *Metascience*, 11, vol. 22, no. 3, 659–665.

———. (2014) *The Pragmatists and the Human Logic of Truth*. Paris, online edition series « La Philosophie de la connaissance au Collège de France », available at: http://books.openedition.org/cdf/3652.

———. (2015) "Chance, Love and Logic: Ramsey and Peirce on Norms, Rationality and the Conduct of Life," in J. Persson, G. Hermerén and E. Sjöstrand (eds.), *Against Boredom. 17 Essays on Ignorance, Values, Creativity, Metaphysics, Decision-making, Truth, Preference, Art, Processes, Ramsey, Ethics, Rationality, Validity, Human Ills, Science, and Eternal Life to Nils-Eric Sahlin on the Occasion of His 60th Birthday*, Stockholm: Fri Tanke, 221–256.

———. (2016) "In Defense of a Critical Commonsensist Conception of Knowledge," *International Journal for the Study of Scepticism*, vol. 6, 182–202.

———. (2017) "Was Peirce a Genuine Antipsychologist?" *European Journal of Pragmatism and American Philosophy*, R. Calcaterra and R. Dreon (eds.), available at: https://journals.openedition.org/ejpap/1003.

———. (2018) "James and Peirce," *The Oxford Handbook of William James*, A. Klein (ed.), available at: https://www.oxfordhandbooks.com/view/10.1093/oxfordhb/9780199395699.001.0001/oxfordhb-9780199395699-e-21.

———. (2019) *Pragmatism and Vagueness*, The Venetian Lectures, G. Tuzet (ed.), Mimesis International, no. 31.

———. (2020) "Principes métaphysiques, principes théologiques", in A. Declos and J.-B. Guillon (eds.), *Les principes Métaphysiques*, Paris: Collège de France, available at: https://books.openedition.org/cdf/7860.

van Inwagen, P. (2005) "What There Is," *Times Literary Supplement* (25 April), 11–12.

26

PEIRCE, JAMES, AND DEWEY AS PHILOSOPHERS OF SCIENCE

Jeff Kasser

Philosophy of science is philosophy enough.

(Quine 1953: 446)

Introduction

This essay pursues themes in the work of the paradigmatic pragmatists that bears on the philosophy of science. It does not try to isolate their pragmatism and consider its bearing on philosophy of science. I think that Peirce, James, and Dewey are uncommonly interesting and underappreciated philosophers, and I think that careful consideration of their work would enrich many areas of philosophy, none more than philosophy of science. I further think that "pragmatism" tends to generate more heat than light. There is value in trying to determine what pragmatism is and then tracing some of its implications for contested issues in "mainstream" philosophy of science (Almeder 2007). And there is value in explicating pragmatism via its contribution to discussions of realism and its competitors (Pihlström 2008). But I am far more confident that the classical pragmatists have been undervalued than I am about how to decide to what extent Kuhn or Quine or Putnam or van Fraassen counts as a pragmatist.

Each of the exemplary pragmatists was deeply engaged in the practice of science. Peirce worked as a geodesist for more than 30 years, during which he made significant contributions to the theory and practice of measurement, including techniques for correcting errors of observation. He and his student Joseph Jastrow carried out a remarkable experiment on smallest-noticeable sensory differences beginning in 1883. Their experiment clearly anticipates both the notion of and the rationale for randomized experiments, usually credited to R. A. Fisher in the 1920s (Stigler 1999: 192–196). Peirce's friend William James received a medical degree, though he never practiced, and he began his long teaching career at Harvard with a course in physiology. He founded the first psychology laboratory on American soil in the 1870s. Dewey helped set up and run two psychology laboratories (at Michigan and at Chicago), and he founded the groundbreaking laboratory school at Chicago. His creative and careful approach to empirical pedagogical research revolutionized the field. Each of the classic pragmatists, then, participated substantively in the methodologically fraught rise of psychology as a science. Peirce, James, and Dewey also each came of age in an intellectual environment suffused with Darwin and evolution (Pearce 2020). As will hopefully emerge below, Darwin's long shadow combined with the divorce between philosophy and psychology (and the resulting professionalization of each discipline) to generate a fruitful environment for new thinking about science. Peirce, James, and Dewey did important work in philosophy of science just before it emerged as a recognized subfield and started telling its own founding narrative. What follows sketches some of what's been lost from that narrative.

Peirce

Fruitful connections between Peirce and later philosophy of science are so plentiful that one survey of pragmatist philosophy of science finds itself compelled to exclude him on just these grounds (Pihlström 2008: 29). Peirce's most influential papers originally appeared in *Popular Science Monthly* in 1877–1878 and inaugurated a series called "Illustrations of the Logic of Science." As the journal and series titles suggest, these papers provide an appropriate launching point for surveying Peirce's work in philosophy of science.

The first of the "Illustrations" is called "The Fixation of Belief," and it stands as a controversial classic of pragmatism. Its resources seem naturalistic and modest, but its conclusions are ambitious. Peirce aims to vindicate the scientific method of settling belief. That method receives an ecumenical characterization, according to which

> [t]here are real things, whose characters are entirely independent of our opinions about them; those realities affect our senses according to regular laws, and, though our sensations are as different as our relations to the objects, yet by taking advantage of the laws of perception, we can ascertain by reasoning how things really are, and any man, if he have sufficient experience and reason enough about it, will be led to the one true conclusion.
>
> *(W3 254)*

This vindication is to come at the expense of such methods as tenacity, which consists of "taking any answer to a question which we may fancy, and constantly reiterating it to ourselves, dwelling on all which may conduce to that belief, and learning to turn with contempt and hatred from anything which might disturb it" (W3 248–249). Strikingly, Peirce does not complain that tenacity and its cousins produce unjustified or irrational beliefs. He even grants that such methods can successfully settle belief for suitably isolated or unreflective inquirers. "Fixation" claims only that unscientific methods can't be deliberately chosen by sufficiently clear-headed inquirers (where clear-headedness includes knowledge that we influence one another's beliefs).

The argument (if in fact it is an argument; see Short 2000) has received a great deal of critical attention. For current purposes, its characteristic pragmatic style matters more than its success. In a pattern we will see replicated by James and Dewey, something practical, both in the sense of "concerned with thriving" and in the sense of "part of an actual practice," provides a starting point and explanatory framework from which something more intellectual or theoretical gets explicated and vindicated. In the case at hand, the (relatively) unproblematic practical aim of settling belief gets used to draw out and ratify implicit intellectual standards. Indulging in the sort of oversimplification endemic to survey essays like this one, we can see here a reversal of the ordinary practice of philosophers, for whom the true retains robust explanatory priority to the good.

Peirce's doubt-belief theory of inquiry, which is premised in "Fixation," further links scientific inquiry and practical problem-solving. It insists that doubts must be "genuine" or "living," it treats beliefs as resources for inquiry (by, for instance, providing a standard of serious possibility), and it evaluates them in a stability-centered, forward-looking way rather than by their pedigree. This is part of Peirce's attempt to bring "laboratory thinking" rather than "seminary thinking" to bear on philosophy (CP 1.126–129). This aspect of Peirce's work (supplemented by some Deweyan insights) has been developed, principally by Isaac Levi, into a formidable contemporary approach to inquiry. Levi draws heavily on these pragmatists while linking their work to epistemic conservatism, Bayesianism, philosophy of statistics, and decision theory (see, for instance, Levi 1991).

The next paper in the "Illustrations," "How to Make our Ideas Clear," formulates what later became known as the pragmatic maxim:

> Consider what effects, which might conceivably have practical bearings, we conceive the object of our conception to have. Then, our conception of these effects is the whole of our conception of the object.
>
> *(W3 266)*

This principle has often served as the basis for comparisons between Peirce's and James's versions of pragmatism. Within mainstream philosophy of science, it has been used to link Peirce to the verification principle of cognitive meaningfulness that figured centrally in logical positivism and its descendants. Though Peirce shared with the positivists an interest in purging philosophy of metaphysical nonsense, he never succumbed to the notion that metaphysics as such constituted nonsense. Accordingly, Peirce instructs us to clarify the meanings of our terms and statements in accordance with (possible) experience rather than insisting on a criterion of meaningfulness. As Cheryl Misak has argued, Peirce also offers a more generous construal of "experience" than one finds among positivist and post-positivist philosophers (Misak 1995: 106ff). Since experience roughly consists of anything suitably forced upon the inquirer, room is carved out for, e.g. moral and mathematical experience to figure in legitimate inquiry.

Verificationism is typically closely linked to anti-realism in the philosophy of science. Statements about unobservable reality are not directly verifiable and so are semantically and epistemically problematic for positivists. Good scientific theories need not (and generally cannot) describe reality beyond experience. Misak, emphasizing Peirce's affinities with verificationism, classifies him as an anti-realist, where "a realist epistemology has it that truth transcends ... our beliefs about what is true" (Misak 1995: 121).[1] Truth, for Peirce, is properly explicated in terms of best possible belief, and beliefs are constrained by experience. So the truth about a matter is what would be believed about it were inquiry into the matter to be undertaken indefinitely.

However, our quote above from "Fixation" presents a decidedly realist impression with its talk of "regular laws" allowing us to ascertain "how things really are" and leading us to "the one true conclusion." Peirce's views about laws of nature and about abductive inference have made him a favored target of anti-realists (van Fraassen 1989). The fact that Peirce, like his pragmatist successors, precedes the 20th-century debates about realism and alternatives to it like instrumentalism, constructive empiricism, and constructivism means that it is perilous to enlist him straightforwardly on behalf of one position or another. A happier way of making the point is to suggest that Peirce and his fellow pragmatists, precisely because of how old their work is, can offer new takes on these all-too-familiar debates in contemporary philosophy of science. In later work, for instance, Misak argues that Peirce "undermines the very contrast that the instrumentalist position depends upon for its existence: the contrast between beliefs that aim at truth and beliefs that aim at fulfilling the pragmatic virtues" (Misak 2006: 402). If truth is best understood in terms of ideal belief, then virtues like simplicity (in some suitable sense of the term) and explanatory power emerge as epistemic as well as pragmatic virtues.

Explanation looms large in Peirce's projects. Peirce coined the term "abduction" and was arguably the first philosopher to insist on its status as a distinct type of inference (in the last of the "Illustrations," among other places). Some of Peirce's writings on abduction present it as a kind of inference to the best explanation, a way of confirming hypotheses in what many call the context of justification. Others suggest a kind of logic for generating and prioritizing hypotheses within what's often called the context of discovery. For van Fraassen and others, the demand for explanation threatens to undermine empiricism, eventuating in metaphysics in the pejorative sense. And indeed, Peirce's ambitious pursuit of explanation, crystalized in his overriding maxim, "Do not block the road of inquiry!" induced him to insist that even the laws of nature require explanation. In an intriguing synthesis of Darwin and physics, Peirce argued that laws could be accounted for only via an evolutionary process. He was thus led, in the second half of his career, into an evolutionary metaphysics that strikes some contemporary philosophers as excessive and unfortunate. Reconciling Peirce's empiricist sympathies (as reflected in the pragmatic maxim and in many other places) with his metaphysical ambitions is challenging and fascinating. And it must be kept in mind that Peirce's close contact with the science of his day tends to discipline his speculations and often renders them fruitful. The physicist Lee Smolin, for instance, has recently defended a loosely Peircean evolutionary account of physical laws (Smolin 1997).

Peirce was uncontroversially a realist in at least one sense, namely that which contrasts with nominalism in the history of philosophy. He held that generals are real (but not actual), that all dogs have something in common that is not arbitrary or merely in minds or words. Peirce very quickly links this old-fashioned-sounding realism with very contemporary issues about scientific laws, causation, and explanation. He maintains, for instance, that nominalists, including those who adopt simple regularity accounts of scientific laws, are poorly positioned to account for the strength of our confidence that copper will continue to conduct electricity in the future (Legg 1999). Laws, for Peirce, are not actual; one cannot *encounter* them. But they are real because they govern actual occurrences and allow us to predict and explain those events.

The central concept in the "Illustrations," however, is probability. Peirce embodied the 19th-century probabilistic revolution as thoroughly as anyone. Under the guidance of the pragmatic maxim, he developed first a long-term limit frequency approach to the metaphysics of probability and then a propensity approach that anticipated important work by Popper. By 1884, he had provided a philosophical basis for what became the Neyman-Pearson approach to statistical inference (Hacking 1990). On the way to rejecting proto-Bayesian accounts of probabilistic reasoning, he developed two important resources on behalf of Bayesianism: a measure of gross weight of evidence that anticipates Keynes on weight of argument, and a measure of net weight of evidence in terms of the logarithm of the odds ratio, which anticipates work by Turing and Good (Kasser 2016). In addition, Short has argued that Peirce's work yields a statistical, non-mechanistic type of scientific explanation importantly distinct from any currently on offer (Short 2007: 112ff). This Peircean account strikingly illuminates explanatory practices in statistical mechanics/thermodynamics and in evolutionary biology. Finally, Peirce became the first modern philosopher to deny determinism and embrace chance as a real feature of the universe. In doing so, he was influenced by everything from his work as a measurement theorist to his evolutionary metaphysics (Hacking 1990).

Later in life, Peirce began thinking of deduction, induction, and abduction less as distinct forms of inference and more as phases of inquiry. Abduction suggests hypotheses with explanatory and other virtues, deduction derives testable consequences from such hypotheses, and induction determines the extent to which those consequences are borne out. His characteristic synthesis of realist ambition and empiricist modesty again manifests itself when he argues both that (1) scientific inquiry is self-correcting in the sense that, pursued diligently enough and for long enough, it will correct its own mistakes and cannot but arrive at the truth, and (2) induction cannot generally tell us how well supported our scientific claims are; our only security comes from the fact that we are using a method that will correct its own mistakes. The in-principle success of science is assured, but not at all at any actual moment.

Peirce's conception of method is by no means exclusively focused on the indefinite long run, however. His pioneering work on the economics of inquiry is no afterthought; it is a central part of his account of scientific method (Wible 2014). Peirce takes the finitude of scientific resources, the social nature of the enterprise, and the importance of communicative practices within science (Ransdell 2000) exceptionally seriously. Even with so prescient an understanding of scientific practices, however, Peirce denies that science should be characterized methodologically. Instead, he thinks of science in terms of its animating spirit. The genuine inquirer must be willing to subordinate selfish goals to the needs of the scientific community. She keenly desires to see questions settled and to put discoveries to use. But she regards the "diligent inquiry into truth for truth's sake" (CP 1.44) as a qualitatively higher goal, one for which the thoroughly scientific inquirer must be willing to sacrifice tenure, fame, fortune, utility, and confidence in the less-than-long-run (Short 2020).

No philosopher since Peirce, I suggest, combines distinctive positions on technical issues in the epistemology and metaphysics of science with equally detailed reflections on the animating values of scientists and the communal nature of inquiry as he does. If many of Peirce's views look to the future, the systematicity with which he combines them can seem old-fashioned, though not necessarily in a bad way. Peirce is an architectonically minded philosopher, as is reflected in his extensive work on the classification of sciences (using "science" in the sense of *Wissenschaft*). He is deeply concerned about circularity, logic, and scientific method. No methodological claim can be vindicated via factual claims discovered by deploying the method itself. This does not render Peirce a defender of a priori, anti-naturalist philosophy of science. For him, philosophy is a cenoscopic discipline; it is observational but relies only on everyday experiences rather than evidence derived from the special sciences. Those sciences can rely on philosophy, but philosophy can draw on them only for examples, suggestions, or inspiration, not for premises. This fascinating not-fully-immersive naturalism contrasts, as we will soon see, with the work of Peirce's pragmatist successors.

James

In philosophy of science as in other areas, James generally gets contrasted rather sharply with Peirce. He is known for taking pragmatism in a less scientific and less scientistic direction. So James often plays the instrumentalist foil to Peirce's supposed scientific realism. We have touched on the complexity of Peirce's relationships with realism, and James turns out to be equally difficult to pigeonhole. James says things like:

> [A]s the sciences have developed farther, the notion has gained ground that most, perhaps all, of our laws are only approximations. The laws themselves, moreover, have grown so numerous that there is no counting them … investigators have become accustomed to the notion that no theory is absolutely a transcript of reality, but that any one of them may from some point of view be useful.
>
> *(P 33)*

As is often the case, it's worth attending to the similarities between Peirce and James, not just the differences. Peirce enthusiastically shares James's suspicions about scientific claims to perfect precision or to completeness. Neither pragmatist holds out for a "transcript of reality," and neither is blind to science as a lived practice that is useful in any number of ways.

James indeed pushes both pluralism and usefulness farther than does Peirce, insisting that theories are "*instruments, not answers to enigmas*" (P: 32; emphasis in original). Despite genuine similarities between the pragmatic maxim (in both its Peircean and Jamesian forms) and the positivists' verification principle, James's instrumentalism does not rest on a freighted distinction between observational and theoretical vocabulary or between observable and unobservable things. It arises instead from a descriptive claim about how scientists deploy their theories. The Jamesian scientist treats theories the way Emerson suggests one read books; she skims many of them and takes what she can use. And James's instrumentalism contrasts less with scientific realism than with a thesis we might call "scientific totalism." He is anxious to give science its due but even more anxious to deflate any pretensions that it has to exhaust legitimate inquiry. Science

> must be constantly reminded that her purposes are not the only purposes, and that the order of uniform causation which she has use for, and is therefore right in postulating, may be enveloped in a wider order, on which she has no claims at all.
>
> *(PP 2:1179)*

James and Peirce thus exhibit distinct aspects of the overused term "naturalism." Peirce is insistent upon avoiding circularity and developing a method of inquiry without appealing to any deliverances of that method. But scientific inquiry is, for Peirce, uniquely authoritative. It is to have the last word in theoretical matters, even if tradition and instinct often provide superior advice about pressing practical issues. James, on the other hand, demotes science's status to that of one (important) practice among others. At the same time, he anticipates 20th-century developments in naturalized epistemology by treating science, philosophy, religion, and other inquiring practices as themselves objects of scientific investigation. Both philosophy itself and individual philosophers' arguments and conclusions are understood as ways of coping with and adapting to a complex world largely structured by needs. While Peirce engages with philosophy of science thematically and explicitly, James's views about philosophy of science tend to emerge as by-products of his reflections on practicing both philosophy and science.

James's deep interest in the young sciences of evolutionary biology and psychology nurtured a fascination for the processes by which hypotheses get articulated, tested, endorsed, and adopted or rejected. Without resorting to anything as simple-minded as the genetic fallacy, James diagnoses the role of needs, wants, sensibility, and temperament in the evidentialism of the tough-minded or the idealism of the tender-minded. Where Peirce envisions a method of inquiry that ultimately filters out everything idiosyncratic or arbitrary, James insists that science cannot and should not eliminate individual interests. Such interests, he insists, get their say in an ideal last word that is not exhausted by science.

The centrality of coping and temperament for James long predates his public announcement of pragmatism in 1898's "Philosophical Conceptions and Practical Results." These themes loom large in important early papers like "Remarks on Spencer's Definition of Mind as Correspondence" (1878) and in his monumental *The Principles of Psychology* (1890), in which a lot of philosophy of science gets used, if not exactly mentioned. James was clear about the philosophical consequences of his work in psychology. Truth and reality are to be understood in terms of successfully taking-true and thinking-real, and these are, in turn, explained in large part by needs and interests. We adopt certain theories and deploy certain concepts as part of a project of coping and adapting; an explanation of such intellectual activities in terms of correspondence to the way the world is performs at best a secondary function. A thoroughly scientific explanation of science shows why it works so well, but it also shows why it won't work as the be-all and end-all of our intellectual lives.

Principles, with its sustained discussion of introspective psychology, provides an extended example of Jamesian science and philosophy of science. The trail of the human inquirer is over every introspection, but that does not prevent such an approach (suitably supplemented and corrected) from delivering genuine and valuable scientific results. James evidences little concern with the (then and now) dominant question about introspection, viz. its fallibility. Introspection is valuable because it is ready-to-hand, and it combines enough reliability and informativeness to be useful to investigators. Characteristically, he worries not just about its reliability but also about its effects on its practitioners. Does it tend to induce a depression in those who pursue it, and would that depression blind the psychologist to important aspects of experience? He also insists on combining the deliverances of introspection with experimental, physiological, and statistical evidence. James was far from celebrating every quirk of every inquirer; he thought that previous philosopher-psychologists had committed serious errors by, for instance, distorting the continuity of experience and relegating experiences of relations to second-class status. From carefully attended-to and suitably corrected individual experiences, each of which is shaped by needs and interests, evidence can accumulate on behalf of improved theories. "That theory will be most generally believed which, besides offering us objects able to account satisfactorily for our sensible experience, also offers those which are most interesting, those which appeal most urgently to our aesthetic, emotional, and active needs" (PP 2:940). James's task in 1890 did not require him to pronounce theories that meet this desideratum truer than their competitors, and our task does not compel us to engage James's pragmatism about truth. The point is rather that a surprising amount of James's later pragmatism gets illuminated by his thoughts on the theory and practice of science. For more on the topics touched on in this paragraph, see Myers (1986), and for a fascinating and detailed engagement with James's (and Peirce's) thoughts about the boundaries between scientific psychology and the metaphysics of mind, see Klein (2008).

If James's scientific writings attempt to rein in the ambitions of science, his writings about religion reflect an eagerness to give science its due. "The Will to Believe" offers a respectful and traditional characterization of science. Scientific questions are neither forced nor momentous, and so James recommends a Peircean patience with the time horizons of inquiry. James can seem almost Popperian in his willingness to let the norm of error avoidance trump the need to attain the truth when doing science. In the context of discovery, James recommends something of a role for one's passional nature, but he endorses silencing the passions in the context of scientific justification. Though "The Will to Believe" is firmly ensconced in the first tier of outrage-provoking philosophical essays, nothing it says about science would scandalize a logical positivist. *The Varieties of Religious Experience*, a naturalistic study of the boundary between the natural and the supernatural, likewise conveys a real scientific diffidence about its massive and varied subject matter. As James works to develop a "science of religions," his trust in explanatory "overbeliefs" is tentative, and his conclusions remain modest. He amasses a wealth of data (generally in the form of reports of individual experiences) and avows his belief that some religious experiences connect us to a "more" or an "other" that eludes our ordinary conceptual and cognitive resources. But he does not claim to have made a generally convincing case for this conclusion.

It is only in works near the end of his life that James offers bolder metaphysical assertions, and even there he remains keen to render unto science the things that are science's. *A Pluralistic Universe* mixes caution and ambition by proclaiming that religious experiences "point with reasonable probability to the continuity of our consciousness with a wider spiritual environment from which the ordinary prudential man (who is the only man that scientific psychology, so called, takes cognizance of) is shut off" (PU 135). This remarkable, virtually untestable claim gets asserted on the basis of empirical, testimonial evidence and philosophical argumentation, but the assertion is hedged with a modest probabilistic qualifier and the whole argument is forcefully distinguished from science.

Dewey

Unlike his pragmatist predecessors, Dewey lived to see the rise of philosophy of science as a recognized subfield of the discipline. His enthusiasm for the development is likely to have been muted. While, as we shall see, Dewey's work in philosophy of science is more extensive and more impressive than has generally been realized, he shares with James a distaste for what the older pragmatist characterized as the "overtechnicality and consequent dreariness" that set in as philosophy became an academic discipline among others (PU 13).[2] James and Dewey would both insist that philosophy of science has grown drearier and more technical in the intervening decades. Peirce, however, thought philosophy insufficiently technical in its vocabulary and style and urged that it become more like chemistry and less like theology (CP 2.219–2.226). Philosophy of science, it should be said, is no drearier or more technical than epistemology, and Dewey would have much preferred to situate his theory of inquiry with respect to philosophy of science than to the "epistemology industry" (Hickman 1998).

As with James, the early decades of psychology's emergence as a scientific discipline set the stage for Dewey's reflections on scientific methodology. Largely under the influence of *The Principles of Psychology*, Dewey sought to integrate an articulate attentiveness to experience (drawn from the introspective tradition) with the burgeoning field of physiological psychology. He characteristically found the former excessively subjective and the latter excessively mechanical. The subject matter of psychology is not to be found "within" either the mind (as the introspectionists would have it) or the skull (as the physiologists would have it).[3] It is to be found in transactions involving both organisms and their environments, and those two aspects of reality can be separated only somewhat artificially and perilously. In his famous paper on the reflex arc, Dewey finds not organisms awaiting stimuli from the outside world and then responding thereto but instead processes that combine receptivity, spontaneity, and continual mutual adjustments of organism and environment. For the remainder of his long career, Dewey remains committed to biological metaphors, to explanation in terms of functions and processes rather than components and mechanisms, and to context and continuity as tools of analysis.

Like Peirce and unlike James, Dewey talks frequently and substantively about a logic of science. Also like Peirce, Dewey construes logic much more broadly than is common today, including abduction as well as deduction and induction, thus crossing the supposed gulf between the context of discovery and that of justification. Peirce and Dewey part company, however, about how formally logic should be treated. Peirce, as a result of his concerns about circularity and his commitment to a kind of first philosophy, accuses Dewey of psychologism (CP 8.190). For his part, Dewey denies that logic is coextensive with what he calls "formalistic logic." With their unorthodox conceptions of the place of logic in science, Peirce and Dewey each offer distinctive and promising ways forward from the dilemma that Helen Longino says we find ourselves in after positivism and Kuhn. We "seem to be faced with a choice between two unacceptable alternatives: a logical analysis that is historically unsatisfactory and a historical analysis that is logically unsatisfactory" (Longino 1990: 64).

Dewey's way forward is rather, well, Deweyan. He insists that logic and scientific methodology have been excessively and artificially distinguished. Scientific method is not simply applied logic, but that does not imply that logic is irrelevant to scientific practice (Brown 2012: 267). Logic attends to and formulates the abstract principles implicit in inquiry and thus is self-imposed, Kant-style, on inquiry. Dewey would claim to have made headway on Longino's problem; logic so conceived has both descriptive applicability to inquiry and normative force for inquiry (though both "applicability" and "force" require delicate handling here, lest they suggest that logic gets imposed from outside inquiry rather than developing within it). Peirce, Carnap, and company would continue to worry about circularity and psychologism.

Just as methodology is not simply applied logic, engineering, technology, and practical problem-solving are not simply applied science. Though generally a patient man, Dewey grew exasperated when confronting the dominant Greek valorization of a science of demonstration at the expense of craft or technical knowledge. "It is not too much to say," Dewey writes, "that what Greek science and logic rejected are now the head cornerstone of science" (LW 12:96). Post-17th-century scientific knowledge much more closely resembles classical conceptions of the knowledge of artisans than it does demonstrative knowledge. Galileo, to take perhaps the exemplary case, initiated the reconstruction of physics in a Copernican universe by taking careful measurements of motion and time using simple Archimedean machines. Engineering is no more "applied physics" than physics is "applied engineering."

Dewey's hostility to a particular and influential way of elevating the theoretical above the practical does not license an interpretation of him as a vulgar pragmatist who reduces science to the solution of narrowly practical problems. For one thing, problems are not simply given at the outset of inquiry; they emerge as indeterminate situations get clarified (Brown 2012: 274). More importantly, Dewey's "instrumentalism" does not mean that scientific theories are mere tools, either in the sense that they cannot be given a full semantic interpretation or in the sense that they are simply to be judged by their practical usefulness. Science studies the instrumental properties of things, i.e. their relationships with other things, their transformative potentialities. Empirical inquiries into pressing practical concerns share important continuities with scientific inquiries, but they are not themselves scientific. Peter Godfrey-Smith expresses this point well: Dewey

> does not think that decisions *within* science are made on the basis of specific practical goals and demands. He thinks the opposite; empirical inquiry that is directed and constrained by immediate practical goals is not science at all. Science is the study of instrumental properties of things *without* regard for immediate practical uses.
>
> (Godfrey-Smith 2002: S32)

It is as much a mistake to assimilate science to the solution of pressing practical problems as it is to divorce it from problem-solving all together.

Seen through the eyes of Levi and like-minded methodologists, Dewey looks like a close descendent of Peirce, more explicitly making unsettledness a property of a situation rather than of the mental state of an inquirer. Dewey insists that situations rather than inquirers are the proper bearers of properties like doubtfulness; Peirce would likely be puzzled by the contrast, but Dewey is the clearer of the two on this point. The genuine similarities between their logics/theories of inquiry should not be allowed to obscure profound differences between Peirce's architectonic approach to inquiry and Dewey's contextualist approach. The Deweyan slogan "experience as method" means that the concepts and other tools we need to raise and settle questions cannot be determined in advance and imported into a situation. Dewey insists that our concepts, our questions, and our goals get informed and altered by the situations in which we find ourselves and the problems we face. That's partly why a spectatorial or excessively theoretical approach to inquiry is misguided; it deprives the inquirer of needed resources. However, Dewey's theory faces the charge of excessive contextualism. He needs to explain how useful knowledge can transfer from one situation to another without counting as having been "imposed from outside" the new situation. Dewey can draw on such resources as his theory of habits (another theme shared with Peirce and James) to help counter this charge. More broadly, he holds that continued inquiry can sometimes establish the generalizability of originally local results. Though many details will need to be supplied, Dewey offers a new take on classic methodological debates about inductivism and the role of hypotheses going back to Huygens and Newton as well as to Mill and Whewell and Peirce.

Like James, Dewey tends to get assimilated to the anti-realism side of metaphysical debates in the philosophy of science. And as with both Peirce and James, the actual philosopher is more complex and more interesting than the one in circulation. As had James, Dewey begins from a post-Darwinian naturalism that makes the striving and thriving of organisms in environments explanatorily central. Dewey is a realist about this broad framework, though it's perhaps better thought of as a version of naturalism. Within that explanatory structure, he insists that thought mainly functions to transform, not to copy or to report. Problematic situations are resolved when the "existential" aspects of the situation (i.e. those which are taken as granted, and it matters to Dewey that these facts are taken more than given) are made to align with relevant potentialities and prospects. This can sound importantly anti-realist, especially when realism gets cashed out in terms of mind-independence. But the seriousness with which Dewey treats relations and potentialities (a seriousness shared with Peirce and James) cuts against the tempting social constructivist reading. While it's true that, for Dewey, facts are characterized functionally and situationally, their fittingness for various kinds of intelligent treatment is explained by real features of the world. In the contemporary sense of words like "real" and "fact," Dewey holds that science pursues real facts about constraints and possibilities. Thought does and should transform both the constraints and the possibilities, and it further constitutes them as the constraints and possibilities they are, but this is an unmysterious kind of constitution (Godfrey-Smith 2002). The Deweyan conception features ordinary objects and relations situated in ways that foreground their roles in inquiry.

It does not follow that Dewey ontologically downgrades the extraordinary objects of scientific theories. Dewey's contextualism does suggest a distinction between objects of inquiry that can be encountered situationally and those which cannot. But the pragmatists wrote before the distinctions between observable and unobservable objects and between observational and theoretical terms became so central to philosophy of science, and Dewey's "existential" does not line up with the received view's "observable" or "observed." Science develops techniques and technologies that allow us to encounter or manipulate things we cannot directly observe. So an anti-realism like van Fraassen's would seem oddly spectatorial to Dewey; his view is much closer to Hacking's instrumental realism and to Woodward's manipulationism. Here we see yet another pragmatist approach that attractively combines a naturalistic modesty about human investigators with confidence in the epistemological and practical prospects of well-conducted inquiry.

For Dewey, as for Peirce, inquiry ends when the problematic situation is taken to be settled. Peirce thinks of inquiry as concluding with belief, where beliefs are ideally explicated in terms of the conceivable actions to which they would lead. Dewey prefers to say that *judgment* closes off inquiry. As we have seen, Dewey insists that judgments alter the world at least in the sense that they make manifest some potentialities, perhaps at the expense of others. Though this is not an especially provocative constructivist thesis, it does bring the theoretical and practical together in ways that would trouble Peirce as well as most non-pragmatist philosophers of science. Inquiry-closing judgments, for Dewey, aren't just explicated in terms of possible practical differences; they involve what he calls judgments of practice (Welchman 2002). In science, such judgments aspire to a kind of generality that abstracts from immediate practical interests, but they still involve what Longino calls contextual values in addition to the constitutive ones on which Peirce focuses. So, for Dewey, science is thoroughly suffused with value judgments, and scientific inquiry is continuous with inquiry in fields like ethics and aesthetics. Dewey thus anticipates the post-Kuhnian focus on the role of values in science that has loomed so large in contemporary feminist philosophy of science, among other areas. Dewey's conception of democracy as a kind of cooperative inquiry is of special interest here. Comparison with Peirce's pioneering work on science's essentially social, communicative, and cooperative structures is again valuable. For Peirce, it is science's freedom from practical concerns and time horizons that allows its distinctive and overriding norms to operate. Dewey characteristically sees more continuity between scientific inquiry and democratic decision-making. Dewey's work has played an even larger role in Philip Kitcher's influential writings on science, truth, and democracy than Peirce had played in Kitcher's earlier work on scientific progress and objectivity (see Kitcher 1993, 2001, 2012).

Conclusion

As is appropriate for an essay on pragmatism, we should close by facing forward. Pragmatism in the philosophy of science generically conceived is attractive enough. It partakes of naturalism, realism, and empiricism in ways that many philosophers might find congenial. It takes seriously the history of science and the role of values in inquiry. It foregrounds the fact that science consists of socially realized practices. But the temperaments of the classical pragmatists forge these ingredients into (at least) three different compelling visions of science and philosophy. I love philosophy of science, mainstream, pragmatist, and otherwise, but one thing it could use more of is personality, which Peirce, James, and Dewey supply in abundance.

Notes

1 It should be noted that anti-realism in this sense is compatible with realism about an importantly different issue, viz. whether reality is independent of any particular people's beliefs about it. On Peirce's realism about the latter issue, see Lane (2018).
2 Many of Dewey's readers would be more willing to acquit him of overtechnicality than of dreariness.
3 Peirce, whose semiotic conception of the mind has it that we are in thought rather than thought's being in us, would have approved of Dewey's conception of mindedness, if not of his methodological naturalism.

References

Almeder, R. (2007) 'Pragmatism and Philosophy of Science: A Critical Survey,' *International Studies in the Philosophy of Science* 21(2), pp. 171–195.
Brown, M. (2012) 'John Dewey's Logic of Science,' *HOPOS: The Journal of the International Society for the History of Philosophy of Science* 2(2), pp. 258–306.
Godfrey-Smith, P. (2002) 'Dewey on Naturalism, Realism, and Science,' *Philosophy of Science* 69(S3), pp. S25–S35.
Hacking, I. (1990) *The taming of chance*, Cambridge: Cambridge University Press.
Hickman, L. (1998) 'Dewey's theory of inquiry,' in L. Hickman (ed.) *Reading Dewey: interpretations for a postmodern generation*, Bloomington: Indiana University Press, pp. 166–186.
Kasser, J. (2016) 'Two Conceptions of Weight of Evidence in Peirce's *Illustrations of the Logic of Science*,' *Erkenntnis* 81(3), pp. 629–648.
Kitcher, P. (1993) *The advancement of science: science without legend, objectivity without illusions*, Oxford: Oxford University Press.
Kitcher, P. (2001) *Science, truth, and democracy*, Oxford: Oxford University Press.
Kitcher, P. (2012) *Preludes to pragmatism: toward a reconstruction of philosophy*, Oxford: Oxford University Press.
Klein, A. (2008) '*Divide et impera!* William James's Pragmatist Tradition in the Philosophy of Science,' *Philosophical Topics* 36(1), pp. 129–166.
Lane, R. (2018) *Peirce on realism and idealism*, Cambridge: Cambridge University Press.
Levi, I. (1991) *The fixation of belief and its undoing*, Cambridge: Cambridge University Press.
Longino, H. (1990) *Science as social knowledge*, Princeton: Princeton University Press.
Misak, C. (1995) *Verificationism: its history and prospects*, London: Routledge.
Misak, C. (2006) 'Scientific realism, anti-realism, and empiricism,' in J. Shook and J. Margolis (eds.) *A companion to pragmatism*, Malden: Blackwell Publishing Ltd.
Myers, G. (1986) *William James: his life and thought*, New Haven: Yale University Press.
Pearce, T. (2020) *Pragmatism's evolution: organism and environment in American philosophy*, Chicago: The University of Chicago Press.
Pihlström, S. (2008) 'How (Not) to Write the History of Pragmatist Philosophy of Science?' *Perspectives on Science* 16(1), pp. 26–69.
Quine, W. (1953) 'Mr. Strawson on Logical Theory,' *Mind* 62(248), pp. 443–451.
Ransdell, J. (2000) 'Peirce and the Socratic Tradition,' *Transactions of the Charles S. Peirce Society* 36(3), pp. 341–356.
Short, T. (2000) 'Peirce on the Aim of Inquiry: Another Reading of "Fixation,"' *Transactions of the Charles S. Peirce Society* 36(1), pp. 1–23.
Short, T. (2007) *Peirce's theory of signs*, Cambridge: Cambridge University Press.
Smolin, L. (1997) *The life of the cosmos*, Oxford: Oxford University Press.
Van Fraassen, B. (1989) *Laws and symmetry*, Oxford: Clarendon Press.
Welchman, J. (2002) 'Logic and judgments of practice,' in F.T. Burke, D.M. Hester, and R.B. Talisse (eds.) *Dewey's logical theory: new studies and interpretations*, Nashville: Vanderbilt University Press.
Wible, J. (2014) 'Peirce's Economic Model in the First Harvard Lecture on Pragmatism,' *Transactions of the Charles S. Peirce Society* 50(4), pp. 548–580.

27
PRAGMATISM AND LANGUAGE

David Boersema

For philosophical pragmatists, what language is and what language does are inherently and necessarily interrelated. A pragmatist approach to understanding and evaluating language takes language to be a process, not (only) a product – a matter of social negotiation involving intersubjective understandings and an external world; language and linguistic meaning are forms of coping with the world as well as disclosing and shaping the world of our experience.

Imagine that you are having a meal with a friend at a restaurant, and, pointing at your plate, she asks, "Are you going to eat that?" The words that are spoken have meaning that I assume we all comprehend. We understand what has been asked. However, what was said could well be interpreted in more than one way. For instance, you might take the question to mean: "If you aren't, I would like to eat it." Or you might take it to mean: "It looks so disgusting that I can't believe you would really eat that" (or perhaps even something else). That is, you might interpret the question differently depending upon what you think your friend intended when she asked you that. You might think that she was trying to be funny or perhaps that she was trying (and succeeding) in being rude or perhaps something else altogether. Furthermore, suppose this same question ("Are you going to eat that?") was asked, not by your dining companion but by the waiter, who is pointing at your plate (and suppose your companion has finished her meal). So, how you might interpret the meaning of the question could be quite different depending upon who has asked it. What this points to is that language is complex and involves a number of dimensions that are salient when we want to understand it, such as what is said (or meant), how it is said, why it is said, and who says it. And, of course, language is not simply about speech.

Additionally, we know that we often use language in non-literal ways. Metaphors are commonly used by us all the time. For example, in *Romeo and Juliet*, when Romeo remarks that Juliet is the sun, no one would seriously take that to mean that he was saying that she is a gigantic ball of fiery gas. Or when someone remarks that "man is a wolf," we understand that to mean not that humans are the same species as wolves but rather that humans can and do act viciously at times. Another non-literal use of language that is commonplace for us is humor and sarcasm. We can all get the following linguistic element in this joke: A termite walks into a saloon and asks, "Is the bar tender here?"

Likewise, we all recognize the sarcasm in the following: Your friend shows up, stumbles, and falls flat on the ground; you remark, "Nice play, Shakespeare."

While the examples above might seem a bit frivolous, we know that language has important social and moral significance, for example, with racial or ethnic slurs. The terms we use to talk about the world can have important implications and consequences with respect to matters of

DOI: 10.4324/9781315149592-31

social justice and status, with individual and group identity and dignity, as well as social and political experience. The terms we use to talk about the world do not merely reflect how things are, but they can also create how things are. They can and do shape how we live.

This issue of language as not merely reflective, but creative, was of central importance to the American pragmatist philosopher, George Herbert Mead (1863–1931). Mead became known primarily for his work on the social nature of persons, especially the notions of personal identity, consciousness, and language. Throughout his writings, Mead emphasized the social development, even the social construction, of a person's "self."

One of the most fundamental ways in which humans "acquire" a self (or become who they are) is through interactions involving language. On an everyday, commonsense level, people recognize this in their lives. For example, people understand, both in terms of thinking and of feeling, how linguistic labels can affect them. When people use, say, racial slurs, they are meant to shape and characterize someone's identity. People even can come to see not only others, but themselves, in terms of those linguistic labels. This is why racist and sexist language is seen as being hurtful; it is because it can have the power to shape, or at least, influence, people's identities. But even deeper than that, Mead said, that language is fundamental to human experience, not just some accidental feature. Our most basic experiences are ones involving communication; that is how we interact in the world. For humans, communication means using language. In particular, for Mead, language is verbal gestures and language evolved out of physical gestures. In a sense, gestures were the beginning of social activity and relations. Indeed, speech – that is, articulate sounds that have meaning – arose out of interactions with others and truly were, for Mead, verbal gestures (rather than silent physical gestures). Language is not primarily just a verbal means of expressing one's thoughts, for Mead, but, instead, is primarily a means of communication. As a result, he claimed that linguistic meaning is primarily social in nature, because it is communication-based. We will see later in this chapter another American pragmatist philosopher, John Dewey (1859–1952), echo and expand upon this view.

Language, then, is fundamental to our lives, and in this chapter, we will consider how philosophers have investigated and analyzed multiple dimensions of language, and in particular how pragmatists have approached addressing concerns about language. To appreciate how pragmatism has focused its approach to language, it helps to see two philosophical traditions that provide the context in which pragmatists worked and how they relate to those traditions (both in the sense of drawing on them and critiquing and supplementing them). The traditions are often referred to as analytic philosophy of language and semiotics.

Analytic philosophy of language

Analytic philosophy is a broad area of philosophical thought and practice, mostly associated with philosophers in England and the United States in the 20th and 21st centuries. What characterizes analytic philosophy is not so much a set of shared beliefs but a way of approaching philosophy. This approach includes a focus on analysis, a style of argument that emphasizes clarity and precision, and a respectful attitude toward science. Much of 20th-century analytic philosophy concerned language, both as a subject in its own right and as a way of investigating traditional philosophical topics such as knowledge and the good. Analytic philosophy is often contrasted with continental philosophy (roughly speaking, philosophy carried out by philosophers on the European continent in the 20th and 21st centuries).

Traditional analytic philosophy is often regarded in the context of the "linguistic turn" in philosophy, as many prominent analytic philosophers turned their attention to language. Analytic philosophers studied (and continue to study), for instance, not only knowledge and goodness but also the words *knowledge* and *goodness*. They also studied the logical form (or structure) of sentences, as well as parts of speech such as names ("J.K. Rowling," for instance) and descriptions (such as "the author of *Harry Potter and the Deathly Hallows*").

Broadly speaking, linguists and philosophers of language have analyzed language along three dimensions: syntax (or the grammatical structure of language), semantics (or how language connects with the world; how words refer to the world), and pragmatics (or how language gets used to do things). In the dining example above, imagine if, instead of saying, "Are you going to eat that?", your friend had said, "To you going that are eat." Such a remark would be gibberish and have no meaning. That is because the syntax is nonsensical. So, one dimension of meaning is syntactical meaning. Also, if she had said, "Are you going to drink that?" the meaning of that remark would be different than "Are you going to eat that?" because the word *drink* has different semantic meaning than the word *eat*. And, as was already noted, the sentence "Are you going to eat that?" can have a given syntactic and semantic meaning but different pragmatic meanings, depending upon what is intended or understood, or even upon who is saying it. For our present concerns, it is important to note that traditional analytic philosophers focused almost exclusively on semantic meaning in their analyses, as opposed to pragmatic philosophers, as will be seen below.

Analytic philosophers study language for at least two reasons. The first is the belief that there are interesting philosophical issues about language itself. Historically, the work of late-18th-century German philosopher Gottlob Frege (1848–1925) was influential in this respect. A second reason philosophers focus on language is the belief that it is possible to shed light on other philosophical subjects by examining language.

Bertrand Russell's influential theory of descriptions was important in this respect. Russell (1872–1970) believed that ordinary language (everyday, informal language) was flawed, vague, and potentially misleading. By putting the sentences of ordinary language into statements of formal logic, he thought, it was possible to view a sentence's logical form. Viewing the logical form of a sentence, in turn, helped philosophers better understand the meaning of those sentences and avoid potential confusions. An instance of this concerned *true negative existential statements*, that is, statements that accurately say that an object does not exist, such as "Pegasus does not exist." Such a statement is puzzling because, in order for "Pegasus does not exist" to be about anything (and therefore in order for the statement to be true), it seems there would have to be an object Pegasus. If there is an object Pegasus, however, then "Pegasus does not exist" is false. Yet the sentence clearly seems to be true. Some philosophers tried to solve this puzzle by arguing that objects such as Pegasus do exist, but in a special way different than how ordinary objects exist. Pegasus, on this line of thought, *subsists*, but because Pegasus does not exist in the usual way, "Pegasus does not exist" is true.

This proposed solution left many philosophers puzzled, for it was far from clear how an object could both exist in some way and yet not exist. Russell's way of solving this puzzle was to analyze the logical forms of such sentences. Russell argued that, in the logical form of "Pegasus does not exist," the term *Pegasus* does not name an object but instead stands for a description of an object. To analyze the truth or falsity of the sentence "Pegasus does not exist" involves analyzing the components of the description that *Pegasus* stands for. An analysis of those components shows that "Pegasus does not exist" is true. Because the truth or falsity of the sentence does not require an object "Pegasus," this eliminated a need to claim that objects such as Pegasus subsisted in some mysterious way. A close attention to the logical form of language, then, could resolve philosophical disputes.

Together with the influential Austrian philosopher Ludwig Wittgenstein (1889–1951), Russell elaborated the relation between language and reality in the theory of Logical Atomism. According to this view, the structure of language mirrors the structure of the world. Therefore, by analyzing the logical form of language, it is possible to learn the structure of the world itself. According to this view, there are atomic propositions expressed by very basic sentences, such as, "This is red."

In such a sentence, the word *this* picks out an object in the world, and the word *red* picks out a quality that the object has. For each atomic proposition, there corresponds an atomic fact, in this case the fact that something is red. Such propositions are atomic in two senses: first, they cannot be analyzed into further parts; second, the structure of the world is built out of atomic facts. Language, if true, mirrors the world.

In contrast to the formal approach was the school of ordinary language philosophy, advocated by philosophers such as J.L. Austin (1911–1960) and Gilbert Ryle (1900–1976). On this view, ordinary language is not something that needs to be fixed. Rather, ordinary language was therapeutic in the sense that a study of ordinary language could show philosophers not how to *solve* certain philosophical problems but how to *dissolve* those problems by showing that they are not genuine problems at all. We can see by the use of the word *know*, for example, that to have knowledge is not just to have an internal mental state but to behave in certain ways. We say that a person knows how to fix a flat tire on a bike, for example, only if she can manipulate the tire in certain ways, say, by putting a patch on it. So, we do not need to look for some logical form of sentences connected to knowledge; we simply need to look carefully at how we use the concept of knowledge.

Also, in his later work, Wittgenstein pointed out that language has many more uses than offering true or false descriptions of the world. Humans tell jokes, ask questions, make promises, complement each other, and many other things. He spoke of these many uses as being language games, and a complete understanding of language cannot focus solely on how sentences are true or false. Following these lines, philosophers such as Austin and others developed a theory of speech acts, giving an account of the different kinds of things people do with words, a view that pragmatist philosophers endorse and expand upon.

Another important school of thought in analytic philosophy is logical positivism, a view advocated by some members of what was called the "Vienna Circle" and popularized by A.J. Ayer (1900–1989). Logical positivism represents a strain of thought in analytic philosophy that favors, and is based upon, science and that rejects traditional abstract metaphysics. On a logical positivist view, a sentence is meaningful either (1) because it is true or false by definition or (2) because it can be verified through some experience of the empirical world. Sentences that do not meet either of these criteria are meaningless. So, some logical positivists claimed that many statements about metaphysics, aesthetics, ethics, and religion are meaningless.

Among contemporary 20th-century philosophers, the American philosopher W.V. Quine (1908–2000) was crucial both to continuing the tradition of analytic philosophy and to critiquing it. Quine continued the traditions of analytic philosophy insofar as his work often focused on language and the relation between language and reality; like analytic philosophers before him, Quine drew lessons about ontology (the study of existence) in part from studies of both formal logic and language.

Semiotics

Semiotics (from the Greek word *semeion*, meaning "sign") is the study of signs. Signs are anything that can represent something. While most people think of signs as things such as stop signs on street corners or "No exit" placards in buildings, practically anything can function as a sign. For example, there are natural signs, such as smoke being a sign of fire or dark clouds being a sign of (impending) rain. These are said to be natural signs because they are not the result of human actions or decisions. There are also conventional signs, that is, things that represent something on the basis of human actions and decisions, such as stop signs or "No exit" signs, or even language.

The study of signs goes as far back as the beginning of human history. Certain natural events have been understood as signs for eons. For example, the changing color of leaves on trees or patterns of birds migrating at different times of the year have long been treated as signs of present or upcoming changes in the seasons. Likewise, tracks left by animals on the ground have been treated as signs of the presence or movements of those animals. Ancient physicians from Egypt, Babylonia, China, indeed across the world, interpreted physical signs on people's bodies as evidence or symptoms of some disease or other. Early Greek physicians, such as Hippocrates, wrote works on various signs of different ailments and injuries.

The modern study of semiotics is usually associated with the work of two thinkers, Ferdinand de Saussure (1857–1913) and Charles S. Peirce (1839–1914). Saussure was a Swiss linguist, writing at the start of the 20th century. He is most known for introducing the terms *signifier* and *signified*. A signifier is a word or phrase that is used as a sign of something else. For example, the word *cat* is a signifier of an animal. A signified is the object that the signifier relates to. One thing that Saussure emphasized about signs is that they have no meaning in themselves but only in relation to other signs. The word *cat* does not have any significance by itself, but it does, he claimed, in terms of its relationship to other signs. In particular, he said, it has whatever meaning it has by being different than other signs, so *cat* is not *car* or *cut* or *rat*. What is crucial is that signs have meaning in terms of their differences from other signs, a point that later philosophers elaborated upon.

Peirce, an American pragmatist philosopher, wrote a great deal on signs. He claimed that there are three fundamental aspects to any sign. There is the sign itself, the object that is represented, and what he called an interpretant. The sign itself is whatever is used to represent something else. For instance, if smoke is a sign of fire, then smoke is the sign; if *cat* is a sign of some animal, then the word *cat* is the sign. The object is the thing that is represented (fire, in the first case, and some animal in the second case). The interpretant, for Peirce, was that which relates the sign and the object. It is what gives the sign its representative power. We usually think of interpretants as being people (so, to an English speaker, *cat* is a sign of cats, but to a non-English speaker, *cat* might mean nothing at all). However, for Peirce, interpretants do not have to be people. For instance, to a cat, certain sounds might be a sign of a mouse scurrying by or the hiss of another cat might be a sign of warning.

A major contribution of Peirce was in introducing three basic terms that identify three different kinds of signs or ways that signs can in fact be signs and function as signs. These terms are *icon*, *index*, and *symbol*. An icon is something that represents something else in virtue of having a close resemblance to that thing being represented. Most people know icons as portraits of religious figures. For Peirce, what made something an icon, and why a portrait is a sign, is because it (the icon) shares relevant features with the object. Quite simply, a portrait represents someone in particular because it looks like that person. An index is something that represents something else because of a causal connection to that thing. For example, a scab is a sign of a wound or injury, not because a scab looks like a wound but because it is caused by the healing process connected with a wound. Or a fingerprint is an index because the print is caused by oils of a particular person's skin that are deposited on a surface; the touching of that surface causes those oils to be there. A symbol represents something, and so is a sign of it, on conventional grounds. Again, the word *cat* is a sign of cats, but there is nothing about the particular sounds (or, in the case of writing the word, nothing about the particular squiggles of ink) that connects to that kind of animal. In Spanish, the word *gato*, not the word *cat*, is the conventional sign for cats. Almost all language is a matter of symbolic signs.

There is not a sharp distinction among these various kinds of signs. For instance, some symbolic signs, such as words, are used conventionally because there is a causal (indexical) connection to the object. A chickadee (a species of bird) is called that (at least in English) because the sound it makes sounds very much like "chickadee." Likewise, there is not necessarily a sharp distinction between an icon and an index; although a portrait is an icon (because it looks like the person who is pictured in the painting), there is clearly a causal connection to how that person looks and how the portrait looks. In addition, there is not a sharp distinction between natural signs and conventional signs. For example, some colors are often associated with danger, such as red being associated with blood, which is why many conventional signs that are intended as warnings are made red. Those who study semiotics claim that signs are everywhere and that everything is at least potentially a sign. For instance, in the Sherlock Holmes story, "Silver Blaze," the detective claims that the fact that a dog did not bark in the night was evidence of (i.e., a sign of) the identity of the criminal; the absence of barking signified to Holmes that the dog recognized the person. So, even the absence of something can be a sign.

Most philosophers have focused on two sub-fields within semiotics: social semiotics and linguistic semiotics. Social semiotics is concerned with signs with respect to their social significance. There are many non-verbal symbols of social status. For example, different styles of clothes represent different kinds of jobs (such as a nurse's uniform or a police officer's uniform) or different "levels" of jobs (such as a suit and tie vs. jeans and tee shirts or the different number of stripes on a military uniform). In addition, even something like the relative placement of people in a gathering can be a sign of relative importance. For instance, who is standing closest to the President during an official ceremony often signifies the importance of that person. Likewise, the very way that people carry themselves (upright vs. slumping) is taken as representing various things. Many social scientists, as well as social philosophers, investigate and analyze such social signs, often with respect to underlying issues of social status and power.

Besides social semiotics, philosophers have focused on linguistic semiotics. In particular, philosophers have looked very closely at the issue of meaning, especially as it relates to language. Much of the area of philosophy of language is the study of various aspects of conventional symbolic signs. This includes semantics (how words or phrases relate to objects) as well as pragmatics (how people use language for various purposes and in various ways). Among the semantic concerns is the issue of what is called "the locus of meaning." That is, obviously, people can and do mean certain things when they speak. For example, when someone says, "It is really hot today," those words have meaning, but where is the meaning? Is the meaning in the words (signs) themselves or in the "heads" of the speakers or somewhere else? In addition, although the words have meaning, in different contexts those same words might carry different meanings. In one context, those words might mean: Please open a window; it's really hot in here! In another context, those same words might mean: It is too uncomfortable for me to get any work done today! Besides the broad issue of meaning, there is even semiotic importance to timing or intonation. With respect to written signs, in an effort to capture spoken signs, we use different fonts or marks. For example, there is different meaning to *HELP!* as opposed to *help*, or to *Oh!* versus *Oh?* versus *Oh*, versus *Oohhhh!*

These concerns overlap with those of social semiotics. For instance, both social semioticians and linguistic semioticians look at what is involved with forms of address, for instance, calling people by first names versus nicknames versus titles, as well as who is expected and allowed to use certain forms of address and in what contexts. Some philosophers, such as Jacques Derrida (1930–2004) and Roland Barthes (1915–1980), have argued that people live only in a world of signs; in Saussure's terms, there are only signifiers that refer to other signifiers, never to signifieds, because everything that we encounter is immersed in meanings that are just signs.

Pragmatism on language

Having surveyed analytic philosophy of language and semiotics, we are now in a better position to approach and understand a pragmatist perspective on language. As noted at the start of this chapter, pragmatist philosophers of language did not operate in a philosophical vacuum; they were aware of and worked alongside philosophers in the analytic and continental philosophical traditions. As will be seen below, they drew on those traditions and certainly agreed with some of the analyses that emerged from them. However, they also critiqued and supplemented them, in the sense of identifying what they saw as both mistakes committed by those traditions (for instance, particular stances on what makes linguistic statements true) and omissions in those traditions (for instance, not being concerned with social and creative aspects of language).

From our earliest experiences in life, we learn that there are different kinds of things in the world, and we use language to talk about them. A fundamental component of that is learning and using names for things. Grammatically speaking, we learn the meaning of nouns. For example, we learn to pick out Mama and Papa, or we learn to refer to our pet as doggy or kitty. We have names for many types of things, not only proper names to identify and distinguish particular individuals (so, "Shakespeare" refers to someone different than "Mister Magoo") but also kinds of things, like dogs or cats or trees or clouds, etc. Besides names for objects, we have names for events, particular events, such as "The Battle of Waterloo," and even kinds of events, such as sneezing. Again, we identify and distinguish different kinds of things (so, the Battle of Waterloo is not the same event as the 23rd Super Bowl, and sneezes are not the same things as hiccups). The importance of this is that we use names (nouns) to engage in the world, but we don't have names for everything. Names are a basic way that we understand and operate in the world. For this reason, philosophers of language have spent considerable energy in analyzing and evaluating names and how they are to be understood. For our present concerns, this also provides a lead in to seeing how pragmatists approach and treat not just names but language generally. Earlier in this chapter, we noted the work of George Herbert Mead, and while there are many philosophers who identify themselves as pragmatist and have variations in their specific commitments to what pragmatism espouses, probably the clearest and most detailed pragmatist account of this is in the work of the American pragmatist John Dewey. For our present concerns, then, he will be taken as the primary spokesman for a pragmatist perspective on language.

Noting that one basic kind of linguistic sign is names, Dewey remarked,

> We take names always as namings: as living behaviors in an evolving world of men and things. Thus taken, the poorest and feeblest name has its place in living and its work to do, whether we can today trace backward or forecast ahead its capabilities; and the best and strongest name gains nowhere over us completed dominance.
>
> (LW.16.7)[1]

This quotation points to a number of features not only of Dewey's view of names but also of his view of language in general, a view that underlies his understanding of names. First, by claiming that names are namings, it points to his insistence that "things" be understood as processes, especially when those "things" (such as language) are human creations and artifacts. Second, as living behaviors they are organic reactions to the world, they are part of our engagements as agents in the world. Third, names, language, behaviors, all have "work to do." They are the result of, and also a means, of our inquiry (which is never disinterested) and purposive agency. Fourth, names have capabilities. They are, as for Dewey all languages are, tools. We use them in the context of our purposive conduct. And, again, this is not just about names but also and more importantly about language and meaning in general.

These four features highlight Dewey's overall understanding of language, which, in *Knowing and the Known* (1949), he defines as:

> Language: To be taken as behavior of men (with extensions such as the progress of factual inquiry may show to be advisable into the behaviors of other organisms). Not to be viewed as composed of word-bodies apart from word-meanings, nor as word-meanings apart from word-embodiment. As behavior, it is a region of knowings. Its terminological status with respect to symbolings or other expressive behaviors of men is open for future determinations.
>
> (LW.16.266)

Besides being a process, language, for Dewey, is also an organic reaction to the world. It is worth quoting Dewey at length on this point. In his book *Experience and Nature*, he claims:

> Language is a natural function of human association; and its consequences react upon other events, physical and human, giving them meaning or significance. Events that are objects or significant exist in a context where they acquire new ways of operation and new properties. Words are spoken of as coins and money. Now gold, silver, and instrumentalities of credit are first of all, prior to being money, physical things with their own immediate and final qualities. But as money, they are substitutes, representations, and surrogates, which embody relationships. ... Language is similarly not a mere agency for economizing energy in the interaction of human beings. It is a release and amplification of energies that enter into it, conferring upon them the added quality of meaning. ... Gestures and cries are not primarily expressive and communicative. They are modes of organic behavior as much as are locomotion, seizing and crunching. Language, signs and significance, come into existence not by intent and mind but by over-flow, as by-products, in gestures and sound. The story of language is the story of the *use* made of these occurrences; a use that is eventual as well as eventful. ... The heart of language is not "expression" of something antecedent, much less the expression of antecedent thought. It is communication; the establishment of cooperation in an activity in which there are partners, and in which the activity of each is modified and regulated by partnership.
>
> (LW.1.138–141)

In terms of language having "work to do," Dewey is explicit about this when he speaks of naming:

> Naming does things. It states ... Naming selects, discriminates, identifies, locates, orders, arranges, systematizes. Such activities as these are attributed to 'thought' by older forms of expression, but they are much more properly attributed to language when language is seen as the living behavior of men.
>
> (LW.16.134)

This view of naming, and of language generally, is certainly reminiscent of (although it predates) Wittgenstein's notion of language games as well as of the speech-act views of Austin and Searle, noted above in the section on analytic philosophy of language.

The fourth feature of language noted above, that of language as a tool, is ubiquitous throughout Dewey's writings. As a tool, language is purposeful; it is used to accomplish something. In addition, like tools, language has physical features that are separable from their use in the purposive behavior of the tool-user.

Dewey's emphases, then, when focusing on language, are varied. One emphasis is on the origins of language. As his remarks about gestures and cries noted above reveal, he takes language to be one of many organic behaviors that emerged out of our interactions with our natural and social environments. In his 1922 *Human Nature and Conduct*, he reiterates this position:

> Men did not intend language; they did not have social objects consciously in view when they began to talk, nor did they have grammatical and phonetic principles before them by which to regulate their efforts at communication. These things came after the fact and because of it. Language grew out of unintelligent babblings, instinctive motions called gestures, and the pressure of circumstances. But, nevertheless, language once called into existence is language and operates as language. It operates not to perpetuate the forces which produced it but to modify and redirect them.
>
> (MW.14.56)

Besides pointing to the origins of language, Dewey, as already noted, emphasizes the sociality of language, both in its genesis and in its functions, especially its communicative function. Both these emphases are captured by his characterization of language as a behavior in "the region of knowings," and this is why commentators have insisted that his understanding of language be placed in the context of his theory of inquiry.

Another emphasis of Dewey is on the publicity of language, indicated in his definition mentioned above as having a terminological status "open for future determination." This points to publicity in two respects. First, "terminological status" refers to its terminal status, that is what, say, a name will ultimately mean or refer to. Second, "terminological status" refers to a word's status as a linguistic term, say, as a name. In both senses, what Dewey is getting at is that what counts as being a linguistic term and what is its reference is a matter of future determination, i.e., by its public, social functioning in the interactive discourse of language users. It is not a matter of simply what a single person means "in his head" so to speak, nor is it a matter of simply what the given semantic content or role that it has had in the past.

Given this brief outline of Dewey's remarks, an overall pragmatist perspective on language, meaning, and interpretation emerges. Meaning is future oriented although connected to past and present interpretations and understandings; meaning involves not only syntax and semantics but also necessarily pragmatics; meaning is a process, not (only) a product; meaning is a matter of social negotiation; meaning is a relation involving agents, intersubjective understandings, and an external world; the "lived nature" of language points to interpretations and understandings fitting into working categories, not simply onto a given set of states of affairs; language and linguistic meaning are forms of coping with the world as well as disclosing and shaping the world of our experience.

Note

1 All references to Dewey's works are from *John Dewey: Collected Works, 1882–1953* (37 volumes), edited by J.A. Boydston, Carbondale: Southern Illinois University Press, 1976–1990. Specific references are to the Early Works (EW), Middle Works (MW), or Later Works (LW). So, for example, LW.16.266 stands for volume 16, page 266 of the later works.

Further reading

Bacon, Michael. *Pragmatism: An Introduction*. Cambridge: Polity Press, 2012.
Brandom, Robert B. *Between Saying and Doing: Towards an Analytic Pragmatism*. Oxford: Oxford University Press, 2010.
Dewey John. *John Dewey: Collected Works, 1882–1953* (37 volumes), edited by Jo Ann Boydston. Carbondale: Southern Illinois University Press, 1976–1990.
Glock, Hans-Johann. *What Is Analytic Philosophy?* Cambridge: Cambridge University Press, 2008.
Hall, Sean. *This Means This, This Means That: A User's Guide to Semiotics*. London: Laurence King Publishers, 2007.
Lee, Barry (ed.). *Philosophy of Language*, second edition. New York: Bloomsbury, 2020.
Martinich, A.P. and Sosa, David E. *A Companion to Analytic Philosophy*. Malden: Blackwell Publishers, 2001.
McGinn, Colin. *Philosophy of Language: The Classics Explained*. Cambridge, MA: The MIT Press, 2015.
Misak, Cheryl. *The American Pragmatists*. Oxford: Oxford University Press, 2013.
Sebeok, Thomas A. *Signs: An Introduction to Semiotics*. Toronto: University of Toronto Press, 1994.
Spencer, Albert R. *American Pragmatism: An Introduction*. Cambridge: Polity Press, 2020.

28
PRAGMATISM IN THE PHILOSOPHY OF MIND

Aaron Zimmerman

Pragmatic Theses in the Philosophy of Mind

Everyone is talking about pragmatism in the philosophy of mind (Engel et al. 2013; Menary 2016). This should come as no surprise once we recall the origins of pragmatism in Alexander Bain's naturalistic theory of belief and the application of that theory to central semantic and methodological questions by C.S. Peirce and William James. I review this history below, before turning to ongoing debates in the philosophy of mind in which pragmatist positions have been articulated and evaluated.

Contemporary theorists have developed distinctively pragmatic accounts of content, consciousness, and agency: the three core subjects in the philosophy of mind. First, contemporary pragmatists have joined William James in arguing that we can believe for practical reasons and that we have a kind of control over our beliefs denied by their evidentialist opponents. Second, while some philosophers of mind reject mental representations altogether, and some adopt a naïvely realistic interpretation of folk psychology according to which there is always a fact of the matter to be discovered as to what a person perceives, thinks, or remembers, pragmatist views of mental representation have drawn attention to the indeterminacy that marks the content of biologically real beliefs and experiences. Pragmatists allow that constraints on assignments of content can be extracted from the cognitive sciences, but they insist we cannot ignore the role our social needs play in further specifications of what we perceive, think, and intend. Finally, pragmatists who approach the philosophy of mind through a biological lens have developed theories of subjectivity or consciousness that emphasize the functional purpose of experience and its gradual emergence over the course of various animal lineages, as organisms adapt to secure their innately specified ends in the face of disequilibria.

Though philosophers continue to debate the meaning and nature of pragmatism, and we remain to some degree divided over the consequences of adopting a pragmatic approach to central debates in the philosophy of mind, an ideology emerges from this survey of the literature. On the whole, pragmatists are committed to an evolutionary psychology that views the human mind as continuous with the psychologies of other animals; pragmatists see the historical (phylogenetic and ontogenetic) method of explanation as an essential component of theorizing about the mind; pragmatists see integration with biology in general and neuroscience in particular as an important desideratum on theorizing about the mind; and, when it comes to evaluating the different theories of agency, consciousness, and mental representation in play, and the integration of the cognitive sciences with the "folk psychology" we use to frame our social interactions, pragmatists highlight

Bain's Account of Belief and the Genesis of Pragmatism

It is arguable that pragmatism began its life as a view about the nature of belief. Recollecting the origins of the movement, C.S. Peirce described Bain's theory of belief as the "axiom of pragmatism," claiming that pragmatism's many doctrines were all just "corollaries" that emerged from Nicholas St. John Green's relentless applications of Bain's definition of belief to issues of philosophical significance during discussions of the ironically named "Metaphysical Club."

> Nicholas St. John Green was one of the most interested fellows, a skillful lawyer and a learned one, a disciple of Jeremy Bentham. His extraordinary power of disrobing warm and breathing truth of the draperies of long worn formulas, was what attracted attention to him everywhere. In particular, he often urged the importance of applying Bain's definition of belief, as 'that upon which a man is prepared to act.' From this definition, pragmatism is scarce more than a corollary; so that I am disposed to think of him [i.e. Green] as the grandfather of pragmatism.
>
> (Hartshorne and Weiss, 5: 1)[1]

the real practical consequences that attend the results of this endeavor and the relevance of practical consequences to theory choice (Zimmerman 2018).

Should we trust Peirce's recollection? Max Fisch (1954) provides ample evidence in favor of the account, ably documenting the explicit role Bain's theories played in the work of James and Peirce during pragmatism's first wave, and the unacknowledged debt owed to Bain's theory of mind by Dewey and those intellectuals who took up the pragmatist mantle in the years to come. My aim here is just to provide a barebones account of the initial stages of this history and the philosophy of mind at its core, which is perhaps not as well known as it should be.

Bain begins his analysis of the mind in a naturalistic vein by observing the primacy of movement. People and other animals are born in motion, having already practiced their motor routines in utero. Surveying the course of our development in light of the neuroscience of his day in *The Senses and the Intellect*, Bain concluded that motor cognition is both developmentally and psychologically primary. Our most basic conceptions of space are representations of our own bodies in motion in preparation for perception and the adaptation of motor performance to sensation. Space, and the actions that unfold within it, precede the representation of space, despite the pretensions of transcendental philosophy. Competence precedes comprehension.

But Bain's philosophy of mind was best known for his functionalist theory of belief, which is biological in orientation insofar as it incorporates developmental and evolutionary analyses granting beliefs to bees and infants. (Bain was a contemporary of Darwin and is cited as much as any other author in Darwin's *Descent of Man*.) According to Bain, the identification of means to ends is the "primordial" form of belief. An organism generates beliefs when its innate or instinctive routines fail to secure its innately specified ends or goals, and it must cast about for information that will enable it to adapt its behavior to this contingency. Since the identification of means preceded the evolution of sentential language and the kinds of discursive reasoning human language enables, Bain rejected Cartesianism and felt compelled to offer an analysis of belief that eschewed anything uniquely human. On this issue and many others, Bain sided with Hume over Reid.

The theory of belief is relegated to the second of Bain's two masterworks, *The Emotions and the Will*, because, according to Bain, belief is best defined in terms of action and classified under will, in partial contrast with states of sense or intellect or emotion. We believe that upon which we are prepared to act.

> It will be readily admitted that the state of mind called Belief is, in many cases, a concomitant of our activity. But I mean to go farther than this, and to affirm that belief has no meaning, except in reference to our actions; the essence, or import of it is such as to place it under the region of will. We shall see that an intellectual notion, or conception, is likewise indispensable to the act of believing, but no mere conception that does not directly or indirectly implicate our voluntary exertions, can ever amount to the state in question.
>
> *(1859: 568)*

Bain's Pragmatic Axiom: A belief is a representation poised to guide an animal's voluntary actions or exertions.

The consequences of this definition are manifold. First, it follows that skepticism about the external world is fake unless it blocks expectations and memories, which are beliefs in that world. Even skeptical attacks on the truth or objectivity of morals are insufficient; if the audience continues to enact a morality, doubt is feigned.[2] In every area of thought, belief is the default. Indeed, it is implied as soon as animals identify means to their ends, as when finding water to drink.

> The animal that makes a journey to a pool of water to relieve thirst believes that the object signalized by the visible appearance of water quenches thirst.
>
> *(1875: 506)*

Relatedly, doubt is the true psychological "opposite" of belief according to Bain, rather than belief in the contrary (or disbelief), as real doubt inhibits those actions or dispositions to action which are constitutive of belief. So radical doubt cannot be sustained over time, as Hume insisted. It's radically maladaptive.

Moreover, since the evolved aim of belief is solving a practical problem, the essential function of belief is not accuracy but success. Belief does not begin with the search for truth but the struggle for existence. Of course, to construct science, we must enforce norms of truth. But these norms are only necessary because epistemic virtue is unnatural. In an Aristotelian vein, Bain argues that epistemic virtue is a mean between extremes of skepticism and delusion, where the latter pole is the rule. Most of us just believe the first thing that works and hang on to our dogma until we are forced to adapt.

Bain labels his teachings on this subject "the principle of primitive credulity" and applies this Spinoza's dictum to a number of different issues, including the "foundation" of our belief in the regularity of nature. Do our experiences provide the premises for an inductive inference, which then fixes a belief in the regularity of nature: a general expectation that the future will resemble the past and the unobserved resemble what has been perceived to date? Bain rejects this account as untenable and not because of traditional worries over the circularity of "justifying" our inductive inferences, which are often attributed to Hume. Instead, Bain insists that belief in uniformity comes first in the order of development and that regularities in experience merely enable this assumption to persist.

> It is not proceeding from the right end, to say that the extended knowledge that enables us to substitute sure uniformities for hasty assumptions is the cause or essence of our believing disposition; it is rather the pruning operation that saves it from destructive checks.
>
> *(1875: 516)*

Bain published several editions of the two works cited above, the third and final edition of the *Emotions* appearing in 1876, and these volumes were read, argued over, notated, and lectured from by James and Peirce in the subsequent decades. (Consult Fisch [1954] and Zimmerman [2022] for details.) Twenty years would pass, as James and Peirce marinated in Bain's "action first" views on the development and functioning of human psychology, before James would first use the term "pragmatism" as a description of a philosophy or worldview in an address delivered before the Philosophical Union at the University of California, Berkeley, on August 26, 1898.

> Years ago ... direction was given to me by an American philosopher whose home is in the East, and whose published works, few as they are and scattered in periodicals, are no fit expression of his powers. I refer to Mr. Charles S. Peirce, with whose very existence as a philosopher I dare say many of you are unacquainted. He is one of the most original of contemporary thinkers; and the principle of practicalism—or pragmatism, as he called it, when I first heard him enunciate it at Cambridge in the early '70s—is the clue or compass by following which I find myself more and more confirmed in believing we may keep our feet upon the proper trail.
>
> *(1898: 290)*

What was this philosophy? We have already seen Peirce's acknowledgment that its axiom was Bain's theory of belief as wielded by Green. But Peirce developed the theory further. His earliest contributions are perhaps clearest in "How to Make Our Ideas Clear," the essay James named above when introducing "pragmatism" to the world. For after rehashing Bain's conceptions of belief, doubt, and inquiry in that essay, Peirce explicitly uses them to derive a semantic corollary. Since belief is an action-guiding or potentially action-guiding representation, there can be no difference in belief without an at least potential difference in action. And since the value of a statement or the meaning of an expression can be equated with the beliefs it is used to communicate, there can be no difference in meaning between statements that are identical in their implications for action.

> The essence of belief is the establishment of habit, and different beliefs are distinguished by the different modes of action to which they give rise. If beliefs do not differ in this respect, if they appease the same doubt by producing the same rule of action, then no mere differences in the manner of consciousness of them can make them different beliefs, any more than playing a tune in different keys is playing different tunes. ... Thus, we come down to what is tangible and practical, as the root of every real distinction of thought, no matter how subtle it may be; and there is no distinction of meaning so fine as to consist in anything but a possible difference in practice.
>
> *(Peirce 1878: 135–137)*

> **Peirce's Semantic Corollary**: Statements, theories and other representations only differ in meaning if their acceptance (i.e., belief in their contents) would introduce different voluntary habits or actional dispositions.

What then of James's contribution? In his 1898 address, James develops Peirce's semantic proposals by applying Bain's theory of belief to the act of philosophizing itself, which can be understood, in a pragmatic way, as communicative actions mounted with a significant point to make or thesis to defend. If we think of the adoption of a philosophical theory as itself a change of belief in the sense defined by Bain, then philosophical disputes are not genuine unless they have some practical difference: the very thesis James defends in his lecture. There is no significance to debates about God's existence, if our actions or actional dispositions would not be altered by conversion to our opponents' stated view on the matter.

Stewart and Brown, James Mill, John Mill, and Bain, have followed more or less consistently the same method; and Shadworth Hodgson has used it almost as explicitly as Mr. Peirce. These writers have many of them no doubt been too sweeping in their negations; Hume, in particular, and James Mill, and Bain. But when all is said and done, it was they, not Kant, who introduced the critical method into philosophy, the one method fitted to make philosophy a study worthy of serious men. For what seriousness can possibly remain in debating philosophic propositions that will never make an appreciable difference to us in action? And what matters it, when all propositions are practically meaningless, which of them be called true or false!

(James 1898: 308)

James' Methodological Corollary: A philosophical dispute is only real (i.e. not merely a matter of words) if changing sides would entail a change in behavior or introduce different actional dispositions.

It would seem, then, that according to its progenitors, pragmatism began in discussions centered on the philosophy of mind in general and the nature of belief in particular, even if the philosophy soon came to encompass distinctive semantic and methodological (or meta-philosophical) doctrines as well.[3]

Doxastic Pragmatism in Contemporary Philosophy

Though more than a century has passed since James introduced pragmatism to the world, the top philosophy journals continue to feature debates over the nature and evaluation of belief, pitting "pragmatists" against their rivals. According to the dominant strain in this discourse, the pragmatist is someone who thinks that we can have practical reasons (e.g. moral or prudential reasons) for believing what we do (Maguire and Woods 2020; McCormick 2015; Rinard 2019a, 2019b; Roeber 2019). They are opposed in this view by the "evidentialists," who think that we can only believe for theoretical reasons or seemingly epistemological or alethic (truth-centric) reasons. According to many self-styled opponents of pragmatism, the only possible grounds for belief are "evidence" or "seeming evidence" in some sense of these terms (Berker 2018; Boyle 2011; Hieronymi 2008; Parfit 2011; Way 2016).

Reasons Pragmatism: people sometimes believe what they do for prudential or moral reasons.

The debate over pragmatic reasons for belief is perhaps more a matter of epistemology or the ethics of belief than it is a matter for philosophers of mind. But its sources are diffuse and hard to pin down. Do the disputants disagree over psychological doctrines or critical norms? The proposition that x is or is not a reason to do or think something is fairly clearly normative or extra-scientific. It can only be be settled by frankly normative arguments, grounded in values shared by disputants. (We would seem to require for the resolution of disagreements about what is a reason for what something beyond experimentation and good faith attempts to provide the best explanations of the observed results of experiments.) But a philosopher might allow that the prudential benefits of adopting a belief do provide someone with a reason for adopting it and yet insist that as a psychological matter, we are unable to believe for these reasons. And that would shift the debate from epistemology to philosophy of mind.

Thus, the contemporary debate over the possibility of believing things for pragmatic reasons is intimately caught up with long-standing debates among philosophers over the control people have over their beliefs and the question of whether people can decide to believe, or choose to believe, or perform acts of mind (by, e.g. the drawing of conclusions or the reaching of judgments), which have belief as a direct consequence.

And there are additional connections between debates over what we have reason to believe and debates over the psychological or metaphysical nature of belief. For if one concedes that we do have significant control over our beliefs, it is hard to defend evidentialism against the kind of reasons pragmatism defined above. For it is unlikely that every person, no matter her level of epistemic virtue, would willingly forgo love or money to retain accuracy in her beliefs were the choice hers to make. (As Bain observed, people need to be coerced into scientific habits of thought. Control enables perversion.) So the evidentialist will feel pressure to admit that some people (if only the epistemically "vicious") would believe for pragmatic reasons if only they could manage it. Conversely, philosophers have argued that evidentialist norms are vacuous or epiphenomenal unless we have some control over our beliefs. How is a subject supposed to respond to the evidentialist injunction that she apportion her credence to the evidence, if it is not up to her whether she follows the order? Evidentialist norms might be used to evaluate or criticize people for believing what they do, but they could not be utilized as a guide to belief were belief wholly passive (cf. Dennett 1981).

Given these connections, "pragmatism," as it is now used in debates over the kinds of reasons we can have for belief, stands or falls with a view on our control over our beliefs that was first sketched by James in "The Will to Believe," an essay published in 1896, two years before the lecture quoted above, and so two years before James first used the term "pragmatism" in print.

Pragmatic Voluntarism: people sometimes exercise discretion over their beliefs by believing what they could have refrained from believing or by refraining from believing what they might have.

Though the pragmatist admits that many of our beliefs are unavoidable or passively held, she identifies a range of important cases in which it is up to us to determine what we believe. You can choose to trust a friend, or withhold judgment against an adversary, when you could have indulged in an inclination to believe the worst. Or you might convince yourself that you will succeed at an important task, when a more dispassionate evaluation of your prospects would have left you in doubt (McCormic 2015; Roeber 2019). And what are we to say about the use of moral considerations to condition belief? Basu (2019) describes cases in which people are morally criticized for believing what they do, quoting former President Barack Obama on his reaction to being mistaken for waitstaff at a black-tie dinner. Suppose Bob has seen a lot of Black men at his country club over the years, and Obama is the first to enter who is not waitstaff. If Bob can see that Obama is a Black man but has no other clue to his identity, the induction to waitstaff would appear as straightforward as any other were it not for the racial context. If the epistemic dispositions that generate such inductions were truly irresistible, would we have grounds on which to criticize Bob's insulting assumption?[4] Basu tries to reconcile her view on which beliefs can morally wrong with doxastic involuntarism by insisting we can blame Bob for his doxastic dispositions even if he cannot resist them in the moment. But the voluntarist has an easier time with the case. Bob should consider the potential moral insult of assuming a Black man at a formal dinner is waitstaff, and he should on these grounds remain agnostic until decisive evidence settles the matter. Moral prohibitions (or considerations of mutual respect) seem to many of us to trump epistemic permissions in cases of this kind and to therein provide a practical reason to withhold judgment.

To retain their commitment to evidentialism, philosophers do fancy things to redescribe the phenomenology that attends all of these phenomena.[5] The evidentialist might allow that a representation of the potential benefits of belief plays a causal role in generating that belief without "interfacing" with the subject's decision-making faculties (Mele 2001). Perhaps our desires can affect our beliefs "unconsciously" without implicating anything deemed essential to "exercising discretion" over belief in the sense at issue. But there are also times when it seems to us as though we are withholding judgment, when we could have drawn conclusions. Is there a reason to deny the appearances?

There are in fact a wide range of normative debates distinct from (if related to) the debate over whether we do or can believe for practical reasons that have consequences for the philosophy of mind. For example, as a normative matter, contemporary pragmatists tend to describe themselves as "permissivist" in their habits or policies of epistemic criticism (Schoenfield 2014). A permissivist typically allows that respect for truth and evidence (however defined) are important values, essential to science, history, journalism, and those forms of life that we see strained by the palpable dishonesty that marks much of our communication in the digital age. But the permissivists refuse to fetishize "the truth," and they point to a number of distinct areas in which belief can be rationally held on an at least partially practical basis. In morality, religion, art, friendship, and self-assessment, pragmatic considerations can guide belief in unobjectionable ways. Echoing Bain's Aristotelian conception, virtue is a mean between extremes of pure fantasy and scientism.[6]

> **Pragmatic Permissivism**: There are a range of beliefs people can adopt at any given time without vulnerability to legitimate criticism. A person's set of evidence constrains what she can rationally believe, but it does not determine a single mindset set she must adopt on pain of irrationality.

In contrast, the evidentialists are more rigid and exacting in their requirements. Most do not follow Plato in calling for a ban on the poets. But they insist we must never go beyond or beneath the evidence in fixing our beliefs, utility be damned (Adler 2002). And if it comes down to utility, they can point to the role played by myths in the fragmentation of societies as a cautionary tale.

Of course, debates about the control we have over our beliefs and the ways in which we evaluate people for believing what they do are closely connected to those debates over the nature of belief that gave rise to pragmatism as a philosophical movement. Descartes equated belief with a kind of inner affirmation of a statement or proposition. Indeed, he denied the other animals belief precisely because they are unable to construct sentences in speech or thought. In more recent work, Blake Roeber (2019) takes a similar view (citing Williamson [2000] as inspiration) and argues on its basis that we directly control or will our beliefs. I can assert at will. So if belief is essentially inner assertion, I can believe at will.

Those theorists attracted to a broadly Cartesian metaphysics of belief who want to avoid drawing voluntarist consequences typically build in some reference to truth to complicate matters (as does Roeber). It would be tough to motivate a picture on which there are no false affirmations, so Cartesian theorists, as a first step, tend to equate belief with *sincere* affirmation. But this risks circularity. If "sincerity" is defined as asserting what you in fact believe, and belief is defined as sincere assertion (or a disposition to such), little has been gained. But perhaps this worry can be massaged by analyzing sincerity in less cognitive terms, as, say, automatic or spontaneous assertion. Even so, the move is insufficient, as we often sincerely express false, irrational, and unjustified beliefs, which are neither true nor adequately grounded in evidence. To avoid embracing voluntarism without contradicting this truism, theorists have resorted to metaphor, suggesting that belief is an affirmation that is in some sense "aimed at" the truth. Perhaps belief is an attempt at reporting or describing or predicting the truth, even when it very clearly falls short of this aim (Bratman 1992; Williams 1973; Shah and Velleman 2005).

But pragmatists continue to reject a telos of truth in their definitions of belief for much the same reasons Bain adduced in the 19th century. From a biological perspective, belief fulfills its function when it contributes to the functioning of the animal whose belief it is. Belief only "aims" at truth to the extent that belief in the truth in question augments that animal's fitness. Maladaptive levels of accuracy are not functional in this sense, and their persistence must therefore be explained by something beyond individual selection. (Again, science is a cultural achievement.)

An even more common metaphysics of belief treats inner affirmation as a symptom of belief or a symptom of certain beliefs: i.e. those the subject is positioned to articulate. This metaphysics is accompanied by (and perhaps motivated by) an account of introspective self-knowledge on which we attribute beliefs to ourselves when we have affirmed their contents to ourselves. Against Descartes, we can still insist that functional connections to behavior and inference are necessary for belief and are presupposed in self-ascription, where this is thought to explain the reliability of our views on our own beliefs. If it is more often true than not that one will act on a representation when one is disposed to assert its content and therein self-ascribe belief in what one has said, introspection will prove largely reliable. But introspection is fallible, as these connections between speech and non-communicative action are contingent. When a person talks the talk without walking the walk, she will erroneously self-ascribe the belief that p on the basis of her inner affirmation of p, an affirmation which is misleading precisely because she lacks the behavioral dispositions necessary for belief in p. Self-ignorance and self-deception are often epistemically blameless (Peacocke 1999).

Alternatively, one might argue that saying something to yourself gives you no real reason to believe that you believe it unless you are prepared to act on that information when it proves relevant. An agent's epistemic authority over her mind (her "first-person privilege") is in part constituted by her authority over her behavior, or her capacity to act from those beliefs she self-ascribes when she focuses on the task at hand (Zimmerman 2006, 2018).

Pragmatic Theories of Mental Representation

Belief is not the only state of mind that has been subjected to pragmatic analysis. If a philosopher acknowledges the holistic nature of folk psychology, and she adopts the pragmatic analysis of belief, she will have to process its ramifications for her conceptions of perception, memory, intention, emotion, and so on. We have already indicated the important conceptions of doubt and inquiry that Bain and the early pragmatists extracted from Bain's analysis of belief, but its effects on their views of perception were just as significant. For example, if we conceptualize belief in Bain's terms, we must allow that an animal's perceptions and memories *constitute* beliefs in their own right, unless she takes steps to block their influence on her actions and plans. In almost every instance, perception is a sensorially rich experience which involves (as a component) believing that something is happening, whereas recollection and expectation are more sensorily anemic ways of believing that something has happened or will happen. Indeed, though Bain argues that the emotions we experience when we read dramatic fiction do not involve belief in what is depicted (1875: 505), he describes cases in which an emotion is itself a belief.

> The soldier in a campaign, cherishing and enjoying life, is unmoved by the probability of being soon cut off. If, in spite of the perils of the field, he still continues to act in every respect as if destined to a good old age, his conviction is purely a quality of his temperament, and will be much less strong at those moments when hunger and fatigue have depressed his frame, or when the sight of dying and dead men has made him tremble with awe. ... Under this hypothesis of no positive evidence, elevation of tone and belief of good to come, are the same fact. Where the acquired trust in evidence does not find its way in any degree, belief is the same thing as happy emotion.
>
> *(1875: 524)*

A further consequence of Bain's axiom concerns the determinacy of mental representations. For if we accept this conception of the relationship between our beliefs and our perceptual, memorial, and emotional experiences, we are naturally led to a view on which the contents of belief are indeterminate as between various discursive articulations or specifications of them. Does my visual experience of the water represent the same thing as the sentence "There is water before me," or does it represent the same thing as "That is water"? Or, if what appears to be water is just a mirage, might my experience have an alternative content (e.g. a "gappy" content) that is neither demonstrative nor existential in nature? Does my experience tell me to believe what it is "saying"? Does it have an imperatival content (Tye 2017)?

If belief requires judgment, as Descartes supposed, and judgment requires inner affirmation, we can ask which content is judged and acted on by the animal who is perceiving or seemingly perceiving the body of water in question. But if the visual experience itself constitutes a perceptual belief in contexts in which the animal is acting on what it sees (or is prepared to do so), we must think of ourselves as theorists as *assigning* contents to the experience in question. We created the distinction between "There is water before me" and "that is water" when we constructed these sentences (or their semantically equivalent predecessor expressions). So neither content need enter into our account of perception and the actions the animal takes that are informed by its perceptions. Pragmatists are prepared to say that cognitive science, even when it is coupled with practical considerations, might fail to yield the kinds of constraints on assignments of content that would warrant a determinate answer to the kinds of questions that confront more "realist" philosophers of mind (e.g. Burge 2010; Tye 2017).

In recent work, Frances Egan (2014) defends this position by distinguishing Fodor's "hyperrealist" account of perceptual content from Chomsky's "ersatz representationalism," and advocating as a middle ground a "deflationary" account of mental representations, which couples a realist construal of representational vehicles with a "pragmatic" account of their content (cf. Coelho Mollo 2020; Egan 2020). Cognitive scientists construct a "function-theoretic" description of perception or skilled movement, which provides an "environment-neutral, domain-general characterization of [a] mechanism" implicated in the animal's output (2014: 123). Of course, at some point, the scientist needs to show how computing these functions enables the animal to see or move if he is to "address the explanandum with which he began," but "no naturalistic relation is likely to pick out a single, determinate content" (2014: 123–124). As Millikan (1989) maintains when arguing that representations can be inaccurate as a rule, and so needn't carry information in Dretske's (1994) sense, to the extent that content is determinate, this is the result of the *uses* to which the organism has put the mechanism. According to Egan, any further determinacy than this must be motivated by "pragmatic considerations" arising from the scientist's need to communicate and utilize her theory of the neurocognitive mechanism in an effective manner (2014: 125).

The Evolution of Subjectivity

Resistance to this pragmatic conception of mental content comes from disparate sources. For example, Horgan and Graham (2012) appeal to phenomenology to attribute greater determinacy to our experiences than Egan allows. But as a general matter, phenomenologists and the heirs to their methodology have been wary of positing representations to account for experience (Hutto and Myin 2013). Using the language of Husserl, for example, Thompson (2007) suggests that perception is "presentational" and that talk of "representation" is only appropriate when theorizing about memory and the phenomena in which it is implicated (2007: 25). And Dewey embraced a relational account of experience that might seem at odds with Egan's acceptance of representational vehicles. But if we need to posit these vehicles to construct effective sciences of perception and motor activity, we lack pragmatic grounds for skepticism.

Indeed, while there are obvious affinities between pragmatism and "enactivist" approaches to the philosophy of mind, the connection is often difficult to ascertain. Gallagher (2014) surveys the enactivist accounts of perception advanced by Thompson (2007), Varela (1991), O'Regan and Noe (2001), Noe (2004), and Hutto and Myin (2013) and finds it surprising that "the philosophical tradition of pragmatism is hardly ever mentioned" in this work (2014: 111).[7] And this is indeed surprising, given the biological orientation shared by the pragmatists and the enactivists and the action-first approach that Dewey (unknowingly) inherited from Bain. Gallagher quotes Dewey's views on the primacy of motor cognition in this vein.

> It is the movement which is primary, and the sensation which is secondary, the movement of body, head and eye muscles determining the quality of what is experienced.
> *(1896: 358–359; quoted from Gallagher 2014: 112)*

Enactivists also argue that movement precedes sensory perception, but they find inspiration for this thesis in Merleau-Ponty's (1942/1963) work and the phenomenological approach to philosophy of mind he helped develop (Thompson 2007: 47–48).

But Dewey's views on the evolution of experience have explicitly inspired a great deal of recent work on sensation, experience, and the emergence of phenomenal consciousness. For example, Peter Godfrey-Smith describes Dewey as his favorite philosopher (2002: 25) and has tried to rescue Dewey's philosophy of science from anachronistically "constructivist" readings. As a counterweight, Godfrey-Smith emphasizes the causal nature of most of those dependencies Dewey posits between mind and world, and the "extrinsic" quality of those cognitive effects which are supposed to be instantaneous and hence non-causal in nature (Godfrey-Smith 2016). At the same time, Godfrey-Smith (2019) has been using contemporary accounts of the origins of embodied, object-oriented, spatial cognition (e.g. Trestman 2013) to bridge the so-called explanatory gap between our knowledge of an organism's biology and our hypotheses about its subjective experiences. If we join the pragmatists in viewing the mind as a control system and a problem-solving device with a natural history, and we view consciousness as a state or property of this system, we can sketch an account of phenomenal consciousness that avoids both panpsychism and "biopsychism," on the one hand, and Cartesianism or human exceptionalism, on the other (cf. Feinberg and Mallatt 2020). This kind of work on the "hard problem of consciousness" is pragmatic in both inspiration and methodology.

Conclusion

In *The Language of Thought 2*, Jerry Fodor tried to defend a Cartesian account of beliefs and other mental representations against a "pragmatist" opponent, describing as "absolute" the disagreement between pragmatism "and the kind of cognitive science that LOT 2 has in mind" (2008: 12). The chasm Fodor saw between these two approaches is worth quoting at length.

> Cartesians think that thought is prior to perception (because perception is, inter alia, a kind of inference). Pragmatists think the opposite. Cartesians think that concepts are prior to percepts (because inference requires, inter alia, subsuming a percept under a concept). Pragmatists think the opposite. Cartesians think that thought is prior to action (because acting requires planning, and planning is a species of reasoning). Pragmatists think the opposite. Cartesians think that concept individuation is prior, in the order of analysis to concept possession. Pragmatists think the opposite. Cartesians think that action is the externalization of thought. Pragmatists think that thought is the internalization of action. In effect, pragmatism

is Cartesianism read from right to left. There are lots of issues that a sufficiently shameless philosopher of mind can contrive to have both ways; but not the issue between pragmatists and Cartesians.

(2018: 12)

Though Fodor was wrong about a lot of things, his characterization of the issues that divide pragmatist and Cartesian approaches to the mind is deeply insightful and tracks the disagreements highlighted in this essay. Of course, Fodor would immediately go on to say, "and pragmatism can't be true; not, at least, the kind of pragmatism according to which the distinctive function of the mind is guiding action" (2008: 13), and he would proceed to advance a flurry of (decidedly a priori) arguments meant to establish this critique. But Fodor's diatribe has proved unpersuasive to date, as pragmatic approaches to the mind have continued to gain traction. As this review of the contemporary philosophical literature demonstrates, the death of pragmatism remains beyond belief.

Notes

1 Peirce gave a similar account in an undated letter to the editor of the *Sun*. "Green was especially impressed with the doctrines of Bain, and impressed the rest of us with them; and finally the writer of this paper brought forward what we called the principle of pragmatism" (Weiner 1946: 223). See, further Bain (1855).
2 See, e.g. J.S. Mill's (1998/1863: 81–86) admission that the non-instrumental value of happiness is not a "matter of fact" that can be confirmed via observation or reasoning from observation but that it is still rational to believe in it.
3 This interpretation of the origins and structure of philosophical pragmatism is spelled out in more detail in Zimmerman (2022). Cf. Fisch (1954).
4 I put to the side the natural criticism we would have of Bob for frequenting a racist club. Perhaps we can assume that he has been regularly dragged there against his will by his racist parents.
5 Does the evidentialist cling to her theory for practical reasons and therein refute herself? I leave consideration of this possibility as an exercise for the reader.
6 This characterization is obviously complicated if one adopts an anti-realist account of truth in one's analysis of a certain discourse. Can't the pragmatist insist that we ought always to say or believe what is true in this relativistic sense (Capps 2020)? I have argued otherwise. Suppose Harry and his friends define truth with regard to claims of handsomeness in terms of a certain aesthetic, or the opinions of the athlete they idolize, or whatever. It is not bizarre to assume that Harry will be more attractive if he avoids a debilitating form of self-loathing, which he can only avoid by falsely believing that he is handsome in the sense at issue. If you think the practical benefits Harry gains from the confidence instilled by his false belief can't compensate for the crime of self-deception, you are to that extent abandoning a pragmatic orientation toward the ethics of belief (Zimmerman 2020). Note too that Bain was in fact an evidentialist in his normative epistemology, which may explain James's complaint (quoted in the text above) that Bain was "too sweeping in his negations."
7 See too Williams (2018), who evaluates the pragmatist elements of "predictive processing" models of perception.

References

Adler, J. (2002) *Belief's Own Ethics*, Cambridge: MIT Press.
Bain, A. (1855) *The Senses and the Intellect*, 1st ed., London: Longmans Green and Company.
Bain, A. (1859) *The Emotions and the Will*, 1st ed., London: Longmans Green and Company.
Bain, A. (1875) *The Emotions and the Will*, 3rd ed., London: Longmans Green and Company.
Basu, R. (2019) 'What We Epistemically Owe to Each Other', *Philosophical Studies*, 176, pp. 915–931.
Boyle, M. (2011) 'Making Up Your Mind and the Activity of Reason', *Philosophers' Imprint*, 11/16, pp. 1–24.
Berker, S. (2018) 'A Combinatorial Argument against Pragmatic Reasons for Belief', *Analytic Philosophy*, 59/4, pp. 427–470.
Bratman, M. (1992) 'Practical Reasoning and Acceptance in a Context', *Mind*, 101, pp. 1–14.
Burge, T. (2010) *Origins of Objectivity*, Oxford: Oxford University Press.

Capps, J. (2020) 'Pragmatic Accounts of Belief and Truth', *William James Studies*, 16/1, pp. 39–56.
Coelho Mollo, D. (2020) 'Content Pragmatism Defended', *Topoi*, 39, pp. 103–113.
Dennett, D. C. (1981) 'How to Change Your Mind', in D. Dennett (ed.), *Brainstorms: Philosophical Essays on Mind and Psychology*, Cambridge, MA: Bradford/MIT, pp. 300–309.
Dewey, J. (1896) 'The Reflex Arc Concept in Psychology', *Psychological Review*, 3/4, pp. 357–370.
Dretske, F. (1994) 'If You Can't Make One, You Don't Know How It Works', in P. French, T. Uehling, and H. Wettstein (eds.), *Midwest Studies in Philosophy*, 19, pp. 468–482.
Egan, F. (2014) "How to Think About Mental Content', *Philosophical Studies*, 170/1, pp. 115–135.
Egan, F. (2020) 'A Deflationary Account of Mental Representation', in J. Smortchkova, K. Dolega, and T. Schlicht (eds.), *What Are Mental Representations?* New York: Oxford University Press, pp. 26–53.
Engel, A. K., Maye, A., Kurthen, M., and König, P. (2013) 'Where's the Action? The Pragmatic Turn in Cognitive Science', *Trends in Cognitive Science*, 17, pp. 202–209.
Feinberg, T., and Mallatt, J. (2020) 'Phenomenal Consciousness and Emergence: Eliminating the Explanatory Gap', *Frontiers in Psychology*, 11/1041, pp. 1–15.
Fisch, M. H. (1954) 'Alexander Bain and the Genealogy of Pragmatism', *Journal of the History of Ideas*, 15/3, pp. 413–444.
Fodor, J. (2008) *LOT2: The Language of Thought Revisited*, Oxford: Clarendon Press.
Gallagher, S. (2014) 'Pragmatic Interventions into Enactive and Extended Conceptions of Cognition', *Philosophical Issues*, 24, pp. 110–126.
Godfrey-Smith, P. (2002) 'Dewey on Naturalism, Realism and Science', *Philosophy of Science*, 69/S3, pp. S25–S35.
Godfrey-Smith, P. (2016) 'Dewey and the Question of Realism', *Noûs*, 50/1, pp. 73–89.
Godfrey-Smith, P. (2019) 'Evolving Across the Explanatory Gap', *Philosophy, Theory, and Practice in Biology*, 11/1, pp. 1–13.
Hartshorne, C., and Weiss, P. (eds.) (1934) *C. S. Peirce: Collected Papers*, Cambridge, MA: Harvard University Press.
Hieronymi, P. (2008) 'Responsibility for Believing', *Synthese*, 161/3, pp. 357–373.
Horgan, T., and Graham, G. (2012) 'Phenomenal Intentionality and Content Determinacy', in R. Schantz (ed.), *Prospects for Meaning*, Boston: De Gruyter, pp. 321–344.
Hutto, D., and Myin, E. (2013) *Radicalizing Enactivism: Basic Minds without Content*, Cambridge, MA: MIT Press.
James, W. (1898) 'Philosophical Conceptions and Practical Results', *Berkeley University Chronicle*, 1/4, pp. 287–310.
Maguire, B., and Woods, J. (2020) 'The Game of Belief', *Philosophical Review*, 129/2, pp. 211–249.
McCormick, M. S. (2015). *Believing against the Evidence: Agency and the Ethics of Belief*, New York: Routledge.
Mele, A. (2001) *Self-deception Unmasked*, Princeton: Princeton University Press.
Menary, R. (2016) 'Pragmatism and the Pragmatic Turn in Cognitive Science', in K. Friston, A. Andreas, and D. Kragic (eds.), *Pragmatism and the Pragmatic Turn in Cognitive Science*, Cambridge, MA: MIT Press, pp. 219–236.
Merleau-Ponty, M. (1963) *The Structure of Behavior*, A. Fisher (trans.), Boston: Beacon Press; originally published in 1942.
Mill, J. S. (1998) *Utilitarianism*, R. Crisp (ed.), Oxford: Oxford University Press; originally published in 1861.
Millikan, R. G. (1989) 'Biosemantics', *Journal of Philosophy*, 86, pp. 281–297.
Parfit, D. (2011) *On What Matters*, Vol. 1. Oxford: Oxford University Press.
Peirce, C. S. (1878) 'How to Make Our Ideas Clear', *Popular Science Monthly*, 12, pp. 286–302.
Roeber, B. (2019) 'Evidence, Judgment and Belief at Will', *Mind*, 128, pp. 837–859.
Rinard, S. (2019a) 'Equal Treatment for Belief', *Philosophical Studies*, 176/7, pp. 1923–1950.
Rinard, S. (2019b) 'Believing for Practical Reasons', *Noûs*, 53/4, pp. 763–784.
Schoenfield, M. (2014) 'Permission to Believe: Why Permissivism is True and What It Tells Us about Irrelevant Influences on Belief', *Noûs*, 48/2, pp. 193–218.
Shah, N., and Velleman, D. (2005) 'Doxastic Deliberation', *The Philosophical Review*, 114/4, pp. 497–534.
Trestman, M. (2013) 'The Cambrian Explosion and the Origins of Embodied Cognition', *Biological Theory*, 8, pp. 80–92.
Tye, M. (2017) 'How to Think about the Representational Content of Visual Experience', in C. Limbeck-Lilienau and F. Stadler (eds.), *The Philosophy of Perception: Proceedings of the Austrian Ludwig Wittgenstein Society*, Berlin and Boston: De Gruyter, pp. 77–94.
Way, J. (2016) 'Two Arguments for Evidentialism', *The Philosophical Quarterly*, 66/265, pp. 805–818.

Weiner, P. P. (1946) 'Peirce's Metaphysical Club and the Genesis of Pragmatism', *Journal of the History of Ideas*, 7/2, pp. 218–233.

Williams, B. (1973) 'Deciding to Believe', in B. Williams (ed.), *Problems of Self*, Cambridge: Cambridge University Press, pp. 136–151.

Williams, D. (2018) 'Pragmatism and the Predictive Mind,' *Phenomenology and the Cognitive Sciences*, 17, pp. 835–859.

Williamson, T. (2000) *Knowledge and Its Limits*, Oxford: Oxford University Press.

Zimmerman, A. (2006) 'Basic Self-Knowledge: Answering Peacocke's Criticisms of Constitutivism', *Philosophical Studies*, 128, pp. 337–379.

Zimmerman, A. (2018) *Belief: A Pragmatic Picture*, Oxford: Oxford University Press.

Zimmerman, A. (2020) 'Pragmatism, Truth, and the Ethics of Belief', *William James Studies*, 16/1, pp. 82–93.

Zimmerman, A (2022) 'Bain's Theory of Belief and the Genesis of Pragmatism', *Transactions of the Charles S. Peirce Society*, 57/3, pp. 319–340.

29
PRAGMATISM AND COGNITIVE SCIENCE

Shaun Gallagher

In recent years numerous researchers have discussed a pragmatic turn in cognitive science (Crippen and Schulkin 2020; Engel, Friston and Kragic 2015; Engel et al. 2013; Gallagher 2014; Johnson 2016; Madzia and Jung 2016; Menary 2015). The general consensus is that this turn, or return to pragmatism, is closely tied to the turn within cognitive science to non-representational embodied cognition, sometimes referred to as 4E (embodied, embedded, extended, and enactive) cognition. Thus, Engel et al. (2013: 202) write:

> In cognitive science, we are currently witnessing a 'pragmatic turn', away from the traditional representation-centered framework towards a paradigm that focuses on understanding cognition as 'enactive', as skillful activity that involves ongoing interaction with the external world.

In some regards the pragmatic turn just is this turn to embodied action-oriented cognition that came to the fore starting in the 1990s. Still, I'll argue that this is an oversimplification in a number of ways. First, in regard to the timing; second in regard to how pragmatism may have already been influencing mainstream cognitivists even prior to the turn to embodied cognition (EC); and third, in regard to how pragmatism relates, somewhat unevenly, to the variety of EC theories. Furthermore, we should be careful not to conflate pragmatism with 'the pragmatic turn', even if these two movements are conceptually linked.

Timing and connections with pragmatism in mainstream cognitive science and philosophy of mind

First, in regard to timing, anything that resembles a self-conscious pragmatic turn in the EC camp doesn't correlate to the emergence of EC approaches. If we review the early works on EC, there is rarely a mention of the pragmatists. Varela, Thompson and Rosch (1991: 30–31), for example, make a general reference to pragmatism but leave it undeveloped and do not directly relate it to the embodied-enactivist aspects of their work. Likewise, even in later texts associated with enactivism, for example, in Noë (2004), Thompson (2007), or Hutto and Myin (2013), there is no mention of pragmatists like Peirce, Dewey, or Mead. When James is mentioned, it's not James the pragmatist but James's *Principles of Psychology* (1890; see, e.g., Varela 1999: 266). Although Clark (2008) begins his book on the extended mind with an invocation from Dewey, he fails to exploit any further connection with pragmatism. Significant mentions of the pragmatists in these contexts, however, do start just around this time. Menary (2007), for example, does exploit pragmatist themes to frame his understanding of the extended mind. He appeals to Dewey's notion of organism-environment transactions to work out a characterization of how embodied cognitive processes incorporate the environment, and to Peirce's 'continuity principle', to counter the idea of a metaphysical discontinuity between the mind and the world (2007: 129). Steiner (2008, 2010) also points out the affinity between Dewey's account of experience and externalist approaches, including extended mind and enactivist accounts of cognition. Johnson (2008: 274) also celebrates Dewey and equates the latter's notion of transaction with enaction. The association between EC and pragmatism starts to gain traction in Chemero (2009), as well as in Menary (2011), and with further detail in Gallagher (2014), Johnson (2016), and Di Paolo, Buhrmann and Barandiaran (2017). Accordingly, if the pragmatic turn is measured by mention and use of the pragmatists, the turn is made only some 15 years after the initiation of EC in the early 1990s.

Second, there are elements of pragmatism that influence (implicitly at least) debates about cognition that predate EC and, indeed, from the very beginnings of the classical cognitivist regime. These influences remain implicit in the sense that they stay, for the most part, in the background. The psychologist, James Gibson (1977), whose ecological psychology and theory of affordances (later taken up by EC) were influenced by pragmatism (Burke 2013; Chemero 2009; Heft 2001; Rockwell 2005), was portrayed as the opposition to what Fodor and Pylyshyn (1981) defined as the establishment view of cognitivism. At the same time, Fodor and Pylyshyn admit that a 'conciliatory reading' of some of Gibson's points might be possible. According to them, Gibson's notion of direct perception as information pickup is overly promiscuous since it implies that we would pick up anything that catches our eye. Perception requires constraints – it needs to be selective – and for Fodor and Pylyshyn, constraint comes by way of internal representational, inferential processes. Although this is a rejection of the fully 'direct' nature of perception, the notion that there must be both some direct pickup (transduction) plus some selectivity may be the conciliatory point. We'll see later that inferential processes, understood as a kind of Peircean abduction, can fit with a pragmatic conception of what drives the cognitive process. The question is whether we should think of the required abductive selectivity process (conceived as active orienting, exploring, investigating, attuning, etc.) to be built into the affordance structure of the organism-environment relation (as Gibson and the pragmatists would hold), or to be the product of an internal representational process working with impoverished 'premises' generated by transducers more narrowly construed (as Fodor and Pylyshyn would have it). In this respect, however, it's clear that the pragmatic elements of Gibson's theory are, for Fodor and Pylyshyn, not the points of conciliation.

Like Gibson, the developmental psychologist, Jerome Bruner (1966), who helped to launch the field of cognitive psychology, and who proposed the concept of enactive (action-based) representation, was influenced by pragmatism, and specifically by Dewey (see Bruner 1961). In an essay on 'Language and experience', Bruner reflects on the revolutionary work of Chomsky and notes that Chomsky overlooks some of the more important aspects of language – 'those precisely that were context dependent'. This context dependency of language and the idea that linguistic practices 'necessarily reflect the circumstances' (Bruner, Caudill and Ninio 1977: 12) are ideas that Bruner finds in Dewey. For Bruner, this issue came back into focus at a conference in London in 1975 (the Third International Child Language Symposium). He is led to emphasize language as performance and the importance of intersubjective interaction for the child's acquisition of speech.

> As Dewey says, communication by itself does not accomplish anything. In so far as the dialogue between mother and infant succeeds in getting the child to fill his role in exchange … the child is in fact learning not so much a language, as how to proceed in achieving certain ends by the use of language. The input is not a corpus; the output is not a grammar.
>
> *(1977: 19)*

It is notable that Bruner cites Grace de Laguna's (1927) work on speech and, in so doing, gestures (perhaps unknowingly) to deeper philosophical connections with pragmatism that continued to hover in the background, shaping the philosophy of science that was immediately informing the cognitive revolution.[1] De Laguna, along with her co-author (and husband) Theodore de Laguna, had early on engaged critically with pragmatism (de Laguna and de Laguna 1910). The pragmatist view is central to their evolutionary epistemology and confirmation holism: 'concepts, apart from the conduct which they prompt, mean nothing' (1910: 206). If they were not pragmatists in this regard, they were the first neo-pragmatists. Importantly, it also seems quite possible that their work informed another quasi- or neo-pragmatist, Willard V. O. Quine. Joel Katzav (2018) makes the connection clear. Both Grace de Laguna and Quine contributed papers to the 1950 APA Eastern Division symposium, which were subsequently published in *The Philosophical Review*. Quine's paper, 'Two dogmas of empiricism' (1951), in which he suggests, there is 'a shift towards pragmatism', turned out to be highly influential. Like the de Lagunas, Quine also defends confirmation holism, the idea that no concept or theory can be verified in isolation since it is embedded in a background or web of other concepts and beliefs, including other scientific theories. 'Our statements about the external world face the tribunal of sense experience not individually, but only as a corporate body' (Quine 1953: 41).

Katzav (2018) argues that an even more sophisticated version of this idea is to be found in the de Laguna critique of pragmatism although in a formulation that incorporates the pragmatist linking of concept and conduct.

> Our thoughts direct our conduct, and it is in this service that their meaning ultimately consists; but every concept means both more and less than any particular application of it contains.
>
> *(1910: 206)*

Likewise, with respect to the analytic-synthetic distinction, de Laguna holds that part of the meaning of any concept involves a reference to experience ('along the edges' or on the periphery of the web), such that the concept is judged by its ability to control conduct, although concepts of logic have a greater autonomy from experience (1910: 206–207, 210, 212; see Quine 1953: 42).

> We have always a multitude of general beliefs in accordance with which we interpret each new matter of fact; and though any one of these beliefs may at some time be called into question, this is always on the supposition of the acceptance of a host of others. Science, accordingly, can never be a system of judgments with one way relations of implication. Our judgments support one another. And when, as occasionally happens, they contradict one another, there is no ultimate standard of imperishable truth by which they can be tested.
>
> *(de Laguna 1930: 404)*

Theodore de Laguna explicitly states that his views were directly influenced by James and Dewey (de Laguna 1930: 406; see Ben-Menahem 2016 for the James-Quine relation).

Whether we consider Quine a pragmatist or not (see Haack 2004; Koskinen and Pihlström 2006), Quine's naturalism and empiricism, views widely shared by cognitive science, mean there is no hard line between science and philosophy. He attributes this to the influence of Dewey (Quine 1969: 26):

> Philosophically I am bound to Dewey by the naturalism that dominated his last three decades. With Dewey I hold that knowledge, mind, and meaning are part of the same world that

they have to do with, and that they are to be studied in the same empirical spirit that animates natural science. There is no place for a prior philosophy.

(also see an interview with Quine in Bergström and Føllesdal 1994)

As Hilary Putnam puts it, 'like Quine, the classical pragmatists do not believe that there is a "first philosophy" higher than the practice that we take most seriously when the chips are down' – i.e., the practice of science (Putnam 1994: 154). This part of pragmatism is imported directly into the cognitivist camps of cognitive science. Patricia Churchland, for example, follows Quine on this point: 'philosophy at its best and properly conceived is continuous with the empirical sciences' (1986: 2).

It also leads to debates, which again hover in the deep philosophical background of cognitive science. Starting in the 1970s, for example, a debate between Putnam and Rorty about the relevance of pragmatism (and neo-pragmatism) ran simultaneously with debates about functionalism in the philosophy of cognitive science (Putnam 1975; Rorty 1972, 1982). Putnam, a central figure in the philosophy of mind who had a significant influence on the development of the concept of functionalism in cognitive science in the 1960s, started to explore the ideas of pragmatism in James and Peirce. This included Peirce's idea that a concept or a belief was not something one simply entertains in one's mind, in a propositional form, but something on which one is prepared to act, or something equivalent to the consequence of the habit that it is calculated to produce – a view that is not alien to the more recent embodied action-oriented views of cognition.[2] By the time that Putnam gives up his own earlier functionalist view, that is, just around the time that EC is getting started, the pragmatism he studied throughout the 1980s remains implicit but discernible in his emphasis on the role of the physical and social environment in cognition (Putnam 1992a; also see 1992b, 1995).

Let me say what many others have said (starting perhaps with Lovejoy 1908, if not Peirce himself; also see Haack 2004; Koskinen and Pihlström 2006; and discussions between Putnam and Rorty about whether Quine could be considered a pragmatist): it is difficult to define pragmatism or to say who counts as a pragmatist. In this regard there are more issues than we can address in this short chapter, but let me suggest that just as sure as Jerry Fodor never was and never could be considered a pragmatist, one could make a good argument that Dan Dennett is a pragmatist. There is, perhaps, a vague line of pragmatism that runs from William James' essay, 'Does consciousness exist' (1904), directly to Dennett's *Consciousness Explained* (1991). The invocation to one of the central chapters in the latter is a quote from James' *Principles of Psychology*, although, to be sure, there are only scattered mentions of James and no mention of other pragmatists. The point, however, is not about lineage but about strategy. Bjørn Ramberg (2004) makes a clear argument about this, which helps to explain why *Consciousness Explained* appears to some as 'consciousness explained away' (e.g., Lowe 1993).

> Dennett is motivated by the diagnosis that the folk-notion of consciousness keeps us wedded to a set of interwoven descriptions of mind and self that inhibit our susceptibility to the naturalizing influence of science on our self-image. This set of descriptions is what we gesture at with the notion of the subjective. The sense that the notion of the subjective is a rich and bona fide mine of philosophical problems and insights is an explicit target of Dennett's seditious account of mind. Dennett's view is that the linguistic practices in which our notion of consciousness is embedded (the vocabulary from which the philosophical invention 'qualia' takes its intuitive power ...), are practices we would do well, if we want to naturalize our conception of ourselves, to alter. But this, any pragmatist knows, we can do only in so far as we are able make satisfying alternative descriptions available.
>
> *(Ramberg 2004: 5)*

In other words, Dennett uses a pragmatist's strategy, the 'interpretivist strategy', following Quine. Pragmatist interpretivism regards ontological intuitions about the mind as vocabularies to be deconstructed.[3] 'Naturalistic pragmatists [like Quine and Dennett] are proposing ways to describe ourselves as thinkers and agents that make the philosophical contrast between mind and matter seem to be without any particular ontological point' (Ramberg 2004: 3). Dennett is forever trying to undercut our armchair intuitions about the mind, mainly by appealing to empirical science (and sometimes humor). This interpretivist strategy makes him an unheralded pragmatist.

The point of this quick and incomplete sketch of how some elements of pragmatism already inform discussions in cognitive science and philosophy of mind prior to the emergence of EC is intended to qualify any strong claim about a pragmatic turn in connection with EC. In some ways, cognitive science has always turned on some very basic pragmatist concepts that continue to inform the philosophy and practice of studying the mind.

How pragmatism relates to the variety of EC theories

I return now to the pragmatic turn. I've noted first, in terms of timing, there was not a complete coincidence between the development of EC approaches and any turn to pragmatism. Second, I've provided a brief sketch of some of the elements of pragmatism to be found in cognitive science and philosophy of mind prior to the advent of EC. My third task is to indicate how pragmatism relates to the variety of EC theories.

Weak EC

Mark Johnson and Tim Rohrer (2007) focus on the rejection of mind-body dualism and trace this rejection from pragmatists like James and Dewey into the recent EC theorists. They take this rejection as also a rejection of the representationalist theory of mind and the establishment of a pragmatically centered cognitive science. In place of representation, the pragmatists offer an emphasis on action, treating cognition as a kind of action as a response to problematic situations.

Johnson and Rohrer emphasize three features that derive from pragmatism to inform EC. First, in Dewey's terms, the idea that the explanatory unit is organism-environment rather than brain. As Dewey suggests, 'to see the organism in nature, the nervous system in the organism, the brain in the nervous system, the cortex in the brain is the answer to the problems which haunt philosophy' (Dewey 1925: 198). Second, a principle of continuity, concerning which they again cite Dewey: 'there is no breach of continuity between operations of inquiry and biological operations and physical operations. "Continuity" ... means that rational operations grow out of organic activities, without being identical with that from which they emerge' (Dewey 1938: 26). Johnson and Rohrer suggest that this fits well with the concept of autopoiesis – the self-organizing, self-producing system discussed by Maturana and Varela (1980), which informs the enactive branch of EC. On this account, no internal representations are needed for intelligent action, and indeed, even single-cell organisms are capable of engaging in sensorimotor coordination in response to environmental changes. Johnson and Rohrer marshal significant scientific evidence to support various notions of organism-environment coupling, and they emphasize brain plasticity and adaptability in this context.

The third feature relevant to cognition is the fact that in speaking about the environment, EC, like pragmatism, acknowledges that the environment is not just physical but also social, and this means that multiple organisms can cooperate in response to current or anticipated problems. Once again, pragmatists such as Dewey (1925) have taken this complex kind of social interaction as 'emblematic of cognition par excellence' (Johnson and Rohrer 2007: 18).

It is useful to distinguish between different versions of EC to understand how Johnson and Rohrer put these principles to work. As a first distinction, consider the difference between 'weak' EC and 'strong' EC (Alsmith and de Vignemont 2012). Strong EC (which would include enactivist views – see below) endorses a significant explanatory role for the (non-neural) body itself, and usually for the environment, in cognitive processes. According to weak EC, in contrast, the significant explanatory role is given to 'in the head' neural processes, or what are variously called body or B-formatted (neural) representations, understood as simulations of bodily functions in the brain (e.g., Goldman 2012, 2014; Goldman and de Vignemont 2009). In this regard, weak EC remains close, and consistent with, classic cognitivist-representationalist conceptions of cognition. Thus, Goldman and de Vignemont assume that almost everything of importance for human cognition happens in the brain, 'the seat of most, if not all, mental events' (2009: 154). They discount both the non-neural body (the role of anatomy and body activity, such as actions and postures) and the environment as significant contributors to cognition. They are thus left with, as Goldman and de Vignemont put it, 'sanitized' body or 'B-formatted' representations (2009: 155). Such representations are computational, even if they are not propositional or conceptual in format; their content may include the body or body parts, but also they may include action goals and the bodily-motoric means to achieve them. Barsalou's notion of grounded cognition also suggests that cognition operates on reactivation of motor areas, simulations that 'can indeed proceed independently of the specific body that encoded the sensorimotor experience' (2008: 619; see Pezzulo et al. 2011).

The processes involving B-formatted neural representations are models or maps internal to the brain. Such processes may be the product of evolutionary 'reuse' (Anderson 2010), i.e., the idea that neural circuits originally established for one use can be redeployed for other purposes while still maintaining their original function. An example can be found in linguistics. Pulvermüller's (2005) language-grounding hypothesis shows that language comprehension of action words involves the activation of action-related cortical areas. This suggests that higher-order symbolic thought, including memory, is grounded in low-level simulations of motor action (Barsalou 1999; Casasanto and Dijkstra 2010; Goldman 2014; Glenberg 2010). Goldman (2014) considers Lakoff and Johnson's (1999) body-related metaphors to be a good example of B-formatted representations.

Although Johnson and Rohrer associate much of what they propose as pragmatic EC with enactivist views, they cash out most of these pragmatic principles in ways that remain close to weak EC. They link their analysis to image-schemas and neural maps as a way to explain higher-order thought. They argue, however, that these neural maps are not representations but are formed as different sets of neurons compete to become topological neural maps driven primarily by regularities in the environment. Brain plasticity allows for the reorganization of such maps in response to changing environments. The maps code perceptual space in a topological fashion but also increasingly allow for abstract topological structure so that 'we live in the world of our maps. Topologically speaking, our bodies are in our minds, in the sense that our sensorimotor maps provide the basis for conceptualization and reasoning' (Johnson and Rohrer 2007: 10). They reject a strict (Fodorian) representationalist interpretation of these maps but argue that 'actual neural representations are perpetually situated in dynamic organism-environment interactions'. Lakoff (1987) and Johnson (1987) propose the term 'image-schemas' to explain the cognitive relevance of these neural maps, and they cite evidence from psychology and linguistics that these image-schemas are neurally embodied as patterns of activation in and between topological neural maps. Image-schemas are involved in the simulations of abstract concepts (Gallese and Lakoff 2005), form the basis for Conceptual Metaphor Theory (Lakoff and Johnson 1999), and help to explain the neuroimaging findings of Pulvermüller (2005), as mentioned by Goldman.

Strong, enactive EC

Strong EC includes what is sometimes called 4E cognition. It maintains that cognition is not just a brain activity and that in regard to evolutionary claims, one has to understand the significance of the fact that the brain and body co-evolved. Consider, for example, the hypothetical case in which humans evolved without hands. Not only would our brains be different, if this were the case, but we would perceive the world differently. On enactivist and ecological accounts, our perception is action oriented, and we perceive the world pragmatically, in terms of affordances, i.e., in terms of what we can do with the things around us and how we can interact with other agents. Both physical and social affordances would be different for an organism without hands.

If, as Johnson and Rohrer have shown, pragmatism can inform weak versions of EC, Engel et al. (2013) propose that it can also inform strong EC. They emphasize the action-oriented perspective, moving away from representationalism and suggesting 'that cognitive processes are so closely intertwined with action that cognition would best be understood as "enactive"' (2013: 202), or 'pragmatic'.

> The term 'pragmatic' is used here, first, to highlight our conjecture that cognition is a form of practice. Second, we introduce the term to refer to action-oriented viewpoints, such as those developed by the founders of philosophical pragmatism [they cite Dewey and Mead], albeit without suggesting a return to exactly the positions put forward by these authors.
>
> *(2013: 202)*

They cite empirical evidence, specifically brain imaging studies, to show that concepts are closely tied to action, specifically that 'object concepts in semantic memory do not only rely on sensory features but, critically, also on motor properties associated with the object's use' (2013: 204). Similar evidence exists for attentional and decision-making processes (Ibid).

Like Johnson and Rohrer, Engel et al. also highlight the brain plasticity of cortical maps, and they point to a great deal of empirical evidence to show how such maps 'critically depend on sensorimotor interactions and active exploration of the environment' (2013: 203). I think it is also clear, that, like Johnson and Rohrer, Engel et al. would eschew any kind of strict representationalism, and if neuroscientists have license to refer to 'neural representations', at most we should think of them as simple co-variations that exist between brain processes in a body that is coupled to an environment. In this respect, Engel et al. point to Andy Clark's concept of action-oriented representation (sometimes called minimal representation) (Clark 1997; Gallagher 2008), the idea that 'brain states prescribe possible actions, rather than describing states of the outside world' (Engel et al. 2013: 206).

Whether there is a better way to think about how the brain actually works in an enactive system is still a matter of debate (Gallagher 2020; Gallagher et al. 2013; Hutto and Myin 2013). Pragmatism, however, does have at least three things to offer in this regard. First, with respect to the question of representation, Menary (2015) reminds us that Peirce proposed a developed account of representation that is neither of the strict Fodorian kind nor of the minimal action-oriented kind. Peirce does not conceive of representation as a vehicle carrying information or semantic content, or as a stand-in for an object. Rather, he emphasizes 'continuous dynamical interpretation' (Menary 2015: 222). Peirce's notion of representation involves 'sign action'. In the simplest terms, something (a sign) is produced by a mechanism or agent, some other mechanism or agent (a consumer) attunes to it, thereby accomplishing something (leading to some end). In that case the sign counts as a representation in Peirce's sense. Menary explains that this representational process 'requires the coordination of producer and consumer mechanisms for some further end; therefore it requires either the coordination of mechanisms within the organism or the coordination of a mechanism in the organism with a mechanism in that organism's environment' (2015: 224). Furthermore, this idea of representation 'makes no commitment as to whether sign action must be internal, external, or distributed across brain, body, and world' (Ibid.).

Second, I think that Engel et al. come close to what might count as a pragmatic solution, at least for basic perception-action cognition. They discuss 'dispositions for action embodied in dynamic activity patterns' rather than representations. Such dispositions are not simply neural events but include extra-neural patterns enacted by the body as it couples to the environment. Accordingly, '[k]nowing what an object is does not mean to possess internal descriptions of this object, but to master sets of sensorimotor skills and possible actions that can be chosen to explore or utilize the object' (2013: 206).

Third, Dewey offers an excellent clue that has nothing to do with representations or maps.

> The advance of physiology and the psychology associated with it have shown the connection of mental activity with that of the nervous system. Too often recognition of connection has stopped short at this point; the older dualism of soul and body has been replaced by that of the brain and the rest of the body. But in fact the nervous system is only a specialized mechanism for keeping all bodily activities working together. Instead of being isolated from them, as an organ of knowing from organs of motor response, it is the organ by which they interact responsively with one another. The brain is essentially an organ for effecting the reciprocal adjustment to each other of the stimuli received from the environment and responses directed upon it. Note that the adjusting is reciprocal; the brain not only enables organic activity to be brought to bear upon any object of the environment in response to a sensory stimulation, but this response also determines what the next stimulus will be.
>
> (1916: 336–337)

Both Engel et al. (2013) and Menary (2015) suggest that pragmatism is consistent with embodied, extended, and enactive versions of EC, and we have seen, above, that it may also be consistent with weak EC, and perhaps even in some respects with aspects of more traditional forms of cognitive science. The final question that we will consider is whether pragmatism is consistent with predictive processing (PP) approaches. Engel et al. (2013) hint, and some of the essays in Engel et al. (2015), including Menary (2015), argue that pragmatism, especially in its conception that we are fallible cognitive agents who actively explore their environments, may provide some guidance for Bayesian PP.

Peircean and neuro-Peircean predictive processing

To provide a pragmatist interpretation of PP, Menary (2015) focuses on the Peircean concepts of 'exploratory inference', abduction, and their similarity to Bayesian or PP accounts of active inference. Pragmatist versions of PP, like enactivist versions (Gallagher and Allen 2019), would steer us away from internalist and avowedly non-pragmatist models of the sort found in Hohwy (2013), in which the brain, isolated from the world, tries to infer the causes of its sensory input while minimizing prediction errors, with the aim of maintaining a veridical representation of the world. Pragmatist versions rather push toward a more externalist conception of active inference according to which action and perception are both needed to minimize prediction errors by optimizing the states of brain-body-environment. Peirce's notion of abduction, 'the process of forming explanatory hypotheses' (CP 5.172), fits this externalist model. Abduction is a fallible, self-corrective process that allows the human agent to explore its environment in a hands-on fashion. In the process of problem solving, we form hypotheses primarily through action (rather than by formulating propositions in-the-head), and we test them out in a science-like exploration of the environment. Menary takes this to be a process similar to PP accounts of active inference, which facilitates the development of priors and of stable predictions. This is a version of PP that emphasizes the Free Energy Principle (Friston 2010), roughly, the idea that the organism, by means of active inference, tends to remain in a range of states, close to homeostasis, allowing it to reduce surprise (prediction error or variational free energy, understood in an information theoretic way) and thereby survive or avoid entropy.

Menary highlights the potential of this model in non-representationalist, strong-EC, terms:

> [In] at least some sensorimotor cases, the conjecture and test may be based on motor activity rather than on beliefs or representations. Therefore, it is possible to give a non-representational account of active inference, and this would be entirely consistent with the likely evolutionary origin of those inferences in sensorimotor interactions with the environment. This interpretation is consistent with the reflex arc concept developed by Dewey ... as a matter of sensorimotor coordination.
>
> *(2015: 230)*

Again, in contrast to Hohwy's (2013) internalist version of PP, in which perceptual inference is accomplished entirely in an isolated brain, with active inference merely serving such central processes, the pragmatist model works out in the open and emphasizes interaction between organism and environment. This view sits well with Clark's (2016) conception of PP, where 'active inference and cultural props help to minimize prediction errors ... and [where] there is a deep continuity between mind and world mediated by active inference and the cultural scaffolding of our local niche' (Menary 2015: 232; also see Williams 2018).

For Friston (2015), PP is based on a Bayesian mathematical formalism that is fully consistent both with the pragmatism of strong (externalist) EC, of the sort championed by Clark or Menary's Peircean model, and with Hohwy's non-pragmatist internalism, the idea that the brain, understood 'as a statistical organ that generates hypotheses', is doing most of the work. In the latter respect, Friston and Stephan (2007: 433) put it simply: 'sustained exposure to environmental inputs causes the internal structure of the brain to recapitulate the causal structure of those inputs. In turn, this enables efficient perceptual inference'.

Engel et al. (2013) stake out a center position, which might be called a neuro-Peircean model that frames the connection between pragmatic conceptions of action and PP in terms of very basic motor control processes (brain-based forward models), which run predictions about the sensory outcome of movement. Engle, Friston and Kragic (2015) interpret this in terms of the predictive aspects of sensory-motor contingencies (SMCs) as this idea has been developed in enactive accounts by Noë (2014) and O'Regan and Noë (2001).

> SMCs are defined as law-like relations between movements and associated changes in sensory inputs that are produced by the agent's actions. Once acquired, an agent can use these SMCs to predict consequences of its own actions.
>
> *(Engle, Friston and Kragic 2015: 176)*

Engle, Friston and Kragic, however, take a very broad view of what counts as pragmatic EC, to include versions of weak EC (discussed above). Specifically, SMCs play a role in the formation of object concepts by means of neural simulation (or what Johnson and Rohrer called image-schemas), as well as in speech perception and language comprehension (Pulvermüller 2015; see Pezzulo 2015: 26ff).

Conclusion

In mapping out the influence of pragmatism on cognitive science, it becomes clear that there is a wide spectrum of pragmatic ideas that inform a diversity of models for studying the mind. The central idea that seemingly characterizes the pragmatic turn and that unites, or at least draws together, these various, and somewhat diverse, embodied approaches in cognitive science – weak or grounded EC, the SMC approach, some versions of active inference/PP, Gibson's affordance theory, and enactive perspectives – is the focus on the idea that cognition is action oriented. Dewey's emphasis on organism-environment coupling clearly shows up in the more embodied EC approaches. More generally, the pragmatist views on confirmation holism in philosophy of science, the tight conjunction of philosophy and science, and the interpretivist strategy have played a continuing role in the development of cognitive science.

Notes

1 He also cites David McNeil (1975), whose theory of gesture figures into later discussions of EC, enactivism, and extended mind, in Gallagher (2005) and in Clark (2008).
2 Zimmerman (this volume) makes clear Peirce's debt to Alexander Bain for this view ('belief [has] no meaning, except in reference to our actions' [Bain 1859: 568]) and Bain's insistence on the primacy of movement.
3 This way of putting it is more Rorty than Dennett; Rorty (1982b) sees Dennett as a pragmatist in just this way and approves; see Dennett's (1982) response – he agrees ±72.4%. Thanks to Aaron Zimmerman for alerting me to these two essays.

References

Alsmith, A.J.T. and De Vignemont, F. (2012) 'Embodying the mind and representing the body', *Review of Philosophy and Psychology*, 3(1), pp. 1–13.
Anderson, M.L. (2010) 'Neural reuse: a fundamental reorganizing principle of the brain', *Behavioral and Brain Sciences*, 33, pp. 245–266.
Bain, A. (1859) *The emotions and the will*. London: Longmans Green and Company.
Barsalou, L.W. (1999) 'Perceptual symbol systems', *Behavioral and Brain Sciences*, 22, pp. 577–660.
Barsalou, L.W. (2008) 'Grounded cognition', *Annual Review Psychology*, 59, pp. 617–645.
Ben-Menahem, Y. (2016) 'The web and the tree: Quine and James on the growth of knowledge', in Janssen-Lauret, F. and Kemp, G. (eds.) *Quine and his place in history*. London: Palgrave Macmillan, pp. 59–75.
Bergström, L. and Føllesdal, D. (1994) 'Interview with Willard Van Orman Quine in November 1993', *Theoria*, 60(3), pp. 193–206.
Bruner, J. (1961) 'After Dewey, what?' in Bruner, J. (ed.) *On knowing: essays for the left hand*. Cambridge, MA: Harvard University Press, pp. 113–131.
Bruner, J. (1966) *Toward a theory of instruction*. Cambridge, MA: Harvard University Press.
Bruner, J., Caudill, E. and Ninio, A. (1977) 'Language and experience', in Peters, R.S. (ed.) *John Dewey reconsidered*. New York: Routledge. Reprint 2010, pp. 12–22.
Burke, F.T. (2013) *What pragmatism was*. Bloomington: Indiana University Press.
Casasanto, D. and Dijkstra, K. (2010) 'Motor action and emotional memory', *Cognition*, 115(1), pp. 179–185.
Chemero, A. (2009) *Radical embodied cognitive science*. Cambridge, MA: MIT Press.
Churchland, P.S. (1986) *Neurophilosophy*. Cambridge, MA: MIT Press.
Clark, A. (1997) *Being there: putting brain, body, and world together again*. Cambridge, MA: MIT Press.
Clark, A. (2008) *Supersizing the mind: reflections on embodiment, action, and cognitive extension*. Oxford: Oxford University Press.
Clark, A. (2016) *Surfing uncertainty: prediction, action, and the embodied mind*. Oxford: Oxford University Press.
Crippen, M. and Schulkin, J. (2020) *Mind ecologies: body, brain, and world*. New York: Columbia University Press.
De Laguna, G. (1927) *Speech: its function and development*. New Haven: Yale University Press.
De Laguna, T. (1930) 'The way of opinion', in Adams, G.P. and Montague, W.P. (eds.) *Contemporary American philosophy: personal statements*. London: George Allen and Unwin Ltd. and the MacMillan Company, pp. 401–422, https://archive.org/details/contemporaryamer01unse/page/402/mode/2up.

De Laguna, T. and de Laguna, G.A. (1910) *Dogmatism and evolution: studies in modern philosophy*. London: MacMillan Company.

Dennett, D.C. (1982) 'Comments on Rorty', *Synthese*, 53(2), pp. 349–356.

Dennett, D.C. (1991) *Consciousness explained*. Boston: Little, Brown, and Co.

Dewey, J. (1916) *Essays in experimental logic*. Chicago: University of Chicago Press.

Dewey, J. (1925) 'Experience and nature', in Boydston, J. (ed.) *J. Dewey, The later works, 1925–1953*, vol. 1. Carbondale: Southern Illinois University Press, 1981.

Dewey, J. (1938) *Logic: the theory of inquiry*. New York: Holt, Rinehart and Winston.

Di Paolo, E., Buhrmann, T. and Barandiaran, X. (2017) *Sensorimotor life: an enactive proposal*. Oxford: Oxford University Press.

Engel, A.K., Friston, K.J. and Kragic, D. eds. (2015) *The pragmatic turn: toward action-oriented views in cognitive science*. Cambridge, MA: MIT Press.

Engel, A.K., Maye, A., Kurthen, M. and König, P. (2013) 'Where's the action? the pragmatic turn in cognitive science', *Trends in Cognitive Sciences*, 17(5), pp. 202–209.

Fodor, J.A. and Pylyshyn, Z.W. (1981) 'How direct is visual perception? some reflections on Gibson's "ecological approach"', *Cognition*, 9(2), pp. 139–196.

Friston, K. (2015) 'The mindful filter: free energy and action', in Engel, A.K., Friston, K.J. and Kragic, D. (eds.) *The pragmatic turn: toward action-oriented views in cognitive science*. Cambridge, MA: MIT Press, pp. 97–108.

Friston, K.J. (2010) 'The free-energy principle: a unified brain theory?' *Nature Reviews Neuroscience*, 11(2), pp. 127–138. https://doi.org/10.1038/nrn2787.

Friston, K.J. and Stephan, K.E. (2007) 'Free energy and the brain', *Synthese*, 159, pp. 417–458.

Gallagher, S. (2005) *How the body shapes the mind*. Oxford: Oxford University Press.

Gallagher, S. (2008) 'Are minimal representations still representations?' *International Journal of Philosophical Studies*, 16(3), pp. 351–369.

Gallagher, S. (2014) 'Pragmatic interventions into enactive and extended conceptions of cognition', *Nous – Philosophical Issues*, 24(1), pp. 110–126.

Gallagher, S. (2020) 'The brains behind radical ecological and enactive approaches to cognition', in Malaforis, L. (ed.) *Beyond Biology and Culture*. Balzan Papers, vol 3. Florence: Olschki Publications, pp. 355–367.

Gallagher, S. and Allen, M. (2018) 'Active inference, enactivism and the hermeneutics of social cognition', *Synthese*, 195(6), pp. 2627–2648. doi:10.1007/s11229-016-1269-8.

Gallagher, S., Hutto, D., Slaby, J. and Cole, J. (2013) 'The brain as part of an enactive system', *Behavioral and Brain Sciences*, 36(4), pp. 421–422.

Gallese, V. and Lakoff, G. (2005) 'The brain's concepts: the role of the sensory-motor system in conceptual knowledge', *Cognitive Neuropsychology*, 22, pp. 455–479.

Gibson, J.J. (1977) 'The theory of affordances', in Shaw, R. and Bransford, J.(eds.) *Perceiving, acting, and knowing*. Hillsdale: Lawrence Erlbaum, pp. 67–82.

Glenberg, A.M. (2010) 'Embodiment as a unifying perspective for psychology', *Wiley Interdisciplinary Reviews: Cognitive Science*, 1(4), pp. 586–596.

Goldman, A.I. (2012) 'A moderate approach to embodied cognitive science', *Review of Philosophy and Psychology*, 3(1), pp. 71–88.

Goldman, A.I. (2014) 'The bodily formats approach to embodied cognition', in Kriegel, U. (ed.) *Current controversies in philosophy of mind*. New York and London: Routledge, pp. 91–108.

Goldman, A.I. and de Vignemont, F. (2009) 'Is social cognition embodied?' *Trends in Cognitive Sciences*, 13(4), pp. 154–159.

Gopnik, A. and Meltzoff, A.N. (1997) *Words, thoughts, and theories*. Cambridge, MA: MIT Press.

Haack, S. (2004) 'Pragmatism, old and new', *Contemporary Pragmatism*, 1(1), pp. 3–41.

Heft, H. (2001) *Ecological psychology in context: James Gibson, Roger Barker, and the legacy of William James's radical empiricism*. Mahwah: L. Erlbaum.

Hohwy, J. (2013) *The predictive mind*. Oxford: Oxford University Press.

Hutto, D. and Myin, E. (2013) *Radicalizing enactivism: basic minds without content*. Cambridge, MA: MIT Press.

James, W. (1890) *Principles of Psychology*. New York: Dover.

James, W. (1904) 'Does consciousness exist?' *The Journal of Philosophy, Psychology and Scientific Methods*, 1(18), pp. 477–491.

Johnson, M. (1987) *The body in the mind: the bodily basis of meaning, imagination, and reason*. Chicago: University of Chicago Press.

Johnson, M. (2008) *The meaning of the body: aesthetics of human understanding*. Chicago: University of Chicago Press.

Johnson, M. (2016) 'Pragmatism, cognitive science, and embodied mind', in Madzia, R. and Jung, M. (eds.) *Pragmatism and embodied cognitive science: From bodily intersubjectivity to symbolic articulation.* Berlin: Walter de Gruyter, pp. 101–125.

Johnson, M. and Rohrer, T. (2007) 'We are live creatures: embodiment, American pragmatism, and the cognitive organism', in Zlatev, J., Ziemke, T., Frank, R. and Dirven, R. (eds.) *Body, language, and mind*, vol. 1. Berlin: Mouton de Gruyter, pp. 17–54.

Katzav, J. (2018) 'Grace and Theodore de Laguna, and the making of Willard V. O. Quine', *Digressions and Impressions*, 05/04/2018. https://digressionsnimpressions.typepad.com/digressionsimpressions/2018/05/grace-and-theodore-de-laguna-and-the-making-of-willard-v-o-quine-guest-post-by-joel-katzav.html.

Koskinen, H.J. and Pihlström, S. (2006) 'Quine and pragmatism', *Transactions of the Charles S. Peirce Society*, 42(3), pp. 309–346.

Lakoff, G. (1987) *Women, fire, and dangerous things. what categories reveal about the mind.* Chicago: University of Chicago Press.

Lakoff, G. and Johnson, M. (1999) *Philosophy in the flesh: the embodied mind and its challenge to western thought.* New York: Basic Books.

Lovejoy, A.O. (1908) 'Pragmatism and theology', *The American Journal of Theology*, 12(1), pp. 116–143.

Lowe, E.J. (1993) 'The causal autonomy of the mental', *Mind*, 102(408), 629–644.

Madzia, R. and Jung, M. eds. (2016) *Pragmatism and embodied cognitive science: From bodily intersubjectivity to symbolic articulation.* Berlin: Walter de Gruyter.

Maturana, H.R. and Varela, F.J. (1980) *Autopoiesis and cognition: the realization of living.* Dordrecht: Reidel.

McNeill, D. (1975) 'Semiotic extension', in Solso, L.E. (ed.) *Information processing and cognition.* Hillsdale: Lawrence Erlbaum. np.

Menary, R. (2007) *Cognitive integration: mind and cognition unbounded.* London: Palgrave-Macmillan.

Menary, R. (2011) 'Our glassy essence: the fallible self in pragmatist thought', in Gallagher, S. (ed.) *The Oxford handbook of the self.* Oxford: Oxford University Press, pp. 609–632.

Menary, R. (2015). 'Pragmatism and the pragmatic turn in cognitive science', in Engel, A.K., Friston, K.J. and Kragic, D. (eds.) *The pragmatic turn: toward action-oriented views in cognitive science.* Cambridge, MA: MIT Press, pp. 219–236.

Noë, A. (2004) *Action in perception.* Cambridge, MA: MIT Press.

O'Regan, K. and Noë, A. (2001) 'A sensorimotor account of vision and visual consciousness', *Behavioral and Brain Sciences*, 23, pp. 939–973.

Peirce, C.S. (1931–1935, 1958) *Collected papers of C.S. Peirce.* Hartshorne, C., Weiss, P. and Burks, A. (eds.). Cambridge: Harvard University Press (abbreviated: CP followed by the conventional '[volume].[page]'-notation).

Pezzulo, G., Barsalou, L.W., Cangelosi, A., et al. (2011) 'The mechanics of embodiment: a dialogue on embodiment and computational modeling', *Frontiers in Psychology*, 2, pp. 1–21.

Pulvermüller, F. (2005) 'Brain mechanisms linking language and action', *Nature Reviews Neuroscience*, 6, 576–582.

Pulvermüller, F. (2015) 'Language, action, interaction: neuropragmatic perspectives on symbols, meaning, and context-dependent function', in Engel, A.K., Friston, K.J. and Kragic, D. (eds.) *The pragmatic turn: toward action-oriented views in cognitive science.* Cambridge, MA: MIT Press, pp. 139–158.

Putnam, H. (1975) *Mind, language and reality.* Cambridge: Cambridge University Press.

Putnam, H. (1992a) 'Why functionalism didn't work', in Earman, J. (ed.) *Inference, explanation, and other frustrations: essays in the philosophy of science* (Vol. 14). Berkeley: University of California Press, pp. 255–270.

Putnam, H. (1992b) *Realism with a human face.* Cambridge, MA: Harvard University Press.

Putnam, H. (1994) *Words and life*, ed. J. Conant, Cambridge, MA: Harvard University Press.

Putnam, H. (1995) 'Pragmatism', *Proceedings of the Aristotelian Society*, 95, pp. 291–306.

Putnam, H. and Putnam, R.A. (2017) *Pragmatism as a way of life.* Cambridge, MA: Harvard University Press.

Quine, W.V.O. (1951) 'Two dogmas of empiricism', *The Philosophical Review*, 60(1), pp. 20–43.

Quine, W.V.O. (1953) *From a logical point of view.* Cambridge, MA: Harvard University Press.

Quine, W.V.O. (1969) *Ontological relativity and other essays.* New York: Columbia University Press.

Ramberg, B. (2004) 'Naturalizing idealizations: pragmatism and the interpretivist strategy', *Contemporary Pragmatism*, 1(2), pp. 1–66.

Rockwell, W.T. (2005). *Neither brain nor ghost: a nondualist alternative to the mind-brain identity theory.* Cambridge, MA: MIT Press.

Rorty, R. (1972) 'Functionalism, machines, and incorrigibility', *The Journal of Philosophy*, 69(8), pp. 203–220.

Rorty, R. (1982a) *Consequences of pragmatism: essays, 1972–1980.* Minneapolis: University of Minnesota Press.

Rorty, R. (1982b) 'Contemporary philosophy of mind', *Synthese*, 53(2), pp. 323–348.
Steiner, P. (2008) 'Sciences cognitives, tournant pragmatique et horizons pragmatistes', *Revue de sciences humaines*, 15, pp. 85–105.
Steiner, P. (2010) 'Philosophie, technologie et cognition: état des lieux et perspectives', *Intellectica*, 53(54), pp. 7–40.
Thompson, E. (2007) *Mind in life: biology, phenomenology and the sciences of mind.* Cambridge, MA: Harvard University Press.
Varela, F.J. (1999) 'The specious present: a neurophenomenology of time consciousness', in Petitot, J., Varela, F.J., Pachoud, B. and Roy, J.-M. (ed.) *Naturalizing phenomenology: issues in contemporary phenomenology and cognitive science.* Stanford: Stanford University Press, pp. 266–314.
Varela, F.J., Thompson, E. and Rosch, E. (1991) *The embodied mind: cognitive science and human experience.* Cambridge: MIT Press.
Williams, D. (2018) 'Pragmatism and the predictive mind', *Phenomenology and the Cognitive Sciences*, 17, pp. 835–859.

30
KNOWLEDGE-PRACTICALISM

Stephen Hetherington

Being a Pragmatist about Knowing: Some Motivation

Shh! Please don't tell the authorities: I like to translate "epistemology" as *knowledgeology* rather than the official *theory of knowledge* (Hetherington 2019: 13). The ancient Greek word "episteme" is usually translated as "knowledge". But ever since that word appeared in the main texts – from Plato – that gave rise to Western philosophy, we have met *many* theories of knowledge – its nature, what distinguishes it from not knowing. So, "theory of knowledge" is misleading: epistemology is many theories. It is also true that epistemology is not always discussing *knowledge*, let alone the question "What is knowledge?" Still, no matter how far its scope has extended beyond knowing's nature, epistemology began with that focus (in the *Meno* and *Republic*, most famously), and we should be aware of epistemology's possibly not having really (rather than apparently) progressed as far *beyond* that initial moment as is generally assumed. A "Back to (Epistemological) Basics" move remains apt.

That is a provocative thought. Here is a pertinent datum.

Epistemologists in general are confident that, thanks to Edmund Gettier (1963), they know a previously overlooked aspect of *what it is* to know. Can we accurately (even if generically) delineate knowing from not knowing, courtesy of the following definition? "S knows that p =df S has a belief that p, it is true that p, and S has good (rationally supportive) justification for the truth of a belief that p". Is that conceptual equation true? Not according to Gettier, or to the legion of post-1963 epistemologists who agree that he told two wholly convincing tales of someone's having a justified true belief that fails to be knowledge. What, then, *is* it to have knowledge? Answering this fully is a challenge. But its epistemological motivation is secure. In asking it, we are building upon *firm* epistemological knowledge, thanks to Gettier, of an aspect of what knowledge is not.

Or so we are standardly told.

But might that view of Gettier's supposed insight, and correlatively of knowledge's nature, be more sensibly *questionable* than epistemologists believe it is? Indeed so: *yes*.

Or so I have long argued (most recently: 2016).

And if I am correct, maybe epistemologists have been *quite* mistaken in their correlative conceptions of knowing. (Given how confident most have been about what Gettier had established about knowing's nature, they have presumed that his supposed insight must somehow be accommodated within any subsequent account of what knowledge is.) In which event, we may explore new epistemological vistas: the door should be opened upon some *alternative* conceptions of knowing.

And pragmatist conceptions satisfy that description, not being *at all* the norm in contemporary epistemology. This chapter points toward what I see as a tempting pragmatist knowledgeology.

Being a Pragmatist about Knowing: Some Possible Forms

Let us begin with a few features that might be apt for a pragmatist conception of knowing.

Spirit. Ideally, we will develop our pragmatist conception in a pragmatist *spirit*. We will regard it as a tool, an implement, perhaps a story or picture to be *used* – and to be used with humility, aware that it might be mistaken or misleading, aware that it might need to be replaced by a better tool or implement. We may deem it a theory that *is* true – yet (like any epistemological theory) an interpretive tool or implement that *might* not be true. "Let us apply it, always testing it, staying open to improving it, even to discarding it". Can you hear the *fallibilist* spirit in that stance? Hopefully so (and, for more on adopting a fallibilist stance toward oneself, see Hetherington 2021b).

Content. Appropriately, we will blend some epistemology with some metaphysics. A pragmatism about knowledge's nature is ultimately a metaphysical thesis, in the literal sense of aiming to reveal something of *what it is to be* an instance of knowing.

Of course, any metaphysical theory is more, or less, "concrete" and substantial – less, or more, deflationary and minimalist – in what it highlights within reality. How is that manifested within a pragmatist knowledgeology? On any such venture, *practicality* is the key aspect of reality that is accorded a presence – a role – within any instance of knowing.

Recent epistemology has emphasized two ways of instantiating that generic idea; this chapter develops a further way. The result will be the following hypothesis:

> Knowledge *is* power – literally, within its nature; for this is *all* that knowledge is.

Before engaging with that idea, though, we should survey those two other pragmatist perspectives on knowing's nature. We can approach them gently, after reminding ourselves of an "everyday" claim that we might have thought would exhaust the philosophical potential of a pragmatist stance on knowing.

Knowledge as power-giving. Everyone has heard the aphorism "Knowledge is power", usually said to be Francis Bacon's. Was it pointing toward knowing's inner nature? No, as Stephen Gaukroger (2001: 17, 18) explains.

> Bacon's claim ... is widely treated as a provocative claim about knowledge, as if it were on a par with claims that knowledge is a grasp of Forms or universals. But it should in fact be read as a claim about power, about something practical and useful, telling us that knowledge plays a hitherto unrecognised role in power. The model is not Plato but Machiavelli.
>
> His point is not to redefine epistemology but to underpin the responsible use of power.

Can we enlarge the scope of Bacon's claim, beyond the political? Yes, but we will still have a claim falling short of describing knowing's inner nature. Enlarge the Baconian claim, and what are we being told? Simply this: have knowledge, will travel; or at least be able to travel, being well-informed on where to go and how to get there. But, even when granted this larger scope, the aphorism is saying just that to have knowledge is to have a power, an ability to do something – *as a causal consequence* of having the knowledge. Metaphysically speaking, that power is *extrinsic* to the knowing: to know is to be in a position to use that knowledge – which has already and independently been constituted by whatever is intrinsic to its being knowledge. The power is not literally *part* of the knowing; it is *merely* a result of the knowing. Various worldly benefits might accrue by having the knowledge. Yet this power *in having* the knowledge is not literally a power *in the knowledge itself.*

We may regard that line of thought as an "everyday" – a not-clearly-philosophical – pragmatism about knowing. It observes that knowledge can be useful – which is hardly a surprising thesis. Metaphysically, this is a *shallowly* pragmatist conception of knowledge's nature. Our epistemological aims should reach further.

Knowledge as interests-sensitive. So, let us meet a more philosophically challenging idea – the thesis of *pragmatic encroachment* (as epistemologists call it). Typically, this is motivated by comparing a "high stakes (in knowing)" version of a possible situation with a "low stakes (in knowing)" version, as follows.

Do I know that my aged mother is alive? I think so. Suppose (counterfactually) that I suddenly feel a powerful desire for her to know, before she dies, that I like and love her. At this moment, experiencing that desire, do I *still* know that my mother is alive? Suddenly I am not sure. If I *am* now to have that knowledge, must I call her, hearing her voice as extra evidence of her being alive? It can now *matter* to me, while previously it did not, that she is alive. And suddenly I am not confident of knowing – really knowing, we might say – that she is alive, simply *because* it now matters to me whether she is, hence whether I know that she is. The practical *stakes* of knowing are now heightened for me. So, I need much better evidence of my mother's being alive if I am to see myself as knowing that she is. "Low" stakes for me in knowing this have been supplanted by "high" ones. What *had* sufficed for my knowing of my mother's being alive no longer does so. With this raising of the stakes, even *whether* I knew has been affected, retrospectively. At this "high stakes" moment, I might see myself as not having *really* known, when the stakes were "lower".

Along such lines, it has seemed to some epistemologists (e.g. Cohen 1988; DeRose 1992; Fantl and McGrath 2009, 2014; Hawthorne 2004; Stanley 2005) that practical factors – interests, stakes, personal significance – can *encroach into* the knowing: practical significance becomes *part* of the knowing, intertwined with the need – which that practical significance has created – for improved evidence, say, to be part of the knowing. Motivations or personal ends can thus be *embedded within* knowing – not merely associated with, or depending causally upon, the knowing, but a constitutive part of it.

This is a substantive metaphysical move, allowing for a pragmatically constituted range of "standards" across different instances of knowing. Everyday knowing includes everyday stakes, thereby an everyday quality of evidence. More "extreme" cases of knowing include "higher" stakes, thereby a stronger quality of evidence.

Knowledge as norm. Spurred on by Timothy Williamson's (2000: ch. 11) view of knowing as a norm of assertion, some philosophers (e.g. Fantl and McGrath 2014; Hawthorne and Stanley 2008) advocate treating knowledge as a norm of action more generally. Just as asserting something should reflect one's knowing, the proposal is that actions should occur only upon a basis of congruent knowledge – motivating or corroborating knowledge. Such actions can be bodily and externally observable, or they might be mental and only "privately" observable. In either event, the idea is that – grounded in knowing – they express or manifest agency, one's *being* an agent. Knowing functions as a normative constraint upon, or regulative standard for, agency.

This is not a consequentialist tale of whatever worldly power one gains by knowing: one might express agency properly without gaining worldly benefit. This story is more substantively metaphysical. It shares with the interests-sensitive form of pragmatism the idea that part of the *point* of, or *value* in, knowing is a *link* with something practical – be this interests or stakes, or properly conducted agency.

As I said earlier, though, those forms of knowledge-pragmatism are not this chapter's focus. You are about to meet what I call knowledge-*practicalism*, perhaps an even more substantively metaphysical knowledge-pragmatism. Knowledge is indeed power (it says) but not because knowing *gives* one power. Rather, knowing *is* power – within itself, part of its nature. To know is to have various abilities or skills. This is not because knowing *gives* one these abilities or skills, nor is it because knowing is a normative *ground* of actions arising through exercising such abilities or skills. Again, no, my account will be even more inherently and fully pragmatist: knowing *is* – nothing but – one's having abilities or skills.

Why so? How so? This is, in my view, an elegantly inviting picture. But it is also heterodox. Added care will be required in presenting it.

Knowledge-That and Knowledge-How

I begin by distilling the picture in more overtly epistemic terms, parsing its talk of abilities and skills with the idea of *knowing-how*:

> All knowing is knowing-how – that is, knowing how to do X (for some specific or generic action "X").

This is an initial formulation of knowledge-practicalism; the next section will add details. Before then, preparatory points are needed.

First, I will not argue that knowing-how to do X equals, fully and precisely, having an ability to do X or a skill in doing X. My quest is a practical equivalence – the thesis that, at least typically, abilities and skills constitute knowing-how.

Second, in appreciating practicalism's heterodoxy, we should notice its main conceptual competitors.

- *Knowledge-Dualism*. All knowing is either knowing-that or knowing-how – two fundamentally different forms that knowing can take. Neither's being knowledge is due to a role played by the other.
- *Knowledge-Intellectualism*. All knowing, including all knowing-how, is fundamentally some or other form of knowing-that.

Knowledge-Dualism is Gilbert Ryle's (1946, 1971 [1949]). Epistemologists have traditionally sought to understand several supposedly distinct kinds of knowledge – perceptual, memory, testimony, *a priori*, etc. – all of these being knowledge-*that*. However, Ryle argued for knowledge-*how*'s being a fundamentally distinct kind of knowledge: its being knowledge is not due at all to the presence of knowledge-that. His argument centered upon "intelligent actions": these are not mere movements; they express or manifest knowledge-how. How so? When you open a door deliberately, you manifest knowledge-how. Is this your having in mind a description of how to perform the action – a description that you know to be true, and one that, in moving your hand, you apply knowingly? Not according to Ryle: you might have such knowledge-that; it is not *needed*, though, as part of opening the door; hence, this knowledge-that ("D is a way to open a door") is a *further* epistemic aspect of you at that moment. You know-how, *and* you know-that. The former need not be a way merely of implementing the latter in a practical way, with this being *why* the former is epistemic – a form of knowledge – at all. (The implementing would itself need to be an "intelligent" action. Would *more* knowledge-that therefore be required to implement that implementing? Yes; and so on, endlessly.) We should thus allow for there being these conceptually separable forms of knowledge – knowledge-that *and* knowledge-how. (In a slightly ironic spirit, then, we may speak here of Rylean dualism – "ironic", given Ryle's fame as a debunker of any Cartesian dualism positing a non-physical mind as accompanying one's physicality, in explaining one's being a person.)

What of practicalism's other main competitor – *intellectualism* – about knowledge-how's nature? This is prominent in recent epistemology, thanks mainly to Stanley and Williamson (2000) and Stanley (2011). It seeks to counteract Ryle's denial that knowing-how's being knowledge is partly due to some knowledge-that's presence. On intellectualism, when you deliberately open the door in a way that manifests knowledge-how, you *are* applying knowledge-that possessed by you (such as "D is a way to open a door"). This need not be your consciously thinking "D is a way to open a door". But knowing how to open the door has to include one's somehow having some such knowledge-that – with the latter being knowledge in a way that is conceptually prior to its helping to constitute knowledge-how (such as of how to open a door).

Knowledge-dualism and knowledge-intellectualism have been discussed in detail (e.g. Bengson and Moffett 2011; Carter and Poston 2018; Fridland 2012, 2013; Pavese 2015a, 2015b, 2017). I will therefore devote the rest of this chapter to knowledge-practicalism (see also Hetherington 2011a: ch. 2, 2011b, 2013, 2015a, 2015b, 2017, 2018a, 2020b, 2021a). Like intellectualism, practicalism rejects knowledge-dualism. But it reverses intellectualism's conceptual ordering of knowing-how and knowing-that: we aim to understand knowing-that's being knowledge at all in terms of knowing-how, rather than the more widely accepted converse ordering of ideas.

We might also parse that explanatory aim in manifestly metaphysical terms: is one of knowledge-that and knowledge-how a *species* of the other? Practicalism replies with this metaphysical ordering: "Yes, even knowledge-that's apparently *purely intellectual* relation to a fact or truth is better understood as a species of knowing *how to do* something".

How can that be so?

Introducing Knowledge-Practicalism

You were born in … let us call it country C. You know this about yourself. Which means … what, exactly? On practicalism, what might it involve?

It involves, above all else, the presence of knowing-how. In knowing that you were born in C, you have some knowledge-how. How much? And what form does it take? We need not delineate it narrowly; flexibility abounds here. You have some complex knowledge-how encompassing *further* instances of knowledge-how, each of which can, at least in theory, be manifested or exemplified by associated actions. Those actions *also* thereby (by transitivity) manifest or exemplify your knowing that you were born in C.

What might be some of those subordinate instances of knowing-how ("subordinate", in that each functions here as a species within the genus that is your knowing that you were born in C)? Here are some possible "contained" pieces of knowing-how.

> You *know how* to say and to write an accurate response to "Where were you born?", such as in conversation or on a government form. You *know how* to "form" and "see" the answer "in your mind". You *know how* to find the answer from some official source(s). (In such ways, you know how to find, and convey, the answer.) You also *know how* to infer both from, and to, the thought that you were born in C: you can embed this thought within an inferential web (since the concepts involved are not abstruse).

I call that collection (just one shape that might be taken by) an *epistemic diaspora* for your knowing that you were born in C. On a given occasion, the knowledge's epistemic diaspora is *whatever* such grouping is present for you on that occasion: today, your knowledge is constituted by one such grouping; next month, the grouping might have changed, as you know in a *different* way how to manifest or express your knowledge of where you were born. So, the knowledge can take different shapes while remaining your having various skills or abilities – various pieces of knowledge-how bearing, more or less directly, upon your knowing where you were born. The knowledge is a shape-shifting bundle, a complex of pieces of knowledge-how, a complex that is thereby knowledge-how itself. To know even a simple truth *is* to have some more, or less, complex knowledge-how.

That is the basic idea of my preferred form of knowledge-practicalism.

Refining Knowledge-Practicalism

Like any fundamental philosophical proposal, though, the basic practicalist idea needs to be refined – extended, sharpened, tested – before it will feel true. Here are a few such moves.

Generality. The concept of an epistemic diaspora for any knowing that p (for a specific "p") is welcomely flexible, with explanatory power within the epistemic world.

For example, as the shape-shifting potential of an epistemic diaspora implies, a person can know the same fact in different ways at different times (even while having the same evidence, perhaps). Equally, different people can know that same fact in different ways even at the same time (and, again, even while having the same evidence for that fact's obtaining).

Accordingly, in representing the epistemic diaspora for knowing at time t that p, this is the schema that I favor:

> At time t, the person has one *or* more of knowledge-how$_1$, knowledge-how$_2$, ..., knowledge-how$_n$ – these being (I will suppose, for simplicity) all of the possible instances of knowledge-how that are ways of knowing that p. At different times, the person might have different such instances of knowledge-how.

Also allowed by this schema is the possibility of describing knowledge as *improved* – or, for that matter, *weakened*. You could know *better* today than yesterday that you were born in C, if now – *unlike* yesterday – you know how to find this information online. This *knowledge-gradualist* interpretive possibility is ignored by most epistemologists, but I have argued elsewhere (2001, 2011a: secs 2.6–2.8, 5.9–5.13) that it should not be. Practicalism accommodates it naturally. There is also the supportive point that abilities and skills – instances of knowledge-how – within a given epistemic diaspora can themselves be more or less developed, more or less complete. And this is *obvious*. So, if knowing is knowing-how, *clearly* it can be better or worse, stronger or weaker. (I am not saying that this is the only way in which a knowledge-gradualism can be derived; it is *a* way, though.)

Epistemic justification. Traditionally, epistemologists have required knowing to include good epistemic justification, such as good evidence. Routinely, they allow that this justification can become weaker or stronger – even once it is better than was *needed* (with all else being equal) for knowing. Do they allow also that the *knowing* – which remains in place throughout that justificatory weakening or strengthening – is thereby weakened or strengthened as knowledge of whatever fact is being known? Generally, no (as I mentioned a moment ago). In this way, epistemology has traditionally been *absolutist* about knowledge – seeing it as only ever *present or not*, with no associated capacity to be present *to some or another degree* or *with some or another variable epistemic quality*.

I cannot prove that such a knowledge-absolutism must be relinquished. But I doubt that it must be adopted. Again a gradualist alternative arises, bespeaking knowledge-practicalism's interpretive generality. Traditional epistemology accepts that one's evidence, say, for p can be strengthened or weakened, even as one continues knowing in an *un*weakenable and *un*improvable way that p.

In contrast, a gradualist can interpret this as the knowledge *also* becoming stronger or weaker. And practicalism has a ready explanation of this, focusing on the *point* of having evidence, say, as part of knowing: the point is the knowing's being able to include pertinent *justificatory skills and abilities*. Note that I write "being able" rather than "having". As part of some knowing, one might know *how to use* related justification, such as evidence; one would not *simply have* the justification. If justification-within-knowledge is to be *worth* having, it should be present – whenever it is – as an expression or manifestation of knowing *how to use* it. For instance, you might know how, if asked, to direct a questioner to an authoritative database that houses the information that you were born in C. Or you know how to extract such information from your parents. In such ways, the knowing can have justificatory *power* – literally, as part of itself. Otherwise, the knowing includes justification – yet not justificatory *power*. To require justification within knowing, therefore, without treating this as the presence of a justificatory *power*, is literally to settle for potentially power*less* justification. I see little point – little value – in that.

I diverge from traditional knowledgeology, too, in not regarding even justificatory skills as *required* within knowing. Instead, they are among the many optional kinds of know-how that can, but need not, be part of the epistemic diaspora for a given instance of knowing. Does this mean accepting the possibility of some knowing *lacking* justification within itself? It does, as I explain elsewhere, both in practicalist terms (2011a: ch. 4) and non-practicalist ones (2001: ch. 4, 2018b, 2020a). I have no view on how likely this disconnection is in practice. But I allow its conceptual possibility, in a continued spirit of potential explanatory reach.

Belief. Something similar can be said about another element of traditional knowledgeology – the usual claim that, necessarily, knowledge is a kind of *belief*. Here are the two main interlinked motivations for that traditional view.

- Any instance of knowledge has to be some *thing*. (Any case of know*ing* must be the possessing *of* that thing.)
- What *kind* of thing might knowledge be? Is it a species of a revealing genus? We often talk of knowing as not *merely* believing. A natural reading of such talk is that knowing is epistemically *strengthened* or *augmented* believing.

That picture seems to reflect some simple metaphysics: the belief that is the knowledge amounts to a kind of *substance*, to which adhere those *properties* that constitute its epistemic augmenting or strengthening – those properties thereby "turning the belief into" the knowledge. Knowledge is thus conceived of in *substance-attribute* terms – a metaphysically venerable way to conceive of individuals. ("Again, what else *is* an instance of knowledge, other than an individual thing?")

But metaphysics offers alternatives to that form of picture. This is good news for knowledge-practicalism, which treats any instance of knowing as more akin to a *bundle* of properties. Practicalism regards any instance of knowledge as an *epistemic diaspora*, which (we saw) is a complex of skills or abilities – instances of knowledge-how – coalescing, forming the knowledge that *p*, for a particular *p*; whereupon nothing further is needed, in the metaphysics of the knowledge's *being* knowledge. I am not saying that this practicalist picture is inescapable. It is available, though – and not clearly *less* likely to be true as a view of knowing than is a substance-attribute conception.

We can continue asking whether the latter conception is mandatory, by reflecting on what is needed to explain the following two cases. (I also welcome, without discussing, Kenneth Sayre's [1997] thesis that knowledge and belief are metaphysically distinct, having quite different objects – respectively, facts and propositions.)

- Some instances of knowing are beliefs that do not "feel" like beliefs: no "active and felt believing" is occurring. This might be so, for example, when the belief is present as a disposition.

- Sometimes, that "belief feeling" *is* present, such as when you say that a propositional content is "before your mind, feeling true": you have it in mind, actively.

The first case fits easily into a practicalist account. Insofar as a belief is a disposition, it is a capacity to respond in a range of ways to a pertinent range of circumstances (e.g. Cohen 1992). But this capacity is itself knowledge-how, since the responses are to be systematic, not random. To have a belief that p – which is not a belief that q, or that r, etc. – is to have *this* range of responses – not *that* one, or that *other* one, etc. So, when knowing includes this form of belief, it includes knowing-how – as practicalism expects.

As to the second of those cases, a practicalist need not dismiss your *having in mind* – consciously, actively – a representation with the content "I was born in C", for example. But this is not your *needing* the representation as part of knowing where you were born. Again, practicalism can explain the situation: the representation's arising for you would be just one possible kind of *manifestation* or *expression* of some of those instances of knowledge-how nestled within the associated epistemic diaspora that amounts to the "larger" knowledge-how that is your knowing that you were born in C. But the representation's arising is not a *required* manifestation or expression.

Still, it might be a useful epistemic *tool* or *implement*. On practicalism, to know is *to know how to act* (in ways commensurate with the specific case). Such actions might be bodily, readily observable in the public world. Or they might be mental moves occurring only within the knower's "inner" world. In either event, the practicalist moral persists: the belief's *presence* is not what matters deeply for the knowing; what most matters for the knowing is what *actions* are associated with the belief, either actually or potentially. On practicalism, the epistemic point of your having a belief as part of an instance of knowledge would only be your *knowing how to do relevant things with* the belief. Once again, practicalism makes us mindful of the need for the belief, if present within an instance of knowing, to be a tool, an epistemic implement. Indeed, on practicalism this is *all* that the belief is, whenever it is present within some knowing. The belief *is not* the knowledge. It is just a tool that the knowledge can – if this is apt – *use*.

Truth. We have seen how practicalism permits two of knowledge's traditionally deemed essential elements, belief and epistemic justification (such as evidence), to be *inessential* to knowing. This is not to banish them from ever being part of any actual instances of knowing: your knowing where you were born might happen to include some believing skills and some justificatory skills. But this would not entail that these skills had to be present; the epistemic diaspora for your knowing that p could have comprised alternative relevant skills. *A fortiori*, neither a belief, nor justification, is essential – even when present – to knowing, on a practicalist view. What would be helping to constitute an instance of knowing is only any associated skills – that is, whatever one *knows how to do* with the belief and/or the justification. The *presence* as such of the belief and the justification is explanatorily epiphenomenal; what contributes to the knowledge's being knowledge worthy of the name is what one *knows how to do* with the belief and the justification. Neither belief *per se* nor justification *per se* is *essential to there being knowing-that at all*.

So, some theses central to traditional knowledgeology can quietly be set aside by practicalism. What of the thesis – also long venerated – that knowledge is only ever of a *fact* or, equally, a *truth*? Can a practicalist do justice to this thesis, even if she does not retain it in its usual form? It is apparent how a belief can be true: its content represents accurately. But we have seen that practicalism no longer requires any instance of knowing to include a belief. Can practicalism nonetheless replicate at least the spirit of knowing's truth requirement?

I believe so. Here is my preferred approach.

Truth is *as truth does*. That is what any pragmatism should say. What knowledge-practicalism, in particular, might say is that truth, when within knowledge, is *as knowledge-how does (succeeds in doing)*. This is to say that truth, within knowing, is *as* the knowledge (of that truth) is *manifested or expressed in action*. On practicalism, any instance of knowledge-that is a complex of instances of knowledge-how. At a given moment, any such instance of knowledge-how is either dormant or being manifested in action. It is in such actions that – in either of two ways, depending on the case – we find the (practicalist version of the) knowledge-that's being true.

Here are those two ways.

Representing. Your knowing that you were born in C might include your knowing how to answer the question accurately of where you were born. Any such action embodies (in the answering) a content that *reflects the fact* of where you were born. This reflecting can include, in a traditionally characterizable way, your answer's *representing* that fact.

Making-true. But sometimes your representing the world is not the mark of the truth present within your knowing. Suppose that you know that you intend to buy aspirin. On practicalism, you thereby know *how to bring it about* that you will buy aspirin. You thus know, in this practical way, that you will buy some. Your knowing this includes your knowing *how to make it true* that you will buy some.

G.E.M. Anscombe (1963) distinguished between what she called "practical knowledge" and "contemplative knowledge". Practicalism treats all knowing, including contemplative knowledge, as knowing-how. But we should preserve, from Anscombe's distinction, there being two ways in which knowledge, even while always being knowledge-how, can have a truth-link with the world. Traditional knowledgeology reports only one of those two – the *representing*, the *reflecting-back* (like a mirror). But sometimes knowing can be *bringing about* what is being known: it can be *making* true. Practicalism does justice to this significant feature of knowledge. (And perhaps not only of *our* knowledge. On how practicalism might model this aspect of God's knowledge, see Hetherington [2019b].)

Practicalism thus imbues knowledge's truth requirement with increased generality and increased explanatory power. There is representing, and there is bringing-about. There is reflecting the world, and there is creating it. Each of these can be a way for it to be true that *p*, *as part of* your knowing that *p*. In this way, practicalism – far from failing to maintain truth as an element in knowing – allows truth an *enhanced* presence within knowing.

That remains so, even if we talk of facts, not truth, as being known. Witness this pertinent observation from Kenneth Olson (1987: 10):

> "'Fact'" derives from the Latin *factum*, meaning "thing done", itself the neuter past participle of *facere*, "to do or make". The English word, too, originally had the sense of deed or action, more often than not criminal. (The word "feat" which is also derived from *factum* – by way of the Old French *fait* – retains the original sense of something done.)

Practicalists should likewise be reassured by this sage point from Sayre (1997: 106):

> truth is the upshot of something's being true, which in turn is the upshot of something done truly. Of these three terms ["truth", "true", "truly"], "truly" is the most concrete, referring as it does to a quality of particular acts (the truly aimed shot that goes straight to the target, the truly spoken words that tell things as they are).

A Precedent? A Competitor?

As I mentioned, practicalism is a heterodox conception of knowing. Is it also new? One might think that it had been anticipated by John Hartland-Swann (1958: 10–14). He does say that knowing is a form of skill, which is one way to expand practicalism's core idea of all knowing being knowledge-how. But Hartland-Swann's focus was not ours: he was not talking about the *state of* knowing. His topic was actions of *claiming* knowledge. I agree that these can be performed more, or less, skillfully, and that they should be manifestations or expressions of skill (which we can expect to find within many epistemic diasporas amounting to knowledge). Still, none of this is knowledge itself being a skill.

Perhaps knowledge-practicalism's closest competitor in contemporary epistemology is John Turri's *abilism*. He reaches this view by subjecting *knowledge reliabilism* (2016: 190) to the demands of experimental philosophy, a methodology that has gained momentum in the past two decades (and of which Turri is a notable proponent). Must knowing that *p* include having reliably acquired, and reliably holding, a belief that *p*? Reliabilism says so; abilism denies so. How are we to resolve this clash? Turri conducts nine experiments, en route to his abilist conclusion (ibid.: 225), which is that "knowledge [is] true belief manifesting the agent's cognitive abilities". This is confronting, because it has this surprising consequence (ibid.): "cognitive abilities can be unreliable in the sense that they needn't produce true beliefs more often than not". Turri thus regards his experiments as supporting this hypothesis: "unreliable knowledge is not only possible but actual and widely recognized as perfectly ordinary" (ibid.: 226). The experiments reveal that this combination is "recognized" – even if not by epistemologists writing in academic settings, at least by people consulted in those experiments.

It is difficult to know how confident we should be about Turri's use here of experimental philosophy, hence about his conclusion. *Does* that method have such power to support a thesis strongly, individually, and philosophically? Or is its proper philosophical role more stage-setting and suggestive, especially when directed (as Turri's "experiments" were) at a specific thesis?

When experimental philosophy sprang into existence, thanks initially to Jonathan Weinberg, Shaun Nichols, and Stephen Stich (2001), its aim was to engage critically with a recurring aspect of contemporary philosophy – the ready recourse to "intuitions" and "what is intuitively clear" in deciding whether to accept a philosophical position. In practice, many philosophers claim that a thesis is "intuitively true" – having consulted nothing more, it seems, than their own "intuitions" and perhaps those of some philosophy peers. I do not do this, having long been skeptical about the epistemic power of "intuition" as a way to engage well with a subtle philosophical thesis. I therefore welcomed experimental philosophy, cautiously, when it entered contemporary philosophy.

It needs to be used carefully, though. Experimental philosophy advised philosophers who reach readily for "intuitions" in support of a thesis to consult a *wider* range of people for "intuitive" reactions to the thesis. What should we infer from the results of doing so, when these seem to reveal philosophers as speaking or thinking differently about an issue than how an alternative, maybe a wider, range of people does so? *Qua* philosophers, what should we infer?

Various options are available. But what I do *not* infer is that enough has been done, by using this experimental method, to establish, for philosophical purposes, a complex and substantial thesis such as abilism. To state my concern bluntly, I am not sure that this method is philosophical enough, at least for some issues, to allow us to do so. What more might be needed?

Speaking for myself, I also want *system-building* to be happening, or at least *system-sensitivity* to be on display and doing some work. When I ask someone whether, in a specific situation, knowledge is present, her answer (the sort elicited in Turri's "experiments") *is* a datum that might be of interest to an epistemologist. It is *only* a datum, though. We should then ask, "Has the person answered with a *theory* of knowledge in mind, perhaps applying or testing it?"

- If "yes", then her response is "just further theory". This is not a failing in itself. But in this case the response should not receive distinctive *praise* from philosophers: it should not be accorded added epistemic weight by them simply because it arises, or is reflected, within an "experiment".

- If "no", we should grant to Turri that he has revealed abilism to be a thesis to which some people are inclined. But how philosophically decisive is that? It can be philosophically valuable in a lesser way, since abilism might not even have *occurred* to most knowledgeologists as something to scrutinize seriously, let alone sympathetically. Still, this is not yet a strong argument for abilism's being true, let alone being both true and deserving to be our favored account of knowledge. We are merely being guided in a professionally unusual direction for the next stage of our philosophical testing; now we need to embed the thesis within a wider philosophical theory, refining and modifying and extending and discarding and, well, you get the idea. This is where the *distinctively* philosophical elements of our enterprise enter the story.

I have never, in this chapter or elsewhere, argued for knowledge-practicalism by claiming that it is "intuitive". I offer it here and elsewhere as a philosophical theory, as a hypothesis to be applied to case after case, with its thereby being *tested* time after time. In that same spirit, I welcome Turri's endorsement of the idea that the *presence of an ability* is a key element within knowing. But I happen to take this idea further, in a more theoretical – and more fully and manifestly pragmatist – way.

References

Anscombe, G.E.M. (1963 [1957]). *Intention*, 2nd ed., Oxford: Blackwell.
Bengson, J. and Moffett, M.A. (eds.) (2011). *Knowing How: Essays on Knowledge, Mind, and Action*, Oxford: Oxford University Press.
Carter, J.A. and Poston, T. (2018). *A Critical Introduction to Knowledge-How*, London: Bloomsbury.
Cohen, L.J. (1992). *An Essay on Belief and Acceptance*, Oxford: Clarendon Press.
Cohen, S. (1988). "How to Be a Fallibilist," *Philosophical Perspectives* 2: 91–123.
DeRose, K. (1992). "Contextualism and Knowledge Attributions," *Philosophy and Phenomenological Research* 55: 913–929.
Fantl, J. and McGrath, M. (2009). *Knowledge in an Uncertain World*, Oxford: Oxford University Press.
———. (2014). "Practical Matters Affect Whether You Know," in M. Steup, J. Turri, and E. Sosa (eds.), *Contemporary Debates in Epistemology*, 2nd ed., Malden, MA: Wiley Blackwell, pp. 84–94
Fridland, E. (2012). "Knowing-How: Problems and Considerations," *European Journal of Philosophy* 23: 703–727.
———. (2013). "Problems with Intellectualism," *Philosophical Studies* 165: 879–891.
Gaukroger, S. (2001). *Francis Bacon and the Transformation of Early-Modern Philosophy*, Cambridge: Cambridge University Press.
Gettier, E.L. (1963). "Is Justified True Belief Knowledge?" *Analysis* 23: 121–123.
Hartland-Swann, J. (1958). *An Analysis of Knowing*, London: George Allen & Unwin.
Hawthorne, J. (2004). *Knowledge and Lotteries*, Oxford: Clarendon Press.
Hawthorne, J. and Stanley, J. (2008). "Knowledge and Action," *The Journal of Philosophy* 105: 571–590.
Hetherington, S. (2001). *Good Knowledge, Bad Knowledge: On Two Dogmas of Epistemology*, Oxford: Clarendon Press.
———. (2011a). *How to Know: A Practicalist Conception of Knowledge*, Malden, MA: Wiley Blackwell.
———. (2011b). "Knowledge and Knowing: Ability and Manifestation," in S. Tolksdorf (ed.), *Conceptions of Knowledge*, Berlin: De Gruyter.
———. (2013). "Skeptical Challenges and Knowing Actions," *Philosophical Issues* 23: 18–39.
———. (2015a). "Technological Knowledge-That as Knowledge-How: A Comment," *Philosophy & Technology* 28: 567–572.
———. (2015b). "Self-Knowledge as an Intellectual and Moral Virtue?" in C. Mi, M. Slote, and E. Sosa (eds.), *Moral and Intellectual Virtues in Western and Chinese Philosophy: The Turn Towards Virtue*, New York: Routledge.
———. (2016). *Knowledge and the Gettier Problem*, Cambridge: Cambridge University Press.
———. (2017). "Knowledge as Potential for Action," *European Journal of Pragmatism and American Philosophy* 9, http://journals.openedition.org/ejpap/1070.
———. (2018a). "Knowledge and Knowledge-Claims: Austin and Beyond," in S.L. Tsohatzidis (ed.), *Interpreting Austin: Critical Essays*, Cambridge: Cambridge University Press.

———. (2018b). "The Redundancy Problem: From Knowledge-Infallibilism to Knowledge-Minimalism," *Synthese* 195: 4683–4702.
———. (2019a). *What Is Epistemology?* Cambridge: Polity Press.
———. (2019b). "Creating the World: God's Knowledge as Power," *Suri* 8: 1–18.
———. (2020a). 'Knowledge-Minimalism: Reinterpreting Plato's *Meno* on Knowledge and True Belief," in S. Hetherington and N.D. Smith (eds.), *What the Ancients Offer to Contemporary Epistemology*, New York: Routledge.
———. (2020b). "The Epistemic Basing Relation and Knowledge-That as Knowledge-How," in J.A. Carter and P. Bondy (eds.), *Well-Founded Belief: New Essays on the Epistemic Basing Relation*, New York: Routledge.
———. (2021a). "Knowledge as Skill," in E. Fridland and C. Pavese (eds.), *The Routledge Handbook of Philosophy of Skill and Expertise*, New York: Routledge.
———. (2021b). "Some Fallibilist Knowledge: Questioning Knowledge-Attributions and Open Knowledge," *Synthese* 198: 2083–2099.
Olson, K.R. (1987). *An Essay on Facts*, Stanford, CA: CSLI.
Pavese, C. (2015a). "Knowing a Rule," *Philosophical Issues* 25: 165–188.
———. (2015b). "Practical Senses," *Philosophers' Imprint* 15: 1–25.
———. (2017). "Knowledge and Gradability," *The Philosophical Review* 126: 345–383.
Ryle, G. (1949). *The Concept of Mind*, London: Hutchinson.
———. (1971 [1946]). Knowing How and Knowing That, in *Collected Papers*, Vol. II, London: Hutchinson.
Sayre, K.M. (1997). *Belief and Knowledge: Mapping the Cognitive Landscape*, Lanham, MD: Rowman & Littlefield.
Stanley, J. (2005). *Knowledge and Practical Interests*, Oxford: Oxford University Press.
———. (2011). *Know How*. Oxford: Oxford University Press.
Stanley, J. and Williamson, T. (2001). "Knowing How," *The Journal of Philosophy* 98: 411–444.
Turri, J. (2016). "A New Paradigm for Epistemology: From Reliabilism to Abilism," *Ergo* 3: 189–231.
Weinberg, J.M., Nichols, S. and Stich, S. (2001). "Normativity and Epistemic Intuitions," *Philosophical Topics* 29: 429–460.
Williamson, T. (2000). *Knowledge and Its Limits*, Oxford: Clarendon Press.

31
PRAGMATISM AND RELIGION

Sami Pihlström

Philosophers associated with the pragmatist tradition have made lasting contributions to virtually all areas of philosophical inquiry. Pragmatists have also been active in a wide range of other disciplines and interdisciplinary fields of study, engaging in the methodological debates over the nature of not only the natural sciences but also the social sciences and humanities, including religious studies and theology. This entry will explore pragmatist approaches to the philosophy of religion and, more generally, the study of religion. While the focus will be on philosophical investigations of religion, no dualism between philosophical and empirical perspectives on this topic is presupposed, as pragmatist accounts are typically (broadly) naturalist and thus willing to avoid such essentialist dichotomies. Indeed, *antiessentialism* can be regarded as one of the main characteristics of pragmatist ways of understanding religion: neither religious phenomena themselves nor the philosophical and methodological principles guiding our study of those phenomena can be narrowly reduced to a single overarching essence. Pragmatism, it may be argued, is valuable for both our understanding of religious faith (or life) itself and for our understanding of how religious faith (and life) ought to be studied, explained, and understood – empirically, conceptually, and philosophically.

The place of the philosophy of religion within philosophy generally is sometimes considered relatively marginal. Clearly, the philosophy of religion is institutionally only a small subfield of philosophy today. However, philosophical investigations of religion may in interesting ways integrate metaphysical, epistemological, as well as ethical ideas and arguments; they may also have highly relevant connections to, for instance, the philosophy of science (or general theory of inquiry) and political philosophy. This fact about the very nature of the philosophy of religion highlights the relevance of pragmatist contributions in the field, as it is entirely natural for pragmatists to abandon not merely the dualism between philosophy and the empirical disciplines studying religion but also strict divisions between the areas of inquiry within philosophy.

Therefore, pragmatist philosophy of religion offers a highly illuminating case study of the distinctive character of pragmatist inquiries more generally. This is because religion is a multiply problematic element of human life, especially in modern liberal societies, with contested relations to scientific rationality as well as ethical and political values. Pragmatism emerges as a dynamically developing approach to both philosophy of religion and what we may call the philosophy of religious studies and theology.

Pragmatic fallibilism between science and religion

Pragmatism was initiated in a culture – late 19th-century America – that was undergoing the deep intellectual and spiritual transition due to the Darwinian theory of evolution and other advances in the natural sciences. Scientific materialism and mechanistic determinism were among the "modern" conceptions of the universe and of humanity that many intellectuals of the day were attracted by. Yet, the quest for the meaning and value of human life in a thoroughly natural world, as well as the need to postulate the freedom of the will as fundamental to morality, was in this context experienced as possibly more pressing than ever. Pragmatist philosophers accordingly took up both the intellectual and ethical challenge of investigating the relations between religion (or, more generally, existential value, meaning, freedom, and spirituality) and scientific rationality. In a sense, pragmatism, thus, established a unique synthesis of the Enlightenment celebration of reason and the more "Romantic" focus on experience and emotions. We may, in extremely general terms, view pragmatist philosophy of religion as an outcome of both Enlightenment rationality and a more romantic aspiration to defend what is distinctive in human life even in the context of scientific naturalism. (On romanticism as a background of pragmatism, see Goodman 1990.)

As many other entries in this *Companion* explain, one of the defining characteristics of pragmatism – in relation to both science and religion, and everything else for that matter – is its resolute *fallibilism*. Pragmatists typically accept no ideological dogmas; on the contrary, any view or idea we may hold can and should be subjected to careful critical scrutiny in terms of its potential experiential outcomes and, if necessary, revised or rejected and replaced by a better one. This dynamic fallibilism, focusing not on particular beliefs but holistically on our system of beliefs inviting continuous critical revision (cf. White 2002), is built into the pragmatist conception of inquiry as a process of moving from whatever initial beliefs we hold through the doubt provoked by the unexpected results the habits of action embodying those beliefs result in, which, in turn, launch an inquiry into those original beliefs, leading to their revision or rejection.[1] It may be unclear how exactly (or whether) the fallibilist principles of scientific (or everyday) inquiry can be reconciled with religious thought, but it is clear that insofar as pragmatists committed to thoroughgoing fallibilism can endorse any religious faith at all, that faith must be open to critical assessment – not necessarily in narrowly understood scientific terms, or based on scientific evidence, but in terms of the conceivable practical results that faith, or its objects, may have in our individual and social lives.

Biblically, it is from the "fruits" of our religious views (or irreligious ones, for that matter) that we really know what those views are all about. When William James (1975 [1907]: 51–56) examines how "God" is pragmatically known, he means that it is in terms of the conceivable practical results in our experience that the significance of religious ideas, just like any other ideas, ought to be investigated (cf. James 1985 [1902]: Lecture XX). Accordingly, religious belief is, for pragmatists of various outlooks, constitutively embedded in our habitual practices that we continuously revise through processes of inquiry – but so are rejections of religious belief, for that matter. It is one of the key insights of classical pragmatism, in particular, that beliefs not merely result in or embody but literally *are* habits of action and as such also need to be constantly subjected to the meta-habit of self-critically ameliorating our habits.

Due to its undogmatic and fallibilistic spirit, pragmatism can, furthermore, be characterized as a species of *critical philosophy* – not quite in Immanuel Kant's original sense of this term but with an important Kantian background nonetheless (cf., e.g., Pihlström 2020). While pragmatist philosophy of religion investigates the meaning and value of religious (as well as anti-religious) ways of thinking by taking seriously their embeddedness in human practices of purposively engaging with the world – instead of understanding religious outlooks as static systems of propositional beliefs taken to represent a standing reality "out there" independent of those beliefs – it also analyzes, in a quasi-Kantian manner, the practice-laden conditions for the possibility of critical discussion of religious beliefs. These conditions themselves enable us to continue the self-critical exploration and renewal of our habits of thinking and acting, both religious thinking and academic inquiry seeking a deeper understanding of religion.

The classical pragmatists on religion

While the pragmatist tradition should definitely not be reduced to its three best-known classical figures, in this entry I can only comment, among the classics, on Charles S. Peirce's, William James's, and John Dewey's views on religion (see also, e.g., Pihlström 2013; Slater 2014; Zackariasson 2015). There are numerous other historical pragmatists and thinkers close to pragmatism who developed original ideas relevant to the philosophy of religion: for example, Jane Addams's social thought and practical social work were shaped by her early Christian influences, while Josiah Royce's "absolute pragmatism" incorporates Hegelian themes into the pragmatist tradition.[2]

Among the classical pragmatists, Peirce (whose fallibilism was already briefly introduced above) was – despite being a highly original scientific and mathematical genius – most clearly a conservative Christian theist. He argued for the significance of sentiment and instinct, rather than critical scientific or philosophical inquiry, in "matters of vital importance", that is, ethics and religion, and in a late 1908 essay he defended a "neglected argument for the reality of God" (Peirce 1992–1998: vol. 2, 434–450). The human mind is, according to Peirce, instinctively tuned into the postulation of a divinity. Instead of subjecting our theistic worldview to any pragmatist analysis, then, we should trust our sentimental "musement" revealing a divine presence in the cosmos. Peirce's "agapistic" metaphysics of evolutionary love as a cosmic principle is an important element of this overall conception.[3] Accordingly, while Peirce was, along with James, the founder of pragmatism, he somewhat paradoxically did not maintain that we should use the Pragmatic Maxim, according to which we should determine the meaning of our concepts and conceptions with reference to the conceivable practical effects of their objects (ibid.: vol. 1, 132), for ascertaining how exactly religious concepts ought to be used; instead, we should conservatively rely on our sentiments in this area.

James, often portrayed as the most religiously and/or spiritually inclined among the great old pragmatists, was actually the one who was most painfully and anxiously concerned with "existential" religious issues. In contrast to Peirce, he saw it as a chief difficulty of the scientific age to reconcile religious sentiments with the advancements of rational inquiry. His *"will to believe" argument* (presented in the title essay of James 1979 [1897], presumably the most widely read piece of pragmatist philosophy of religion) has often been criticized for licensing wishful thinking, but for James himself the difficulty was rather the impossibility of ever fully embracing religious belief. According to his famous, or notorious, argument, we have the right to actively exercise our will to believe in a "religious hypothesis" that cannot be settled on evidential grounds, if (and only if) it is, for us, a "genuine hypothesis" in the sense of being "live", "forced", and "momentous" (ibid.: 15–18). In the late 19th century, James directed his words to academically educated people who were tempted to embrace the rapidly developing scientific worldview while still hoping to maintain their religious faith in some form. The will to believe strategy can also be employed to defend other existential or worldview-related ideas, such as the freedom of the will. In developing the will to believe argument, James attacked evidentialists like W.K. Clifford, whose "ethics of belief" urged us to never believe anything upon insufficient evidence. James argued that there are cases in which we have to use our active will to believe "ahead of evidence" in order to avoid losing a truth that we might be able to find. It has been debated whether James's view leads to irrationalism by harboring the epistemic vice of credulity or whether it actually develops a sustainable virtue-epistemological approach to developing our doxastic practices and characters.

In addition to "The Will to Believe", James's main works exploring religion include his psychological study of religious experience, *The Varieties of Religious Experience* (1985 [1902]), aptly subtitled "A Study in Human Nature", and his engagements with the "pragmatic method" and the so-called pragmatist conception of truth and its applications to religion in *Pragmatism* (1975 [1907]) and other late writings. *The Varieties* comprehensively describes individual religious experiences, including mystical experiences, as constitutive of religiosity, emphasizing the experiential rather than institutional or dogmatic basis of religion. Hence, James's empirically informed work on religious experience can in important ways be seen as a precursor of modern interdisciplinary studies of religion within a broadly understood methodological naturalism (cf., e.g., Proudfoot 2004, 2018). At the same time, it is radically *individualist*: personal experience is the core of religion (see especially James 1985 [1902]: Lecture II).

It may be speculated whether James, despite the anti-evidentialist tone of "The Will to Believe", actually came to believe in the possibility of embracing religious faith on the basis of evidence, after all, insofar as mystical or supersensory "evidence" produced by religious experiences can be included (see Putnam and Putnam 2017). In his late reflections on pragmatism, pluralism, and truth, James also examined the senses in which religious beliefs may be pragmatically true by yielding valuable and satisfactory experiential results (cf. James 1975 [1907]: Lecture VIII). However, neither in the context of religion nor elsewhere was his goal to *replace* the correspondence theory of truth by a rival pragmatist theory reducing the truth of a belief to its useful consequences (as his view has often been misleadingly portrayed) but to give a pragmatist elucidation of what the key notion of the correspondence theory, viz., "agreement" between a belief and the reality it is about, actually means in concrete experiential terms.[4]

Dewey, along with Peirce, is usually regarded as one of the more scientifically minded pragmatists, developing his own version of a pragmatist theory of experimental inquiry that had been initiated by Peirce and James. His *pragmatic naturalism* leaves little room for traditional religion as such. However, in *A Common Faith* (1934; see LW9) – a slim volume known as his only work primarily focusing on religion – he defended the value of the "religious qualities" in experience, arguing that such qualities may be manifested in a wide variety of different types of experience, including political, ethical, aesthetic, educational, and even scientific. Like so many other works by Dewey, *A Common Faith* is ultimately a defense of democracy as a way of life. Dewey was deeply suspicious of any supernaturalist metaphysics, but he suggested that we may use the concept of God to refer to the "*active* relation between ideal and actual" (LW9:34, original emphasis) experientially pursued in, e.g., democratic and educational processes of human "growth".

In the history of pragmatism, Dewey stands out more importantly as a naturalist critic of (especially organized and institutionalized) religion than as a philosopher of religion as such. His secularized reconceptualization of the divinity is shared by few, but his account of the religious qualities in experience as something that need no reference to supernaturalism contributed, beyond pragmatism proper, to the development of the tradition known as "religious naturalism".[5] Dewey's democratic focus on religious qualities in experience as a social force also highlights an important tension within pragmatist philosophy of religion: James's individualism celebrating the uniqueness of each and every individual religious experience is replaced by an emphasis on religious faith as something "common" to human beings in their experiential lives.

Neopragmatists on religion

Another interesting tension regarding religion can be perceived between the two most influential neopragmatists who were active in the late 20th and early 21st centuries, viz., Richard Rorty and Hilary Putnam. Rorty (e.g., 1989, 1999, 2007), on the one hand, insisted on drawing the full consequences from thoroughgoing atheism, rejecting the philosophical realists' notions of a mind-independent world and objective, discourse-independent "Truth" as mere remnants of a divinity to which human beings were in the past thought to be "answerable". On the other hand, he defended a "pragmatic polytheism" or pluralism based on a sharp distinction between the public and the private, also elaborating on the fully secular transcendence of "social hope" (see Rorty 1999: Chapter 10). Our liberal hope for a better future should not be thought of as depending in any way whatsoever on metaphysical or transcendent speculations about the ahistorical essence of humanity, correspondence-theoretical truth, or God; these are all versions of the Platonic quest for something superhuman and can be dropped in a humanistic culture finally realizing that human beings are answerable only to each other. In their private lives, people can believe in what they wish – this is a central tenet of liberalism – but in mature liberalism, public use of reason should remain thoroughly secular (or "anti-clerical", as Rorty put it in some of his late writings).

Putnam (e.g., 1992, 1997a, 1997b), in turn, approaching issues in the philosophy of religion from the standpoint of his pragmatist rejection of what he called metaphysical realism and the fact/value dichotomy, arrived at a much more positive assessment of religious faith than the radically secularist Rorty.[6] Putnam was in his religious thinking heavily influenced by his Jewish background (see especially Putnam 2008), in addition to his continuous appropriation of the classical pragmatists, particularly James (cf. also Putnam and Putnam 2017), as well as Ludwig Wittgenstein. For Putnam, religious perspectives on reality may be acceptable among the plurality of conceptual schemes by means of which we categorize the world we live in. Rejecting metaphysical realism postulating an "absolute conception of the world" (to be ideally captured by most advanced science), he affirmed, with James and Wittgenstein, the possibility of employing religious concepts and "pictures" as fundamentally constitutive of one's individual life orientation, despite their remaining beyond scientific evidential assessment.

Rorty and Putnam agreed that we need to reject metaphysical realism in favor of something like *pragmatic pluralism* in order to arrive at any adequate understanding of the ways in which we, through our epistemic and ethical practices, engage with the world we live in. They disagreed, however, about the ways in which religious practices may still be relevant to modern people in modern societies. For Rorty, religion is only legitimate in the private sphere, while Putnam was more sympathetic to the religious perspectives people may find valuable in seeking meaning in their lives. In this sense, Putnam – at least the late Putnam, self-identifying as a "practicing Jew" – became closer to the kinds of positions developed (independently of, yet in dialogue with, pragmatism) by philosophers like Charles Taylor and Jürgen Habermas seeking to understand the complex interplay of secularized and religious outlooks in the modern age.[7]

Thus, we may view Rorty as a humanist and naturalist secularist essentially continuing the Deweyan strand of pragmatism critical of (especially organized) religion. At the same time, we may view Putnam as an equally humanist and non-reductively naturalist pragmatist who nevertheless genuinely sought to accommodate the possibility of religious outlooks within a pluralism of rationally acceptable world descriptions. Rorty's atheist pragmatism is, at any rate, to be acknowledged by pragmatist philosophers of religion also for its explicit emphasis on the significance of the concept of God: it is with reference to religious responsibility to something non-human that we may view what, according to Rorty, goes wrong in the realist theories of truth and objectivity forgetting that human beings are *only* responsible to other human beings.

Other late-20th- and early-21st-century scholars with serious interest in issues in both pragmatism and the philosophy of religion have included, among others, European thinkers like Hans Joas, Eberhard Herrmann, and Dirk-Martin Grube. Joas (2017), for example, draws from Jamesian pragmatism, in particular, in his reassessment of the history of "disenchantment" (*Entzauberung*) going back to Max Weber; his historical sociology of religiously relevant concepts is strongly informed by the pragmatist tradition. Grube (e.g., 2004; see also Grube and Van Herck 2019) can be seen as a Jamesian pluralist in his defense of epistemic religious diversity, while Herrmann's (e.g.,

2003) work has significantly contributed to invigorating a Putnamian pragmatic and pluralist realism in the philosophy of religion and theology in the Nordic countries (cf. also, e.g., Zackariasson 2015). In the younger generation of pragmatists, Ana Honnacker (2018) has interestingly defended a basically Jamesian orientation of pragmatist humanism (see also below).

Pragmatism and contemporary philosophy of religion

These and many other philosophers influenced by and further developing pragmatist themes in the philosophy of religion should not be too simply classified as "pragmatists", because many of them have also been equally at home in other philosophical traditions. This is an important aspect of pragmatist philosophy of religion today: scholars engaging with pragmatist thinkers and themes may also contribute to other philosophical orientations. Therefore, in addition to investigating the pragmatist tradition as such, it is illuminating – for both pragmatism and other schools of thought – to place pragmatist philosophy of religion in a critical comparison with other recent and current approaches. This demonstrates both the distinctive character of pragmatism and its versatility in creating dialogues between rival philosophical outlooks.

First, pragmatism can, to be sure, be clearly distinguished from the heavily evidentialist trends of mainstream *analytic philosophy of religion*, both theist and atheist. Pragmatism can, however, also be taken to renew and rearticulate a liberalized form of evidentialism by taking into account other types of evidence than the standard intellectual evidence one typically exclusively refers to when either defending or attacking the idea that religious beliefs can be "rational" or epistemically justified in the same sense in which scientific beliefs are. Analytic philosophers of religion rarely refer to pragmatism, though. Furthermore, the recent interest among analytic philosophers of religion in what has come to be called "analytic theology" – that is, philosophical theology analyzing theological (e.g., Christian) doctrines and ideas by means of methods drawn from analytic philosophy – seems to be far from pragmatist approaches to religion, because pragmatists rarely focus on specific religious dogmas but have, characteristically, wanted to keep their inquiries open to a wide range of religious (and non-religious) outlooks. In any case, pragmatists certainly need not reject analytic philosophers' general concern with clear conceptualization and argumentation; on the contrary, pragmatist classics like Peirce and Putnam are also major figures of the analytic tradition, and the interpenetration of these approaches should be appreciated in the philosophy of religion, too.

Second, more specifically, the *reformed epistemology* defended by "Christian philosophers" like Alvin Plantinga – a key orientation within analytic philosophy of religion today – superficially resembles pragmatism in its emphasis on the relative autonomy of our "doxastic practices". It might sound like a pragmatist view to maintain that such practices embody "properly basic" beliefs that cannot be justified on a neutral ground. However (*pace* Slater 2014), at a deeper level of investigation, Christian philosophy embodies the kind of ideological attachment to a form of exclusivism that may be regarded as foreign to the tolerant and open-minded spirit of pragmatists which can hardly accept the idea of advocating Christianity from explicitly Christian premises. While reformed epistemologists are typically also considerably stronger realists than pragmatist philosophers of religion, the main problem is their employment of analytic philosophical argumentation in the service of specific Christian positions, also drawing from theological and Biblical sources. Pragmatists generally view such endeavors very critically.

Third, *Wittgensteinian philosophy of religion* is perhaps closer to pragmatism than the analytic tradition, though in a broad sense it also belongs to analytic philosophy of religion. Putnam is a key figure for this comparison, because his philosophy of religion was equally strongly informed by James and Wittgenstein (as well as by Jewish thinkers like Emmanuel Levinas; see again Putnam 2008). While pragmatists may characterize religion in terms of religious practices, Wittgensteinians usually refer to religious ways of using language ("language-games") within human forms of life, though Wittgenstein himself did not explicitly speak about "religious language-games". Wittgenstein's (1969) understanding of "certainties" as practice-laden basic convictions grounded in our habitual action comes, suitably interpreted, very close to the pragmatist (e.g., Jamesian) view that religious faith can be maintained independently of evidence insofar as it is a genuine option for us, defining our entire life. Yet, both Wittgenstein and the pragmatists would insist on the contingent and revisable nature of even our most deeply entrenched commitments: the ways we use religious language and engage in religious activities are, while not either defended or criticized in scientific terms, nevertheless historically mutable and constantly under critical evaluation – or tested in the "laboratory of life", as Putnam (1997a: 182–183) aptly puts it. For both Wittgensteinians and pragmatists, our practices of life – or the natural development of our "forms of life" – may lead us to use religious concepts, or out of such concept use.[8]

Fourth, many "Continental" philosophers of religion – not to be examined here in any detail, of course – have defended ideas that may to some extent be familiar with pragmatism as well. This is clear in the phenomenology of religion, for instance, relying on conceptions of religious experience that may be comparable to both James (individualism) and Dewey (shared common experience). Furthermore, Richard Kearney's (2010) "anatheism" rejects both metaphysical realism and "theodicist" construals of the problem of evil and suffering (cf. below) – more or less like James's pragmatism does. It is typical of postmodern philosophy of religion to dispense with the metaphysically realist picture of God assumed by classical philosophical theism; indeed, what postmodernists often call (following Martin Heidegger) "onto-theology" has also been firmly rejected in Rorty's resolutely antirepresentationalist pragmatism.

Postmodern "Continental" philosophy of religion, Wittgenstein-inspired investigations of religious uses of language, and (neo)pragmatist (e.g., Rortyan) critiques of realism and representationalism may all agree that the relation between religious language and "religious reality" (whatever that is) should not, and cannot, be understood in terms of any straightforward representational relations between language and the world, based on the idea of our language (or the human mind) functioning as a "mirror of nature" (cf. Rorty 1979), aiming at an accurate depiction of extra-linguistic reality. While inspired by all these currents of thought, a less radical pragmatist like Putnam may, however, insist that it is impossible for us to give up representational relations between language and reality altogether. Pragmatist philosophy of religion can thus seek to critically inquire into the distinctive ways in which religious discourse may seek to represent reality rather than taking any pre-understood conception of representation for granted.

One of the tensions defining the philosophy of religion today is the one between the realistic, representationalist tradition of metaphysical and epistemological explorations of *theism* and *atheism*, on the one side, and critical semantic considerations of the criteria for *cognitively meaningful religious language*, on the other. The latter kind of issues were strongly present in early analytic philosophy of religion, while the metaphysical turn of analytic philosophy toward the end of the 20th century led to an increasing focus on classical philosophical theism. Pragmatism should arguably remain fallibilist and undogmatic here, too. While many have been suspicious of traditional agnosticism – for example, James (1979 [1897]) found it unstable, as it in a sense practically reduces to atheism – it may be suggested that agnosticism need not merely apply to the theism versus atheism issue (at an epistemic level) but can also be applied to the meta-level question concerning the meaningfulness and truth-aptness of religious discourse (cf. Pihlström 2021: Chapter 6). It may remain an undecided issue for pragmatists whether, and how exactly, we can "make sense" of religious language, and taking this issue seriously could be a way of emphasizing the distinctiveness of such language use. Pragmatist philosophy of religion generally seeks to maintain a critical perspective enabling the continuous amelioration of our practices of inquiry, both at the "first-order" level of inquiry into the metaphysics and epistemology of theism versus atheism and at the meta-level addressing the cognitive significance (or lack thereof) of religious language.

Moreover, pragmatism need not make any exclusive choice between investigating religion with reference to either "religious language" or "religious experience", although it might be tempting to portray the difference between "post-linguistic-turn" neopragmatism and experience-focused classical pragmatism in terms of this distinction. Both language and experience are arguably equally important for an adequate understanding of religious beliefs and practices. Insofar as there are religious experiences worthy of philosophical attention, they need to be articulated in language, but religious uses of language can also be expected to have a genuine experiential grounding in individuals' life practices. At the same time, the *limits* of language capable of articulating, say, mystical experiences of the transcendent can be pragmatically investigated.

Furthermore, in inquiring into religious practices of using language and religious experiences, pragmatism can fruitfully collaborate with a range of empirical disciplines studying religious phenomena, e.g., multidisciplinary religious studies integrating psychological, anthropological, and social-scientific approaches to the investigation of religion and religiosity. The methodological debates over the nature and aims of such fields – for instance, their focus on external causal explanations of the emergence and spreading of religious beliefs *versus* more "participatory" understanding of religious ways of life "from within" – can also benefit from the non-reductively naturalistic insights of the pragmatists. In addition to pragmatist philosophy of religion, it is possible to develop a pragmatist philosophy of the study of religion (including theology and comparative religion), analogous to pragmatist philosophies of the various special sciences and disciplines (such as pragmatist philosophy of social science, for instance).[9]

Summarizing the role played by pragmatism in contemporary philosophy of religion, we may return to James's view of pragmatism as a *critical middle path* between various philosophical extremes. James (1975 [1907]: Lecture I) maintained that pragmatism mediates between what he called the "tough-minded" and "tender-minded" philosophical temperaments. The former is, among other things, empiricist, materialist, and determinist, whereas the latter leans toward rationalism, idealism, and indeterminism (ibid.: 13). The tough-minded thinker is scientifically oriented, while the tender-minded takes religious sentiments seriously. Now, while pragmatist philosophy of religion can, even today, be claimed to continue to mediate between science-based naturalist critiques of religion, on the one side, and religious or spiritual sentiments, on the other, it also acts as a mediator at a more philosophical meta-level. Pragmatism seeks to integrate the most plausible elements of, e.g., theological and/or religious realism, on the one side, and constructivist critiques of realism, on the other. Accordingly, pragmatist philosophers – classical and recent – are in most cases moderate realists about religious discourse, maintaining that religious language use does purport to refer to reality that is not just linguistically constructed while also maintaining, with constructivists of various stripes (including postmodernists), that there is no "ready-made" world out there that would be simply "given" to us in an ontologically pre-categorized form. Any reality that human beings can be meaningfully claimed to engage with is, though not simply a human construct, in an important sense, dependent on our categorizing activities – and this applies to the "reality" categorized in religious and theological discourses, too.

In addition, as already suggested, pragmatism offers a middle ground option between *evidentialism*, according to which we can only be justified in holding beliefs that are supported by the kind of evidence that we would require our scientific (or, by extension, everyday) beliefs to be supported by, and *fideism*, which makes a sharp distinction between reason and faith, suggesting that religious faith is not, and should not be, based on evidential considerations. In particular, Jamesian pragmatism can be taken to articulate a novel critical version of evidentialism, which includes among relevant pieces of "evidence" the ways in which religious beliefs *function* in the believer's overall practical world-engagement. This does not entail fideism, according to which we cannot rationally judge religious beliefs at all but can only adopt them on the basis of an act of faith (perhaps comparable to "the will to believe"), but it borrows from fideism the idea that we cannot subordinate religious beliefs to evidential criteria in the same way we evaluate scientific theories by means of evidence.

Finally, pragmatism clearly avoids both fundamentalist religious doctrines and equally fundamentalist and dogmatic (and, arguably, anti-philosophical) versions of "New Atheism",[10] both of which seek a super-objectivity beyond our human reach, a kind of "God's-Eye View" or an absolute conception of reality. By so doing, pragmatism need not, however, argue for the simplified idea that the "rationality" of religious thought (if there were such a thing) is merely some kind of practical rationality instead of theoretical rationality comparable to the rationality of scientific inquiry. On the contrary, pragmatism may seek to reconceptualize the very idea of rationality in terms of habitual practices, and thereby it can reconceptualize the notions of truth and objectivity as well. Truth, objectivity, and rationality are, then, in pragmatist philosophy of religion (and of religious studies) understood as deeply *practice-embedded*: far from being neutral to human practices, they emerge from our reflective engagements in our practices.

Special themes in pragmatist philosophy of religion: evil, death, and religious diversity

In addition to metaphysical and epistemic issues concerning the rationality of believing in the reality of God, semantic ones concerning the meaningfulness or truth-aptness of religious language, and methodological debates on the nature of rational inquiry into religious phenomena, pragmatists have made important contributions to a number of special topics that are highly central in the philosophy of religion both today and historically.

One such topic is *the problem of evil and suffering*, traditionally discussed in terms of "theodicies". While theodicies seek to excuse God for the evil and suffering we find in the world around us, thus responding to the atheist "argument from evil" asking how an omnipotent, omniscient, and absolutely benevolent deity could possibly allow our world to contain so much apparently meaningless evil and suffering, pragmatists generally either reject the problem of evil altogether or defend an "antitheodicist" approach to this issue. There is little in the pragmatist tradition that could be regarded as an explicit attempt to *solve* the problem of evil and suffering, or as a direct response to the "argument from evil" questioning the justification of theism with reference to the empirical reality of suffering. From a pragmatist point of view, the contemporary analytic discourse on theodicies problematically emphasizes the purely epistemic, intellectual dimension of the problem while neglecting the need to respond to concrete suffering ethically and politically. The problem of evil as a key issue in the philosophy of religion thus highlights the entanglement of the epistemic and the ethical in pragmatism more generally (cf. Pihlström 2020).

The philosopher most acutely aware of the existential dimensions of this problem in the pragmatist tradition was clearly James, who framed his pragmatic method with discussions of evil (see James 1975 [1907]: Lectures I and VIII). According to James, it would be crudely insensitive to the individuals experiencing their suffering as absurd and meaningless to claim to find some divine purpose or meaning in that suffering. Hegelian as well as Leibnizian metaphysical and/or teleological theodicies must therefore be firmly rejected (ibid.: 19–22); it would be a serious misunderstanding of pragmatism to suggest that absurd suffering would also have to serve some pragmatic purposes in an allegedly divine scheme of things. Just as pragmatist philosophy of religion in general can be understood as continuing the project of Kantian critical philosophy, James's criticism of theodicies can be placed in the context of what may be called "Kantian antitheodicism" (Kivistö and Pihlström 2016). Jamesian antitheodicism refuses to instrumentalize individual human beings' sufferings in the service of any divine purpose, real or imagined (or any secular proxy thereof, for that matter).

Another issue that some pragmatists have been preoccupied with is the traditional religious topic of *death and the afterlife*. James, in particular, formulated his own theory of "human immortality",[11] while more secular pragmatists have focused on death and mortality as inescapable features defining the horizons of our experience and finite practices (cf. Pihlström 2016). It might be thought that speculations on the afterlife are only available to Jamesian pragmatists "willing to believe" in religious hypotheses not backed up by evidence; yet, Peirce, despite his strong natural-scientific orientation, speculated about a kind of immortality based on his doctrine of "synechism", the metaphysical view that everything is continuous with everything else.[12]

Finally, the contemporary debates over *religious diversity* (and more generally the diversity of worldviews in liberal multicultural societies) are to a certain degree defined by the traditional opposition between religious *exclusivism* and *inclusivism* (see, e.g., Jonkers and Wiertz 2020). Here, pragmatist philosophy of religion can hardly endorse any form of exclusivism but begins from an inclusivist attempt to put a plurality of religious and/or spiritual resources into work in the amelioration of the human condition generally. Instead of succumbing to the temptations of exclusivistically seeking "the" correct religious outlook (or seeing one's own as such), pragmatism is, arguably, primarily a form of *meliorism* assessing religion not from the perspective of its (exclusively) getting things right but from the perspective of its enabling human beings to cope with the world they live in, including the sufferings they must endure, through their various individual religious outlooks. The rejection of exclusivism also follows from the fallibilist character of pragmatism: our own beliefs are never immune to criticism, however firmly we may be committed to them. Pragmatist philosophy of religion is generally open to otherness – and thus also to others' ways of viewing the world religiously, even though those viewings may be very different from our own.

Conclusion: pragmatism as humanism

Pragmatist philosophy of religion can ultimately be regarded as a *critical examination of the human condition* and on how to make this condition better in its irreducible diversity. Therefore, pragmatism is, finally, a form of *humanism*. As such, pragmatist philosophers have generally avoided reducing the human condition to something that can be simply described and explained in terms of the natural sciences while nevertheless finding human life as fully natural. Even our experiences of the supernatural can be studied, echoing the subtitle of James's *Varieties*, as features of "human nature". Furthermore, the pragmatist antitheodicist approach to the problem of evil and suffering (as very briefly described above) is also, most importantly, a humanist response to this problem, suggesting that suffering is, for us, an inescapably ethical issue that cannot be captured from a metaphysical or theological point of view referring to any super-human divine harmony (or an atheist rejection of any such harmony).

Despite some pragmatists' (e.g., Rorty's) strong secularism, "humanism" here does not simply mean secular humanism but can encompass a wide range of meanings (for recent articulations of pragmatist humanism, see, e.g., Honnacker 2018; Jung 2019). Its core for pragmatist philosophers of religion is the pursuit of ever deeper understanding and ever more sensitive amelioration of the human condition. Whatever positive role religion may play in human lives and societies, this role should be pragmatically investigated within this general humanist outlook. But so must, obviously, the negative functions of religion. For the pragmatist philosopher of religion, only the critical path of inquiry is open. Peirce's rule, "do not block the way of inquiry", thus applies to pragmatist philosophy of religion as directly as to any other field of philosophy. Finally, it is important to apply this principle of critical inquiry to the challenges pragmatism itself faces. The pragmatist philosopher of religion must continue dialogue with rival approaches, taking seriously its critics' worries that, for instance, pragmatist accounts of truth and rationality are too relativistic, the will to believe strategy is too irrational, or the conception of non-reductive naturalism remains vague. Pragmatism is never completed in religion – or religious studies – but must be constantly renewed.

Notes

1. This doubt-belief theory of inquiry is primarily indebted to Peirce's writings on the scientific method. See, e.g., the classical essays "The Fixation of Belief" (1877) and "How to Make Our Ideas Clear" (1878), available in Peirce (1992–1998: vol. 1, 109–123 and 124–141). (These famous essays can also be found at W3: 242–257, 257–276.)
2. See other relevant entries in this volume for discussions of these and other classical pragmatists.
3. Peirce's essay, "Evolutionary Love", is also available in Peirce (1992–1998: vol. 1, 352–371). See Atkins (2016) for a detailed and comprehensive analysis of Peirce's philosophy of religion. Peirce's semiotics has also been proposed as a source of theological insights and inquiry: see Raposa (2020).
4. For recent interpretations of James's views on religion, including the will to believe argument and the pragmatist conception of truth, see, e.g., Slater (2014); Campbell (2017); Putnam and Putnam (2017); Pihlström (2013, 2020, 2021).
5. On Dewey's pragmatic naturalism about religion, see, e.g., Rockefeller (1991); Pihlström (2013: Chapter 2). For an influential attempt to develop a naturalistic pragmatism that is largely Deweyan in spirit today, also with regard to the critique of supernaturalist religion, see Kitcher (2012).
6. For a critical comparison between Rorty's and Putnam's legacies in pragmatist philosophy of religion, see Pihlström (2013: Chapter 3).
7. Habermas's (e.g., 2010) late views on religion have to some extent been influenced by pragmatism. Taylor (2002) insightfully comments on James's religious thought while also criticizing its excessive individualism.
8. Putnam's (1992, 2008) writings on religion move smoothly from Wittgensteinian to pragmatist considerations and back again. For an attempt to incorporate Wittgensteinian ideas to pragmatist philosophy of religion, see Pihlström (2020: Chapter 5).
9. See the essays collected in Bagger (2018) for perspectives on pragmatism and naturalism in the study of religion.
10. This militant form of atheism is associated with radical critics of religion such as Richard Dawkins and Sam Harris.
11. James's 1898 essay with that title is available in James (1982).
12. Peirce's 1893 essay, "Immortality in the Light of Synechism", is available in Peirce (1992–1998, vol. 2: 1–3).

References

Atkins, Richard (2016). *Peirce and the Conduct of Life*. Cambridge: Cambridge University Press.
Bagger, James (ed.) (2018). *Pragmatism and Naturalism: Scientific and Social Inquiry after Representationalism*. New York: Columbia University Press.
Campbell, James (2017). *Experiencing William James: Belief in a Pluralistic World*. Charlottesville and London: University of Virginia Press.
Dewey, John (1934). *A Common Faith*. In Jo Ann Boydston (ed.), *The Collected Works of John Dewey: Later Works*. Vol. 9 (LW9). Carbondale: Southern Illinois University Press, 1967–1986, pp. 1–57.
Goodman, Russell B. (1990). *American Philosophy and the Romantic Tradition*. Cambridge: Cambridge University Press.
Grube, Dirk-Martin (2004). "Refuting the Evidentialist Challenge to Religion: A Critique Inspired by William James". *Ars Disputandi* 4, www.arsdisputandi.org.
Grube, Dirk-Martin and Van Herck, Walter (eds.) (2019). *Philosophical Perspectives on Religious Diversity*. London: Routledge.
Habermas, Jürgen (2010). *An Awareness of What Is Missing: Faith and Reason in a Post-Secular Age*. Cambridge: Polity Press.
Herrmann, Eberhard (2003). "A Pragmatic Realist Philosophy of Religion". *Ars Disputandi* 3, www.arsdisputandi.org.
Honnacker, Ana (2018). *Pragmatic Humanism Revisited: An Essay on Making the World a Home*. Basingstoke: Palgrave Macmillan.

James, William (1975–1988). *The Works of William James*. 19 vols. Eds. Frederick H. Burkhardt, Fredson Bowers, and Ignas K. Skrupskelis. Cambridge, MA and London: Harvard University Press. (Contains, e.g.: *The Will to Believe and Other Essays in Popular Philosophy* [1897/1979]; *The Varieties of Religious Experience: A Study in Human Nature* [1902/1985]; *Pragmatism: A New Name for Some Old Ways of Thinking* [1907/1975]; *Essays in Religion and Morality* [1982].)

Joas, Hans (2017). *Die Macht des Heiligen: Eine Alternative zur Geschichte von der Entzauberung*. Frankfurt am Main: Suhrkamp.

Jonkers, Peter and Wiertz, Oliver J. (eds.) (2020). *Religious Truth and Identity in an Age of Plurality*. London and New York: Routledge.

Jung, Matthias (2019). *Science, Humanism, and Religion: The Quest for Orientation*. Basingstoke: Palgrave Macmillan

Kearney, Richard (2010). *Anatheism*. New York: Columbia University Press.

Kitcher, Philip (2012). *Preludes to Pragmatisms*. Oxford: Oxford University Press.

Kivistö, Sari and Pihlström, Sami (2016). *Kantian Antitheodicy: Philosophical and Literary Varieties*. Basingstoke: Palgrave Macmillan.

Peirce, Charles S. (1976–). *Writings of Charles S. Peirce: A Chronological Edition*. Ed. Nathan Houser et al. Bloomington: Indiana University Press.

Peirce, Charles S. (1992–1998). *The Essential Peirce*. 2 vols. The Peirce Edition Project. Bloomington: Indiana University Press.

Pihlström, Sami (2013). *Pragmatic Pluralism and the Problem of God*. New York: Fordham University Press.

Pihlström, Sami (2016). *Death and Finitude: Toward a Pragmatic Transcendental Anthropology of Human Limits and Mortality*. Lanham, MD: Lexington Books.

Pihlström, Sami (2020). *Pragmatic Realism, Religious Truth, and Antitheodicy: On Viewing the World by Acknowledging the Other*. Helsinki: Helsinki University Press.

Pihlström, Sami (2021). *Pragmatist Truth in the Post-Truth Age: Sincerity, Normativity, and Humanism*. Cambridge: Cambridge University Press.

Proudfoot, Wayne (ed.) (2004). *William James and the Science of Religions*. New York: Columbia University Press.

Proudfoot, Wayne (2018). "Pragmatism, Naturalism, and Genealogy in the Study of Religion". Bagger 2018: 101–119.

Putnam, Hilary (1992). *Renewing Philosophy*. Cambridge, MA and London: Harvard University Press.

Putnam, Hilary (1997a). "God and the Philosophers". *Midwest Studies in Philosophy* 31: 175–187.

Putnam, Hilary (1997b). "On Negative Theology". *Faith and Philosophy* 14: 407–422.

Putnam, Hilary (2008). *Jewish Philosophy as a Guide to Life*. Bloomington: Indiana University Press.

Putnam, Hilary and Putnam, Ruth Anna (2017). *Pragmatism as a Way of Life*. Ed. David Macarthur. Cambridge, MA: The Belknap Press of Harvard University Press.

Raposa, Michael L. (2020). *Theosemiotics*. New York: Fordham University Press.

Rockefeller, Steven C. (1991). *John Dewey: Religious Faith and Democratic Humanism*. New York: Columbia University Press.

Rorty, Richard (1979). *Philosophy and the Mirror of Nature*. Princeton, NJ: Princeton University Press.

Rorty, Richard (1989). *Contingency, Irony and Solidarity*. Cambridge: Cambridge University Press.

Rorty, Richard (1999). *Philosophy and Social Hope*. London: Penguin.

Rorty, Richard (2007). *Philosophy as Cultural Politics*. Cambridge: Cambridge University Press.

Slater, Michael R. (2014). *Pragmatism and the Philosophy of Religion*. Cambridge: Cambridge University Press.

Taylor, Charles (2002). *Varieties of Religion Today: William James Revisited*. Cambridge, MA and London: Harvard University Press.

White, Morton (2002). *A Philosophy of Culture: The Scope of Holistic Pragmatism*. Princeton, NJ: Princeton University Press.

Wittgenstein, Ludwig (1969). *On Certainty*. Eds. G.E.M. Anscombe and G.H. von Wright. Trans. Denis Paul. Oxford: Basil Blackwell.

Zackariasson, Ulf (2015). "Religion". In Sami Pihlström (ed.), *The Bloomsbury Companion to Pragmatism*. London and New York: Bloomsbury.

32
PRAGMATISM AND THE MORAL LIFE

Diana B. Heney

The subject of this chapter is the engagement with ethics that permeates pragmatism. My tasks are to identify contributions that pragmatists have made to philosophical ethics, to identify some challenges that may be associated with taking a pragmatist ethics on board, and to gesture toward possible next steps for pragmatist ethics.

The tasks just described, arranged in that way, have a kind of intuitive order – what's on offer; what's trouble; what's next? Despite the plausibility of proceeding in the intuitive way, I think there is something to be gained by starting with the troubles, for in doing so we can perhaps begin to clear the way for making good use of pragmatism's contributions in our own times and on our own terms.

As a final prefatory remark, because my task is to survey the relationship between a philosophical way of thinking and a particular domain, I give short shrift to biographical considerations. I defer the reader interested in knowing more about the thinkers behind the thoughts explored in this chapter to the entries of my colleagues in Part I of the handbook.

Three Challenges

There are three main challenges to exploring pragmatist ethics that it is helpful to delineate before setting out.

The first is to determine which thinkers we would need to consider carefully to get a grip on what pragmatist ethics is and what it could be. Celestial navigation is the ancient art of using objects in the night sky to orient oneself in space. To do so, sailors can rely on the most visible such objects: the brightest stars. History of philosophy proceeds in a similar way – contemporary students and readers often orient themselves with respect to the most well-known philosophers of an era.

This presents a challenge with respect to pragmatism because there are numerous thinkers worthy of serious study, especially when we focus on contributions to moral philosophy. It is common for introductory courses or volumes on pragmatism to focus exclusively on three thinkers: Charles S. Peirce (1839–1914), William James (1842–1910), and John Dewey (1859–1952). But those who want to engage the possibilities of pragmatist ethics seriously will benefit from going beyond Peirce-James-Dewey treatments of the tradition and including the thinkers with whom the "big three" stand in constellation. This could include any of the following philosophers – some quite bright stars in the history of pragmatism already, some whose place in the firmament has not been so well appreciated: Josiah Royce (1855–1918), Jane Addams (1860–1935), George

Herbert Mead (1863–1931), George Santayana (1863–1952), Ella Lyman Cabot (1866–1934), W.E.B. Du Bois (1868–1963), Clarence Irving Lewis (1883–1964), and Alain Locke (1885–1954). To complicate matters further, we could go beyond pragmatism's first 100 years and consider the work of self-identified pragmatists active in moral philosophy today, but I take it that the point is made even without the full pragmatist family tree.

The second challenge is closely related to the first, and it is the sheer volume of work to consider. With a constellation of thinkers, many of whom were prolific writers, some of whom lacked judicious editors, you get a lot of books. In the case of the pragmatists, we have much more than books – we have encyclopedia entries, magazine articles, autobiographies, correspondence, and great heaps of unpublished literary remains. As just one example, consider Clarence Irving Lewis, the subject of Chapter 8 of this handbook. His relevant contributions include the following essays and books: "Judgments of Value and Judgments of Fact" (1936), "The Objectivity of Value Judgments" (1941), *An Analysis of Knowledge and Valuation* (1946), "The Empirical Basis of Value Judgments" (1950), "The Rational Imperatives" (1951), "Subjective Right and Objective Right" (1952), "The Individual and the Social Order" (1952), "Turning Points of Ethical Theory" (1954), *The Ground and Nature of the Right* (1955), and *Our Social Inheritance* (1957). Consider that this volume of work is fairly typical for any one of these writers, and the scope of the commitment required to fully engage the tradition begins to come into view.

Although the magnitude of materials can be daunting, this is actually a good thing. Unlike some schools of ethical thought, such as Epicureanism, which philosophers have had to reconstruct on the basis of textual fragments, we have a lot to work with in understanding how historical pragmatists engaged the ethical. Further, seeing the richness and variety in classical pragmatism's moral philosophy supports our imaginative engagement with the projects those interested in experimenting with pragmatist methods might take up.

Our first and second challenges arise from what I have suggested is an appropriate expansiveness with respect to *writers* and *writings*. The third challenge relates to a different kind of expansiveness, which is with respect to what should be considered under the heading of *ethics*. One common approach is to see ethics as divided into three main branches: metaethics, normative ethics, and applied ethics. Metaethics is the most abstract and broadly concerns the preconditions and presuppositions of moral thought, talk, and practice. When we ask questions like "are any moral judgments true?" or "what, if anything, do moral claims refer to?", we are in the realm of metaethics. Normative ethics concerns good conduct and asks such questions as "what makes good acts good?" or "what is the mechanism for identifying morally good actions?" Applied ethics is the most concrete, as it asks questions that bring the theoretical elements of ethics into contact with real and pressing problems, asking such questions as "should people be allowed to receive medical assistance in dying?" or "does allowing carbon offset purchase by the wealthy do anything to help those most harmed by climate change?"[1]

This separation can promote clarity in our thinking about what a philosopher is most centrally doing in a given text or argument. But that clarity is in part achieved through a kind of distortion. In reality, the subfields of ethics make frequent contact and grow together. While our contemporary distinctions for carving up the space of academic moral philosophy are prominent, with many historical texts, the lines are blurry at best.

What all this means is that there are works by pragmatist thinkers that may not look so obviously like ethical treatises but which are shot through with themes that bear on our attempts to model – and to live – moral lives. As an example, consider Jane Addams's *The Long Road of Women's Memory*, published in 1916. This book recounts the season when women of diverse backgrounds came to the Hull House settlement founded by Addams to investigate rumors that a "Devil Baby" had been born there. The book is part memoir, part reflection on the puzzles and promises of democracy, part anthropology, part feminist epistemology, part argument for the importance of narrative in understanding others, and part ode to meeting others in community even – perhaps, especially – when we have no idea what they are talking about.[2]

In sum, a full study of pragmatist ethics has to chart a path through these three challenges: whose work to engage, from which of a myriad of sources, and on how wide an understanding of "moral philosophy".

Our task is not the full study. Instead, my approach in what follows is to simplify matters by identifying entry points that are intelligible and interesting without taking on the tradition in its entirety. I will focus on three contributions to moral philosophy found in classical pragmatism: its distinctive moral epistemology, its treatment of habit, and its abundant advice on norms to live by.

From Methodology to Moral Epistemology

In his 1903 Harvard Lectures, Peirce described pragmatism as "a principle and method of right thinking", that principle being that to understand our concepts requires us to interrogate them where we actually find them: in use, playing their parts in the cognitive life of human beings. He described pragmatists as "laboratory philosophers", animated by "the impulse to penetrate into the reason of things" (CP 1.44) and by the conviction that "philosophy is a science based upon everyday experience" (CP 8.112). In her *Newer Ideals of Peace*, Addams articulates the pragmatist orientation in terms directly connected to moral philosophy. She declares that "It is necessary from the very beginning to substitute the scientific method of reason for the *a priori* method of the school men" – at least, it is necessary "if we would deal with real people and obtain a sense of participation with our fellows" (*NIP* 18). Hilary Putnam remarked that the distinctive balance of pragmatism as a method is between being fallibilist and being anti-skeptical: we must be prepared to change our beliefs when faced with appropriate evidence and experience, but beliefs as habits for actions are not undermined arbitrarily or by "paper doubts". Further, pragmatism as a method places human practice always at the center of the philosophical enterprise: 'the emphasis on the primacy of practice', he says, is 'perhaps *the* central' emphasis of pragmatism (1995: 52). While the application of the pragmatic maxim to the concept of truth may be its most (in)famous iteration, pragmatist methods in general are open to reiteration. In fact, they demand reiteration as human practices, institutions, and needs evolve. This is why pragmatism can be construed as a metaphilosophical program rather than a set of substantive first-order views.[3]

At the center of pragmatism as practice, we find its naturalized account of inquiry. This model is most famously articulated in a pair of papers by Peirce, "The Fixation of Belief" (1877) and "How to Make Our Ideas Clear" (1878). Peirce's presentation is striking because it considers how an individual is moved to, and through, inquiry. It also locates human animals as knowers not in isolation but in community – where what we often need to progress in the pursuit of our own settled beliefs is the experience and expertise of others. The earliest pragmatists took on the English psychologist Alexander Bain's precept that beliefs are habits for action. Bain framed belief as active and predictive, the "expectation of some contingent future about to follow on our action" (1859: 568). Our beliefs are formed not apart from life but in the living of it. Once formed, they are not inert contents tucked away in the storehouse of the mind but real commitments that guide our conduct. Just as we are not isolated and unchanging, our concepts – including our moral concepts – are not static placeholders in thought. Rather, they are instruments for classification, comparison, and inference, for solitary and shared deliberation.

Genuine inquiry begins when belief is undermined by with doubt. As beliefs are the cognitive backstop for practical reasoning and our confident action, the perturbance of a belief we had relied on brings us up short. In order to find our way back to action, we must replace or update the previously held belief. The only reliable way to do this is inquiry, which is characterized by its cognitive aspirations: when we begin an inquiry, we take ourselves to be in the truth business. Peirce also calls this "the method of science" (CP 5.384), but this need not raise connotations of hadron colliders or hazmat suits. Rather, as Cheryl Misak puts it, a commitment to inquiry amounts to aiming at "empirical adequacy, predictive power, understanding the way things work, understanding ourselves" – that is, aiming at the truth (Misak 2000: 1).

This account of inquiry carries over to pragmatist moral philosophy. Dewey frames how extensive the scope of this method of belief fixation is in his *Logic: The Theory of Inquiry*, where he declares that "Inquiry is the life-blood of every science and is constantly employed in every craft, science, and profession" (1986 [1938]: 12). Whether we articulate our cognitive aspirations to ourselves deliberately and declare our purpose explicitly, or whether we experience ourselves as flummoxed and simply trying to find a way forward that won't give way beneath our feet, our desire for secure and stable belief shows us as committed to inquiry in practice. This is no less true of our beliefs about moral matters. While pragmatism portrays our approach to ethics as scientific – informed by evidence and experience – it is not scientistic, for it is still human experience that moves inquiry forward.

It might be thought that pragmatism's focus on experience is radically subjective in a way that leads straightforwardly to moral relativism. This need not be so. As James puts it, "when as empiricists we give up the doctrine of objective certitude, we do not thereby give up on the quest or hope of truth itself"; rather we see ourselves as engaged in the process of pursuing it – and "we gain an ever better position towards it by systematically continuing to roll up experience and think" (1979 [1896]: 23–24). Systematically rolling up experience can be done at both the individual level and the communal level. It also requires us to consider experiences broadly. When Peirce says that "Ethics as a positive science must rest on observed facts" (CP 8.158), this is a capacious understanding of "observation" that includes consideration of our reactions to thought experiments or other modes of imaginative engagement. Also central to the idea of treating ethics as a normative science is the rejection of a dichotomy between facts and values. All inquiry is goal directed, and such goals express values – whatever facts are turned up cannot be considered apart from the aims that led to their discovery.

Rather than taking this as a strike against science, we can take it as a point in favor of the overlap in method between science and ethics. The antidote to subjectivity of the solipsistic kind is not in an objective sense of morality available through reason alone but shared objectives. This is how pragmatism positions inquiry as available to the laboratory scientist and the ethicist alike: we must conceive investigation into questions of good and bad conduct as moral inquiry, "aimed at finding the right answer and improving our beliefs through considering more evidence, argument, and perspective" (Misak 2000: 85).

Treating moral life as amenable to investigation is also a kind of democratization of access to the truths of morality, which follows from the wide range of whose experience is salient. Rather than a few sages or oracles who speak the truth to them any, pragmatism frames the pursuit of moral knowledge as group work. Participation in moral evaluation arises in much the same way that participation in inquiry does: from a natural inclination toward one's own benefit to an appreciation of the contributions others may make to one's understanding and pursuit of one's own benefit and arrival at the necessity of the knowledge of others to best serve that pursuit. Lewis maintains that "to seek the good and to avoid the bad is the basic bent of conscious life" (1957: 83). Humans ascend from this "basic bent" to full-blown normative notions of right and wrong because we are capable of "the extension of this process to experiences distant in time through deliberation and self-criticism" (Murphey 2005: 382). The capacity for judgment, furnished with experience, makes it possible for a person to govern their actions in ways that produce good results for themselves and others – to adopt effective norms.

Habit Revisited

The improvement of our habits of mind is a central goal of pragmatist inquiry. Because of the active and predictive nature of beliefs on the pragmatist model, it will come as no surprise that habits of conduct should follow.

Of course, the idea that habit is an important focal point in trying to model a morally good life and in trying to live one is not a Pragmatist™ idea. Indeed, it is an ancient idea. Much of Aristotle's *Nicomachean Ethics* is devoted to explaining how the interested student can cultivate moral virtues through processes of habituation – and much of the advice is practice oriented. For example, in discussing how to cultivate particular virtues of character, Aristotle advises that self-knowledge is crucial: if we can identify which vice we are inclined toward, we can adjust our "aim" in action accordingly to hit the mean. Steering toward the vice contrary to the one we naturally lean toward is a mechanism for addressing the propensity leading us toward a bad habit.

Similarly, Epictetus's *Handbook* stresses self-knowledge when considering taking on a project or path forward in life, focusing at least in part on one's capacity to become habituated in the ways necessary to complete that project or walk that path. As he demands of his reader,

> Just you consider, as a human being, what sort of thing it is; then inspect your own nature and whether you can bear it. You want to do the pentathlon, or to wrestle? Look at your arms, your thighs, inspect your loins. Different people are naturally suited for different things. Do you think that if you do those things you can eat as you now do, drink as you now do, have the same likes and dislikes?
>
> *(§29, White 1983: 20)*

If we assess our conduct and find it wanting, we cannot simply continue to do as we now do and hope improvement somehow occurs. This is not because self-improvement is hopeless, but because in order for us to earn *rational* hope, we must replace our problematic beliefs – and their resultant habits – with something stable.

Pragmatism contributes to our understanding of habit as an evaluative focal point for moral philosophy by giving an account of how that stability can be secured. When habits of mind – beliefs – are disturbed by experiences that give rise to genuine doubt, it is back to inquiry that we turn. Becoming accustomed to *turning back* as part and parcel of doing ethics converts that initially unpleasant experience of doubt into a learning opportunity. The function of inquiry, recall, is to fix beliefs not merely as cognitive units of semantic content but in connection with our intellectual and moral habits and in connection with one another.

> Mead articulates this interplay between the individual and the social as follows:
> It is by means of reflexiveness – the turning back of the experience of the individual upon himself – that the whole social process is thus brought into the experience of the individuals involved in it; it is by such means, which enable the individual to take the attitude of the other toward himself, that the individual is able consciously to adjust himself to that process, and to modify the resultant of that process in any given social act in terms of his adjustment to it.
>
> *(2015 [1934]: 134)*

Updating our beliefs while participating in our collaborative social life is the mechanism for avoiding calcification of those habits. But to observe that humans run on habits is only part of the picture; pragmatists are deeply invested in habit as the site for individual and social improvement.

It is in this spirit that James describes habit as "the great fly-wheel of society" (1981 [1890]: 121). He adds that, "Could the young but realize how soon they will become mere walking bundles of habits, they would give more heed to their conduct while in the plastic state" (1981 [1890]: 127). Anyone who has undertaken a failed self-improvement program may find solace in a pair of remarks made by Peirce. On the one hand, "Where hope is unchecked by any experience, it is likely that our optimism is extravagant" (CP 5.366); on the other hand,

> The trial of this method of experience ... encourages us to hope that we are approaching nearer and nearer to an opinion which is not destined to be broken down – though we cannot expect ever quite to reach that ideal goal.
>
> *(CP 5.384)*

While certainty in ethics is not offered by pragmatist moral epistemology, it is nonetheless melioristic. Terence MacMullan points out that this measured hopefulness permeates pragmatist thought and takes various forms in the problem-engaged theorizing of multiple figures (2013).

The question that arises next is a practical one: what ought we do to give appropriate heed to our conduct? Or in other words, through the lens of what ideal should we strive to improve our habits?

Abundant Advice

Pragmatists offer a refreshing diversity of ideals as candidates for the human good. Normative ethics, as we saw earlier, asks precisely what it is that makes good acts good. Once a criterion is established, it serves as a guiding normative notion: an ideal and focal point, a standard held up as a key piece of conceptual architecture in modeling moral life. The history of normative ethics in Western philosophy has furnished us with prominent examples, including virtue, duty, utility, and care.

A number of the most prominent views handed down in the canon of Western philosophy are monistic about normative ideals: they either explicitly endorse just one guiding notion or reduce an apparent plurality to a singularity. One question engagement with the classical pragmatists urges us to consider is whether we might get closer to understanding or accurately modeling the aspirations of human morality if we are more expansive. In response to the search for a criterion of good conduct, pragmatists offer a plethora of guiding normative notions. Here, I introduce four candidates: growth, loyalty, harmony, and culture. I shall not strive to adjudicate between them but only to suggest that these notions are worth considering alongside duty, virtue, utility, and care, that we might enrich contemporary conversations about what makes a life go well in the moral and prudential senses.

First, growth. Considering humans as a life form, some pragmatists forward a guiding normative notion of growth as the highest human good. All living things experience growth, but what is intended here is no merely biological process. Dewey focuses on growth in his philosophy of education, where he argues that education is characterized not a singular overarched end but by the principle of growth (Dewey 2008 [1916]). Sidney Hook notes that

> it is obvious that for Dewey growth is an inclusive and not a single exclusive end. It embraces *all* the positive intellectual, emotional, and moral ends which appear in everybody's easy schedule of the good life and the good education – growth in skills and powers, knowledge and appreciation, value and thought.
>
> *(1959: 1013)*

As Hook adds, Dewey does not simply offer an objective-list theory of well-being but stresses the activity requisite to "bring these ends into living and relevant relation to the developing powers and habits and imaginations of the individual person" (Hook 1959: 1014).

Peirce's account of growth is similarly expansive and also encompasses community-level growth. He suggests that the summum bonum of human life, "the highest of all possible aims", is "to further concrete reasonableness" (CP 2.34). Growth in reasonableness that renders it "concrete" demands of us more than an individually cultivated disposition. As Aaron Massecar has argued, Peirce's ideal requires us "to ensure that we use our reason to address the concerns of our community" (2016: 17). Reasonableness is made concrete not only in well-formed habits but also in practices and norms that govern our public life and shared institutions.

Second, we have loyalty. The ideal of loyalty as the highest good is introduced by Royce and affirmed by Cabot. In his lectures published as *The Philosophy of Loyalty*, Royce develops an account that is structurally similar to Aristotle's virtue ethics, which presents a plurality of good moral qualities united under a single, overarching virtue. For Aristotle, all other intellectual virtues and moral virtues are arranged under the virtue of practical wisdom; for Royce, all other good moral qualities are arranged under the principle of loyalty to loyalty. In describing this structure, he asserts that "You can truthfully center your entire moral world about a rational conception of loyalty. Justice, charity, industry, wisdom, spirituality, are all definable in terms of enlightened loyalty" (1908: 160).

That loyalty should be enlightened requires a commitment to the ideal of loyalty itself. Royce describes how this leads to an imperative that informs self-cultivation:

> Find your own cause, your interesting, fascinating, personally engrossing cause; serve it with all your might and soul and strength; but so choose your cause, and so serve it, that thereby you show forth your loyalty to loyalty, so that because of your choice and service of your cause, there is a maximum of increase of loyalty amongst your fellow-men.
>
> *(1908: 909–910)*

The final clause is meant to help guard against the objection that some people will choose bad causes to be loyal to; such a choice would not inspire in others an increase of loyalty in others and thus is not enlightened loyalty.

Cabot shares Royce's interest in loyalty and connects the ways in which he frames it as providing normative structure to a life with her own philosophical work – especially her reflections on ethics education. In describing her pedagogy, John Kaag notes that she identifies "The loyalties of childhood to gangs, to friends, to family members and local communities", which "are usually exclusive and remain antagonistic to diversity and difference" (Kaag 2011: 67). Cabot believes that from an early age, moral questions "are as close to our life as the air we breathe", and that "anyone who has any interests whatsoever is concerned with ethical problems" (1910: 1). Education can begin from that interest in a variety of ways. She sees clearly that not all have the same opportunities to engage firsthand with problematic situations in a way that would be edifying. It is on that account that she develops a necessary place for both sympathy and imagination. Sympathy, Cabot claims, "is intricately bound up with interest" and "like interest, grows with knowledge". The more we know of someone else's problems, the more equipped we will be to learn from and with them. But knowing facts is insufficient. We must "make real to ourselves" their struggles and solutions, and for that, imagination is required: "imagination is kindled and kept shining by sympathy" (1910: 202). To move from exclusive and antagonist loyalties to mature and worthy ones is the work of becoming educated for the world in which we live.

Third, harmony. This ideal is hinted at by a number of pragmatists and made explicit by both James and Santayana. In keeping with the radical empiricism that James develops to consider experiences of all kinds, he remarks in *Varieties of Religious Experience* that "Were one asked to characterize the life of religion in the broadest and most general terms possible, one might say that it consists of the belief that there is an unseen order, and our supreme good lies in harmoniously adjusting ourselves thereto" (1985 [1902]: 51). Santayana connects the notion of harmony directly with the aspiration to live a morally good life, remarking that "All that morality can require is the inward harmony of each life" (SB 134). Achieving harmony requires both knowledge of self and knowledge of the world; this is why Santayana connects the possibility of living harmoniously with wisdom. We sometimes find the following quote in isolation: "To be happy, we must be wise" (EGP 152). But what immediately precedes this is an important piece of context. For wisdom here is earned through self-knowledge: "You must have taken the measure of your powers, tasted the fruits of your passions and learned your place in the world and what things in it can really serve you". Martin Coleman and Herman Saatkamp have described this process of coming to know oneself in relation to the world as an individual one but also as egalitarian in spirit:

> No one can claim a central place above others. But each entity also has an embodied set of values, and the art of life is to structure one's environment in such a fashion as to best realize those embodied values, i.e., to place in harmony the natural forces of one's life and one's environment.
>
> *(Saatkamp and Coleman 2012: 2002)*

Fourth, and finally, we have culture. Alain Locke develops "an ethics of culture", in which he argues that the "highest intellectual duty is to be cultured" (1991 [1923]: 435). Speaking in an address to students at Howard University, Locke exhorts them to take a wider view of education than "the necessary hardship that is involved in preparing to earn a better living" (436). He states that the word "culture" represents a "higher function of education, the art of living well" (436). It is higher, on his analysis, because it is that part of education that is self-administered. Locke duly considers objections against the idea that we have a duty to become cultured – that it is artificial, useless, selfish, or snobbish – by calling his students back to their own investment in becoming educated. Crucially, that investment is not merely economic but personal. But an investment in one's continued education, in learning to live well, is not made selfish by being personal: "culture", Locke says, "even when it is rich and mature, gives only by sharing" (438).

Locke offers practical advice to his students about how to become cultured, which goes some distance further to illuminating how he understands this task as an ethical duty and not a mere instrumental good to the individual. He recommends that we refine our habits of consumption, cultivate an intelligent appreciation of one of the "great human arts",[4] and constantly practice that in which we have an interest. Locke connects his ideal with the Socratic dictum "Know thyself" by marking culture out as a constitutive end of education: "the goal of education is self-culture, and one most hold it essential even for knowledge's own sake that it bs transmuted into character and personality" (441). In connecting culture with education and growth, both individual and societal, there is a clear resonance here between Locke's view and those of Peirce and Dewey. There is also some overlap with Santayana's account of harmony (SAF 223). While culture seems highly specific, it may actually be the pragmatist normative notion that could most easily accommodate the insights of the others.

Whether to try to synthesize these ideas of morally worthy goals or allow them to stand separately is a question at the point of putting pragmatist ethics into action. If the list of options just rehearsed feels daunting, there may be some (cold) comfort in the doubt generated by having so much to choose from. For choose we can, and we must: to be human is to be the kind of thing that places oneself under norms, and such norms express themselves in our unconscious habits and our conscious choices. As Lewis puts it,

> To act, to live, in human terms, is necessarily to be subject to imperatives; to recognize norms. Because to be subject to an imperative means simply the finding of a constraint of

action in some concern for that which is not immediate; is not a present enjoyment or a present suffering. To repudiate normative significances and imperatives in general, would be to dissolve away all seriousness of action and intent, leaving only an undirected floating down the stream of time; and as a consequence to dissolve an significance of thought and discourse into universal blah.

(1946: 481)

Far from encouraging such "blah", organizing ourselves under norms derived from ideals of growth, loyalty, harmony, or culture offers many possible routes away from doubt and indecision.

How could we decide which to enact? For the pragmatist, the test of any principle is what Dewey, following John Stuart Mill, called experiments in living. Presumably some version of this thought is what informs the popular event Stoic Week. Per the Modern Stoicism website, "Stoic Week is an annual event that invites you to 'live like a Stoic for a week'. It is run online and is completely free. Since 2012 over 20,000 people have signed up for Stoic Week".[5] The obvious challenge of "Pragmatism Week" as a competitor is that, as we've just seen, the pragmatists differ on what they emphasize in the pursuit of a good – socially engaged, morally upright – human life. But the mechanism of experimentation is the same for any candidate normative principle: we just have to try what the pragmatists offer to see what works. It is up to the student of ethics to determine which of the candidate norms sketched here can be lived out fully and whether in the living we would find it vindicated in practice.[6]

Conclusion

We began by considering some challenges to getting serious about the study of pragmatist ethics and then turned to a consideration of pragmatism's contributions to moral philosophy. What could be next is roughly that our understanding of pragmatist ethics has the potential to go both backward and forward.

By "going backward", I do not mean "diminishing" – rather, I mean going back to those pragmatists whose work has not been widely read and engaged despite their contributions to moral philosophy. To go backward in this sense, we would need to meet – and not simply steer around – the challenge of meeting the pragmatist tradition in ethics in a more complete way than has usually been attempted.[7]

To go forward, we need to consider to what purposes and projects the elements of pragmatist ethics might be put: what effects, that might conceivably have practical bearings, we conceive pragmatist ethics to have.

Whatever steps we take toward improving our understanding of moral life, pragmatism tells us to take them together. The experience of others is an essential and everyday corrective; deliberation on the pragmatist model is not solitary a priori reasoning but an activity where we have company. In a pluralistic society, that company will be mixed: people with different ends in view must find ways to get along and go along. In order to do so, we must adopt Peirce's recommendation and "rate [our] own powers of reasoning at the very mediocre price they would fetch if put up at auction" (1.673). To know better, and do better, a person must make the most of being one among a community of knowers. The moral life is a shared life.

Notes

1 As Andrew Altman has argued, once pragmatism as a method is taken on board, we can see that applied philosophy has "a principal role" to play (1983: 227). There is an abundance of work being done by pragmatists in applied ethics, which has not been my topic here. This is not because pragmatist applied

2   Other examples will spring to mind for those familiar with pragmatism, so I add just one more here: James's *Principles of Psychology*, principally intended as a textbook for the emerging science of the mind, is replete with metaethical observations.
3   For the argument that metaphilosophical pragmatism is the only pragmatism worth having, see Aikin and Talisse (2018).
4   Locke's examples are "literature, painting, sculpture, music, or what not" (439). "What not" leaves considerable room for expansion and interpretation, which is important to Locke's connection between authenticity and projects of self-cultivation.
5   Stoic Week is a yearly event. For details, see https://modernstoicism.com/stoic-week/.
6   For more on the standard of pragmatic vindication, see Wiggins (1990–1991).
7   Shannon Dea (2017) recounts the experience of developing a more expansive syllabus for an undergraduate course in classical pragmatism. As she observes, "In classic pragmatism courses, the philosophers who are usually covered are Peirce, James, Dewey, etc. If any non-White male philosopher is included, it is typically Jane Addams." Describing herself as "determined to teach a more pluralistic canon", Dea's syllabus began as follows: "In this seminar, we will survey classic pragmatism, a distinctively American philosophical movement that spanned about 1870 to the 1930s. We will read and discuss representative works by central figures in the movement and by other authors working in a similar idiom in the period". Dea's survey course included Peirce, James, Dewey, Addams, and Locke, along with Anna Julia Haywood Cooper (1858–1964), W.E.B. Du Bois (1868–1963), and Mary Parker Follett (1868–1933). Students were tasked throughout the term with three questions: "what is pragmatism?"; "what have historically been the grounds for inclusion into (or exclusion from) the canon of classic pragmatism?"; and "what, if anything, is distinctively American about classic pragmatism?" Further thought has been put into criticizing the narrowness of and expanding the pragmatist canon by Raposa (forthcoming).

Recommended Reading

For alternate surveys of similar terrain, I recommend the following three works by way of introduction:

Andrew Sepielli (2017). "Pragmatism and Metaethics", in Tristram McPherson and David Plunkett (Eds.), *The Routledge Handbook of Metaethics*, New York: Routledge: 582–594.

J.E. Tiles (1998). "Pragmatism in Ethics", in E. Craig (ed.), *Routledge Encyclopedia of Philosophy,* Vol. 7. London: Routledge, 640–645.

Juan Pablo Serra (2010). "What Is and What Should Pragmatic Ethics Be?: Some Remarks on Recent Scholarship", in *European Journal of Pragmatism and American Philosophy*, II (2): 1–15.

References

Addams, Jane (2007 [1906]). *Newer Ideals of Peace* (NIP), with an introduction by Berenice A. Carroll and Clinton F. Fink. Indianapolis: University of Illinois Press.

Altman, Andrew (1983). "Pragmatism and Applied Ethics", *American Philosophical Quarterly*, 20(2): 227–235.

Aristotle (1999). *Nicomachean Ethics*, 2nd edition, trans. and ed. Terrence Irwin. Indianapolis, IN: Hackett Publishing Company.

Bain, Alexander (1859). *The Emotions and the Will*. London: Parker and Son.

Cabot, Ella Lyman (1910). *Everyday Ethics*. New York: Henry Holt and Company.

Curry, Tommy J. (2009). "Royce, Racism, and the Colonial Ideal: White Supremacy and the Illusion of Civilization in Josiah Royce's Account of the White Man's Burden", *The Pluralist*, 4(3): 10–38.

Dea, Shannon (2017). "Deep Pluralism and Intentional Course Design: Diversity from the Ground Up", *Rivista di Estetica*, 64: 66–82.

Dewey, John (2008 [1916]). "Democracy and Education", in *The Middle Works of John Dewey 1899–1924*, ix, ed. Jo Ann Boydston. Carbondale, IL: Southern Illinois University Press.

———. (1986 [1938]). "Logic: The Theory of Inquiry", in *The Later Works of John Dewey 1925–1953*, xii, ed. Jo Ann Boydston. Carbondale, IL: Southern Illinois University Press.

Hook, Sidney (1959). "John Dewey—Philosopher of Growth", *The Journal of Philosophy*, 56(26): 1010–1018.

James, William (1975–1988). *The Works of William James*, 18 volumes, eds. Frederick H. Burkhard, Fredson Bowers, and Ignas K. Skrupskelis. Cambridge, MA: Harvard University Press.

———. (1979 [1896]). "The Will to Believe", in *The Works of William James*, vi: *The Will to Believe and Other Essays in Popular Philosophy*, in *The Works of William James*, 18 volumes, eds. F.H. Burkhard, F. Bowers, and I.K. Skrupskelis. Cambridge, MA: Harvard University Press.

———. (1981 [1890]). "The Principles of Psychology", in *The Works of William James*, xiii–x, eds. F.H. Burkhard, F. Bowers, and I.K. Skrupskelis. Cambridge, MA: Harvard University Press.

———. (1985 [1902]). "The Varieties of Religious Experience", in *The Works of William James*, 18 volumes, eds. F.H. Burkhard, F. Bowers, and I.K. Skrupskelis. Cambridge, MA: Harvard University Press.

Kaag, John (2011). "Narrative and Moral Psychology in the Philosophy of Ella Lyman Cabot", *The Pluralist*, 6(3): 64–79.

Locke, Alain (1991[1923]). "The Ethics of Culture", reprinted in Leonard Harris (ed.), *The Philosophy of Alain Locke*. Philadelphia, PA: Temple University Press.

MacMullan, T. (2013). "The Fly Wheel of Society: Habit and Social Meliorism in the Pragmatist Tradition", in *A History of Habit: From Aristotle to Bourdieu*, eds. Tom Sparrow and Adam Hutchinson. Lanham, MD: Lexington Books: 229–253.

Massecar, Aaron (2016). *The Fitness of an Ideal: A Peircean Ethics*. Lanham, MD: Lexington.

Mead, George Herbert (2015 [1934]). *Mind, Self, and Society: The Definitive Edition*, eds. Charles W. Morris, annotated by Daniel R. Huebner and Hans Joas. Chicago, IL: University of Chicago Press.

Misak, Cheryl (2000). *Truth, Politics, Morality: Pragmatism and Deliberation*. New York: Routledge.

Murphey, Murray G. (2005). *C.I. Lewis: The Last Great Pragmatist*. Albany: State University of New York Press.

Peirce, Charles Sanders (1931–1958). *Collected Papers of Charles Sanders Peirce*, i–vi, eds. C. Hartshorne and P. Weiss, vii and viii, ed. A. Burks. Cambridge, MA: Belknap Press. Cited as *CP* plus volume and paragraph number.

Putnam, Hilary (1995). *Pragmatism: An Open Question*. Cambridge: Blackwell.

Raposa, Michael (forthcoming). "Peirce and Racism: Biographical and Philosophical Considerations," 2021 Presidential Address, Charles S. Peirce Society.

Royce, Josiah (1908). *The Philosophy of Loyalty*. New York: The Macmillan Company.

Saatkamp, Herman and Martin Coleman (2012). "George Santayana", The Stanford Encyclopedia of Philosophy (Fall 2020 Edition), ed., Edward N. Zalta, https://plato.stanford.edu/archives/fall2020/entries/santayana/.

Santayana, George (1916). *Egotism in German Philosophy* (EGP). New York: Charles Scribner's Sons.

———. (1923). *Scepticism and Animal Faith: Introduction to a System of Philosophy*. New York: Dover.

———. (1955 [1986]). *The Sense of Beauty* (SB). New York: Dover.

Serra, Juan Pablo (2010). "What Is and What Should Pragmatic Ethics Be?: Some Remarks on Recent Scholarship", *European Journal of Pragmatism and American Philosophy*, II (2), 1–15

Tiles, J.E. (1998). "Pragmatism in Ethics", in E. Craig (ed.), *Routledge Encyclopedia of Philosophy*, Vol. 7. London: Routledge, 640–645.

White, Nicholas (1983). *Handbook of Epictetus*. Indianapolis, IN: Hackett Publishing Company.

Wiggins, David (1990–1991). "Moral Cognitivism, Moral Relativism and Motivating Moral Beliefs", *Proceedings of the Aristotelian Society*, 91: 61–85.

33
ARTWORLD PRACTICE, AESTHETIC PROPERTIES, PRAGMATIST STRATEGIES

Robert Kraut

These, then, were some of my broodings as I looked at [Jasper] Johns's pictures. And now I'm faced with a number of questions, and a certain anxiety.
What I have said—was it *found* in the pictures or read into them? Does it accord with the painter's intention? Does it tally with other people's experience, to reassure me that my feelings are sound?
—Leo Steinberg, "Contemporary Art and the Plight of Its Public"

Steinberg's anxiety involves a familiar contrast within metaphysical and art-critical discourse: between the found and the made, the discovered and the invented, and—more generally—between various strains of realism and constructivism. One wishes to know whether interpretation of artworks discloses determinate, objective features of artworks—there to be discerned by perceptive viewers and/or listeners—or whether interpretation somehow projects subjective constructs onto artworks. An ongoing inquiry in aesthetic theory involves the nature of the contrast—and, more broadly, the determinants of correct interpretation (if such there be) and the locus and nature of artistic meaning. Pragmatism offers an illuminating perspective on these inquiries—and, in its more radical moments, seeks to dismiss them as riddled with illegitimate assumptions.

Pragmatism is not domain specific: its methods, perspectives, and doctrines are applicable within any philosophical inquiry, regardless of topic. The general goal here is to explore the impact of pragmatist doctrines and methods upon issues arising in philosophical reflections upon the arts—more specifically, upon the ontology of aesthetic properties and the nature of interpretive content. The inquiry ramifies to concerns with the legitimate place (if any) of metaphysics in aesthetic theory. These issues bear no special relation to pragmatism: each could be explored from a variety of perspectives. But focus upon these areas highlights the relevance of pragmatism to inquiries into description, evaluation, and interpretation in the arts and also highlights controversial aspects of the pragmatist agenda.

Pragmatism and Aesthetics

"Pragmatism" is said in many ways. Within philosophy, the term designates themes associated with Dewey and Peirce, visible in Heidegger and the later Wittgenstein, and moving through Sellars, Quine, Davidson, Rorty, Brandom, and others. Such themes include the following:

1. doubts about the explanatory role of semantic notions (*meaning, reference, truth*);
2. deflationist theories of truth and reference; rejection of truth-as-correspondence-to-reality; minimalism about semantic discourse; the relation between truth and warranted assertibility;
3. the significance of justificatory holism (Rorty 1979);
4. doubts about the viability (and utility) of the bifurcation thesis: viz., the (alleged) distinction between declarative sentences that describe the world and those that, though meaningful, manifest non-cognitive attitudes or express commitments (Kraut 1990);
5. Peirce's Pragmatic Maxim: "Consider what effects, that might conceivably have practical bearings, we conceive the object of our conception to have. Then, our conception of those effects is the whole of our conception of the object" (Peirce 1902);
6. (derivative from 5): the idea that conceptual content is best understood in terms of human practices and activities: differences in meaning are constituted by differences in how people behave; ("Meaning ... is not a psychic existence; it is primarily a property of behavior" [Dewey 1925: 179]);
7. the philosophical value of a non-circular "genealogy or anthropology ... about how [a given] mode of talking and thinking might come about, given in terms of the functions it serves" (Blackburn 2009: 2).
8. (derivative from 7): anti-representationalism: a shift from concern with the representational properties of a discourse to concern with the role(s) played by the discourse within the larger sphere of human goals and practices;
9. (derivative from 8): shift of emphasis from denotational semantics to speech-act theory and linguistic pragmatics;
10. (derivative from 9): shift in emphasis from "What is truth?" to "What role is played by the truth predicate in natural language?"; and from "What is knowledge?" to "What role is played by epistemic attributions in natural language?"; and from "What is morality?" to "What co-ordinating/regulatory role is played in human affairs by moral discourse and thought?"; and parallel shifts in other domains. In each such case, the representational function of the problematized language/concepts is downplayed;
11. frequent deployment (with approval and occasional obscurity) of the phrase "primacy of practice";
12. rejection of the idea that normativities sustained within social practice depend upon metaphysical facts; skepticism about grounding institutional practices in ontological facts;
13. rejection (and frequent derision/ridicule) of metaphysics;
14. suggestion that the notion of *objectivity* is defective and should be replaced with the notion of *social solidarity* (Rorty 1989).

Call this the *canonical list* (hereafter CL). Some entries are variants of others; some are consequences of others; some are inconsistent with others. Caution is required. If the pragmatist agenda is formulated too broadly—e.g., in terms of "the primacy of practice"—then virtually *any* adequate aesthetic theorizing falls under the pragmatist rubric: that is because any such theory worthy of acceptance must accommodate actual goings-on among consumers and critics of art.

Yet the history of aesthetics offers theories that diverge from pragmatist methodology even broadly construed: philosophical accounts of art that ignore actual practices upheld within the artworld. The sociocultural realities fall by the wayside. Clive Bell, e.g., offers a "formalist" aesthetic theory built upon the assumptions that there exists "[a] peculiar emotion provoked by works of art" and that art "transports us from the world of man's activity to a world of aesthetic exaltation" (Bell 1913). Both are doubtful. And his "isolationist" conception of art and aesthetic experience, viz.

> to appreciate a work of art we need bring with us nothing from life, no knowledge of its ideas and affairs, no familiarity with its emotions ... For a moment we are shut off from human interests; our anticipations and memories are arrested; we are lifted above the stream of life.
>
> (Bell 1913: 27)

is out of phase with artworld realities. Most people regard artworks as essentially embedded in cultures and histories; were Bell to have taken account of actual artworld practice, his theory would have taken a different shape.

Other notable disconnects of theory from practice come to mind. Leo Tolstoy's "infection" theory of art—which insists upon the transmission of sincerely experienced feelings and the lowering of ego boundaries between artist and audience—disenfranchises an enormous number of generally accepted artists and artworks of Tolstoy's time; his theory finds insufficient ground in actual artworld practice (Tolstoy 1896). Likewise, R.G. Collingwood offers a theory of art in terms of emotional expression, where this latter concept is understood by Collingwood as emancipation from the helpless state of not knowing one's own mind (Collingwood 1938: 109). Collingwood ties genuine art ("art proper") to this self-clarificatory process: objects/events that fail to conform to this model fall by the wayside as either non-art or "counterfeit art." But privileging this self-clarificatory enterprise as an essential condition of artistic production results in a theory that disenfranchises much of what routinely qualifies as art: Collingwood's aesthetic theory, like those of Tolstoy and Bell, loses touch with actual artworld practice.

It is perhaps unfair to construe these theorists as seeking (and failing) to accommodate and explain actual practices of their time; perhaps their goal was not descriptive but normative: aiming to legitimize (or impugn) those practices in light of metaphysical, epistemic, or semantic theories. Bell, for example, deployed his theory of art as a platform from which to condemn critics' preoccupation with realistic representation: offering recommendations about how they should proceed and aspects of artworks to which they should attend. Similarly, Tolstoy offered a theory that served to delegitimize much of the putative art celebrated by his aristocratic contemporaries as elitist non-art. It appears that Bell and other theorists offered normative recommendations about proper evaluation, interpretation, and/or experience art, given the (alleged) metaphysical facts about what artworks are.

But here the pragmatist demurs: for the very idea of assessing the legitimacy of a practice relative to standards grounded in metaphysical, epistemic, or semantic theories is deemed problematic. If there is tension between theory and practice, it is theory—not practice—that demands revision. The "primacy of practice" insisted upon by the pragmatist consists not in treating social behavior as beyond reproach—massive errors are possible—but rather in its serving as the ultimate tribunal to which philosophical theories must answer. Put another way: there is a class of phenomena which constitute the data for an aesthetic theory to explain; the pragmatist focuses, first and foremost, upon institutional phenomena.

Not all aesthetic theories run afoul of pragmatist scruples. George Dickie's "institutional" theory of art is paradigmatically consistent with pragmatist methods (Dickie 1974, 1983, 1984). Eschewing traditional efforts to define "art" in terms of imitation, representation, emotional expression, formal composition, or other features traditionally cited as essential characteristics of art, Dickie foregrounds a social-institutional framework—dubbed "the artworld"—and characterizes art and related aesthetic notions in terms of practices sustained within that framework:

> A work of art in the classificatory sense is (1) an artifact, (2) a set of the aspects of which has had conferred upon it the status of candidate for appreciation by some person or persons acting on behalf of a certain social institution (the artworld).
>
> (Dickie 1974: 34)

Dickie nowhere avows explicit commitment to pragmatism; nevertheless, his strategy comports with pragmatist sentiments, insofar as social practices are regarded as essentially constitutive of the contrast between art and non-art. If art does indeed have an essence, it is found—according to Dickie—in the position of an object or event relative to a normatively constrained, informally characterized institutional framework: the artworld.

Dickie's theory is not beyond reproach. Despite his insistence that the artworld is a loosely organized, "informal" structure, it is not clear that such a structure actually exists in sufficiently robust form to do the work he enlists it to do. Here is an especially critical assessment by Richard Wollheim:

> Does the art-world really nominate representatives? If it does, when, where and how do these nominations take place? Do the representatives, if they exist, pass in review all candidates for the status of art, and do they then, while conferring this status on some, deny it to others? What record is kept of these conferrals, and is the status itself subject to revision? If so, at what intervals, how, and by whom? And, last but not least, Is there really such a thing as the art-world, with the coherence of a social group, capable of having representatives, who are in turn capable of carrying out acts that society is bound to endorse?
>
> (Wollheim 1987: 15)

Other writers voice similar misgivings about the existence of institutional structures and patterns of representative authorization sufficiently robust to do the work required by Dickie's theory.[1]

Whatever the prospects for Dickie's theory, it is a pragmatist-friendly story about the property specified by the predicate "x is a work of art". Other aesthetic properties involve additional complexities.

Art and Metaphysical Questions

There could be a dispute—among metaphysicians and ontologists of art—about the nature of aesthetic properties: the coherence of a symphony, the mystery of a painting, the energetic tension of an improvised jazz solo, the tranquility of a cinematic moment, the funk of an R&B performance, the spiritual devotionalism of a poem.[2]

Perhaps such properties depend upon subjective matters of taste: projected onto the works rather than discovered in them. Questions arise as to whether there can be substantive disagreement and/or dispute about the presence of such properties, whether and why some consumers of art have more refined taste than others, whether assessment of aesthetic merit is based upon sentiment rather than rational judgment, whether there are facts of the matter concerning the melancholy of a Futurist painting, and whether Hume was right about the standards by which conflicting tastes can be assessed and some sentiments prioritized over others (Hume 1757).

Or perhaps aesthetic properties are objective: discernible by viewers and listeners equipped with adequate perceptual acuity and selective sensitivity. Various possibilities fall into this "realist" camp:

1 Aesthetic properties are response-dependent—analogous to Lockean secondary properties—constituted by dispositions to induce specific responses within a certain class of observers. Questions arise as to which responses—among which people—are relevant, whether some responses are more appropriate than others (and, if so, the grounds of such appropriateness), and so on. As with any account of dispositional properties, complexities abound.

2 Aesthetic properties closely resemble semantic properties of linguistic items: heavily dependent upon context and causal etiology, but real properties—there to be discovered. Questions

arise about their supervenience upon more fundamental structural properties, their dependency upon broader cultural phenomena, whether such properties are grounded in artists' intentions, the relationship between language comprehension and artistic understanding, and so on. Questions also arise within the aesthetic domain that parallel those encountered in semantic theory: concerning indeterminacy of translation, referential inscrutability, rule-following considerations, and the contrast between genuine and merely verbal disagreement.

A metaphysics of aesthetic properties is surely relevant to understanding artworld practice. It is, in part, because of what pictorial horror *is* that certain responses to Francis Bacon's paintings are appropriate; it is, in part, because of what doom and despair *are* that Rothko's paintings express them (or not). Analogies in other areas are plentiful: it is because of what the natural numbers *are* that the Dedekind/Peano axioms are the appropriate theory for arithmetic, it is because of what persons *are* that they should be treated as ends rather than means, and so on. Practice is normatively constrained by ontology. And so it goes in the artworld, where correct descriptive, interpretive, and evaluative practices mirror the ontology of aesthetic properties. Metaphysics has both explanatory and justificatory purchase.

That's one view. Another is that metaphysics is an illegitimate enterprise. Humean condemnations of metaphysics as "sophistry and illusion," positivist rejections as "literal nonsense," and postmodern rhetoric about contingency and social construction, all conspire to suggest that the practice of metaphysics is somehow defective.

Pragmatism tends toward rejection of metaphysics—at least, certain strains of it. Thus:

> It was in the earliest seventies that a knot of us young men in Old Cambridge, calling ourselves, half ironically, half defiantly, "The Metaphysical Club"—for agnosticism was then riding its high horse, and was frowning superbly upon all metaphysics—used to meet, sometimes in my study, sometimes in that of William James.
>
> *(Peirce 1904: 191)*

Here is Cheryl Misak's gloss on Peirce:

> Metaphysics, "in its present condition" [1904] is a "puny, rickety, and scrofulous science," but it need not remain so. The pragmatic maxim [see CL5] will sweep "all metaphysical rubbish out of one's house. Each abstraction is either pronounced gibberish or is provided with a plain, practical definition".
>
> *(Misak 2013: 43)*

Pragmatist rejection of metaphysics continues to the present. Huw Price, a strong contemporary voice of pragmatism, rails against " metaphysics, or at least a distinctively misguided and self-inflicted kind of metaphysics, to which philosophy has long been subject" (Price 2011: 312, and see Price and MacArthur 2007). Simon Blackburn echoes the sentiment:

> [Huw Price and David Macarthur] say that for the pragmatist the crucial thing is not to answer questions about the function of language in ways that encourage metaphysics. On this I am entirely at one with them.
>
> *(Blackburn 2007: 2)*

It takes considerable work to articulate the precise basis for the rejection and to determine whether it is justified—even by pragmatist standards. One strain of opposition involves the idea that ontology is a shadow cast by social practice—not a causal determinant or a normative ground of the practice—and thus cannot, on pain of circularity, do the explanatory or justificatory work it is enlisted to do (Kraut 2012). Another strain, closely associated with Richard Rorty and inspired by Wilfrid Sellars's attack on the "Myth of the Given" (Sellars 1997), challenges the very notion of *appropriate response to the world*. Although the world exerts causal force, it does not exert normative force; thus, any effort to ground the correctness of response to an artwork in the metaphysical nature of its properties must be abandoned. Norms sustained within a practice do not have metaphysical foundations; the grounds of normativity lie elsewhere.

Note that the target of the rejection of metaphysics is not the ontological respectability of certain entities; it is, rather, what Brandom calls "the metaphysical strategy for grounding normative appraisals of different forms of life" (Brandom 2009: 134). Pragmatism opposes the idea that legitimacy of institutional practices consists in their conformity to requirements issuing from objects, properties, and/or states of affairs that demand compliance. It is people, not objects, who make demands and whose patterns of behavior constitute the ultimate grounds of normativity. As Michael Williams puts it, "Pragmatism is 'anti-metaphysical' in its hostility to postulating supernatural entities to guide human practices" (2010: fn.4). The key word here is "guide": the pragmatist alleges that metaphysical entities fail to confer normative status upon our practical interests or activities. But aesthetic properties, whatever their nature, are not "supernatural entities." Care is thus required to determine which sorts of metaphysics fall under the pragmatist's condemnations.

Despite the misgivings, metaphysical explanations are a familiar trapping of human attempts to understand the world and our place in it—including participation in the artworld. Perhaps such explanations are misguided, but perhaps, when properly understood, they are not.

Objectivity and Artworld Practice

One possible conception of aesthetic properties is summarized by Jerrold Levinson:

> [Alan] Goldman's central idea is that aesthetic attributions are irreducibly evaluative, that they necessarily reflect the differing attitudes and sensibilities of critics, and thus that any attempt to construe them as signifying aesthetic properties, even if construed dispositionally and indexed to standard conditions and qualified perceivers, is illegitimate.
>
> *(1994: 351)*

Call this the "subjectivist" view—here invoked to capture, in coarse grain, the side of Steinberg's musings whereby experienced aesthetic properties are read into Johns's artworks rather than found there. Such subjectivism is susceptible to countless variations.

The alternative conception—intended to capture the "found" side of Steinberg's musings—is broadly realist: aesthetic properties are determinate realities of artworks, there to be right or wrong about, there to be targets of disagreement or consensus, there to be discovered (or overlooked) by audiences. Thus construed, there are facts of the matter as to whether Ornette Coleman's solos are forceful, eager, and spacious, whether Edvard Munch's *Starry Night* (1893) has a mysterious and poetic strength, whether Duchamp's *Nude Descending a Staircase* presents shattered picture planes and neo-Dada humor, whether T.S. Eliot's *Four Quartets* are meditations on time, redemption, and eternity, and so on. As with the subjectivist view, a realist conception of aesthetic properties is susceptible to countless variations.

Thus, we have a venerable metaphysical contrast: between subjective and realist conceptions of aesthetic properties. They cannot both be right, any more than J.J. Thomson's "plumb pudding" model of the atom and Rutherford's "planetary" model can both be right.

It follows from Peirce's Pragmatic Maxim (CL5) that the contrast between subjectivist and realist pictures of aesthetic properties must, if meaningful, bear upon what people do in connection with artworld activities: description, evaluation, interpretation, and/or commodification of artworks. But it is not clear how the allegedly contrasting metaphysical theories hook onto practice or how the contrast does any work at all. A difference, to be a difference, must make a difference. But what kind of difference?

It need not be a difference in sensory-perceptual experience. When Carnap despairs of the contentfulness of specific metaphysical contrasts, he is almost certainly moved by an unduly restrictive empiricist epistemology:

> I cannot think of any possible evidence that would be regarded as relevant by both philosophers [viz. the realist and the nominalist concerning the existence of numbers], and therefore, if actually found, would decide the controversy or at least make one of the opposite theses more probably than the other.
>
> *(Carnap 1950: 219)*

But—as Quine often stressed—a wider notion of evidence is required. Pragmatism is not verificationism. The contrast between subjectivist and realist conceptions of aesthetic properties involves differences not in experiential evidence but in explanatory power.

Stephanie Ross identifies the relevant explananda:

> Aesthetic disputes are commonplace; arguing about art is a major source of entertainment and edification. Moreover there is a clear justificatory structure in place for such exchanges. Yet there seems to be a problem about the objectivity of aesthetic claims. ... Absent laws of taste, how can we secure the objectivity of aesthetic claims? The worry pertains not only to summary verdicts and evaluations, but also to the assignment of any and all aesthetic properties to works of art. ... Is there any way to secure at least some degree of aesthetic realism?
>
> *(Ross 2014: 590)*

Ross identifies two phenomena worthy of explanation: the presence of aesthetic disputes—and thus *norms of dispute* within aesthetic discourse—and the *objectivity* of aesthetic claims. Her description of artworld practice is not unassailable: the alleged disputes might not be real disputes—only conflicts of attitude that appear to be disputes; moreover, seeking to secure "the objectivity of aesthetic claims" is ill-advised if, in fact, no such objectivity obtains.

The plausibility of Ross's descriptions is complicated by the fact that identifying genuine disagreement (or dispute) is difficult; nor is it obvious what objectivity amounts to. If aesthetic judgments are matters of taste, then prima facie conflicting aesthetic claims might not constitute disagreement. Thus:

> if something were true only for him who held it to be true, there would be no contradiction between the opinions of different people. So to be consistent, any person holding this view would have no right whatever to contradict the opposite view, he would have to espouse the principle: *non disputandum est* ... even if his utterances had the form of assertions, they would only have the status of interjections—of expressions of mental states or processes, between which and such states or process in another person there could be no contradiction.
>
> *(Frege 1979: 233)*

John MacFarlane notes that "The concept of disagreement has been the crux of recent debates between contextualists and relativists about epistemic modals, simple evaluative predicates like 'tasty', and terms of epistemic assessment like 'knows'" (MacFarlane 2009: 1). Upshot: whether Siskel and Ebert are engaged in genuine disagreement—rather than a clash of sentiments easily mistaken for disagreement—is a complex conceptual matter. And there are, as MacFarlane notes, varieties of disagreement (MacFarlane 2014).

That's one problem. Another is that the concept of *objectivity* is difficult to pin down. Frequently, it signals aspects of a mind-independent world that "push back" on our conventions and institutional practices; issues emerge about whether such pushback is purely causal, or rather a normative, authoritative force to which we are answerable. Pragmatists might wish to switch the question from "What is objectivity?" to the expressivist question "What are we *doing* when we attribute objectivity?". And here there is room to maneuver: perhaps talk of objectivity signals the legitimacy of giving and asking for reasons and justifications as part of an ongoing effort to achieve consensus in the face of diverse perspectives. Or perhaps such talk plays a quite different role. At the extreme, one might question—as Rorty is alleged to have done—the very legitimacy of the concept of objectivity and/or the utility of its continued deployment (Rorty 1989). Perhaps we do better with a notion of social solidarity. But perhaps not.

Thus, a methodological quandary: the putative aspects of artworld practice identified by Ross might or might not be real. Nevertheless, there is *apparent* dispute and disagreement, and *apparent* objectivity. For the present we take Ross's description of artworld practice at face value and ask what reality must be like for these manifest features of the artworld to be possible; we look to the metaphysical nature of aesthetic properties in answering the question.

Artworld participants might regard Munch's *The Scream* (1893) as representing an existential crisis. Some may resist. There will be vigorous (ostensible) dispute: reasons and justifications will draw upon Munch's biography, sociocultural environment, historical context, the iconography of his artistic peers and ancestors, and other factors possibly deemed relevant. If this be disagreement, it is not "faultless" (an alleged feature of much evaluative discourse): there are aesthetic facts of the matter to be right or wrong about. That is the view from within artworld practice: a thoroughly realist view. A pragmatist committed to the primacy of practice deems the view worthy of explanation rather than dismissal: it is, after all, part of the data to be explained. Criticizing it—perhaps advocating its revision or elimination—would be as misguided as a jungle linguist correcting her native informants' verbal behavior.

Note that Ross's sense of dialectical perplexity results from preoccupation with the general framework set out by Hume, wherein attributions of aesthetic properties are thought to depend upon matters of taste, and aesthetic predicates are regarded as having an evaluative as well as a descriptive dimension. That framework can be avoided. An alternative framework—which provides immediate answer to Ross's query—embraces the idea that the foreboding melancholy and existential solitude of Alberto Giacometti's sculptures, e.g., are no less real than the spin of a microparticle or the height of a person; the properties are there to be discovered and there to be disputed. Aesthetic realism restored. All that's required is the correct metaphysics of aesthetic properties.

Perhaps. But pragmatist insistence that non-social reality contains no guiding norms fosters skepticism about whether the alleged explanation in terms of the metaphysics of aesthetic properties is a real explanation. According to the pragmatist, no genuine explanatory progress has been achieved: only a redescription—in lofty metaphysical terms—of the institutional phenomena to be explained. The prima facie relevance of the metaphysics of aesthetic properties to explanations of artworld practice thus hangs in the balance.

Pragmatism as a Method

In a discussion of objectivity in mathematics, Penelope Maddy highlights the phenomenology of mathematical practice—the way it *feels* to do mathematics:

> One of the many things about the practice of mathematics that makes the philosophy of mathematics so difficult ... arises from the pure phenomenology of the practice, from what

> it feels like to do mathematics. Anything from solving a homework problem to proving a new theorem involves the immediate recognition that this is not an undertaking in which anything goes, in which we may freely follow our personal or collective whims; it is, rather, an objective undertaking par excellence. Part of the explanation for this objectivity lies in the inexorability of the various logical connections, but that can't be the whole story.
>
> (Maddy 2010: 1)

There is something it feels like to participate in mathematical practice—that's the data; such phenomenology is susceptible to explanation. The theme is less familiar in the philosophy of art but equally applicable: there is a phenomenology of engagement with artworks and artistic events, something it's like to connect with paintings, sculptures, dance recitals, and music performances. One feels that one can *understand* an artwork—or fail to understand it—and that attributions of aesthetic properties are constrained by the possibility of getting things right or wrong.

Not all consumers of art are prone to such experiences of objectivity: it depends upon background and levels of artworld sophistication (no easy task to explicate the requisite notion of "sophistication"). But argumentation is misplaced here: the phenomenology of objectivity is sufficiently prevalent among artworld participants to motivate present discussion. Grant that experience and criticism of art are, as in mathematics, "objective undertakings"; how is such objectivity best explained?

Options abound. In the mathematical domain a familiar pattern of explanation is realism about ontology: numbers, sets, and functions are real entities with real properties about which one can be correct or mistaken; there is, as Maddy puts it, "a world of abstracta that we're out to describe." An alternative approach, which downplays the explanatory power of ontology, insists that

> our mathematical activities are constrained not by an objective reality of mathematical objects, but by the objective truth or falsity of mathematical claims, which traces in turn to something other than an abstract ontology.
>
> (ibid.: 4)

The present task—with strong echoes in the philosophy of mathematics—is to explain the appearance of objectivity in the artworld. One strategy—the correlate of Platonism in mathematics—treats aesthetic properties as determinate realities: present in artworks, there to be "found" rather than "read into" the works, thereby providing grounds for the sense of objectivity in engagement, description, interpretation, and criticism. An alternative explanatory strategy—favored by the pragmatist—seeks grounds of objectivity elsewhere: viz. in the structure of conceptual and discursive artworld practices. This latter strategy dictates that norms of correctness sustained within the artworld are shaped by physiology, psychology, history, economics, and countless other factors and that those norms and their institutional foundations—rather than an ontology of aesthetic properties—are to be foregrounded in the explanatory endeavor. The ontology of aesthetic properties is a mere shadow cast by these institutional and normative forces.

So says the pragmatist, thereby reversing the more familiar order of explanation: ontology, rather than providing the grounds of normativity, consists of reifications of that normativity and thus cannot—on pain of circularity—be invoked to explain it. Norms enjoy explanatory priority over ontology (Kraut 2010). Herein lies the essence of "the primacy of practice."

In current philosophical climate, it is unwise to tie norms of correctness to a realist ontology. Pragmatists of various stripes—often focusing upon "expressivist" and "nondescriptivist" semantic explanations of a given region of discourse—claim that the concepts of *truth*, *correctness*, and *objectivity* can be given a foothold without a grounding in realism. Thus, e.g., Simon Blackburn's "quasi-realist" program purports to explain and accommodate the prima facie realistic trappings of a discourse without any concession to interpreting it in realist, representational terms (Blackburn 1993, *passim*). Norms of correctness do not require a realist ontology.

The dialectic is now clear. Recall critic Leo Steinberg's pondering whether his interpretations of Jasper Johns's work consist in finding aesthetic features in the work or reading them in. Call the "finding" option *aesthetic realism*: a position which, to paraphrase Gideon Rosen's description of moral realism, might be identified thus:

> [the insistence] that true aesthetic statements *describe a genuine domain of aesthetic fact*; that aesthetic knowledge *represents* what is *already* the case; that aesthetic inquiry aims to *discover* the aesthetic properties and relations of things and in no way serves to *construct* or *create* them; that aesthetic truth is altogether *independent* of us; etc.
>
> (Rosen 1998: 386)

If this be the view from within artworld practice, the pragmatist claims to accommodate and explain it without recourse to an "ontology of aesthetic properties" (no wonder doubts about "the ontology of art" occasionally surface [Kraut 2012]). Whether the pragmatist's preferred order and method of explanation are advisable, superior to alternatives, and even possible involves profoundly complex philosophical issues in metaphysics, semantic theory, and theories of normativity. Pragmatism is not a theory; it is a set of directives for construction of theories. The pragmatist insists that institutional practices—artworld and otherwise—are not "guided" by determinate ontological realities: normative constraints are instituted by communities, not forced upon them by ontological realities that demand compliance. There are no non-social realities that guide us. Therein lies the heart of pragmatism.

Pragmatism and Artworld Practice

In his 1976 William James Lecture, Michael Dummett raises questions often avoided in polite philosophical conversation:

> We also face another and greater difficulty: to comprehend the content of the metaphysical doctrine. What does it mean to say that natural numbers are mental constructions, or that they are independently existing immutable and immaterial objects? What does it mean to ask whether or not past or future events are *there*? What does it mean to say, or deny, that material objects are logical constructions out of sense-data? In each case we are presented with alternative *pictures*. The need to choose between these pictures seems very compelling; but the non-pictorial content of the pictures is unclear.
>
> (Dummett 1991: 10)

We wish to know whether the aesthetic features attributed by Steinberg to Jasper Johns's pictures are there to be found. Dummett's remarks prompt the question: "What would it *be* for the features to be there? What precisely is the contrast between the *found* and the *read into*?" The contrast between finding and making is often reasonably clear, but humility is required in admitting that in the context of realist versus subjectivist pictures of aesthetic properties, we might not know what we are talking about. What role is played by the contrast between the found and the projected? What is Steinberg *doing* in wondering whether the problematized features of Johns's artworks are found or read into them?

College-level introductory courses in the Philosophy of Art provide suggestive clues. A split emerges, early in the semester, between those students who regard artworks as things about which one can say virtually anything (phrases such as "eye/ear of the beholder" and "matter of taste" quickly emerge) and those who think otherwise. Members of the latter group—frequently majoring in Art History, Art Education, Music, Dance, or Graphic Arts—regard artworks as things about which one can engage in dispute, things which can be experienced appropriately (or not), things which—with proper training and sensitivity to the medium and genre—can be properly understood. They regard students in the former group as artistically illiterate philistines: analogous to those who cannot grasp the meaning of an inscription and deny that there is anything there to grasp. The instructor's charge in such courses is, inter alia, to facilitate realization that artworks, like texts, jokes, or persons, can be understood or not: there is something it's like to get them right.

The instructor displays a slide of Mark Rothko's *Untitled, 1960*, asking students to engage the artwork and provide description, interpretation, and evaluation. Some regard the assignment as daunting; others treat it as a playful exercise no more demanding than that of discerning patterns in cloud formations. One student claims to find tragedy, ecstasy, and doom in the artwork; others discern no such features, denying that anything lurks there other than somber, dark hues and mesmerizing patches of color.

The inevitable dialectic ensues: attributions of tragedy, ecstasy, and doom are susceptible to challenge; the claimant might undertake the task of providing reasons and justifications for her attributions. Her justificatory strategies might draw upon aspects of Rothko's biography—including his suicide in 1970—the history and evolution of his work, the general trappings of abstract expressionism, and/or other artworld realities deemed relevant. But it is clear that she has—as Wilfrid Sellars would have put it—taken up a position in the space of reasons (Sellars 1997): she has undertaken responsibility for justifying her interpretive attributions. It is precisely her doing so which is expressed in her sense that the problematized aesthetic features are found, rather than read in.

But are those features really *there* rather than projected by the student? The pragmatist invites us to ask what it would *be* for those features to be really there and has transformed the problem into one of identifying the functional role of the very contrast between the found and the projected. It is that contrast—between those claims for which one undertakes responsibility to provide reasons and justifications and those for which one does not—expressed in the metaphysical contrast between the found and the made.

Thus, the question about artworld ontology turns into questions about artwork practice, discursive norms, and dispositions to respond to dialectical challenges. These are the favored ingredients of pragmatist theories: institutional transactions, normativities, and communal dispositions. Whether Leo Steinberg found the problematized aesthetic features in Jasper Johns's artworks or read them into the works is a matter of dispositions and commitments to engaging with one posture rather than another as a participant in art-critical practice. His musings manifest ambivalence about whether to occupy a position in the space of reasons or whether to plead "subjectivity" and "matters of taste" and thus immunity to challenges.

The dialectical position Steinberg chooses to occupy is neither dictated nor legitimized by the ontology of aesthetic properties: that picture would, according to the pragmatist, embrace the wrong order of explanation and/or a misguided notion of the sources of normativity. Steinberg's posture regarding his interpretive claims about Johns's work is not to be assessed by conformity to rules emanating from the ontological realities of aesthetic properties: there are no such rules. It is to be assessed, rather, by its capacity to facilitate productive and valuable communal practice (e.g., ongoing critical conversation about artistic works and performances).

The ontology of aesthetic properties is a shadow cast by artworld practice rather than a source of constraints upon that practice. Rather than explaining and/or justifying, ontology codifies and reifies possible positions within the space of institutional norms. This, at any rate, is the picture favored by the pragmatist. Whether the picture is superior to the alternatives—in the artworld or elsewhere—remains the essential problematic of pragmatism.

Problems remain:

1. If metaphysics—more specifically, ontology—serves to codify institutional realities, rather than explain and/or legitimize them, it is unclear that customary pragmatist denunciation of metaphysics is warranted.
   This encourages the question of genealogy: how and why the modes of thinking characteristic of "the ontology of art" might come about, in terms of the functions they serve, and why various aspects of discursive engagement come to be reified in metaphysical constructions. More must be said about what we are *doing* when practicing ontology. Should pragmatism encourage engineering in our metaphysical concepts—e.g., replacement of the concept of *objectivity* with that of *social solidarity*—the costs and benefits of such revision must be assessed.

2. If the sense of objectivity admits of degrees, it is likely weaker in the context of experience and interpretation of artworks than in the context of mathematical practice. Thus, the proffered analogy is likely to be controversial. There is, moreover, the additional challenge of distinguishing objectivity from intersubjectivity and the attendant phenomenological contrasts. Engaged artworld participants see themselves as answerable not to their neighbors—or a sophisticated subgroup thereof—but to the artworks and performances themselves. The current discussion seeks to address that phenomenology while admitting its susceptibility to challenge.

These are difficult problems. It is no mark against pragmatism that it fails to provide solutions. It is, rather, a mark in its favor that it foregrounds the problems and provides some indication of how they are best approached.[3]

Notes

1. See, e.g., Khatchadourian (1979). Dickie's theory evolves in response to such objections; later versions abandon the notions of *status conferral* and *acting on behalf of* as "too formal." See Dickie (1983).
2. Delineating the class of *aesthetic properties* is no easy task; see Sibley (1959, 1965, 1968). Evaluative properties—e.g., *beauty*—are not among those considered in the present discussion.
3. I am grateful to Abe Wang, Ali Aenehzodaee, Cheryl Misak, Stewart Shapiro, Tristram McPherson, Penelope Maddy, and William Lycan for helpful suggestions and critical discussion.

References

Bell, Clive. 1913, 1958. *Art*. New York: Capricorn Books.
Blackburn, Simon. 1993. *Essays in Quasi-Realism*. New York: Oxford University Press.
Blackburn, Simon. 2007. "Pragmatism: All or Some?," lecture presented at a conference on *Expressivism, Pragmatism, and Realism*; Sydney, August 2007.
Blackburn, Simon. 2009. "The Steps from Doing to Saying," *Proceedings of the Aristotelian Society*, 110. 1–13.
Brandom, Robert. 2009. *Reason in Philosophy: Animating Ideas*. Cambridge, MA: Harvard University Press.
Carnap, Rudolf. 1950. "Empiricism, Semantic and Ontology," in *Meaning and Necessity*. Chicago, IL: University of Chicago Press: 205–221.
Collingwood, Robin George. 1938. *The Principles of Art*. London: Oxford University Press.
Dewey, John. 1925. *Experience and Nature*. La Salle, IL: Open Court.
Dickie, George. 1974. "What Is Art? An Institutional Analysis," in *Art and the Aesthetic: An Institutional Analysis*. Ithaca, NY: Cornell University Press.
Dickie, George. 1983. "The New Institutional Theory of Art," *Proceedings of the 8th Wittgenstein Symposium*, Vol. 10: 57–64.
Dickie, George. 1984. *The Art Circle: A Theory of Art*. New York: Haven Press.

Dummett, Michael. 1991. *The Logical Basis of Metaphysics: The William James Lectures 1976*. Cambridge, MA: Harvard University Press.
Frege, Gottlob. 1979. *Posthumous Writings*. Chicago, IL: University of Chicago Press.
Hume, David. 1757. "Of the Standard of Taste," in *The Philosophical Works of David Hume*, edited by T. H. Green and T. H. Grose. Volume 3. London: Longman, Green. 248–275.
Khatchadourian, Haig. 1979. "Review of George Dickie, Art and the Aesthetic, an Institutional Analysis," *Nous*, Vol. 13: 113–117.
Kraut, Robert. 1990. "Varieties of Pragmatism," *Mind*, Vol. 99, No. 394: 157–183.
Kraut, Robert. 2010. "Universals, Metaphysical Explanations, and Pragmatism," *Journal of Philosophy*, Vol. CVII: 590–609.
Kraut, Robert. 2012. "Ontology: Music and Art," *The Monist*, Vol. 95, No. 4: 684–710.
Levinson, Jerrold. 1994. "Being Realistic about Aesthetic Properties," *The Journal of Aesthetics and Art Criticism*, Vol. 52, No. 3: 351–354.
MacFarlane, John. 2009. "Varieties of Disagreement." Unpublished text of presentation at University College Dublin and the Institute of Philosophy in London, May 2009.
MacFarlane, John. 2014. *Assessment Sensitivity: Relative Truth and Its Applications*. Oxford: Clarendon Press.
Maddy, Penelope. 2010. "Objectivity in mathematics," unpublished ms. Sixth Annual Thomas and Yvonne Williams Lecture for the Advancement of Logic and Philosophy, University of Pennsylvania.
Misak, Cheryl. 2013. *The American Pragmatists*. Oxford: Oxford University Press.
Peirce, Charles Sanders. 1878. "How to Make Our Ideas Clear," *Popular Science Monthly* 12: 286–302.
Price, Huw. 2011. "One Cheer for Representationalism?," in *Naturalism Without Mirrors*. New York: Oxford University Press.
Price, Huw and Macarthur, David. 2007. "Pragmatism, Quasi-Realism, and the Global Challenge," in *New Pragmatists*, edited by Cheryl Misak. New York: Oxford University Press: 91–121.
Rorty, Richard. 1979. *Philosophy and the Mirror of Nature*. Princeton, NJ: Princeton University Press.
Rorty, Richard. 1989. "Solidarity or Objectivity?" in *Relativism: Interpretation and Confrontation*, edited by Michael Krausz. Notre Dame, IN: University of Notre Dame Press: 167–183.
Rorty, Richard. 1991. *Objectivity, Relativism, and Truth*. Cambridge: Cambridge University Press.
Rosen, Gideon. 1998. "Review of Blackburn's *Essays in Quasi-Realism*," *Nous*, Vol. 32 No. 3: 386–405.
Ross, Stephanie. 2014. "When Critics Disagree: Prospects for Realism in Aesthetics," *The Philosophical Quarterly*, Vol. 64, No. 257 (October 2014): 590–618.
Sellars, Wilfrid. 1997. *Empiricism and the Philosophy of Mind*. Cambridge, MA: Harvard University Press.
Sibley, Frank. 1959. "Aesthetic Concepts," *Philosophical Review*, Vol. 68: 421–450.
Sibley, Frank. 1965. "Aesthetic and Non-aesthetic," *Philosophical Review*, Vol. 74: 135–159.
Sibley, Frank. 1968. "Objectivity and Aesthetics," *Proceedings of the Aristotelian Society Supplemental Volume*, Vol. 42: 31–54.
Steinberg, Leo. 1972. "Contemporary Art and the Plight of Its Public," in *Other Criteria: Confrontations with Twentieth-Century Art*. New York: Oxford University Press: 3–16.
Tolstoy, Leo. 1896 and 1904. *What is Art?*. New York: Funk and Wagnalls Company.
Williams, Michael. 2010. "Pragmatism, Minimalism, Expressivism," *International Journal of Philosophical Studies*, Vol. 18, No. 3: 317–330.
Wollheim, Richard. 1987. *Painting as an Art*. London: Thames and Hudson Ltd.

34
PRAGMATISM AND POLITICAL PHILOSOPHY

Matthew Festenstein

Introduction

The political implications of pragmatism have been highly controversial but have settled into the thought that pragmatist ideas tend to provide a basis for a defence of democratic values. In this chapter, I want to tease out some of the debates underlying this plausible but relatively unexciting conclusion and to show how pragmatists come apart, as well as converge, on what it means.

Thinking about the relationship of pragmatism and political philosophy attracted three general worries from an early stage. As a theory of meaning and/or epistemic justification, distinguished from particular ethical and political commitments that particular pragmatists happen to espouse and perhaps captured by general pragmatic rules of the "judge any idea by its consequences" sort, pragmatism may seem to have "no inherent political valence", as Richard Posner put it (Posner 1990: 1658–1659). This suspicion may be supported by a variety of political positions that pragmatists have adopted. For example, of the canonical founders, C. S. Peirce and William James were both idiosyncratic in their politics. Peirce personally subscribed to the kind of reactionary conservatism for which reasoned inquiry was of little relevance to practical affairs. (This has not stopped later commentators finding in this work the raw materials for a democratic political philosophy, as we'll see.) James combined an elitist liberalism with some sympathy for anarchism and, notably, a powerful opposition to imperialism (Livingston 2016). John Dewey, alongside collaborators such as Jane Addams (1902), identified deep affinities between his pragmatism and what they called democracy as a way of life.

A second worry is that pragmatism supports an all-out assault on rationality. In the early part of the 20th century, pragmatism's rejection of fixed foundations for knowledge and reasoning was sometimes met with an anxious response that, without fixed epistemic and ethical standards, the chaos of relativism, scepticism, and nihilism beckoned. Indeed, sceptical commentators in the 1920s and 1930s observed that some fascists such as Giovanni Papini (and, less credibly, Mussolini himself) had had some intellectual links with James, in particular. The worry that pragmatism might unravel political values and practices has persisted but been overlaid with other interpretations. This is echoed in recent writing that identifies a slippery slope from Richard Rorty's so-called postmodernism to irrationalism and "post-truth" politics.

Third, a prominent early view of the politics of pragmatism appealed to the sense that pragmatism combines general dispositions towards flexibility, relativism about ultimate ends, ambivalence about theory, a practical orientation, and a trust in scientific solutions to social and political problems. In this vein, pragmatism was seen as basically a complacent philosophy, which rested on an unquestioned acceptance of the liberal values of the United States. This isn't an entirely unfair interpretation. Even Dewey and some later writers, such as Rorty in particular, at times expressed their political views both in terms of fulfilment of implicit values of liberalism and indeed of achieving patriotic aspirations (Rorty 1998). This view of his understanding of democracy fuelled a broader interpretation of pragmatist political philosophy as what critics described as an "acquiescent" fig-leaf for power politics (for an overview, see MacGilvray 2000).

Yet prominent elements in pragmatism squarely oppose the revolt against rationality and the idea that political goals are beyond reasoned criticism, alongside suggesting ways of building a bridge between pragmatism and political theory. For pragmatists, the absence of fixed epistemological foundations does not imply a rejection of rationality. Further, ethical and other forms of evaluative inquiry are generally thought to be continuous with empirical inquiry: we use our reflective intelligence to improve our judgement in both, testing our value judgements by seeing their results and revising them in the light of those results. As C. I. Lewis puts it,

> [p]ragmatism could be characterized by the doctrine that [...] there can be ultimately no valid distinction of theoretical and practical, so there can be no final separation of questions of truth [...] from questions of the justifiable ends of action.
>
> *(Lewis 1970: 108)*

While pragmatists reject a transcendental and ahistorical conception of truth, they are fallibilist rather than sceptical and committed to doing justice to the objective dimension of inquiry.

Two components of the canonical expressions of pragmatism are particularly important in thinking about the relationship between pragmatism and politics. The first is the Peircean point that objectivity in inquiry requires a conception of open and unlimited community of inquirers. The second is that in seeking truth, as James puts it,

> [s]he [that is, pragmatism] is willing ... to follow either logic or the senses and to count the humblest and most personal experiences. ... Her only test of probable truth is what works best in the way of leading us, what fits every part of life best and combines with the collectivity of experience's demands, nothing being omitted.
>
> *(James 1907: 44)*

For Dewey, in particular, the thought that the methods of scientific inquiry, understood in these pragmatist terms, should be extended to address social and moral problems, and that this extension requires the democratisation of decision-making in those domains, exerted a strong grip. For instance, he writes in his *Logic: The Theory of Inquiry*:

> Failure to institute a logic based inclusively and exclusively upon the operations of inquiry has enormous cultural consequences. It encourages obscurantism; it promotes acceptance of belief formed before methods of inquiry had reached their present estate; and it tends to relegate scientific (that is, competent) methods of inquiry to a specialist technical field. Since scientific methods simply exhibit free intelligence operating in the best manner possible at a given time, the cultural waste, confusion and distortion that results from failure to use these methods, in all fields in connection with all problems, is incalculable. These considerations reinforce the claim of logical theory, as the theory of inquiry, to assume and to hold a position of primary human importance.
>
> *(Dewey 1938: 527)*

As I'll try to bring out, this orientation provides a critical, rather than just complacent or "acquiescent", conception of political thought, although one that seeks to retain a grounding in what pragmatists call the primacy of practice (Putnam 1995). For pragmatists, this tends to provide materials for a justificatory account of democracy that feeds a critical evaluation of the actual workings of democratic societies. Finally, they try to do so in a way that is sensitive to the plurality of ethical and political beliefs and identities that characterise complex societies. These three foci – the primacy of practice, democratic values, and pluralism – structure what follows. Pragmatists are fissiparous, though, and I'll try to bring out the differences under each of these headings, in particular between Dewey, the most prominent pragmatist social and political philosopher, what I'll call neopragmatism, which takes its cue from Richard Rorty's provocative claims, and the new pragmatism, which takes its cue from philosophers such as Hilary Putnam and Cheryl Misak, who want to provide a clearer account of how the standards and practices of inquiry can be both historical and objective.

This should remind us that the topic of this chapter is rife with definitional diversity and contestation. While pragmatism is not a monolithic bloc, the character and scope of political philosophy and its central concepts are even more contentious and certainly contested by a much wider constituency. As a result, this chapter needs carefully situating alongside complementary chapters in this *Companion*: not only for a fuller sense of the variety of thought that has been gathered under the pragmatist umbrella but also because the political runs into the terrain of gender, race, and ethics, among other topics.

The primacy of practice

At the core of Deweyan pragmatism is a conception of beliefs and judgements as tools or instruments for resolving problematic situations. Inquiry is a problem-solving activity, engaged in by particular agents: agitated by some doubt, finding ourselves, in Dewey's terms, in an "indeterminate situation", we respond with inquiry in order to arrive at beliefs and policies of action that can assuage these doubts. Pragmatists also embrace the historical character of belief and value. In his ethical writings, Dewey elaborates a view of moral theory not as an antecedent constraint on action but rather as a repertoire of conceptual resources deposited by a particular history. Conflict among these approaches cannot be resolved in theory – only in practice, if at all, where an agent must make "the best adjustment he can among forces which are genuinely disparate" (Dewey 1930: 288). Moral theories such as deontological ethics and utilitarianism are seen as historical products, in part expressing features of the societies that produce and sustain them, on this view. We can only begin to reason and deliberate on the basis of the beliefs and practices that we have – we cannot call everything into question all at once. The pragmatist views beliefs both as rooted in history and as subject to rational scrutiny. The criteria for what counts as success or failure in inquiry are not pre-given and external to inquiry but are hammered out through it. Finally, recognition of the historical character of our conceptual resources does not mean we must cleave to tradition. Dewey's extensive writings on liberalism stress the historical sources of notions of the individual, rights, freedom, contract, and so on. In part, the point in each case is to explain how theories that emerge and are fitted for one particular social matrix fail to make sense in a different one and so need to be abandoned or reformed. Ensnared by redundant moral notions, we can fail to perceive and respond to the distinctive needs of the present.

In considering the goal of inquiry, in this sense, then, the pragmatist philosopher doesn't seek to play the role of a philosopher-king, although Dewey himself was a fecund source of political proposals, views, and campaigns that he provided as a citizen to other citizens. Instead, the pragmatist only tells us about what he calls the logic of inquiry: that is, this account of how we deliberate in identifying and addressing problematic situations. The language of "adjustment" as a goal provoked many readers to think of Dewey as basically conservative, but it is an important feature of this logic that in addressing problems, it is not only our habits and beliefs but also the environment that is unfixed and potentially plastic. The closest Dewey comes to identifying a summary term or criterion for a successful solution to a problem is to call it "growth", although this is a vague (and, in Dewey's use, non-moral) term – a source of irritation for readers seeking to work out the specific moral and political implications of this logic (Festenstein 2008).

New pragmatists agree with Dewey on the need to eschew the high road of prescribing to practice from ideal moral theory and that (in a sense) the logic of inquiry is the important strand in pragmatism for political philosophy, but they are clearer in seeking to draw out normative requirements from what they see as the doxastic commitments of actual inquirers in practice. The goal of acquiring and retaining well-grounded beliefs about how to deal with social and political problems entails a set of non-discretionary commitments on the part of anyone who wants their beliefs to be well-grounded. Drawing, in particular, on an interpretation of Peirce's conception of truth, Misak and Talisse argue that no belief is held to be a priori certain or beyond the reach of criticism and revision and that the search for a well-grounded belief involves testing claims against as wide a range of different experiences as possible. We should seek out and attend to different perspectives and arguments, in order to test and, if necessary, revise our current beliefs. Our beliefs and judgements aim at being true, and being true, on this pragmatist account, means fitting with reasons and experience. This apparently innocuous condition has critical bite, since genuine ethical inquiry

> requires that we take our beliefs to be responsive to new arguments and sensibilities about what is good, cruel, kind, oppressive, worthwhile, or just. Those who neglect or denigrate the experiences of others because of their gender, skin colour, or sexual orientation are adopting a very bad means for arriving at true or rational beliefs. They can be criticised for failing to aim at truth properly.
>
> (Misak 2000: 104)

Of course, that isn't all they should be criticised for, and it is unlikely to be the most important thing they can be criticised for, but the point is that pragmatism throws a spotlight on a distinctive epistemic failure on their part.

We can distinguish neopragmatists such as Rorty from new pragmatists such as Misak and Putnam. Rorty undoubtedly did a great deal to revive widespread engagement with pragmatism on the current intellectual scene, but his own version of the doctrine is at least as controversial as James's and attracted a similar uproar. In large part, Rorty's philosophy is negative, attacking foundationalist accounts of knowledge and rationality. Like other pragmatists, he believes that there is no single way of representing the world with absolute certainty, and we should view our beliefs as attempts not to mirror the world accurately but to forge tools to deal with it. Any set of tools may work for a particular group at a particular time, but it can make no claim to represent the way the world really is. In his starkest statements, Rorty claims that what gives a belief the power to justify other beliefs is purely sociological, a matter of what others will let us get away with saying. There is no truth or objectivity to be had, only solidarity or agreement within a community. Instead of seeking to line up our beliefs with the world, we should view ourselves as free to come up with new descriptions and "vocabularies" and to see how these help us achieve our ends and formulate new ends. What I'm calling a new pragmatist here balks at this view, seeing it as undermining reason and misreading the pragmatist tradition. For them, the historical development of standards in inquiry does not impugn their objectivity or make this objectivity simply a matter of "what we do around here".

The Rortyan neopragmatist takes different approach to the primacy of practice. The Western philosophical tradition is dominated by a discourse of objectivity, which, in turn, is understood as the abasement of the merely human before something else – the divine, the rational, nature's own language, or one of the other targets of *Philosophy and the Mirror of Nature*, and subsequent works (Rorty 1979). Our standards of normativity, including epistemic authority, are only those given in the practices of reason-giving and justification of those around us; instead of seeking objectivity in our talk, in the sense of a connection to something external to it, we should aspire to "solidarity", orienting our beliefs and desires to those of others around us. Reasons, authoritative descriptions, common conceptions of salience and naturalness, etc., have normative force within particular, socially located, intersubjective practices, where they may be promulgated and challenged. But there is no standpoint outside those practices where practice-independent reasons and descriptions hold true. In keeping with the negative thrust of his epistemological writings, Rorty rejects the idea that political views, and specifically his own social democratic liberalism, require philosophical justification. Such accounts mistakenly try to justify liberal practices with reference to some universally authoritative standard, and his pragmatism rejects the very idea of such standards. Rorty famously argues that social practices, including practices of inquiry, truth-telling, and reason-giving, aren't subject to external theoretical vindication, and in the essay provocatively called "Postmodernist Bourgeois Liberalism", he claims that the role of philosophy in what he calls "rich North Atlantic democracies" is to "convince society that loyalty to itself is loyalty enough, and that such loyalty no longer needs an ahistorical backup" (Rorty 1991: 198–199). A belief in freedom of speech, for example, should be viewed as a local practice, and there is no neutral standpoint outside the societies that endorse this practice, or which don't, from which to evaluate it. It does not follow that it is impossible to evaluate other worldviews. Indeed, part of what it is to be a liberal is to appraise other worldviews, in particular ways – to condemn governments that suppress their critics, for example.

On the one hand, if there isn't a vantage point from which to assess the norms of a particular community, this seems to leave existing practices of political justification in place, whatever they happen to be. This revives the political worry about complacency that pragmatism is insensitive to the ways in which power creates and gerrymanders consensus on particular practices (cf. Festenstein 2003). The other way of looking at Rorty's account, which he increasingly emphasises, is as itself a revisionary proposal, not for leaving all our practices and ways of thinking as they are but for the benefits of coming to see them as contingent and mutable products: this is what he calls philosophy as cultural politics (Rorty 2007). These benefits include both a sense of liberation, as we don't any longer see ourselves as constrained by the requirement to align practice with philosophically justified principles, and humility, as particular proposals for how to live are presented just as instructive stories rather than grounded in such universally compelling principles. A critic can then legitimately ask whether embracing philosophy as cultural politics in fact possesses these inspirational and humbling transformative effects as well as whether they are superior to the outcomes of following different sets of beliefs.

Another way of developing this practice-based account draws on Rorty's student Robert Brandom's imposing architectonic of "the implicit structure characteristic of discursive practice as such" (Brandom 1994: 374; 2000, 2002). As this is developed (notably by Fossen 2011, 2013), this marks out via media between the new pragmatism of the Peirceans and Rorty. Like the new pragmatists, and a Peirce, it takes its starting point in the idea that we incur non-discretionary commitments in our practices of believing, claiming, asserting, and declaring things. Norms arise within practices of giving and asking for reasons, and in accepting reasons and making claims, participants bind themselves to standards that go beyond their subjective interpretation of their commitments. What it is for us to think of ourselves and others as normative beings is as capable of undertaking commitments, ascribing them to others and accepting responsibility for them. From the perspective of a participant in claim-making, we are held responsible for our stances, and these can be evaluated by others, and in engaging in discursive practice, we distinguish between the commitments that we happen to accept and those that it is appropriate to accept. Judging that a public authority is legitimate or illegitimate is a matter of "taking a stance" in a linguistic practice, attributing various commitments and entitlements to oneself and other participants: it is only "from an engaged standpoint, in virtue of subjects taking stances from different perspectives" that "there such a thing as political legitimacy at all" (Fossen 2013: 442). To take a claim to authority to be legitimate is to accept practical commitments to obey, while to reject it is to accept practical commitments to treat it as a coercive imposition, for example. These commitments evolve and the content of these commitments in any given case isn't pre-given by individual fiat or communal consensus but is itself the challengeable product of a process of social contestation.

Democratic values

When we turn to consider the relationship between pragmatist arguments and democracy, a first point to note about Dewey is that he has an unorthodox view of what democracy consists in. Democracy is not primarily a feature of political institutions but of a society as a whole. One reason Dewey thinks that democracy is social, in his sense, is that he sees relationships of governance arising in many social institutions, including (importantly for him) the workplace and schools. Questions of power and popular rule arise across a range of social locations, not only in a narrowly construed set of political institutions. On top of this, though, Dewey thinks of democracy not only as a form of government that extends across a wider range of social contexts than is often thought but as "more than a form of government; it is primarily a mode of associated living, of conjoint communicated experience" (Dewey 1916: 93). Democracy is the way of life of a community of equal citizens.

Dewey sees democracy as itself an embodying form of social inquiry: "[t]he very heart of political democracy is adjudication of social differences by discussion and exchange of views", he writes in a late essay. "This method provides a rough approximation to the method of effecting change by means of experimental inquiry and test" (Dewey 1944: 273). Democratic societies try to arrive at acceptable decisions and to do so in ways that permit the criteria for an acceptable decision, as well as the decision itself, to be critically reviewed, scrutinised, and revisited. For Dewey, democracy is a form of experimental inquiry that allows for a thoroughgoing questioning of the prejudices and assumptions on the basis of which decisions are made, even if, of course, much of ordinary democratic politics does not involve this kind of unsettling challenge. This aligns pragmatism with the deliberative turn in democratic theory: at the core of this turn is the intuition that democracy comprises more than just a majoritarian procedure of decision-making, or a site for the tug of competing forces, but also consists in those processes of "debate, discussion and persuasion" which Dewey thought raised democracy beyond the "stupidity" of mere majority rule (Dewey 1927: 365). One reason why democracy improves social judgements about what to do is that it allows for the expression of beliefs and interests on the part of all, through both voting and less formal mechanisms of debate, discussion, and persuasion. Democracy involves the expression of interests on the part of voters; the vote helps to protect individuals from putative experts about where the interests of people lie. In the absence of this constraint, a class of experts will inevitably slide into a class whose interests diverge from those of the rest, and it becomes a committee of oligarchs, making poor and unresponsive judgements about what to do.

More generally, Dewey argues that inegalitarian class societies are epistemically defective: "[s]uch social divisions as interfere with free and full intercourse react to make intelligence and knowing of members of the separated classes one-sided" (Dewey 1916: 354). The absence of "free and full intercourse" limits experience and opportunities to learn from one another. Class separation tends to express and reproduce forms of unjustified hierarchy and disadvantage. The inequitable distribution of power excludes many from epistemic resources, an exclusion that may flow from censorship and propaganda but can just as effectively arise from informal market pressures:

> [p]eople may be shut out from free access to ideas simply because of preoccupation of their time and energy [...] because of class barriers and because a limited minority group holds a virtual monopoly of whole ranges of ideas and of knowledge [...] It requires a common background of common experiences and of common desires to bring about this free distribution of knowledge.
>
> *(Dewey 1932: 41)*

Membership of a privileged class is also epistemically distorting for the privileged themselves. An important symptom of this is the ideological justifications generated by dominant groups to explain their superior position. Privilege makes it difficult to resist the temptation to develop and accept as true, self-serving rationalisations of this status:

> [t]he intellectual blindness caused by privileged and monopolistic possession is made evident in 'rationalization' of the misery and cultural degradation of others which attend its existence. These are asserted to be the fault of those who suffer; to be the consequence of their own improvidence, lack of industry, wilful ignorance, etc.
>
> *(Dewey and Tufts 1932: 347–348)*

In this way, Dewey's writings anticipate some of the current thinking about epistemic injustice and the wilful ignorance of privileged groups (Medina 2012). By contrast, democratic society is characterised by free interaction which breaks down "those barriers of class, race and national territory that kept men from perceiving the full import of their activity" and so counters the distortion that flows from social separation by removing the obstacles of class, status, and identity to mutual learning (Dewey 1916: 93, 354–355).

Sceptical contemporaries and later interpreters argue that Dewey has too much faith in individual and collective capacities for critical inquiry. Proponents of Deweyan democracy, who find in his thought a fruitful source of recent deliberative conceptions of democracy, are attracted by the thought that democracy is more than merely a procedural minimum and that Dewey's thought provides a critical perspective on this minimalism. Dewey's rejection of procedural minimalism, however, goes further than pressing the epistemic claim. We shouldn't just view democracy as only instrumentally valuable for keeping officials in line and, more broadly, for mutual social learning, in Dewey's opinion, important though these are. Democracy allows us to arrive at a clearer view of social problems and of possible solutions by subjecting proposals to discussion and scrutiny. The kind of freewheeling civic interaction that democracy expresses is itself valuable for citizens: people can properly express their potential for growth only within a democratic society; that is, where they make decisions with others in terms of equality. Democracy as a social ideal harmonises "the development of each individual with the maintenance of a social state in which the activities of one will contribute to the good of all the others" (Dewey and Tufts 1932: 350). Democratic societies – and particular democratic institutions within them, including the different tiers of government, the workplace, and elsewhere – manifest the promise of universal equal standing.

As we've seen, the new pragmatists locate their defence of democracy on a more focused claim to the effect that "there is a direct connection between deliberative democracy and the pragmatist theory of truth" (Misak 2004: 9). A true belief is one that is responsive to, and best fits with, all reasons, arguments, and experience. An authentic, non-specious believer is committed to testing epistemic claims against as wide a range of different experiences as possible, rendering beliefs responsive to reasons and evidence. In particular, this commitment requires us to seek out and attend to different perspectives and arguments, in order to test and, if necessary, revise our current conceptions. From the fact we need access to evidence, arguments, other forms of information, and processes of reason-exchange, it follows that we need to live in a certain sort of canonical social and political order and should exercise certain epistemic virtues.

In extracting or seeking to extract this epistemic argument from the wider complex of claims in the Deweyan account, the new pragmatists develop a clearer picture of the relationship of pragmatist epistemology and political value. Yet it is one that aims to derive democratic commitments more minimally from doxastic commitments rather than on a fuller vision of a democratic community of equals. From the Deweyan perspective, the requirement to include others in decision-making may appear reinstate the kind of instrumentalism about democracy – in this case, as supporting the conditions for epistemic good practice – which the Deweyan wants to avoid, since to interact with you as an equal citizen is to do something more than view you as a good source of arguments, even if you are also that. I'll develop this debate a little further in the next section, when considering pluralism, but note here that new pragmatists can respond by making clear that this pragmatist argument doesn't amount to the totality of normative arguments that can be made on behalf of democracy in their eyes but that the other arguments that can be invoked don't directly derive from the pragmatist logic of inquiry. Indeed, it is important for this kind of approach that it makes sense to seek for truths across a wide range of evaluative matters, and these can include truths about how to organise society, while the narrower pragmatist argument about democracy is about what flows from the commitment to seek truth in this way.

Rorty's neopragmatism is ambivalent about the project of arguing for democratic values (Festenstein 2001). In keeping with the negative thrust of his epistemological writings, Rorty rejects the idea that political views require philosophical justification. Such accounts make the mistake of trying to justify liberal practices with reference to some universally authoritative standard, while his pragmatism rejects the very idea of such standards. In this sense, Rorty gives *priority* to democracy over philosophy, as he puts it, arguing that the democratic conditions for solidarity matter more than any philosophical account of why democracy is valuable (Rorty 1991). At the same time, he provides a distinctive vision of what political thinking in societies that face up to contingency should be like. The task of the social thinker is to sensitise us towards the suffering of others and widen the circle of those with whom we identify, not to elaborate theoretical justifications for this concern. A figure whom Rorty calls the "liberal ironist" combines an awareness of the historical contingency of evaluative categories with a commitment to promoting solidarity and freedom in the public realm while confining diverse and peculiar projects of philosophical self-realisation to the private sphere.

Pluralism

The third area to consider is how pragmatists address the moral, cultural, and political pluralism of complex societies. As we've seen, at the core of Dewey's general vision of democracy is an idea of participation on terms of equality in a complex society in which we can't assume that we know the needs and interests of others. Dewey tends to express sympathy for the value pluralist idea that different ethical theories and principles embody genuine and conflicting normative pulls on a person (Dewey 1930; Dewey and Tufts 1932). Furthermore, he isn't such a simpleton as to imagine that people can't inquire, solve problems, and "grow" in all sorts of anti-social ways – I can improve as a gangster or corrupt politician, for example. From a democratic perspective, however, my growth is hampered or warped if it takes place at the expense of yours, as in these cases. I can't engage on free and equal terms with everyone. We may ask, though, so what? How does this approach treat the political views of those who don't want to deliberate or to engage constructively with others? Dewey's argument that my growth depends on others and that these should lead me to treat them democratically may not impress me if I think that I can benefit from exploiting this interdependence undemocratically or if I view my growth as pleasantly insulated from the plight of fellow citizens.

As we've seen, Dewey has some instrumental points to make about my character in this case that parallel some recent work in the "epistemology of ignorance": I'll be susceptible to epistemic vices that functionally preserve my sense of independence or superiority, such as weird ideologies that purportedly justify my insulated position, and secession from democratic exchange will tend to exacerbate this vulnerability (Dieleman, Rondel and Voparil 2017; Medina 2012). Dewey doesn't want us to consider these as *merely* instrumental, though, since he views avoiding these pitfalls as inherent in a democratic way of life. It is articulating the content and rationale of this way of life for those who are in that is central to his project rather than seeking out points of critical leverage that will have force on it.

Rortyan neopragmatism threads together two responses to pluralism. One is what Rorty provocatively calls a "frankly ethnocentric" response to non-democratic views. Liberal democrats need not seek to justify their position to other positions since this is not only impossible (there isn't a neutral standpoint for this kind of justification) but also pointless (democracy is prior to philosophy). At the same time, it is internal to this worldview that it asks itself, "[w]hat are the limits of our community? Are our encounters sufficiently free and open? Has what we have recently gained in solidarity cost us our ability to listen to outsiders who are suffering? To outsiders who have new ideas?" (Rorty 1991: 13). This openness flows from a commitment not to betray the community's own interests – not, of course, the interest in getting things right in a framework-independent way but something more like an interest in being the best sort of community we can be, by our own lights. Rorty's dialectic notoriously stops short at this point. Critics press him for some framework-transcending reason why these societies should embrace those values and, further, insist that the relevant epistemic virtues of openness and engagement require a commitment to an external standard. Rorty insists that at this point the spade turns, and there is no deeper justification available.

For new pragmatists, the foundation in the concept of true belief acts as a powerful response to this question that marks it off from the Deweyan and Rortyan approaches (Misak 2000; Misak and Talisse 2014; Talisse 2007; Misak 2013). Since no matter what your approach to ethics and politics, you have an interest in acquiring and sustaining true beliefs, you should sign up to the pragmatist canons of inquiry, and these commit you to a deliberative model of democracy. Political commitments are built into doxastic commitments. The minimalism of this approach, which suggested a difficulty in the previous section – namely, that it casts a political relationship as an entirely epistemic one – may provide a robust way of responding to the pluralist challenge. What importance should we attach to the identification of this inconsistency in the anti-democrat's doxastic and political commitments?

Another way of drawing the line between Dewey and the new pragmatist approach takes its inspiration from some well-known arguments in contemporary liberal political theory. Many societies are characterised not just by a variety of moral and political views but by a variety of reasonable views, this thought runs. (As we've seen, Dewey doesn't quite adopt this way of looking at ethics but is broadly sympathetic to this idea.) Liberal societies tolerate these different reasonable views, which means that the state should not base constitutional principles and public policy on one particular view and so compel all citizens to act in accordance with that view, and in violation of their reasonable alternatives (Rawls 1993). The relevance of this to pragmatism, pressed particularly by Talisse (2007, 2011a, 2011b), is that Dewey's political philosophy may seem to violate this constraint, while the new pragmatist argument doesn't. It seems to be based on a contentious view that democracy is a way of life and on the interdependence of individual and social growth, which may itself be reasonable but which can be challenged by alternative reasonable perspectives. The new pragmatists' Peircean argument, however, is immune from this challenge, since it rests on a generic set of claims that anyone with an interest in forming and sustaining true beliefs (sc. anyone) should commit to. It follows that the new pragmatist argument can and should be used as an organising principle for political institutions and public policy, since it embeds a doxastic commitment that anyone qua reasonable believer can be content with rather than resting on a set of contentious ethical claims that may not be the basis of these arrangements.

This is an important line of inquiry, but we may ask why Peircean (or any other) epistemological strictures on political discourse are less reasonably challengeable than particular claims in moral philosophy, from this perspective, since this epistemological position does not self-evidently form part of the overlapping consensus of modern societies. Here, too, the distinction between what Dewey's position and the new pragmatists may be overdrawn. Democracy is a way of life for Dewey as it doesn't apply only to the polling booth but also to the workplace, family, educational and cultural organisations, etc., since these are also subject to government. Now it is difficult to see why the argument in terms of doxastic commitments seeks to have a narrower scope than this, restricted only to constitutional essentials and the bases of public policy, since our commitments as believers seem to hold across all these social contexts. It is not clear what in the new pragmatist argument allows us to suspend acknowledging the democratic implications of these commitments in the workplace, say, but not in public policy. In this sense, democracy as a way of life may be a banner beneath which both Deweyan and new pragmatists can parade and in the radical direction that Dewey encourages.

References

Addams, J. 1902 [1964]. *Democracy and Social Ethics*. Cambridge: Belknap Press.
Brandom, R. 1994. *Making It Explicit*. Cambridge: Harvard University Press.
———. 2000. *Articulating Reasons: An Introduction to Inferentialism*. Cambridge: Harvard University Press.
———. 2002. *Tales of the Mighty Dead: Historical Essays in the Metaphysics of Intentionality*. Cambridge: Harvard University Press.
Dewey, J. 1916 [1985]. "Democracy and Education". In J. Boydston (ed.), *The Middle Works of John Dewey, 1899–1924*. Carbondale: Southern Illinois University Press, 1981–1992, vol. 9, pp. 1–370.
———. 1927 [1984]. "The Public and Its Problems". In J. Boydston (ed.), *The Later Works of John Dewey, 1925–53*. Carbondale: Southern Illinois University Press, 1981–1992, vol. 2, pp. 235–370.
———. 1930 [1984]. "Three Independent Factors in Morals". In J. Boydston (ed.), *The Later Works of John Dewey, 1925–53*. Carbondale: Southern Illinois University Press, 1981–1992, vol. 5, pp. 279–288.
———. 1932 [1985]. "Politics and Culture". In J. Boydston (ed.), *The Later Works of John Dewey, 1925–53*. Carbondale: Southern Illinois University Press, 1981–1992, vol. 6, pp. 40–48.
———.1938 [1986]. "Logic: The Theory of Inquiry". In J. Boydston (ed.), *The Later Works of John Dewey, 1925–53*. Carbondale: Southern Illinois University Press, 1981–1992, vol. 12, pp. 1–528.
———. 1944 [1989]. "Challenge to Liberal Thought". In J. Boydston (ed.), *The Later Works of John Dewey, 1925–53*. Carbondale: Southern Illinois University Press, 1981–1992, vol. 15, pp. 100–120.
Dewey, J. and Tufts, J.H. 1932 [1985]. "Ethics", 2nd ed. In J. Boydston (ed.), *The Later Works of John Dewey, 1925–53*. Carbondale: Southern Illinois University Press, 1981–1992, vol. 7, pp. 1–438.
Dieleman, S., Rondel, D. and Voparil, C. (eds.). 2017. *Pragmatism and Justice*. New York: Oxford University Press.
Festenstein, M. 2001. "Pragmatism, Social Democracy and Political Argument". In M. Festenstein and S. Thompson (eds.), *Richard Rorty: Critical Dialogues*. Oxford: Polity Press, pp. 203–222.
———. 2003. "Politics and Acquiescence in Rorty's Pragmatism". *Theoria*, 50, pp. 1–24.
———. 2008. "John Dewey: Inquiry, Ethics and Democracy". In C. Misak (ed.), *The Oxford Handbook of American Philosophy*. Oxford: Oxford University Press, pp. 87–109.
Fossen, T. 2011. "Politicizing Brandom's Pragmatism: Normativity and the Agonal Character of Social Practice". *European Journal of Philosophy*, 22 (3), pp. 371–395.

———. 2013. "Taking Stances, Contesting Commitments: Political Legitimacy and the Pragmatic Turn". *Journal of Political Philosophy*, 21 (4), pp. 426–450.
James, W. 1907 [1975] *Pragmatism: A New Name for Some Old Ways of Thinking.* Cambridge: Harvard University Press.
Livingston, A. 2016. *Damn Great Empires! William James and the Politics of Pragmatism.* New York: Oxford University Press.
MacGilvray, E. 2000. "Five Myths About Pragmatism, Or, Against a Second Pragmatic Acquiescence". *Political Theory*, 28 (4), pp. 450–508.
Medina, J. 2012. *The Epistemology of Resistance: Gender and Racial Oppression, Epistemic Injustice and Resistant Imaginations.* Oxford: Oxford University Press.
Misak, C. 2000. *Truth, Politics, Morality: Pragmatism and Deliberation.* London: Routledge.
———. 2004. "Making Disagreement Matter". *Journal of Speculative Philosophy*, 18, pp. 9–22.
———. 2013. *The American Pragmatists.* Oxford: Oxford University Press.
Misak, C. and Talisse, R. 2014. "Pragmatist Democracy and Democratic Theory: A Reply to Eric MacGilvray". *Journal of Political Philosophy*, 22 (3), pp. 366–376.
Posner, R. 1990. "What Has Pragmatism to Offer Law". *Southern California Law Review*, 63 (6), pp. 1653–1670.
Rawls, J. 1993. *Political Liberalism.* New York: Columbia University Press.
Rorty, R. 1979. *Philosophy and the Mirror of Nature.* Princeton: Princeton University Press.
———. 1989. *Contingency, Irony, and Solidarity.* Cambridge: Cambridge University Press.
———. 1991. *Objectivity, Relativism and Truth: Philosophical Papers*, vol. 1. Cambridge: Cambridge University Press.
———. 1998. *Achieving Our Country: Leftist Political Thought in Twentieth Century America.* Cambridge: Harvard University Press.
———. 2007. *Philosophy as Cultural Politics: Philosophical Papers*, vol. 4. Cambridge: Cambridge University Press.
Talisse, R. 2007. *A Pragmatist Philosophy of Democracy.* New York: Routledge.
———. 2011a. "A Farewell to Deweyan Democracy". *Political Studies*, 59 (3), pp. 509–526.
———. 2011b. "Toward a New Pragmatist Politics". *Metaphilosophy*, 42 (5), pp. 552–571.

35
PRAGMATISM AND METAPHILOSOPHY

Scott Aikin

Metaphilosophy and the Creep Problem

One does metaphilosophy when one asks and tries to answer the philosophical question of what philosophy is. Many concepts are the sites of debate and contestation, and so questions of *What is justice? What is beauty? What is truth?* and *What is happiness?* are all points of divergence between philosophers. And the question *What is philosophy?* has similar divergence. And just as when we wrangle over the question of what justice is, we must distinguish between what merely passes for justice and what really is just; the same goes for the question of philosophy. That is, we are, when philosophizing about these concepts, trying to get a normative theory of how to use the concepts and an explanatory account as to why we have the diversity of views we have. So we ask, as philosophers, what philosophy is, why we philosophize, whether we can make progress in philosophy, which questions are philosophical and which are not, and what it might be to do philosophy better.

Metaphilosophy has two important uses when philosophizing. The first is that it functions for philosophers as a kind of North Star when they feel lost or in the weeds in a debate or they are unsure of what's worth thinking through. A metaphilosophy, as a normative concept, provides us with direction and even methodological criteria for what's worth doing and how to do it. And metaphilosophical approaches have explanatory uses, too. When one is in the midst of a debate with other philosophers over what beauty is, for example, one can not only look at the first-order philosophical reasons the parties give on the matter but one can also look at their second-order metaphilosophical reasons and orientations to help make sense of what is at stake for them. So metaphilosophy offers a greater understanding of first-order philosophical debates in terms of why they have the structures they do, why they developed as they did, and even why they seem so often deadlocked.

Metaphilosophy, from this initial take, has significant value. It has an inward-regarding power, since it provides an agenda for work to be done, clarity of objectives and appropriate methods, and even some good demarcation for what proper questions, training, and even success looks like. It also has outward-looking power, since it provides philosophers with explanations for why some (or most) debates go the way they do, and it offers paths for understanding broader philosophical dissensus. The hope is that these two orientations, the inward- and outward-looking, have a convergence that yields better work on the first order. With more shared understanding of our respective approaches to philosophy generally, we can have better exchange of reasons about philosophical questions.

DOI: 10.4324/9781315149592-39

The problem is that the hopeful convergence yielding that later good, that of fecund exchange, is regularly disappointed. Generally this disappointment arises because clashes on the first order drive us to the second order, and this makes the second-order assessments into clashes, too. So, debates on free will, for example, can go metaphilosophical and bring in considerations of method or what the objectives are between the two sides. But instead of clarifying a path forward with the first-order issues, the metaphilosophy made the debate wider and more unruly. And notice that once this global version of the problem arises, it is hard to go back to the first-order problem without seeing it as a mere instantiation of the global clash of metaphilosophical orientations. Call this the problem of *metaphilosophical creep* – once one sees first-order philosophical problems through the lens of metaphilosophy, they become tokens of a broader type of metaphilosophical clash.

The creep problem has a further consequence, since now the question is how to decide between metaphilosophies. In philosophy, argument is the prime mover, but the norms for argumentative success themselves are what we are debating when doing metaphilosophy. Inward-looking metaphilosophical norms make conditions for proof clear, but they will not work in cross-traditional contexts, since they themselves are in question. Disagreements that range over not only particular matters but also decision criteria for those matters are called *deep disagreements*, and they have a structure to that of the skeptic's *problem of the criterion*.[1] And once we've seen that metaphilosophy induces *creep*, philosophical disagreements themselves become *deep*.

The stories of metaphilosophical programs are narrated with the hope that they break the cycle of creep-to-deep. So, when the origin story of philosophy in Greece is told, it is how *logos* triumphs over *mythos*. Or when the Christianization of philosophy occurs at the end of the ancient period, it is about how we reorient ourselves toward the divine over the diversions of the pagans. Or with the Enlightenment, reason provides clarity over superstitious scholasticism. It is not hard to provide many other examples. Every new philosophical program comes with some inward- and outward-looking story, and each proceeds with the presumption that it has broken the creep-deep cycle. And further, on the assumption that they have broken that cycle, they proceed as though the philosophical traditions they have broken with are utterly undone by their critique. There is little for those others to say back that does not beg the question or double down on something hopeless. And so, an illusion befalls those practicing the new program, that of *triumph* over those old, outmoded philosophical views and approaches. The problem, at least as it should be seen in retrospect from the challenge of creep and deep disagreement, is that triumphalism in metaphilosophy is all too clearly self-congratulation and dogmatism. And it makes these new programs all too ripe for budding critics to triumph over later.

The metaphilosophical progression from creep to deep disagreement to triumphalist illusion seems quite clear. And perhaps it also explains some of the problems with self-serving stories of philosophical progress. In fact, once seen clearly, one should ask how pervasive the phenomenon is and how we can break its spell. It makes one ask the meta-metaphilosophical question: *What was the point of metaphilosophy in the first place, and should we have resisted the temptation?* Unfortunately, the deed is done. We have eaten from the apple and cannot go back. The question, then, is, *How do we manage this unhappy metaphilosophical situation?*

Pragmatist Metaphilosophies

Pragmatism is a practical philosophy. In fact, it is even more than that, as it is also a practical metaphilosophy. Pragmatists not only approach philosophical questions with matters of practice at the fore of their minds but they also hold that this practical orientation is how one ought to do philosophy. Programs that fail to have the practical perspective as their point of emphasis are deficient by pragmatist lights. On this broad point, most in the pragmatist tradition agree, but the matter between pragmatists is how to put that thought into practice. Once we see the work of clarifying that question in motion, we see a familiar problem emerge.

Charles Peirce's "How to Make Our Ideas Clear" and "The Fixation of Belief" are not only first-order philosophical essays in semantics and epistemology but they also pose metaphilosophical points as a consequence. The *Pragmatic Maxim* is designed to clarify "hard words and abstract concepts" (CP 5.464), and it is stated with the injunction:

> Consider what effects, that might conceivably have practical bearings, we conceive the object to have. Then our conception of these effects is the whole conception of the object.
> *(CP 5.402)*

The result of following this line of clarification yields a "prope-positivism" and "retention of a purified philosophy" with which "so much rubbish is swept away" (CP 5.423). As long as following the Pragmatic Maxim gives an exhaustive analysis of a claim or a term's meaning, a program of demarcation for philosophically worthwhile research follows.[2] Peirce's example is that of transubstantiation, the process of the substance of bread and wine turning into that of the body and blood of Christ in the midst of a Catholic mass. Peirce says that, given that no practical difference is detectable, the term is "senseless jargon" (CP 5.401). He predicted that most of the alternatives to pragmatist analysis or those that diverged from the norm will come out to be "meaningless gibberish ... or else downright absurd" (CP 5.423).

William James's "The Present Dilemma of Philosophy" opens with the observation that the history of philosophy "is to a great extent that of a certain clash of temperaments" (1975: 11). It merely seems that the debates are between grand theses, but they are really contests of basic temperaments. He argues, then, that in light of this insight, there is "a certain insincerity in our philosophical discussions: the most potent of premises is never mentioned" (1975: 11). The main contrast of temperaments is between *tough-* and *tender-mindedness*, which is behind the free will, rationalism-empiricism, theism-atheism, and idealism-naturalism debates. James's proposal is that given that most people are a little of both temperaments, it is appropriate that we should look to work out mixed views: commitments that appeal to those who have a "hankering for the good things on both sides of the line" (1975: 14). Since we are jumbles of those inclinations, James offers alternatives to the "barbaric disjunction" of thinking, say, that free will and determinism are mutually exclusive. One then is allowed to construct "free-will determinism," "pluralistic monism," and what James pronounces as "practical pessimism maybe combined with metaphysical optimism" (1975: 14). And so, the *pragmatic method* is a strategy of interpreting each notion by tracing it to its practical consequences and stakes informed by our antecedent interests. James promises, thereby, "a method of settling metaphysical disputes that otherwise might be interminable" (1975: 28). So, the famous hikers who debate whether the man chasing the squirrel around the tree also goes around the squirrel can be answered if we clarify what the question means first (and both, it turns out, can be right) (1975: 27). The result, James says, is that philosophical views should be seen as various rooms in a metaphorical grand hotel, each pursuing their own individual programs on their own. But all are connected by their relevance to practice. Pragmatism, then, is where all philosophies meet. Pragmatism is the "corridor" through which one must pass to reach one room or the other or to explain the significance of any particular program (1975: 32).

John Dewey's primary metaphilosophical thesis is that most philosophy in the traditional vein is in the service of insulating inherited values from science's progress.[3] Darwinism's emphasis on continuity and change undoes so many of the dualisms and essentialisms that undergird prescientific worldviews (MW 4:3). And the insights of properly informed social theory shows them to be obsolete (MW 4:12). Traditional philosophy is a rear-guard action of pre-modern, undemocratic, unscientific, and inegalitarian cultural interests. So, the spectator theory of knowledge, the doctrine of essences, the opposition of mind and body, and the distinction of reality and appearance are all consequences of these deeply retrogressive social inclinations. What's necessary, then, is a *reconstruction* of philosophy, not as one preserving those old notions but as one that ushers in a democratic and scientifically literate culture of inquiry. Philosophy, then, "gets over" traditional problems (MW 4:14). We philosophize and inquire to "escape from peril" (LW 4:3), to find comfort in turning indeterminate situations into determinate. Philosophy plays this role as a response to particular doubts, and the doubts and indeterminacies that arise for us in contemporary circumstances are vastly different from those faced by those previously. The proper attitude, then, for the pragmatist toward philosophical programs is to ask whether they are "fruitful" or "defined by remoteness from the concerns of everyday life" (LW 1:18). Philosophy is at its heart a critical enterprise, and for Dewey, pragmatism is not just critical but a "criticism of criticisms" (LW 1:298). The result is that pragmatism "provides the way, and the only way ... [to] freely accept the standpoint and conclusions of science ... and yet maintain cherished values" (LW 1:4). From Dewey's perspective, all programs external to the pragmatist orientation are at best uninformed about the things that matter or are complicit with regressive cultural forces at worst.

Richard Rorty's metaphilosophical pragmatism is a radicalized version of Dewey's program of getting over philosophical problems. Rorty holds that "a philosophical 'problem' was the product of the unconscious adoption of assumptions built into the vocabulary in which the problem was stated" (1979: xxxi). Rorty capitalizes on the Wittgensteinian quip that philosophical problems arise "when language goes on holiday." In this approach, the deeper normativity of philosophical work, that of distinguishing between true and false applications of terms such as "justice," "beauty," or even "truth," is deflated. "[T]he only criterion we have for applying the word 'true' is justification, and justification is always relative to an audience" (1998: 4). Instead of holding our vocabularies as accountable to something external to them, we should think of them as accountable only to us, our interests, and what we wish to be.

> It is impossible to step outside our skins – the traditions, linguistic and other, within which we do our thinking and self-criticism – and compare ourselves to something absolute.
> *(1982: xix)*

We should be asking whether our concepts are more useful to us to coordinate and establish "intersubjective agreement." Pragmatism, then, is a "philosophy of solidarity" instead a despairing objectivism (1991: 33). Consequently, philosophers should disavow the transcendental project, caught up with theological yearnings, and instead they should turn toward becoming a cadre of "all-purpose intellectuals" (1982: xxxix) and "ironists" (1989: 73). Instead of grounding our intellectual cultures (whether it be science, logic, mathematics, or aesthetics), philosophy is a kind of epiphenomenon or mere commentary running behind the culture. And philosophy's norms are reflections of the culture. So philosophers must be "ethnocentric" in their approaches, practicing and privileging the norms of their own groups, even without grounds for them (1991: 29). From this metaphilosophical view, there are not real first-order philosophical problems; there are only complications in a group's vocabulary. Or friction with another's. Philosophy is then merely a "kind of writing," a genre no different in kind from poetry or self-help books (1982: 92). And the objection that it must be more is but a kind of intellectual bad faith or "philosophical fantasy" (1982: 89).

Robert Brandom's metaphilosophical program is grounded in his view that philosophy is an extension of our linguistic sociality. Language works on a "reciprocal recognitional model" (2009: 89), and our practices of "giving and asking for reasons" are the core of this mutual recognition. We keep track of what we can say (and do) in light of these coordinated endorsements (1994: 474). Brandom calls this orientation "*fundamental* pragmatism," which he holds he shares with the broader pragmatist tradition (2011: 9). Since reasons animate this activity, the capacity for making how reasons work explicit is of utmost importance. "Logic is the organ of semantic self-consciousness" (1994: xix; 2009: 11). But logic in this guise has its central term, "truth," functionalized. Brandom notes that "I think it is a fundamental mistake to think that what is important is the possession of beliefs with a metaphysically weighty property: being true" (2009: 157). This is because "is true" is functionally equivalent to merely endorsing a commitment. "Truth" is just a term for beliefs we think should be shared (1994: 515). As a result, Brandom holds that "truth is not important to philosophy" (2009: 156). In its stead, he holds, philosophy must tell "Whiggish histories" of our conceptual development – that our concepts are better at doing what we wish them to do. We coordinate with each other, express ourselves, and make decisions. The objective is to see our linguistic arrangements as something being honed and perfected, "as the gradual unfolding of explicitness of commitments that can be seen retrospectively as always already having been implicit in it" (2009: 112). Thereby, philosophy enacts a kind of "expressive freedom" in bringing those genealogies to our attention and endorsing them and their direction (1979: 194; 2009: 150). The traditional alternative, Brandom holds, is "deeply confused and almost totally wrong" (2009: 160).

Triumphalist metaphilosophical programs tell a story about philosophy's past so that the current trends are a culmination of that past's incipient norms. On this story, the intellectual world was waiting with bated breath for the good news of the movement. Thus, Brandom's "Whiggish history" and James's "corridor" metaphor, among all the other pragmatist self-regarding tropes, are part of the triumph. When Dewey dismisses the "industry of epistemology" (MW 7:49) and enjoins returning to "the problems of men" (MW 10:46), he's taken the pragmatist program to have not only cornered the market on relevance but to have shown the pointless busyness of the alternatives. The pragmatic idiom is rife with the temptation to regard non-pragmatist approaches as mired in confusion and mendacity. And it even makes internal debates between pragmatists hard to see as anything short of expressions of contempt, as those retrenching to the classical idiom all too frequently dismiss the analytic brands of pragmatism as "conservative" and "insufficiently innovative" (Rockmore 2005: 270). The question is whether pragmatists can do better, or is this insular and dogmatic fate just what they have to make "work"?

Minimalist Metaphilosophy?

Pragmatist metaphilosophy has come under critical scrutiny. The criticisms have been mostly external. One line has been what one might call *full-frontal assault*. The pragmatist program has been criticized as being based on a theory of meaning that confuses acts of speaking with what is said (Fodor 1975: 71 and Fodor and Lepore 2010: 185). Pragmatism is criticized for promising clarity in forms of practical difference, but it is not yet clear how practical differences are measured or articulated. Pragmatists proceed as though the practical part is obvious, but it in fact needs all the same philosophical work that pragmatists had said was otiose (Reed 2014: 49; Russell 1966: 119). Others have objected that pragmatism is simply an anti-intellectual tradition, one that refuses to take matters seriously – calling the basic idea behind it "paleo-pragmatism" (as Williamson hints 2007: 278).

Alternately, pragmatist metaphilosophy is criticized for being a kind of performative contradiction. Rorty's ironist program about our concepts requires that we still live committed to them, but he proposes we hold them ironically. But can one live ironically committed to these things (Flanagan 2016: 214; Talisse 2009: 116)? One cannot have a critical program without concepts such as *truth* playing substantive roles (Aikin 2016: 226; Heney 2015: 510; Price 2003: 187). And metaphilosophical pragmatism is judged as a kind of self-refuting position with regard to how philosophy is even possible, since it cannot see itself as a response to something properly philosophical (Okrent 1989: 190).

The preceding criticisms are mostly *external* to the pragmatist program, even if many are posed by pragmatists. Either they undercut the pragmatist reconstruction of philosophy in key areas or they wall off parts of traditional philosophy from that reconstructive project. But I believe that a critical program *internal* to the pragmatist tradition is appropriate and likely to yield results pragmatists can recognize as worth acknowledging and internalizing. The problem with pragmatist metaphilosophy, as I think the overview from the previous section showed, is that the slide from creep, to deep disagreement, to triumphalism is tempting for the pragmatists. It is because there is something about the tradition that it tells about itself, namely, that it is *exceptional*. This is not an unusual feature of philosophical traditions, as what makes them all distinctive is what they also take to make themselves exceptional, too. In fact, from the outside, pragmatism is among a staggering variety of takes vying for primacy, to set the stage for debates, to identify questions worth asking. And pragmatists do not have exclusive rights on claiming relevance to life or having experience on their side (as noted forcefully by Hacking 2007: 37). In fact, the Jamesian insight that pragmatism is a *new name for old ways of thinking* (as the subtitle to his Pragmatism lectures runs) is that pragmatist themes antedate the statement of the pragmatist program and inform the very traditions pragmatism stands to oppose or reconstruct. Pragmatist triumphalism, from this perspective, is rhetoric for a frenzied mob without a view of whom it really is about. But in cool reflection, pragmatism stands more in conversation with its competing philosophical schools. And any grown-up who has ever had a critical conversation knows that one must toggle between seeing oneself and one's commitments from the inside and seeing them from the outside. On the one hand, one is in the right and standing up for what's good, but, on the other hand, we might just be wrong, and those other folks may have a point. And so, grown-ups know how to straddle those perspectives, and pragmatists should, too. In fact, given that the metaphilosophical illusions of creep and triumphalism are themselves shared between traditions, pragmatism fits quite well into traditional philosophy and needs the grown-up perspective in order to not simply be the sophomoric radical who consistently overstates their insights.

So, how to manage the slide from creep, to deep disagreement, to triumphalism? I think a form of *metaphilosophical minimalism* is the best answer. It is, mind you, not a solution to the problem but rather a management strategy. The metaphilosophical minimalist proceeds according to the following three axioms:

#1: Problems first. Try to address first-order philosophical problems before going meta-.
#2: Mixed bag. Acknowledge that most programs of addressing philosophical questions have significant trade-offs.
#3: Modest progress. Approach philosophy with thought that better run debate is progress of sorts.

These are interlocking and mutually supporting strategies, but I will try to address them individually. **#1: Problems first** is that our defaults should be to focus on the philosophical problems as tightly as we can. The less often we are out to speak *as pragmatists* (or whatever school, for that matter) on the issue, the better. And this makes us better practicing pragmatists. One reason why is that school-identifications, when made prominently salient, have more likelihood of polarization than understanding in critical exchanges. "Here we go again," thinks the audience of the person who invokes pragmatism, just as one might do when hearing a colleague invoke a philosophical tradition one holds in contempt. We should resist the inclination to go meta- because we've already seen the problems and diversions it creates. Moreover, we know enough about the history of philosophy to know that interesting things emerge from the alchemy of philosophers who meld traditions or think in ways informed by one tradition and are open to the insights from any and all. Cicero took pride in the eclecticism his practicing Academic approach allowed, and Seneca famously quoted Epicurus, even as a Stoic, holding that truth is not the property of the schools. In this regard, Rorty's "all-purpose intellectual" approach is right, but it is wrong in what he excludes to get there. The objective is to remain capacious and open to varieties of lines of thought.

Hilary Putnam's view of his own philosophical work and that even of the so-called revolutionaries shows how this approach has more promise than expected. Seeing ourselves as striving to provide relevant and useful additions to the ongoing conversation about particular problems and issues will have both more uptake and appropriately modest promises than our pronouncements of triumph (Putnam 1995: 31 and 2008: 113).[4] In fact, if we do so, we might even learn something from the folks who we'd thought were so wrong.

We inquire in philosophy in piecemeal fashion, so most of our results will yield what I've termed **#2: A mixed bag**. If we are interested in a variety of questions or see them connected, we see that there are trade-offs for consistency everywhere. And if we respond to where doubts arise and are interested in responding to criticisms from various perspectives (as Rule #1: Problems first runs), our solutions and working hypotheses will likely begin to aggregate in a fashion that begins to look systematic. One thing to remember is that systems deserve to be preserved only if they are good at solving problems in the first place. Of course, this depends on taking them seriously, which is too often what the metaphilosophical approach prevents. In fact, given that so much of the pragmatist outlook is that we are supposed to "get over" philosophical problems when we find out that they are no longer rooted in our lived experience, it should seem strange to them that many problems persist after their debunking has been made clear. At what point does the pragmatist ask whether their diagnosis of a problem as *aufgehoben* gets rethought when it seems so many of their colleagues are still thinking about it? The result is that we have to see our approaches as mixed bags of happy and regrettable commitments. This certainly should be the result of anyone who has seen so much of the pragmatist reconstruction of ethical terms, since the Deweyan analogy between "valuable" and "edible" seems to have solved the problem of making value both objective and motivating but created a new problem of the question of whether some things, though edible, aren't worth eating (LW 4:213; see also Rachels 1977: 161 and Gale 2006: 75). Or another case is worth noting. I'll admit that I, myself, find pragmatism appealing, but I can't kick the thought that the skeptics were really on to something. And the more that the pragmatist tradition dismisses skeptical lines of argument as paper doubts or unserious questions, the more I'm inclined to think that this is the huffing of dogmatists than think this is the proper analysis of a challenge.[5] Announcing that we've "gotten over" a problem that seems to stick around for folks who have listened intently and sympathetically with the therapeutic program seems, well, unpragmatic.

All of this requires what I've called a double-mindedness about our philosophical commitments that itself may not be plausible. Neil Gascoigne and Michael Bacon argue that the creep problem and my objections here (and elsewhere with Talisse in 2017 and 2018) are mere reassertions of a *transcendental urge* that gives rise to an *intellectual model* or taking philosophy seriously. They counter that such an approach I'm proposing will have another, parallel, problem they call "transcendental creep" (2020: 18). They are right about this problem, but this challenge, I believe, is an acceptable trade-off. For sure, many pragmatists have announced that their approaches are consistent with transcendentalist aims (Cooke 2006; Pihlström 2009; Putnam 1981), and even if pragmatism inclines one to look in askance at such programs, it is good argumentative policy to take the pull seriously. It is precisely what drives so much philosophy outside of pragmatism. How can the pragmatist have an argumentatively successful program if they refuse to dialectically take the alternatives seriously?[6]

At a philosophy conference years back, I was confronted by a self-identifying pragmatist for criticizing triumphalism for failing to take other programs seriously. "I thought you were a *pragmatist* – looks like I was wrong" was the zinger. At another time, Robert Talisse and I were criticized for saying that pragmatist reconstructions of too many terms are mere self-serving stipulations. "You'll never understand Dewey" was the response. These kind of replies, by my lights, are best seen as doubling down on triumphalism, and they are then both all-too-pragmatist and yet also unpragmatic. What can pragmatism be if it is not both an ambitious program of reconstructing the philosophical tradition and a form of fallibilism, even about pragmatism itself?

Finally, **#3: Modest progress** is an axiom about what to hope for, as a pragmatist, in philosophical exchange. My thought is that our best hopes for progress in philosophy is to work on norms and conditions for well-run critical dialogue over the big questions. In this regard, finding strategies for clarifying stakes and identifying relevant evidence are particularly useful, and pragmatism is a tradition well placed to not only participate in but lead the way with such stage-setting. So, the Peircean hope with the Pragmatic Maxim and James's "corridor" metaphor are good starting places. The crucial issue, however, is what happens *after* this initial setup. The hope was that properly arranging things will *resolve* or *dissolve* the intellectual tensions, like James's hikers once they got clear on what they respectively meant by "going around the squirrel." But given the fact that most philosophical disagreements are profound, that clarity can make it manifest instead that the issue is deeper than expected. And so, we can see just how intractable the matter is. In fact, wider forms of disagreement are likely, given this setup.[7] The procedural arrangement does not tell us which of the commitments will survive scrutiny, and even the results of our reflections can be contested, because those critical thoughts themselves depend on other challengeable commitments. The clarity with which we proceed does not guarantee consensus but rather a wider, but more commanding, dissensus. Nicholas Rescher's pragmatic approach to metaphilosophy, with the tool he calls *aporetics* as a model for problems and the logical space for solutions, is exemplary of this approach (1985 and 2006). In essence, Rescher sees webs of wider and richer conflicts of commitments arising from these exchanges. And the results, he holds, are both hopeful and tragic. They are tragic in that "no articulation of a philosophical system is free of problems" (2006: 84), but we have hopes for greater clarity and systematic connection going forward. Philosophy, as critical dialogue, is "an ongoing confrontation between competing positions perpetually in conflict, though changing in detail through increasing sophistication and complexification" (2006: 87).[8] In the end, the result is a restatement of the problem of deep disagreement from before, but now we see the commitments in terms of their argumentative and reasoned connections. It is no longer a mere clash but a vast and developing dialectical puzzle.

Minimally metaphilosophical pragmatism is open to plenty of challenges. To start, I've certainly been doing a lot of metaphilosophy while proposing that we should do much less. This sin of hypocrisy is just proof of our deeply unhappy state, of course. Additionally, what I've proposed pragmatist in only a big-tent sense, as it's more a form of philosophical eclecticism (or even modest Academic skepticism). This objection is correct, and I would simply say this mixed result is preferable to the triumphalist alternative. The mixed bag approach is also a mixed bag – not a surprise, really.

Pragmatism is the view that theories are tools, and it should follow that if our theory of theories prevents us from addressing those with whom we disagree, then we should revise the theory. Because it isn't a tool that does what it should be doing. My argument here has been that pragmatism's metaphilosophy as it is widely applied has these unhappy results in the form of the progression from creep, to deep disagreement, to triumphalism. Consequently, pragmatist metaphilosophy must be revised and moderated. This doesn't guarantee smooth work going forward, and it still yields a form of explicitly deep disagreement, but it offers a mitigation of the creep problem and its triumphalist results. This is disappointing, for sure, but it is precisely the limits of what one should expect of a minimalist program.

Notes

1 See Aikin (2021) for an account of the connection between problems of metaphilosophy, deep disagreement, and traditional skeptical challenges. See Aikin and Talisse (2018) for an overview of connections between pragmatist metaphilosophy and philosophical method.

2 See Macbeth (2007), Hookway (2012), Misak (2013), and Heney (2016) for applications of the Pragmatic Maxim for philosophical work.
3 See Gale (2006) and Marsoobian (2007) for accounts of Dewey's program of inquiry as a metaphilosophy.
4 See Crosby (2018) as an extension of the thought that pragmatism can have rich exchange with what are too often dismissed as "traditional" philosophical figures and questions.
5 See Aikin (2018 and 2020) for the case that pragmatism's anti-skepticism is regrettably thin.
6 Langsdorf (1995) argues that the dialectical orientation of pragmatism should be its central virtue.
7 Siegfried (2002) has made the case that this is very good news for pragmatists, that clarity will yield wider opportunities for critique.
8 See also Langsdorf (1995) and Mayorga (2017) for models of pragmatist metaphilosophy and methods as argumentative approaches to philosophy – the objective is to provide grounds for fruitful exchange, which need not always be resolution.

References

Aikin, Scott (2016) "Truth and Brandomian Metaphilosophy," *Al-Mukhabat Journal: Special Issue on Robert Brandom* 16: 217–227.
Aikin, Scott (2018) "Pragmatism, Common Sense, and Metaphilosophy: A Skeptical Reply," *Transactions of the Charles S. Peirce Society* 54: 231–245.
Aikin, Scott (2020) "Does Metaphilosophically Pragmatist Anti-Skepticism Work?" *Logos and Episteme* 11: 391–398.
Aikin, Scott (2021) "Deep Disagreement and the Problem of the Criterion," *Topoi* 40: 1017–1024.
Aikin, Scott and Talisse, Robert (2017) "Pragmatism and Metaphilosophy," in S. Pihlström (ed.) *Pragmatism and Objectivity*, New York: Routledge, pp. 31–49.
Brandom, Robert (1979) "Freedom and Constraint by Norms," *American Philosophical Quarterly* 16: 187–196.
Brandom, Robert (1994) *Making It Explicit*, Cambridge: Harvard University Press.
Brandom, Robert (2009) *Reason in Philosophy*, Cambridge: Harvard University Press.
Brandom, Robert (2011) *Perspectives on Pragmatism*, Cambridge: Harvard University Press.
Cooke, Elizabeth (2006) *Peirce's Pragmatic Theory of Inquiry*, London: Continuum.
Crosby, Pamela (2018) "The Education of Moral Character," in J. Goodson (ed.) *William James, Moral Philosophy, and Moral Life*, New York: Lexington Books, pp. 365–397.
Flanagan, Owen (2016) *The Geography of Morals*, Oxford: Oxford University Press.
Fodor, Jerry (1975) *The Language of Thought*, Cambridge: Harvard University Press.
Fodor, Jerry and Lepore, Ernie (2010) "Brandom Beleaguered," in Bernhard Weiss and Jeremy Wanderer (eds.) *Reading Brandom*, New York: Routledge, pp. 181–193.
Gale, Richard (2006) "The Problem of Ineffability in Dewey's Theory of Inquiry," *The Southern Journal of Philosophy* 64: 75–90.
Gascoigne, Neil and Bacon, Michael (2020) "Taking Rorty Seriously: Pragmatism, Metaphilosophy, and Truth," *Inquiry*. Online first: 1–20. https://doi.org/10.1080/0020174X.2020.1820375
Hacking, Ian (2007) "On Not Being a Pragmatist," in C. Misak (ed.) *New Pragmatists*, Oxford: Oxford University Press, pp. 32–48.
Heney, Diana (2015) "Reality as Necessary Friction," *Journal of Philosophy* 112: 504–514.
Heney, Diana (2016) *Toward a Pragmatist Metaethics*, New York: Routledge.
Hookway, Christopher (2012) *The Pragmatic Maxim*, Oxford: Oxford University Press.
James, William (1975) *Pragmatism and the Meaning of Truth*, Cambridge: Harvard University Press.
Langsdorf, Lenore (1995) "Philosophy of Language and Philosophy of Communication," in L. Langsdorf and A. R. Smith (eds.) *Recovering Pragmatism's Voice*. Albany: SUNY Press, pp. 195–209.
Macbeth, Danielle (2007) "Pragmatism and Objective Truth," in C. Misak (ed.) *The New Pragmatists*, Oxford: Oxford University Press, pp. 169–190.
Marsoobian, Armen (2008) "Metaphilosophy," in J. Lachs and R. Talisse (eds.) *American Philosophy: An Encyclopedia*, New York: Routledge, pp. 500–501
Mayorga, Rosa (2017) "Concrete Reasonableness and Pragmatist Ideals," in S. Pihlström (ed.) *Pragmatism and Objectivity*, New York: Routledge, pp. 150–167.
Misak, Cheryl (2013) *The American Pragmatists*, Oxford: Oxford University Press.
Okrent, Mark (1989) "The Metaphilosophical Consequences of Pragmatism," in A. Cohen and M. Dascal (eds.) *The Institution of Philosophy*, New York: Open Court, pp. 177–198.
Pihlström, Sami (2009) "Pragmatism and Naturalized Transcendental Subjectivity," *Contemporary Pragmatism* 6: 1–13.

Price, Huw (2003) "Truth as Convenient Friction," *Journal of Philosophy* 100: 167–190.
Putnam, Hilary (1981) *Reason, Truth, and History*, Cambridge: Cambridge University Press.
Putnam, Hilary (1995) *Pragmatism: An Open Question*, Cambridge: Blackwell.
Putnam, Hilary (2008) "Philosophers – and Their Influence on Me," *Proceedings and Addresses of the American Philosophical Association* 82: 101–115.
Rachels, James (1977) "John Dewey and the Truth about Ethics," in S. Cahn (ed.) *New Studies in the Philosophy of John Dewey*, Hanover: University Press of New England, pp. 149–170.
Reed, Baron (2014) "Practical Matters Do Not Affect Whether You Know," in M. Steup, J. Turri, and E. Sosa (eds.) *Contemporary Debates in Epistemology*, Malden: Blackwell, pp. 95–105.
Rescher, Nicholas (1985) *The Strife of Systems*, Pittsburgh: University of Pittsburgh Press.
Rescher, Nicholas (2008) *Aporetics*, Pittsburgh: University of Pittsburgh Press.
Rockmore, Thomas (2005) "On Classical and Neo-Analytic Forms of Pragmatism," *Metaphilosophy* 36: 259–271.
Rorty, Richard (1979) *Philosophy and the Mirror of Nature*, Princeton: Princeton University Press.
Rorty, Richard (1982) *Consequences of Pragmatism*, Minneapolis: University of Minnesota Press.
Rorty, Richard (1989) *Contingency, Irony, and Solidarity*, Cambridge: Cambridge University Press.
Rorty, Richard (1991) *Objectivity, Relativism, and Truth: Philosophical Papers I*, Cambridge: Cambridge University Press.
Rorty, Richard (1998) *Truth and Progress: Philosophical Papers III*, Cambridge: Cambridge University Press.
Russell, Bertrand (1966) "William James's Conception of Truth," in *Philosophical Essays*, New York: Touchstone, pp. 112–132.
Siegfried, Charlene H. (2002) "John Dewey's Pragmatist Feminism," in C. H. Siegfried (ed.) *Feminist Interpretations of John Dewey*, University Park: Pennsylvania State University Press, pp. 47–76.
Talisse, Robert (2009) *Democracy and Moral Conflict*, Cambridge: Cambridge University Press.
Talisse, Robert (2013) "Recovering American Philosophy," *Transactions of the Charles S. Peirce Society* 49: 424–433.
Talisse, Robert (2017) "Pragmatism and the Limits of Metaphilosophy," in G. D'Oro and S. Overgaard (eds.) *The Cambridge Companion to Philosophical Methodology*, Cambridge: Cambridge University Press, pp. 229–248.
Williamson, Timothy (2007) *The Philosophy of Philosophy*, Malden: Blackwell.

PART IV

Pragmatism's Relevance

36
PRAGMATISM AND PHILOSOPHICAL METHODS

Andrew Howat

Philosophical methodology is the central focus of pragmatism's founding documents.[1] The early works of Peirce, James, and Dewey examine methodological questions such as "How do we make philosophical ideas clear?", "What is the best method for fixing belief?", and "How do we know whether a philosophical question is answerable?". Thus, many consider pragmatism inherently methodological – as a metaphilosophy, a view about how philosophy should or must be done (e.g. Talisse 2017). Any summary of pragmatist methods is therefore a summary of pragmatism itself. Given such an impossibly broad remit, this chapter does only three things. First, it provides four broad claims common to pragmatist approaches to philosophical methodology, claims reflecting its underlying theory of inquiry. Second, it briefly examines three core pragmatist methods – for conceptual clarification, for fixing belief, and for settling or dissolving philosophical disputes. Third, it briefly describes differences between the Classical figures regarding each method. This is merely a brief sketch – the reader should consider all entries in this volume relevant to pragmatism qua philosophical method.

Pragmatists on Inquiry

Pragmatist ideas about philosophical method emerge from two main sources – its critique of Cartesianism (beginning with Peirce's *cognition* series, CP 5.213–5.317) and its theory of inquiry (Peirce's *Fixation*, CP 5.358, and *Ideas*, CP 5.388). While each of these sources is rich, detailed, and sophisticated, these four claims provide a first approximation:

a   Given that all beliefs are fallible, philosophical theorizing must begin with commonsense beliefs, not with propositions taken or shown to be certain, self-evident, apodictic, etc.
b   Philosophers should question these beliefs only when we encounter legitimate grounds for doing so, not based on the mere possibility of error. Legitimate grounds may take the form of *recalcitrant experiences* (*actual* for James, *actual or possible* for Peirce) or *indeterminate situations* (Dewey).
c   Inquiry is a collective or dialogic long-term effort by an indefinitely large community to arrive at stable beliefs that "work" or are "satisfactory" rather than a solitary inquirer's quest for certainty.[2]
d   Only inquiries concerning propositions with pragmatic meaning – those that pass the pragmatic maxim's test of legitimacy – can be successful. It will be idle and pointless to inquire into the truth or falsity of pragmatically empty and thus illegitimate propositions.[3]

Talisse (2017) argues that there is a marked historical trajectory within the pragmatist tradition concerning the scope or significance of such claims. Talisse calls this trajectory "metaphilosophical creep" and describes it as follows:

> Peirce originally proposed a semantic rule for philosophical enquiry that was transformed by James into a method for re-describing traditional philosophical problems as expressions of psychological differences; Dewey expanded pragmatism further into a full-bore metaphilosophical platform, a comprehensive second-order vision that fixes first-order philosophical views and so ultimately treats all purportedly philosophical disputes as metaphilosophical disputes.

While not all scholars will accept this characterization, it nevertheless captures a crucial point of contention. One way to understand pragmatism is as a method *only* – a way of clarifying ideas that merely facilitates philosophical inquiry. Others consider it more radical – as the attempt to "renew" or "reconstruct" philosophy altogether by identifying and discarding "perennial" problems (e.g. skepticism) and/or entire sub-disciplines (e.g. metaphysics, epistemology), transforming it into a mechanism for democratic or political growth and cultural criticism.[4] This has led some scholars to argue that there are at least *two* pragmatisms (e.g. Misak 2013; Mounce 2002).

Thus, despite deep affinities, pragmatists disagree about some aspects of methodology and sometimes exhibit internal inconsistency or changes of heart. The remaining sections therefore treat the three key figures – Peirce, James, and Dewey – separately, making brief note of some of these tensions.

Peirce on Clarifying Concepts

Peirce wrote that pragmatism is "a method of reflexion having for its purpose to render ideas clear" (CP 5.13, n1, 1906). His first 1878 formulation of the "pragmatic maxim" governing this method reads, "Consider what effects, that might conceivably have practical bearings, we conceive the object of our conception to have. Then, our conception of these effects is the whole of our conception of the object". Thus, as a first approximation, pragmatists claim that clarifying a concept requires identifying the practical effects or consequences that follow from its correct application. Those wishing to *test* a philosophical hypothesis should first deduce from it such practical consequences, just as we might test a scientific hypothesis by similarly constructing a real or simulated experiment.

Peirce's initial formulation of the maxim suggests that practical consequences *alone* matter in conceptual clarification, a suggestion he later corrects.[5] This is because he changed his mind concerning "scholastic realism", or the view that reality contains real generality as well as individuals, which transformed his understanding of the maxim.[6] Thus, I focus instead on this 1903 formulation:

> Pragmatism is the principle that every theoretical judgment expressible in a sentence in the indicative mood is a confused form of thought whose only meaning, if it has any, lies in its tendency to enforce a corresponding practical maxim expressible as a conditional sentence having its apodosis in the imperative mood.[7]

The maxim is a rule for clarifying what Peirce called "intellectual concepts".[8] He argued against the view, popular in the late 19th century, that this merely involved making them "clear and distinct" (a la Descartes and Leibniz). To possess "clearness", Peirce writes, is "merely to have such an acquaintance with the idea as to have become familiar with it, and to have lost all hesitancy in recognizing it in ordinary cases". It is to have "a subjective feeling of mastery" that "may be entirely mistaken". To possess "distinctness" is to be able to give "a precise definition" of the idea "in abstract terms". As Peirce understood them, traditional methods assume that we achieve the maximum level of clarity about an idea by refining distinctness – that is, tinkering with our definition – until either we have a clear apprehension of everything it contains (Leibniz) or it withstands dialectical examination (Descartes). Though he admits that distinctness is "indispensable to exact reasoning", Peirce claims that "Nothing new can ever be learned by analyzing definitions" and suggests that full clarity about an idea requires something more.[9] He argues for a third grade of clarity that relies fundamentally on the concept of reality – scholars typically call this the concept's "pragmatic meaning".

Suppose a student learns to recognize samples of lithium in her high school chemistry class, by reference to its peculiar metallic luster.[10] She becomes confident in picking out samples of lithium, at least in the classroom setting. She thereby attains first-grade clarity about the proposition "this substance is lithium". When she subsequently learns that "lithium" is defined as *the chemical element with atomic number 3*, she thereby achieves second-grade clarity.

Peirce would argue that something vital remains missing from the student's understanding, namely lithium's *practical or experiential* meaning. For Peirce, this is how reality manifests for us in the context of inquiry – as a resistance, compulsion, or surprise that we experience when we try to act in certain ways (e.g. when we conduct experiments).[11] Thus, while she remains unaware of the following "conditional sentence having its apodosis in the imperative mood", the student's grasp of lithium will be inadequate (and dangerously so!): *if you expose lithium to water, expect it to ignite, burn, and possibly explode*. Thus, third-grade clarifications tell us what experiences to expect and what reactions to prepare when we interact with instances of the relevant idea – something that neither clearness nor distinctness supplies.[12]

Peirce's method rests on a controversial assumption about meaning, in this case that *a person who knows that lithium ignites in water understands the meaning of "lithium" better than someone who does not*.[13] More schematically, Peirce is claiming that the practical consequences of a concept's true application are not only an important part of its meaning but in some sense the most fundamental part.[14]

Most semantic theories entail that while true and informative, such a subjunctive conditional nevertheless says nothing about either the *literal meaning* or the essence of "lithium", which are instead exhausted by its proper definition or analysis. Instead, the conditional seemingly expresses a fact about *lithium itself* (the *substance* or *property*), not a fact about our *conception of* lithium (the *word* or *idea*). Thus, Peirce's method will strike many as conflating *words* or *concepts* with the *things* (entities or properties) they pick out.

Some pragmatists concede this point and stipulate that "the meaning in which Peirce is primarily interested is not fixed semantic meaning, but meaning … as it differently informs our practices depending on our context and goals" (Dea 2015: 416).[15] Others seek to refute the objection, typically referencing Peirce's realist metaphysics, his career-long war against nominalism, and his theory of signs. Mounce, for example, writes:

> [Peirce] is accused of attempting to derive conclusions about the world from the study of linguistic or logical forms. But that criticism is based on the very assumption of an absolute gulf between language or logic on the one hand and the world on the other, which it is precisely Peirce's intention to deny.[16]

Peirce on Fixing Belief

Though Peirce sometimes implies the maxim alone *is* pragmatism, he also sometimes recognizes that his views on doubt (an unsettled, dissatisfied state), belief (that it incorporates a stable disposition to act in certain ways), and the transition between them (inquiry) are also crucial.[17] Once we have an idea that is clear at the third grade (for a given context), we still require a method for "fixing our beliefs" about specific hypotheses involving that idea.

Suppose the teacher presents our chemistry student with a sample of what *appears* to be lithium and asks her to determine whether the hypothesis "this is lithium" is true or false. Applying the pragmatic maxim suggests appropriate ways to test the hypothesis, such as placing the sample in water. Knowing the "sensible effects" with "practical bearings" that follow if a hypothesis is true thus facilitates the fixation of belief via the "method of science".

Peirce describes and argues for this method by contrasting it with three historically significant alternatives, which involve fixing belief through tenacious wishful thinking, deference to some comprehensive authority (such as a religious or political authority), and a priori reflection on what is "agreeable to reason". He argues all three methods are inferior in their capacity to fix belief permanently and defines the method of science as resting on a "fundamental hypothesis" that there

> are Real things, whose characters are entirely independent of our opinions about them; those Reals affect our senses according to regular laws, and, though our sensations are as different as are our relations to the objects, yet, by taking advantage of the laws of perception, we can ascertain by reasoning how things really and truly are; and any man, if he have sufficient experience and he reason enough about it, will be led to the one True conclusion. The new conception here involved is that of Reality.
>
> *(CP 5.384)*

James on Dissolving Pseudo-Problems

James hugely admired Peirce's maxim and set out to popularize, apply, and extend it in novel ways. However, in announcing "the pragmatic method" to the world in 1898, James repeated Peirce's initial misleading overstatement of the maxim, essentially eliminating the first two grades of clarity:

> To develop a thought's meaning we need *only* determine what conduct it is fitted to produce; that conduct is for us its *sole* significance. ... To attain perfect clearness in our thoughts of an object, then, we need *only* consider what effects of a conceivably practical kind the object may involve – what sensations we are to expect from it, and what reactions we must prepare. Our conception of these effects, then, is for us *the whole of* our conception of the object, so far as that conception has positive significance at all.
>
> *(1898: 290–291, my emphasis)*

James was also more strident in describing its philosophical significance. He saw its purpose not solely as clarifying concepts and facilitating inquiry but as solving or *dis*solving numerous philosophical (and especially *metaphysical*) (pseudo-)problems.

Peirce himself laid the groundwork for this much bolder interpretation of the maxim when he applied the maxim to various philosophical debates (e.g. transubstantiation, free will), effectively claiming to have reduced them to merely verbal disputes.[18] While Peirce's later change of heart about realism arguably undercuts this early deflationism, James doubles down: "It is astonishing to see how many philosophical disputes collapse into insignificance the moment you subject them to this simple test" (1898: 292). James says of the debate between materialism and theism: "if no future detail of experience or conduct is to be deduced from our hypothesis, the debate… becomes quite idle and insignificant" (2018 [1907]: 395).[19] He also adds a novel element, diagnosing such disputes in terms of competing philosophical "temperaments", seemingly reducing them to mere expressions of psychological differences.

One key reason for this divergence between the two men is that James's interpretation of "practical differences" was significantly narrower than Peirce's.[20] He believed philosophy sought to "find out what definite difference it will make to you and me, at definite instants of our life, if this world-formula or that world-formula be the one which is true" (1898: 292). His most notorious application of this approach was to the hypothesis of God's existence in *The Will to Believe*.[21] Peirce, by contrast, sought only to identify what *possible* differences it could make to some actual *or hypothetical* scenario if a proposition were true, and his conception of truth very deliberately references an indefinitely large community of inquiry working over the indefinitely long run, not any particular individual, group, or time.[22] In addition, James appears to include as genuine practical consequences the effects of *believing* the relevant "formula", allowing the desirability of these consequences to weigh in its favor, at least under certain carefully specified conditions.[23] Many scholars speculate that it was this controversial modification of his maxim that prompted Peirce to rename his position "pragmaticism".

Dewey on Social and Political Philosophy

Dewey too eventually came to greatly admire Peirce – particularly his theory of inquiry, maxim, and scientific metaphysics. Dewey also modified and extended pragmatic methods in important ways, particularly into new areas such as ethics, politics, and social philosophy.[24] Heney (2016) identifies three main methodological continuities between Dewey and his predecessors:

1   a broad conception of experience,
2   the central role of regulative assumptions, and
3   the conception of inquiry (including ethical inquiry) as aiming at truth.

Regarding (1), Dewey embraces Peirce's conception of experience as compulsion, reaction, or "that which surprises", which had enabled Peirce "to count manipulations of mental diagrams and processing of thought experiments as ways of gathering experience".[25] Dewey also emphasizes the important role of the experiences of *others*, extending Peirce's insight about the role of the *community* in inquiry by emphasizing a novel element – the idea "that some of the evidence garnered by taking the experience of our fellows into account is distinctively moral in character and can be put to work in developing better habits."[26] Finally, Dewey emphasizes the ways that experience is *transactional* – that is, it involves a dynamic interaction between an organism and its environment, not a mere passive receptivity of the sort he frequently denounces as a mere "spectator theory of knowledge".[27] This emerges from Dewey's novel analysis of "means" and "ends" in his celebrated early work on "the reflex arc" (which Mounce [2002] calls "one of his finest achievements", p. 131).

Regarding (2), Dewey recognizes the reliance upon *regulative assumptions* – principles that describe our logical habits – as the best alternative to a Cartesian focus on ultimate and invariant first principles. The assumptions are *regulative* because the only vindication they can receive is *practical* rather than metaphysical. Heney writes:

> to consider whether or not a principle is vindicated in pragmatism, look to the consequences of treating that principle as a working hypothesis for action and deliberation. It is not the task of a principle such as that of bivalence to prove itself true; if it proves itself the right kind of instrument for the project of inquiry, no further justification of its adoption can be demanded.[28]

Where Dewey departs from his predecessors is equally important. First, Dewey transcends Peirce and James's approach by extending it into social and political philosophy, developing – with Jane Addams – extraordinarily influential accounts of education and democracy in the process.[29] Forstenzer (2018) argues compellingly that a consequence of this extension is an original form of experimentalism in political philosophy, which rivals the Rawlsian/ideal-theory mainstream. Second, although Dewey may agree with the Peircean idea of inquiry being bound up with truth, he nevertheless largely jettisons the terms "knowledge" and "truth" in favor of "warranted assertibility". It appears that he does so primarily to pre-empt the problematic misunderstandings of knowledge and truth that he sees as part and parcel of "perennial philosophy".[30] But some have argued that in doing so he betrays Peirce's commitment to a genuinely objective conception of inquiry and lays the groundwork for the Rortyan, "post-truth" understanding of philosophical method.[31]

Notes

1. West (1989) describes them as "preoccupied with method" (p. 42).
2. In precisely what sense they must "work" or be "satisfactory" is a point of significant tension and disagreement among pragmatists.
3. To some this claim foreshadows verificationism (Misak 1995), while others argue these affinities are often overstated (Potter 1996).
4. See e.g. Rorty's (2007) vision of philosophy as "cultural politics".
5. CP 8.218, 1910.
6. For a comprehensive treatment, see e.g. Lane (2018).
7. CP 5.18. See Hookway (2012) and Short (2017).
8. It is neither a full-fledged theory of *all* linguistic meaning, nor does it presuppose one (Short 2017). Also, as Almeder (1979) writes:

    [I]n spite of the unfortunate phraseology, the pragmatic maxim, as stated, is not so much a criterion for the meaning of concepts or words as it is a criterion for the meaning of certain propositions or sentences ... [for Peirce] talking about the meaning of a concept or a word is in fact talking about the meaning of a sentence or proposition.

    (p. 4)

9. Given his later change of heart about the maxim, the quoted claim might be an overstatement, though see e.g. Wilson (2020).
10. This example is inspired by De Waal (2013: 4) and CP 2.330, 1902.
11. See e.g. McLaughlin (2009: 404).
12. Peirce's 1878 formulation says that the perhaps infinite list of conditionals describing how lithium behaves in various possible contexts constitutes "the whole of" our conception of lithium. This can give the impression that Peirce was wholly dismissing the significance of the first two grades of clarity, something he later denied (CP 8.218, 1910). The 1903 formulation allows that the other grades still have significance albeit only *derivatively* from those conditionals, that is, from their "tendency to enforce" them. Peirce is probably alluding to final causation here, *per* Hulswit (1997) and Short (2007).
13. See Peirce's response to a dispute between Newton and Kirchhoff concerning force, summarized in Dea (2015: 414–415).
14. In precisely what sense, it is notoriously difficult to say. See e.g. Wilson (2020).
15. See also Hookway (2015: 398).
16. Mounce (2002: 23). This foreshadows the "linguistic turn", per Rorty (1967). On Peirce's realism, see Lane (2018); on his anti-nominalism, see Forster (2011); and on his theory of signs, see Short (2007).
17. See his remarks on Bain at CP 5.12, 1906.
18. On transubstantiation, see Talisse (2017). On free will, see Howat (2018).
19. Peirce also opposes materialism, but it is unclear whether he does so on grounds of simple pragmatic illegitimacy. See e.g. Tiller (2006).
20. James himself says he construes the maxim more "broadly" (p. 291), but what he presumably means here is that on his understanding it has broader *implications*, which only occurs if one narrows the *criteria* so that fewer conceptions and metaphysical hypotheses pass the test of legitimacy.
21. Or so it is popularly assumed. In reality, James defends the claims that *the best things are the more eternal things* and that *we are better off if we believe this fact*.
22. Boncompagni (2016) writes:

    Peirce highlights precisely the exigency of not talking about a single action, nor about a set of actions, as consequences of a concept, but of a habit of conduct, that is of a *general* determination which includes not only what happens but what *could* happen.

    (pp. 146–147)

He also emphasized that our concepts are interconnected so that inquiry will have to be holistic and continuous (see Dea 2015: 410).
23 See e.g. Bacon (2012: 28).
24 On Peirce's difficult relationship with such "vital matters", see Atkins (2016).
25 Heney (2016: 55). Thus, a student conducting employing measurements/diagrams in her mathematics class may also be correctly applying the third grade of clarity, even if the "experiments" she performs take place entirely in her own mind.
26 *Ibid.* p. 56. On the question of how Peirce might have extended his method to incorporate ethical and political inquiry, compare Misak (2000) with Atkins (2016).
27 See e.g. Hildebrand (2003).
28 *Ibid.* p. 57. See also Hookway (1985: 77–79) on Peirce's classification of the sciences, and/or Howat (2015) on "*quietist* grounding".
29 See esp. Dewey (1916). Shields (2017: 23) addresses Addams' role viz. the account of democracy.
30 See Dewey (1938).
31 See Misak (2013) and Forstenzer (2018).

References

Almeder, R., 1979. Peirce on Meaning. *Synthese* 41, 1–24.
Atkins, R., 2016. *Peirce and the Conduct of Life: Sentiment and Instinct in Ethics and Religion.* Cambridge University Press, New York.
Bacon, M., 2012. *Pragmatism: An Introduction.* Polity, Malden, MA.
Boncompagni, A., 2016. *Wittgenstein and Pragmatism. On Certainty in the Light of Peirce and James.* Palgrave Macmillan, London.
De Waal, C., 2013. *Peirce: A Guide for the Perplexed.* Bloomsbury, New York.
Dea, S., 2015. Meaning, Inquiry, and the Rule of Reason: A Hookwayesque Colligation. *Transactions of the Charles S. Peirce Society* 51, 401.
Forstenzer, J., 2018. *Something Has Cracked: Post-Truth Politics and Richard Rorty's Postmodernist Bourgeois Liberalism.* Ash Center Occasional Papers Series.
Forster, P., 2011. *Peirce and the Threat of Nominalism.* Cambridge University Press, New York.
Heney, D., 2016. *Toward a Pragmatist Metaethics*, Routledge Studies in American Philosophy. Routledge, New York.
Hildebrand, D.L., 2003. *Beyond Realism and Antirealism: John Dewey and the Neopragmatists.* Vanderbilt University Press, Nashville, TN.
Hookway, C., 1985. *Peirce.* Routledge, New York.
Hookway, C., 2012. *The Pragmatic Maxim: Essays on Peirce and Pragmatism.* Oxford University Press, Oxford.
Hookway, C., 2015. Comments on Essays from Conference "The Idea of Pragmatism." *Transactions of the Charles S. Peirce Society* 51, 397.
Howat, A., 2015. Peirce on Grounding the Laws of Logic. *Transactions of the Charles S. Peirce Society* 50, 480–500.
Howat, A., 2018. Misak's Peirce and Pragmatism's Metaphysical Commitments. *Transactions of the Charles S. Peirce Society* 54, 378–394.
Lane, R., 2018. *Peirce on Realism and Idealism.* Cambridge University Press, Cambridge, UK.
McLaughlin, A.L., 2009. Peircean Polymorphism: Between Realism and Anti-realism. *Transactions of the Charles S. Peirce Society: A Quarterly Journal in American Philosophy* 45, 402–421.
Misak, C., 2000. *Truth, Politics, Morality: Pragmatism and Deliberation.* Routledge, New York.
Misak, C., 2013. *The American Pragmatists.* Oxford University Press, Oxford.
Misak, C.J., 1995. *Verificationism: Its History and Prospects.* Routledge, London, UK.
Misak, C.J., 2004. *Truth and the End of Inquiry: A Peircean Account of Truth,* Expanded paperback ed., Oxford University Press, Oxford.
Mounce, H., 2002. *The Two Pragmatisms: From Peirce to Rorty.* Routledge, New York.
Potter, V.G., 1996. Peirce's Pragmatic Maxim: Realist or Nominalist?, in: V. Colapietro (ed.), *Peirce's Philosophical Perspectives.* Fordham University Press, New York, pp. 91–102.
Rorty, R., 1967. *The Linguistic Turn: Essays in Philosophical Method.* University of Chicago Press, Chicago, IL.
Rorty, R., 2007. *Philosophy as Cultural Politics: Philosophical Papers.* Cambridge University Press, Cambridge.
Shields, P., 2017. *Jane Addams: Progressive Pioneer of Peace, Philosophy, Sociology, Social Work and Public Administration.* Springer, Dordrecht.
Short, T.L., 2007. *Peirce's Theory of Signs.* Cambridge University Press, Cambridge, UK.

Short, T.L., 2017. The 1903 Maxim. *Transactions of the Charles S. Peirce Society* 53, 345–373.

Talisse, R., 2017. "Pragmatism and the Limits of Metaphilosophy," in Overgaard, S., and D'Oro, G., (Eds.), *The Cambridge Companion to Philosophical Methodology*. Cambridge University Press, Cambridge, UK, pp. 229–247.

Tiller, G., 2006. The Unknowable: The Pragmatist Critique of Matter. *Transactions of the Charles S. Peirce Society* 42, 206–228.

West, C., 1989. *The American Evasion of Philosophy: A Genealogy of Pragmatism*. University of Wisconsin Press, Madison, WI.

Wilson, A.B., 2020. Interpretation, Realism, and Truth: Is Peirce's Second Grade of Clearness Independent of the Third? *Transactions of the Charles S. Peirce Society* 56, 349.

37
PRAGMATISM AND EXPRESSIVISM

David Macarthur

What do the expressivist programs which trace back to logical positivism have to do with the pragmatist tradition of Sellars, Rorty and Brandom? One might think the answer to that is nothing, but Huw Price would like to convince us otherwise. He champions the virtues of a *pragmatist expressivism* which generalizes the expressivist viewpoint, which explains the functions of problematic parts of language in non-representational terms, to the whole of language. In this chapter, I want to examine the merits of Price's position and, in particular, to bring an objection against its instrumental approach to the function of truth, which is, perhaps somewhat surprisingly, also applicable to the instrumental theory of truth propounded by the classical pragmatist William James.

The Placement Problem: Material Versus Linguistic Versions

Wilfrid Sellars bequeathed to philosophy a highly influential version of what has come to be known as the placement problem. In its orthodox formulation one asks, How are we, in philosophical reflection, to 'place' those 'objects' that we recognize in our everyday lives but which apparently do not figure in the scientific image of the world, e.g., moral goodness, numbers, possibilities? Less tendentiously, how are we to 'place' facts about moral goodness, numbers, possibilities (etc.) in a world composed entirely of facts recognized by the sciences? There are four main options for how to answer these questions: two forms of realism- reductionism and non-naturalism; and two forms of antirealism- error theory and expressivism. Reductionists reduce moral goodness and so on to scientific objects or properties, a move which seems to deny or downplay their differences. Non-naturalists simply accept non-natural 'objects' or 'facts' and so can be accused of believing in "a kind of metaphysical fairy story" (Price, 1996, 966). Error theorists think that the language in question refers to various non-existent 'objects', which inevitably leads to the embarrassing implication that ordinary practice is subject to widespread systematic error. Expressivists deny that the relevant language even purports to describe a realm of objects or properties or facts, a denial which seems in sharp tension with the familiar fact that ethical language, say, takes the form of assertions which can be true or false. All options seem blighted.

Huw Prices describes his alternative approach to the placement problem as follows:

> we should begin with expressivism – itself a naturalistic account of why ordinary folk talk in the ways in question – but then show that talk with these origins can properly acquire the trappings of genuine descriptive talk. ... Though it begins with expressivism, its upshot, in effect, is non-naturalism without the metaphysical tears: we explain why we folk talk of

goodness, possibility, chance, causation, or whatever, without compromising our (Humean) naturalistic principles, and without in any way undermining our right to go on talking this way.

(1996, 966)

Classical expressivists such as A.J. Ayer argued that certain discourses, e.g., moral discourse, do not describe any worldly facts but perform a non-descriptive function such as expressing certain attitudes or emotions. If this so-called non-cognitivist position is correct, then the placement problem is dissolved because, despite initial appearance to the contrary, there are no moral 'objects' or 'properties' to accommodate within the scientific image. On this line an utterance of the statement "Murdering the innocent is wrong" is not, as it might at first seem, a description of a certain moral property or relation of moral wrongness attaching to certain acts of killing – for there are no such moral properties or relations, nor are there any moral facts characterizable in terms of them. This utterance serves an entirely different purpose than that of describing such items, viz. expressing a con-attitude or unfavourable emotion towards the murder of the innocent. The placement problem is eliminated but at the expense of raising a serious problem about the form ethical language typically takes: if ethical utterance is non-cognitive, why does it so often take the cognitive (descriptive, fact-stating) form of assertions which are assessable as true or false?

Price's linguistic version of pragmatism builds on the insights and oversights of classical expressivism – hence the name 'pragmatic expressivism'. Like classical expressivism, Price's pragmatism considers metaphysical issues from a deflationary linguistic standpoint. That is, one does not begin philosophizing about the placement problem in the material mode by asking, What is goodness or truth or freedom? Since this orthodox form of the problem is stated in metaphysical terms, it invites a metaphysical answer which attempts to give, in the time-honoured Socratic manner, the nature or essence of goodness, etc. The metaphysically deflationary alternative, practised by expressivism, is to begin in the formal or linguistic mode by asking, how do we use the terms 'good', 'true' and 'free'?[1] Logical analysis of concepts is replaced by an explanation of the functions of concepts and their origins. And, as the previous quote makes clear, it is important to the pragmatist way of seeing things that the answer to the question of linguistic use is consistent with naturalism – according to which philosophy shows a proper deference to the sciences, particularly the successful natural sciences. Pragmatists of Price's stripe want to avoid the metaphysical extravagance of having to posit non-natural objects such as Goodness or Truth or Freedom for our terms to refer to. And they want to avoid the need to engage in the unrewarding project of attempting to provide reductive naturalistic analyses of concepts. These are insights in expressivism that Price means to retain.

The oversight of expressivism is in its handling of the surface linguistic data, in particular, the ubiquity of ethical utterance that takes the form of assertions which are assessable as true or false. Rather than saying that such utterances are not really true or false, or only true or false in a qualified sense, Price aims to provide a non-revisionary naturalistic explanation of the linguistic data (in contrast to, say, reductionism or error theory). As we have seen, Price avoids doing material mode metaphysics by refusing to answer Socratic questions of the form "What is x?". But avoiding metaphysics also requires adopting a strategy for avoiding doing metaphysics in a semantic key by way of representationalist assumptions about how language must operate. For example, it is not to be assumed that in order to countenance ethical truths, we must accept that there are ethical truth-makers, or that ethical terms must have ethical 'objects' or 'properties' as referents. So, in addition to the linguistic starting point, another crucial aspect of Price's pragmatism is *anti-representationalism*, which we can understand in the present context as a limitation of one's explanatory resources for answering linguistic questions about the use of terms by disavowing any substantial semantic vocabulary at the level of theoretical explanation.[2] The upshot is that one is not to explain the use of terms such as 'good', 'true' or 'free' by considering what they *refer* to, or what the *truth-conditions or truth-makers* of the sentences are in which they occur. In this way a metaphysics of language is also avoided.

The classical expressivist conception of the linguistic landscape of non-representational discourses has two main features: (1) a negative claim: that (despite appearances to the contrary) the sentences of the relevant discourse lack truth values, so are not fact-stating or descriptive; and (2) a positive claim: that the relevant sentences typically serve to express pro- or con-attitudes or emotions. Price's pragmatist expressivist program rejects (1) and so acknowledges the ordinary use of the terms 'true' and 'false' in linguistic practice such as that of ethics, mathematics and modality, but Price accepts (2), that there is a theoretical functional characterization of the discourse in anti-representationalist terms (where, let us recall, 'representationalism' is a metaphysical view stating how language *must* operate). Price wants to replace a metaphysics of language with a form of linguistic anthropology according to which one engages in serious scientific study of the uses of concepts and their genealogy. This is what he calls "subject naturalism", contrasting with orthodox "object naturalism", which is typically committed to metaphysics at two levels: material mode metaphysics regarding objects (i.e. answers to Socratic what-is questions) and representationalism regarding language.

With this brief sketch of Price's pragmatist expressivism before us, I'd like to explore in more detail its commitment to a version of linguistic functionalism. The eventual critical target of this inquiry will be Price's instrumental conception of linguistic function as applied to truth.

Price's key innovation is to overcome the distinction between monism and pluralism when considering the question of linguistic function. A monist (say, Davidson, 1984) might say that the primary function of language, to which all others reduce, or on which all others are parasitic, is to make assertions which purport to describe the world.[3] A pluralist (say, Wittgenstein, 2009) might say that language has many diverse functions without there being any core function common to all. Transcending this debate, Price argues that the right picture of linguistic function has both monistic and pluralistic dimensions. He grants to Brandom that language has a "downtown" – namely, making and questioning assertions, the practice of giving and asking for reasons – but notes that that admission is, nonetheless, compatible with accepting a functional pluralism across different discourses or linguistic practices that involve assertions. As Price puts it,

> although assertion is indeed a fundamental language game, it is a game with multiple functionally distinct applications – a multifunction tool, in effect.
>
> *(2013, 33)*

Price's enthusiasm for Brandom's account of assertion depends on two main considerations. Firstly, Brandom shares Price's view that assertion is "a fundamental language game", which is employed in science and ethics alike. This view does justice to the surface linguistic phenomena that posed a serious problem for the non-cognitivist stance of classical expressivism. Secondly, Brandom's inferentialist account of assertion in terms of giving and asking for reasons shares Price's commitment to anti-representationalism (in the sense of a denial of a representationalist metaphysics of language, e.g., neither Price nor Brandom accepts a correspondence theory of truth).[4]

The first thing that is worth noticing about Price's account is that by "function" (or "use") he means *function or use as a theoretical posit within a serious scientific linguistic theory* so not as an explication of what an ordinary language user would say in answer to the question "How are you using that term (expression)?" or "What do you mean by that term (expression)?" (2011, 215). Indeed, Price affirms that a subject naturalist account of the function of a concept is "immodest" in the sense that it does not confer on the theorist who posits that function the ability to use the concept in question in a practical communicative context (2011, ch.10). Paradoxically, Price invokes

Wittgenstein as a forerunner of his pluralist approach to function despite the fact that an explanation of meaning in terms of use for Wittgenstein aims precisely at conferring the ability to use the concept to communicate in practice. In contrast to Wittgenstein's use of "use", Price's functions are, or may well be, *hidden* since they are theoretical posits rather than aspects of practitioner know-how.[5] The disparity with Wittgenstein is clear from remarks such as this:

> Philosophy puts everything before us, and neither explains nor deduces anything. – Since everything lies open to view there is nothing to explain. For whatever is hidden is of no interest to us.
>
> *(2009, §126)*

But for Price it is crucial that philosophy does explain the pragmatic significance of language from a theoretical stance. The significance of these postulated linguistic functions depends upon their being hidden, not something ordinary speakers could be expected to know. To make this point vivid, Price speaks of his program as "dissecting out the hidden functions of language" (2011, 76).

A second aspect of linguistic function is that it is conceived *instrumentally* by which I mean that language is treated for theoretical purposes as a realm of individualizable speech acts and concepts, each serving some isolable function that has some utility when considered from a biological or an anthropological perspective.[6] For example, as a biologist might ask, "Why do gulls have a red spot on their beaks?", Price asks:

> What does assertoric discourse do for us? ... What are the concepts of truth and falsity for? (What function do they serve in the lives of a linguistic community?)
>
> *(2011, 76)*

Assertion plays a prominent role in Price's account given the ubiquity of the truth predicate across our linguistic practices. In view of its central importance, then, I would now like to take up Price's instrumental approach to truth.

A State-of-Nature Genealogy of Truth

Price argues for an anthropological account of the use and genealogy of the concept of truth according to which, "Truth is the grit that makes our individual opinions engage with one another" (2010, 231); or, as he also puts it, truth is what "makes disagreements matter" (241). The aim of this account is to explain the normative function of the concept in our assertoric practices using only naturalistic vocabulary. It is a contribution to 'subject naturalism', a scientific (biological or anthropological) account of the function of our concepts and of the role of the fundamental speech act of assertion.

As a preliminary it is important to note that Price's genealogy of the concept of truth is not proposed as a factual account of its actual origins in human history. It is an imagined history akin to Hobbes' state of nature account of the rise and legitimacy of absolute monarchy. It is a thought experiment which, starting from a basic conception of humans and their needs and interests, imagines a state of things *before* the concept of truth was introduced and then hypothesizes the behavioural difference the concept would make once it is introduced. If successful, this account could plausibly make two important theoretical advances. It could explain our possession of a central normative notion in naturalistic terms without any question of reducing the one to the other. And it could explain in general terms the human motivation for introducing the concept of truth in so far as the practical difference it makes is something that has utility or benefit for us as a community. If successful, this hypothetical genealogy of truth is supposed to show the pragmatic advantages for any human society of employing a concept of truth conforming to Price's conception of its function.[7]

The Function of Truth: The Case of the MO'Ans

Price's thought experiment begins by imagining (or at least attempting to imagine) a human society of MO'Ans whose linguistic practices by hypothesis lack the concept of truth. MO'An discourse is, nonetheless, imagined to be characterized as being committed to the following two norms:

1   Subjectivity Assertibility: A speaker is incorrect to assert that P if she does not believe that P; to assert that P in these circumstances provides prima facie grounds for censure, or disapprobation;
2   Personal Warranted Assertibility: A speaker is incorrect to assert that P if she does not have adequate (personal) grounds for believing that P; to assert that P in these circumstances provides prima facie grounds for censure (2010, 234).

These are essentially norms of sincerity and subjective justification, respectively. Price argues that the first norm need have nothing to do with truth since it is analogous to norms of correctness governing non-truth-apt utterances such as "Coffee please!" where the appropriate norm is, say, "Don't order coffee if you don't want coffee" – perhaps an instance of a more general norm covering orders and ordering. Price treats the norm of subjective justification as a matter concerning the question whether a candidate for belief is justified by one's other current beliefs. Once again, he argues that correctness need not be understood in terms of truth. Coherence with one's own web of current belief is enough. I want to question whether this account of correctness makes sense, but let us continue with the exposition for the present.

In a MO'An community characterized by these two norms, speakers have no motivation to see disagreements as having any motivational force. MO'Ans see no point in arguing with each other to settle disagreements since, by the only lights they recognize and care about – their own internal standards of sincerity and coherence – such disagreements are of no consequence; or, as Price puts it, for them, "disagreements are of a no-fault kind" akin to differences in preferences for such things as chocolate, wine and coffee (2010, 239). Thus, he continues, in order for such disagreements to matter, we have to add a third norm, stronger than both sincerity and subjective justification:

> If we didn't have a normative notion in addition to the norms of subjective assertibility and personal warranted assertibility, the idea that we might improve our commitments by aligning them with those of our community would be simply incoherent.
>
> *(2010, 236)*

The third norm Price proposes is the norm of truth whose role is to "create the conceptual space for the idea of further improvement" in our current commitments (2010, 240). But couldn't one say that all that is required is a communal version of the norm of justification, such as:

> Communal Warranted Assertibility: A speaker is incorrect to assert that P if she does not have adequate communal grounds for believing that P; to assert that P in these circumstances provides prima facie grounds for censure.

This norm makes available a potential normative gap between complying with the first two norms and the correctness of one's assertions. It thus allows for the idea that an individual speaker might improve their views by according not simply with their own current beliefs but by attempting to accommodate the best beliefs of the wider community. With this norm in place, there is a motive to find intra-communal disagreements to be consequential in so far as they represent differences over which opinion coheres best with communal norms.

But the norm of communal warranted assertibility won't do the required work according to Price because it is defined in terms of a *particular* community. Price runs the argument again at the community level. If a MO'An is sincere, is subjectively coherent and conforms to norms of justification within the MO'An community, then inter-communal disagreements don't get any grip. Why should such a MO'An care about disagreements with a member of *another* community? The relation between individuals in the first scenario is mirrored by the relation between communities in the second, but now there is no way of rectifying the problem.[8] One is trained in the norms of one's own community, not those of others. Price concludes that disagreements between communities are no-fault in character, and so we need a further normative dimension to change their character from no-fault to at-fault disagreements.

At this stage in his account, Price wants to call our attention to the fact that this imagined scenario is importantly different from our own practices, since we do, as a matter of fact, take disagreements in beliefs between communities as no less consequential, no less normatively loaded, than those within them. In our linguistic practices, we employ a third norm stronger than communal warranted assertibility that allows "the actual community … to recognize that it may be wrong by the standards of some broader community" (2010, 235).

Price proceeds to argue that the third norm must be understood as a norm of truth, which he characterizes as follows:

> Truth: If not-P then it is incorrect to assert that P; if not-P there are prima facie grounds for censure of an assertion that P.

This is not simply disquotationalism about truth. Even if we grant that the truth of non-paradoxical sentences can be captured in the disquotational schema – 'S' is true iff S – this does not exhaust the function of truth. We can see this by noting that we could add a purely disquotational truth predicate to the MO'An society without that making any significant difference. Saying "That's true" might be used by them to signal the agreement of one speaker with that of another, for example, but no more than that. Price's main claim, then, is that truth is a normative notion which provides disagreements with the right sort of normative impact – the sort characteristic of our own assertoric and doxastic practices. Disagreements are taken to indicate one or other party to the disagreement is mistaken, thus stimulating debate and argument with a view to reasoned resolution of the conflict.

We may summarize Price's account as making out how the norm of truth operates to create the conceptual space for improvement in one's commitments by provoking argument and the sharing of information or evidence. In virtue of this third norm, we can make sense of the possibility of improving our commitments by seeking to align them with those with whom we disagree. This is its passive role. The norm also plays an active role, causally "motivat[ing] speakers who disagree to try to resolve their disagreement" through reasoned argument and so pooling our cognitive resources for collective gain. It encourages speakers to rationally resolve their differences for the benefit of gaining the "community's positive evaluation" (2010, 240–241). Price concludes that the norm of truth is required to convert the MO'An assertoric practice into ours or something close to it.

The Problem of Incoherence

A key question for Price's thought-experiment and the very idea of a distinct genealogy of the concept of truth is whether it is possible to instrumentalize truth. Can we treat truth as an isolable tool or an instrument having an explanatorily substantial function within a scientific theory? In particular, does truth play some distinct instrumental role that we can imagine having done without? I want to argue that the answer to these questions is no.

Price's account relies on the viability of imagining a society of humans (MO'Ans) who have propositional attitudes and the capacity to speak and argue with one another but who, by hypothesis, lack the concept of truth. Given this scenario, he wants to ask what behavioural difference adding truth makes? The assumption is that this will show up by comparing MO'An society with our own society. Here it is of utmost important to notice that Price describes MO'Ans in their supposedly truth-lacking state as having opinions and beliefs, of their making assertions – merely opinionated assertions, hence the name MO'Ans – and of their noticing disagreements in their commitments which they do not care about. The glaring problem is that this is obviously incoherent in so far as all the key terms here – 'opinion', 'belief', 'assertion' and 'disagreement' – presuppose the notion of truth in a circumstance imagined as one in which the concept of truth has not yet been introduced.

Consider the case of opinion or belief. It is important to see that Price is talking about beliefs in the context in which the bearers of the beliefs are rational linguistic agents who can see themselves as believers and who take responsibility for seeing that their beliefs satisfy the norms that are appropriate to them.[9] Belief and truth are conceptually entangled in this sense: to believe that p is to be, and be able to see oneself as, committed to the truth of p. So, too, to assert that p – at least to do so seriously and sincerely – is to express a commitment to the truth of p from a perspective that sees oneself as so committed.[10] In short, to be of the opinion, or the belief, that p is to take p to be true.[11] And to disagree about whether p – again with the requisite seriousness – is to disagree about whether p is true or not from a perspective that sees oneself as not so committed. The moral is clear: without the normative concept of truth, we cannot speak of the MO'Ans as following any norm of MO'An linguistic practice including the first or second norms. Sincerity involves expressing what one takes to be true, and subjective coherence, at a minimum, requires that one's beliefs can all be true together.

Price is aware of the difficulty, wondering whether his account of the third norm "does not viciously presuppose the very notions for which it seeks to account" (2011, 175). It will be instructive to consider his responses, since he has more than one. One Pricean response to the problem of incoherence is to admit that the MO'An thought experiment is impossible but to instead imagine another route to the desired result, "by imposing suitable restrictions on real linguistic practices … [that is], by imagining self-imposed restrictions on what we are allowed to say" (2010, 250). This is to imagine the MO'Ans as something akin to Pyrrhonian sceptics, restricted to saying things like "My own current opinion is that P", without being in a position to flatly assert that P (2011, 172 fn 15). But this does not work either. We can paraphrase this restricted expression as follows:

> I am currently of this opinion: P.

This is equivalent to:

> I am currently of this opinion: "P" is true.

Here both the term 'opinion' and the emphasis on the 1st-p perspective helps to bring out the Janus-faced nature of belief: that it points both back to the believer and his state of believing and, also, towards the world and the subject-matter of the belief. The important point is that even if the believer calls attention to himself and his local epistemic perspective in this way – perhaps by way of a social convention in which this is called for – what he believes is still something which contradicts another who is of the contrary opinion that not-P. And it is still something that there is a certain amount of pressure to argue about on rational grounds – supposing the disputants have the inclination, time and so forth. And such argument, of course, aims at possible resolution.

Price goes on to suggest that the difficulty we have in imagining the MO'An scenario is simply that we have an irresistible, but wholly contingent, urge "to see the situation in terms of our own normative standards" (2010, 250). But it is not a contingent matter that we see someone understandable as a rational self-conscious speaker in terms of having beliefs and making assertions that are assessable as true or false. It is, as Davidson has long argued, a constitutive condition of their being a rational speaker of a language at all (e.g., 1980, 233). In understanding and interpreting the speech of another, we make essential use of the notion of truth whether by way of ascription of the attitude of holding-a-sentence-true or taking certain utterances as the sincere expression of a commitment to the truth of their contents. The capacity to understand and interpret another as a rational agent goes hand in hand with ascribing beliefs to them, understood as truth commitments to various contents; for instance, about the immediate environment.

Price wavers on whether "giving up truth" is incoherent or just contingently very difficult because so deeply ingrained – but he tends to favour the latter view. He writes, "Perhaps a truth-like norm is essential to any practice that deserves to be called linguistic" (2010, 249, fn 16). Yet he goes on to say that if we gave up truth, then that would "reduc[e] the conversation of mankind to a chatter of disengaged monologues". The problem with this is that a monologue is still an intelligible truth-involving use of language even if it does not engage with the monologues of others. Price further claims that "a linguistic practice without truth" would be "radically different", but this falls short of acknowledging that without truth there would be no linguistic practice at all (2011, 166). To attempt to imagine a linguistic practice without truth is akin to attempting to imagine a round square. Without truth there is no dialogue, and no monologue either. The unimaginability of MO'An linguistic practice is not a contingent limitation of our powers of imagining, as Price would have it, but a consequence of the fact that truth is part of a package deal of constitutive concepts, which includes rationality, belief, assertion and linguistic meaning.

As Price sees it, assertive uses of language are "multipurpose tools", which have a two-level functional structure, which combines global uniformity at one level with local diversity at another:

> Global Uniformity: The distinctive and uniform function of truth in the employment of assertoric form "provides a powerful pressure towards alignment of such attitudes across a community, with long-run benefits" (2013, 153). Price spells out the global functional role of truth in terms of its "internal", "in-game" or "i-representational" role within language; one option for which is Brandom's inferentialist treatment of assertion.
>
> Local Diversity: assertions in different discourses have a diversity of local functions that distinguish them from each other. For example, scientific uses of language often have an environment-tracking or 'e-representational' component which is not present in, say, ethical uses of language.

The present criticism of instrumentalizing truth is tangential to the monism versus pluralism question, as applied to linguistic function. What is being denied is that truth can be explained instrumentally without circularity. While it is plausible that a concept must figure in some functional explanations if it plays a meaningful role in the use of language – if it is to do "work" in Wittgenstein's metaphor (2009, §132) – there is no reason to accept that concepts must have isolable functions that figure in substantial scientific explanations. Truth poses a double difficulty in so far as it is a package-deal concept – and so without an isolable function distinct from those of meaning, belief, reason etc. – and a constitutive concept which cannot be explained except in terms that presuppose it. The conclusion is clear: truth cannot be instrumentalized. We cannot strip out the concept of truth from an account of rational agency or our engagement in linguistic practices in order to ask what it does for us without incoherence.[12]

A Problem for Classical Pragmatism

Price's problematic commitment to an atomistic conception of linguistic function arguably also afflicts the pragmatic maxim, one of the two central pillars of classical pragmatism.[13] The pragmatic maxim traces the meaning of a concept to its conceivable practical effects, which, apart from sensations, involves the behavioural difference it makes to our practices. If that is so, then there is more in common between classical pragmatism and the linguistic pragmatism of Price than one might have initially supposed – especially given that Price explicitly sees his own convenient friction account of truth as contrasting with James's view.[14] There are indeed differences, most glaringly, that each has a different focus: for James, experience; for Price, language. But this is arguably more a matter of emphasis than a substantive difference, given that the intersubjective domain of language is contained within James's enriched notion of 'experience', which he uses in a decidedly non-Cartesian sense without any restriction to the subjective standpoint of an individual mind.[15]

In the present context, I cannot explore the Price-James connection in much detail. In particular, I will not broach the similarities and differences in their functional explanations of truth.[16] The aim is only to establish that, notwithstanding their differences of emphasis, there is a close kinship between Price's and James's functionalist approaches to truth even if the results of their inquiries diverge to some extent. What matters for our purposes is that both suffer the same problem of attempting to instrumentalize truth.

As James interprets the pragmatic maxim, the whole meaning of a concept is a matter of its conceivable practical effects:[17] "You must bring out of each [meaningful] word its practical cash-value" (1975, 31). Applying this maxim to truth – as part of an explicit rejection of the correspondence theory – James arrives at what he calls "the instrumental view of truth", according to which "Truth ... becomes a class-name for all sorts of definite working-values in experience" (38). Given that language is part of what James means by 'experience', there is a close analogy with Price's approach, which is based on the following shared commitments: (1) a pragmatic or function-based approach to linguistic or conceptual meaning as opposed to logical analysis; (2) a pluralist conception of pragmatic function, admitting that a single concept can perform a plurality of functions; (3) an atomistic conception of linguistic or conceptual function according to which each word or concept can be studied in isolation; and (4) expecting a fruitful non-representational answer to the philosophical question, 'What does the concept of truth do for us?' – that is, one that does not involve any unique correspondence relation or a category of truth-makers.

Like Price, James treats each word or concept as a specific instrument with a particular practical 'cash-value'.[18] That suggests an atomist conception of pragmatic function and a picture of language according to which it is possible to remove any given conceptual 'atom' while leaving the rest of language relatively intact. Applying that thinking to the concept of truth is to imagine that we can have a working language without the concept of truth. As we have seen, this 'possibility' is explicitly invoked by Price's MO'An thought experiment as well as his explanation of its failure as a matter of contingent inconvenience. James does not explicitly commit to this possibility, but it plausibly follows from his methodological remarks.[19]

By offering a functional genealogical explanation of truth, pragmatism avoids the intractable problems of attempting to analyse normative notions such as truth in non-normative naturalistic terms.[20] But serious problems arise for the instrumental approach to explaining the meaning of concepts. What both Price and James overlook in treating language as an assembly of individual instruments (tools, functions) is that some concepts, of which truth is a leading example, are constitutive of language and reason, constitutive of anything deserving to be counted as language or reason. Since truth is part of the framework of language and rationality, it cannot be regarded as an instrument of language or reason which could be dispensed with under certain circumstances. Pragmatists generally tend to miss this point, I suggest, because they are looking at language from a scientific point of view. Price speaks of viewing language from "an anthropological perspective" (2011, 270) and of his resulting account being a contribution to "serious science" (116). And James's instrumental conception of truth is based on a scientific model of explanation of the following kind:

> to take some simple process actually observable in operation ... and then to generalize it, making it apply to all times, and produce great results by summating its effects through the ages.
>
> *(1975, 34)*

Applied to language this involves the method of breaking down the target phenomenon (language) into its simplest discernible components (e.g., concepts in use), which, once isolated, are explained in causal-functional terms; the results of which are summed or aggregated.

The general lesson for pragmatism is, as Wittgenstein – another thinker who associates meaning and function – puts it, "logic does not treat of language – or of thought – in the sense in which a natural science treats of a natural phenomenon" (2009, §81). Logic treats our employment of language as assessable in normative terms (e.g., correct, incorrect), whereas science treats natural phenomena as explicable in non-normative causal terms. Looking at language from a logical point of view also reveals networks of conceptual connections, which an atomist conception of pragmatic function fails to do justice to. And logic distinguishes framework concepts such as truth for which substantial non-circular explanations are not available, a problem that has no analogue in the causal explanation of natural phenomena. Granting that 'meaning is use' (§43), the function or work performed by a concept-in-use is not a scientific notion explicable in naturalistic (non-intentional, non-normative) terms. From a scientific perspective, the question, 'What is the function of (the concept of) truth?', is ill-posed. Truth cannot be instrumentalized.

Notes

1. On the distinction between material and formal (linguistic) modes of philosophizing, see Carnap (1937).
2. 'Substantial', in this context, means metaphysically inflationary. The non-revisionary nature of Price's proposal depends upon there being no restrictions on semantic vocabulary in non-theoretical contexts.
3. As Rebecca Kukla and Mark Lance note, "analytic philosophers, of any stripe, act as though the most fundamental, important and common thing we do with language is to use it to make propositionally structured declarative assertions with truth-values" (2009, 10). This focus on the indicative or assertoric is part and parcel of what J.L. Austin refers to as "the descriptive fallacy" (1970, 103).
4. Brandom (1994, xvi) differs from Price in allowing a substantial word-world reference relation at the level of theory. Where he diverges from traditional referentialism is in not taking this reference relation as explanatorily basic but deriving it from pragmatic inferentialist foundations.
5. A constraint of Wittgenstein's conception of meaning is that "nothing is hidden" (2009, §435). For a more detailed criticism of this aspect of Price's proposal, see (Macarthur 2014, 77–95).
6. I want to distinguish my use of the terms 'instrumental' and 'instrumentalize' from what is traditionally called *instrumentalism*, the familiar view that x's (e.g., electrons, the average tax payer, moral values) do not exist but we talk *as if* they do for various practical reasons.

7 It is interesting to note that on Price's account, these advantages cannot be thought of as motivations at the person level. His account is narrated at the general level of human communities engaged in social practices, and its posited linguistic functions are motivationally opaque since they are hidden from the perspective of language users.
8 I show there is another way to respond to Price's argument here in (Macarthur 2020). It is also unclear why the move to posit a norm of intercommunal warranted assertibility is ruled out.
9 In particular, he is not talking about animals to whom we ascribe beliefs (in some lesser, perhaps merely analogical, sense) but who cannot see themselves as believers.
10 Here I am treating 'assertions' in a theatrical performance, e.g., as not really assertions at all since the context robs them of the requisite committal character. They are what Frege calls "sham assertions" since they lack "the requisite seriousness" (Frege, 1997, 330).
11 An opinion is a belief except that we tend to use the term 'opinion' when we want to emphasize the differences between the beliefs of different believers.
12 In his reply to Rorty's criticism of the MO'An thought experiment (along similar lines to those sketched here), Price comes close to acknowledging this insight when he remarks, "I think the kind of considerations that show the Mo'ans to be impossible show that nothing can count as an assertoric practice unless the third norm is at least 'on' by default" (2010, 257).
13 There seems a serious tension in Price's thought between sympathy for Brandom's holistic approach to meaning in terms of a network of inferences and his own atomistic approach to meaning as pragmatic function.
14 Price also distinguishes his preferred view of truth from Richard Rorty (2011, 166).
15 This is an important way in which the empiricism of classical pragmatism differs from that of classical empiricism.
16 James writes, "Truth in our ideas means their power to work" (1975, 34). It is important to see that James has a pluralist conception of 'work', which includes theoretical and practical functions. On the theoretical side are included consistency, coherence, conservation of past belief, simplicity, elegance, etc. as well as fitting with observation and making prediction possible. On the practical side, there is the emotional appeal and successful consequences of believing. It is worth noting that James explicitly says that truths "help us get into satisfactory relation with other parts of our experience", which can be understood in terms of the sharing of cognitive resources as in Price's account (1975, 34).
17 As James puts the maxim, paraphrasing Peirce,

> To attain perfect clearness in our thoughts of an object, then, we need only consider what conceivable effects of a practical kind the object may involve – what sensations we are to expect from it, and what reactions we must prepare. Our conception of these effects, whether immediate or remote, is then for us the whole of our conception of the object, so far as that conception has positive significance at all.
>
> (1975, 29)

By 'practical effects' James means the conduct recommended and the sensations expected.
18 This is consistent with Price's example of assertion – the expression of a truth commitment – as a multifunctional tool. Does James also endorse a version of Price's two-level pragmatism? Not if he has the view that all the interesting functional work of truth is done at the local level – the concept doing different work on different occasions within the stream of experience. Although his writings do not clearly decide the question, it is arguable that James does endorse a version of the two-level view. For James could surely accept that, apart from its specialized uses, there are interesting general things to say about the function of truth as a class name, say, in matters of generalization (e.g., "Everything he said today is true") or in argument (e.g., truth is what is transmitted from premises to conclusion in a valid argument).
19 James writes, "The pragmatic method in such cases is to try to interpret each notion by tracing its respective practical consequences" (1975, 28).
20 For discussion of the problem of conceptual normativity in the context of naturalism, see De Caro and Macarthur (2010).

Bibliography

Austin, J.L. (1970). *Philosophical Papers*. J.O. Urmson & G.J. Warnock, eds. (Oxford: Clarendon).
Brandom, Robert (1994). *Making It Explicit* (Cambridge, MA: Harvard University Press).
Carnap, Rudolph (1937). *The Logical Syntax of Language*. Trans. Amethe Smeaton (London: Routledge).
Davidson, Donald (1980). "Mental Events." In *Essays on Action and Events* (Oxford: Clarendon), 207–225.

———— (1984). "Truth and Meaning." In *Inquiries into Truth and Interpretation* (Oxford: Clarendon), 17–36.
De Caro, Mario & Macarthur, David (2010). *Naturalism and Normativity* (New York: Columbia University Press).
Frege, Gottlob (1997). "The Thought: A Logical Inquiry." In Michael Beaney (ed.) *The Frege Reader* (London: Blackwell), 325–345.
James, William (1975; orig., 1907). *Pragmatism: An Old Name for Some New Ways of Thinking* (Cambridge: Harvard University Press).
Kukla, Rebecca & Lance, Mark (2009). *Yo and Lo* (Cambridge, MA: Harvard University Press).
Macarthur, David (2014). "What's the Use? Price & Wittgenstein on Naturalistic Explanations of Language." *Al-Mukhatabat Journal: Special issue on Wittgenstein* 9: 77–95.
———— (2020). "Does Rorty Have a Blindspot about Truth?" *European Journal of Pragmatism and American Philosophy* [Online], XII(1). http://journals.openedition.org/ejpap/1851.
Price, Huw (1996). "Review of Simon Blackburn, Essays in Quasi-Realism." *Philosophy and Phenomenological Research* 56: 965–968.
————. (2010). "Truth as Convenient Friction." In Mario De Caro & David Macarthur (eds.) *Naturalism and Normativity* (Cambridge, MA: Harvard University Press), 229–252.
————. (2011). *Naturalism without Mirrors* (Oxford: Oxford University Press).
————. (2013). *Expressivism, Pragmatism and Representationalism* (Cambridge: Cambridge University Press).
Wittgenstein, Ludwig (2009). *Philosophical Investigations*, 4th ed. Trans. G. E. M. Anscombe, Peter Hacker & Joachim Schulte (Oxford: Blackwell).

38
PRAGMATISM AND NATURALISM

James R. O'Shea

A Legacy of Pragmatism: Reconciling Scientific Naturalism and Human Experience

The classical pragmatists such as Peirce, James, and Dewey each framed their philosophical thought in terms of thoroughgoing scientific naturalist[1] outlooks on the world that they argued could nonetheless, by that very means, provide a more satisfying philosophical account of the irreducible perspective of human agency, experience, and normativity than had hitherto emerged in the wake of the scientific revolution and the Enlightenment period. Not surprisingly, then, neo-pragmatist themes in philosophy today also remain centrally concerned with how to integrate the often seemingly conflicting priorities of both natural science and the perspective of human experience and agency – of both naturalism and free, rational normativity – within one coherent philosophical view of how all things hang together. It is this pragmatist theme that forms the present topic and which I will approach through the following route, as inevitably just one among many such possible routes.

Distinctive of pragmatism since its birth have been variations on the famous *pragmatic maxim* that was initiated by Peirce in the 1870s as a method for clarifying the conceptual content of any of our intellectual ideas. In relation to the meaning of any idea or concept, he wrote, pragmatist thinkers should consider "what effects, which might conceivably have practical bearings, we conceive the object of our conception to have. Then, our conception of these effects is the whole of our conception of the object" (Peirce 1992: 132). Or again, "there is no distinction of meaning so fine as to consist in anything but a possible difference of practice" (Peirce 1992: 131). And in a later formulation in 1903: "pragmatism teaches us [that] what we think is to be interpreted in terms of what we are prepared to do" (Peirce 1998: 142).

The pragmatic maxim has proven to be highly controversial when interpreted in either strongly verificationist terms (attempting, for example, to reduce the meaningfulness of empirical claims to their modes of possible sensory verification) or in the attempt to define or analyze truth in terms of consequences, as opposed to drawing more plausible indirect connections between meaning and truth. However, at the heart of both classical and more recent pragmatist employments of the maxim has been the enduring idea that our concepts or ideas have their meaning in virtue of implicit hypotheses or norms concerning possible and necessary courses of experience and action, as these are reflected in our commitments in practice to the corresponding patterns of thought and rules of action.

Crudely put, our concept of a dog, for instance, would on this pragmatist view be constituted however flexibly by norms pertaining to what must, may, or ought to follow in a possible experience, inference, or action involving an object of that kind, with various approaches being taken to the ostensible circularity in any such summary statement. As James stressed, if you want to ascertain the meaning of a word or image or idea, follow its *function* in our experience and agency.[2] Or as Peirce argued in detail in 1868 (cf. Peirce 1992: Chs. 2–3), the import of all our conceptions is a matter of norms and habits of inference rather than of direct intuition or bare apprehension. In this way Peirce anticipated the substance of Wilfrid Sellars' later pragmatist-inspired rejection of the "myth of the given" in 1956, including the Sellarsian companion conception of knowledge as a normative standing in the "logical space of space reasons" (Sellars 1963: Ch. 5 §36; on Sellars and pragmatism, see O'Shea 2020 and deVries [this volume]). Yet Sellars, too, sought to combine this irreducibly normative inferentialist and broadly pragmatist outlook on mind, meaning, and knowledge with a thoroughgoing scientific naturalist conception of reality as a whole, from top to bottom.

It is this particular line of development of the classical pragmatic maxim into recent normative inferentialist conceptions of meaning and conceptual content that has become particularly influential recently in relation to the overarching pragmatist aim of reconciling the naturalist and the normative dimensions of human experience and agency.[3]

Inferentialism as Pragmatist and as Support for Naturalism

In recent years Huw Price (2011, 2013) has argued influentially in support of a systematic combination of pragmatism and naturalism that springs in part from pragmatist-inspired views from a wide variety of sources, including, in particular, the inferentialist outlooks on meaning and conceptual content in Wilfrid Sellars (1963) and Robert Brandom (1994, 2011). The previous section touched on some of the reasons why inferentialist outlooks in semantics, in general, can be seen as an extension of classical pragmatist views about meaning and thought. But what is the connection between inferentialism and the sorts of philosophical naturalism that have characterized pragmatism throughout its history? Price, Brandom, and other well-known neo-pragmatists, such as Richard Rorty (1982) and Michael Williams (2013), all refer back in this connection to Sellars' (1963: Ch. 5) rejection of the myth of the given, which rests upon Sellars' supporting conception of the irreducibly normative-inferential "space" of reason-giving. These Sellarsian inferentialist views on meaning and knowledge can be seen as rearticulating the classical pragmatists' attempts to forge a reconciling middle way between certain arguably anti-naturalist metaphysical outlooks, on one hand, and implausibly reductive versions of naturalism, on the other. (Compare what follows with James' pragmatist objections to the "copy theory" of knowledge, for instance, or with Dewey on the "spectator theory" of knowledge.) This section will attempt to clarify the connections between inferentialism, pragmatism, and naturalism in Sellars and Brandom in general, followed by some reflections in relation to Price's more explicitly naturalist version of pragmatism.

Sellars saw the myth of the given as coming in both platonistic or rationalist versions and naturalistic or empiricist versions, each of which, he argued, rests ultimately on philosophically problematic presuppositions concerning alleged primitive relations of representation, reference, or direct intuition assumed to obtain between mind or language and reality (cf. Sellars 1963, and see O'Shea 2017 for further references). Rationalist versions, generally speaking, contend plausibly that it is impossible to explain the concepts, principles, and truths pertaining to logic, mathematics, modality, meaning, intentionality, or morality solely in terms of the sorts of natural causal processes recognized in physics and the other natural sciences. Rather, such explanations are argued to require the recognition of various corresponding domains of abstract entities (properties, relations, facts, possibilia) as the semantic "truth-makers" for the claims made in those various rational domains. The Sellarsian pragmatists, however, have offered arguments for the following claims in response to such outlooks: namely, (1) that such quasi-platonist semantic relations are both naturalistically and epistemically problematic; (2) that there is available an alternative, normative-inferentialist semantics that avoids (to use Price's terms) the metaphysical "placement problems" that are arguably inherent in such "big-R Representationalist" approaches; these philosophers offer instead broadly pragmatist approaches to "the problem ... of 'placing' various kinds of truths in a natural world" in cases where, by Representationalist lights, we "seem to have more truths than truth-makers" (Price 2013: 26); and furthermore, (3) that this normative-inferential pragmatist approach also simultaneously enables the rejection of empiricist versions of the myth of the given, since these fail to accommodate the constitutively normative dimensions of such conceptions within the human practice or "logical space" of giving and asking for reasons.

In short, the Sellarsian pragmatists' reconciling middle way or pragmatist "corridor" mediating conception (James 1978: 32) proposes that the relevant irreducibility of the rational and principled to the natural or physical is not of the metaphysical or givenist kinds that are arguably entailed by traditional representationalist and intuitionist views of mind and language, but rather pertains in various subtle ways to the pervasive normative irreducibility of "ought" to "is" throughout human experience and agency.

It is Brandom (1994, 2011) who has done most to offer detailed and systematic views as to how a "normative inferentialist" semantics, built solely on a pragmatist basis of socially norm-instituting attitudes and behavior, and without relying on theoretically primitive or underived notions of reference or representation, can actually seek to accommodate successfully the significant technical and conceptual advances that have been made in philosophical semantics over the last century and a half (and for further contributions, see also Kukla and Lance 2009). A central challenge that such pragmatic-inferentialist views about meaning and truth have perennially faced from the beginning, however, is to account adequately for the *objectivity* of matters of fact, both in ordinary experience and in the stunning success of scientific theoretical developments. Sellars himself was not fully on the pragmatist bandwagon in this respect, at least not in the particular ways in which Rorty and Brandom have developed Sellarsian inferentialist views in relation to the problem of accounting for the objectivity of empirical and scientific matters of fact. The challenge, in a nutshell, is to deliver adequate conceptions of truth and objectivity on the basis of a pragmatist semantics, according to which such notions must be defined in terms of the norms of assertion and practice established within particular social frameworks and communities of inquirers. In particular, the plausible claim that such views seek to address is that our assertions and theories within the normative "space of reasons" are answerable to the world and not just to each other (cf. Bernstein 2010; Levine 2019; McDowell 1996; Talisse and Aikin 2008; Wolf and Koons 2016).

One long-standing pragmatist strategy in Peirce, James, Sellars, Putnam (1981: 56), Rosenberg (1980), and, most recently, Misak (2004) has been to conceive truth in terms of the regulative ideal of the "opinion which is fated to be ultimately agreed to by all who investigate," as Peirce put it (1992: 139). But while this approach to objective truth is worthy of further exploration, such idealizations and extrapolations themselves generate familiar additional challenges to be overcome that I will not pause to reconsider here. Brandom, however, contends that his normative inferentialist account of the necessary "structural objectivity" of any ordinary empirical discourse, including an inferentialist-derived account of de re representation, by itself successfully avoids the collapse of the correctness and truth conditions of assertions to the standards of any particular discursive community or "we." Brandom's proposal is thus to

> reconstrue objectivity as consisting in a kind of perspectival *form*, rather than in a nonperspectival or cross-perspectival *content*. What is shared by all discursive perspectives is *that* there is

a difference between what is objectively correct in the way of concept application and what is merely taken to be so, not *what* it is – the structure, not the content.

(Brandom 1994: 600)

This is based on Brandom's pragmatic account of the complex discursive interplay between our normative commitments and our entitlements to claims, which he argues embodies a principled distinction, in any propositionally contentful discourse, between the truth conditions and the assertibility conditions of those claims, and in particular "between the contents of ordinary empirical claims and the contents of any claims about who is committed or entitled to what" (2000: 201). This structural objectivity could be taken to provide the neo-pragmatist with a plausible via media between Rorty's resolutely anti-representationalist social consensus views on truth and objectivity on the "far left" hand, so to speak, while also resisting on the "far right" hand what Brandom argues is the implausibly strong scientific naturalist view of Sellars himself that "in the dimension of describing and explaining the world, science is the measure of all things" (Sellars 1963: Ch. 5, §41). Since the "small-r" representationalist aspects of Brandom's views are derived entirely from his Sellarsian normative-inferentialist pragmatics, we would avoid the metaphysical "placement problems" that pragmatists from James to Price have argued plague traditional big-R Representationalist or "copy-theory correspondence" views of meaning, truth, and knowledge.

As far as naturalism is concerned, Brandom suggests that "at each stage in the account ... the abilities attributed to linguistic practitioners are not magical, mysterious, or extraordinary" (Brandom 1994: 155–156). The avoidance of metaphysical placement problems would certainly be congenial to naturalists. What remains pragmatically and conceptually irreducible in Brandom's account, as for Sellars, too, is the distinction between the "is" and the "ought": between the normative and the natural. Here, too, however, Brandom's suggestion is that nothing naturalistically mysterious or "supernatural" is required by the account, since the relevant abilities "are compounded out of reliable dispositions to respond differentially to linguistic and nonlinguistic stimuli":

Nothing more is required to get into the game of giving and asking for reasons – though to say this is not to say that an interpretation of a community as engaged in such practices can be paraphrased in a vocabulary that is limited to descriptions of such dispositions. Norms are not just regularities, though to be properly understood as subject to them, and even as instituting them by one's conduct (along with that of one's fellows), no more need be required than a capacity to conform to regularities.

(Brandom 1994: 156)

Brandom does not usually stress the naturalistic dimensions and aims of classical pragmatism in his works on semantics, perhaps due to the overriding tendency of naturalistic outlooks to fail to capture the constitutive normativity that is essential to the Sellarsian pragmatist approach to meaning and intentional agency. In his *Perspectives on Pragmatism* (2011), however, Brandom explores the relationship between pragmatism and naturalism further, making favorable reference to the "subject naturalism" defended in Huw Price's pragmatism: "a naturalism concerning the *subjects* of discursive understanding and agency" rather than the more common naturalist "thesis about the *objects* represented by different potentially puzzling kinds of concepts" (Brandom 2011: 10). He adds that by "contrast to this object naturalism, the American pragmatists were subject naturalists" (ibid.). It will be appropriate to close, then, with some observations on this most recent conception of the relationship between pragmatism and naturalism.

Price's Subject Naturalism, the New Bifurcation Thesis, and Lingering Questions

Price's pragmatism builds on classical Humean *expressivism* as later improved upon in the *quasi-realism* of Simon Blackburn (1993). This view is designed to capture the cognitive idioms pertaining to truth, reasoning, and objectivity as they function in relevant domains that were classically regarded as non-cognitivist. The expressivists argued that "some declarative uses of language were not really doing what philosophers had previously taken them to be doing" (Price 2013: 149). For example, moral or modal statements concerning rightness or necessity might be taken to be representing (or misrepresenting) corresponding properties of independent realities, when in fact their purpose or function is to express certain attitudes toward inquiry or practice held or endorsed implicitly by the subject or speaker. The resulting expressivist *bifurcation thesis* (Kraut 1990: 158) thus sought to distinguish the meaning or functioning of such statements from those that both traditional expressivism and Blackburn's quasi-realism hold are more straightforwardly descriptive of and represent matter-of-factual empirical properties or states of affairs in the external environment. Expressivist quasi-realist accounts are thus taken to hold only for the former domains (i.e., "locally"), while some form of representational realism holds in the latter empirical and scientific domains, on these classical outlooks.

Price argues, however, that the normative-pragmatic inferentialism of Sellars and Brandom discussed in the previous section affords a way of extending the quasi-realist expressivist and pragmatist outlook "globally" to cover all discursive human practices, i.e., on both sides of the classical bifurcations. Furthermore, Price thus defends a pragmatic-inferentialist *"global anti-representationalism"* (in this respect similar to Rorty), but one which in the spirit of both Blackburn and Brandom preserves our conceptions of reasoned objectivity and truth in any discourses in which those concepts can be shown to have a useful "anthropological" function or use. Price's "global pragmatism" is thus designed to steer clear of the "placement problems" that plague correspondence "copy theories" in all domains, including the full spectrum of currently influential "object naturalist" metaphysical accounts of our scientific theorizing.

Instead, Price's "anthropological" subject naturalism attempts, again in the spirit of both Hume and pragmatism, to explain in naturalistically non-mysterious terms the practical functioning of all vocabularies within our norm-governed inferential and other agentive practices (for an incisive analysis of this project, see Williams 2013; for a bifurcationist reply to Price, see Blackburn 2013; and on questions concerning Price's view and "liberal naturalism," see Shapiro, 2022). Within any such inferentialist practice, globally, whether it concerns science, ordinary life, morality, mathematics, or modality, Brandom's pragmatic-inferentialist conception of a "structural objectivity" and Blackburn's "quasi-realist" explanations clarify the grounds for what Price calls our "*i-representation*" of "the world" pertaining to the relevant domain (where "i" is for *internalist* or *inferential*," Price explains, as opposed to "*e-representation*" where "e" is "for *environmental* or *externalist*" [Price 2013: 36n]). The result for Price is a metaphysically and semantically deflated but otherwise realist pluralism of i-represented "worlds" (thus extending a pluralist theme in pragmatism stretching from James to Carnap and Goodman), where such perspectives are however simultaneously non-mysteriously explained functionally in subject-naturalist anthropological terms.

One key question that arises in relation to Price's view has to do with the continually resurfacing plausibility of the bifurcationist view. When in commonsense terms I see and hear the dog barking before me, I do seem to be causally and sensitively related to that object in the environment in systematically dependent ways that are not similarly the case in relation to my claims about numbers and duties, for example (Blackburn 2013: 82–84). Price offers in ameliorating response, however, "a new bifucationism" (2013: 35–39, 149–53). He sees is as having fruitful parallels to Sellars's earlier distinction between (1) truth as "correct semantic assertibility," viewed as characterizing all truth-apt vocabularies globally, though functioning differently in anthropologically investigable ways in different domains (in this respect comparable to Price's inferential i-representations); and (2) what Sellars called basic empirical or "matter-of-factual" discourse, where there can in addition be a very different sort of naturalistic "picturing" or mapping-and-tracking dimension of *e-representation* functioning systematically *within* the i-representational discourses of those particular empirical and scientific kinds (Price 2013: 160–70).

I think Price is importantly right to suggest that this sort of naturalistic, cognitive-scientific investigation into patterns of environmental e-representation can coherently be investigated as systematically generated and integrated within whatever sorts of normative-pragmatic i-representational practices this explanatory stance proves to be a fruitful approach. One of Sellars' overarching aims, I believe, was to show in this way how the insights into the irreducibly normative and social dimensions of the logical space of reasons, derived in part from the pragmatist tradition, are not incompatible with taking scientific-explanatory stances (including e-representational "mapping" theories) on different but systematically related aspects of those same activities. Sellars and Price thus seek to maintain the non-reductive or "liberal" character of the naturalism that has been central to the pragmatist tradition throughout the last century and a half but without diminishing the scientific character of that naturalism when it is applied by taking a different "anthropological" stance on our own rational (and irrational) agency.

Given space limitations, I must leave this particular trajectory of recent "normative naturalist" views in the pragmatist tradition with one central question that such views invariably face, even supposing they are successful in all the respects discussed to this point. What sort of *non-reductive naturalism* do normative-pragmatic inferentialist views of this sort really leave us with? Put another way, how are the irreducibly normative stances that generate the i-representational contents across all domains ultimately related to the scientific-naturalist anthropological stances and in some cases e-representational posits that are supposed to explain in solely scientific-naturalist terms what we are doing at the i-representational level (cf. De Caro and Macarthur 2010; and Shapiro 2022)?

The challenge would be that if the naturalism is as "liberal" as the versions delineated by Strawson, McDowell, Brandom, Price, and others have suggested, the distinctively conceptual and more generally human character of the relevant practices seems for familiar reasons not to be explained by the relevant naturalistic posits and explanations proposed in our scientific stances. Yet neither mere anti-supernaturalism nor mere anti-platonism seems by themselves to capture the scientific naturalist ambitions that characterize so much of the classical pragmatist tradition, if the resulting normative/naturalist dualism of conceptual stances leaves it explanatorily unclear how the two stances are really related to one another within the natural domain. Both Sellars and Price appeal to the distinction between the *use* of, for example, normative vocabulary within our i-representational activities, and the scientific or anthropological *mentions* or targeting of how such vocabulary functions in our doings and sayings. However, delineating the latter itself would seem to require normative stances. Appeal to the natural evolution of cooperative activities and linguistic-conceptual capacities in the first place is a further pragmatist move often made at this point (e.g., Williams 2013: 143–144 on "the *emergence* of norms"; Stovall 2022).). The open question of whether either or both moves would solve or simply reproduce the original conceptual or "stance" dualisms in new contexts is one that I will have to end with here.

Notes

1 Distinctions are both common and needed in discussions of naturalism, but I will raise them only when necessary for immediate purposes. Roughly speaking, "methodological" or "epistemic" naturalism is the view that the methods and modes of justification in philosophy are or ought to be the same as or

continuous with those exhibited in the natural sciences (a classic defense is W.V.O. Quine's "Epistemology Naturalized" (1969); for analysis of the partial agreements and disagreements between Quine's view and classical pragmatism, see Sandra Rosenthal (1996)). "Ontological," "metaphysical," or "philosophical" naturalism is usually stated as the rejection of all "supernatural" or "nonnatural" entities, properties, and relations, though this position is difficult to define in a non-question begging manner both intensionally and extensionally. "Scientific naturalism" often refers to the convergence or overlap of both methodological and ontological naturalism, though I will use the phrase here in a way that leaves open for further debate the possibility of more "liberal" versions of naturalism that might aim to be consistent with scientific naturalism (e.g., Macarthur 2018; Strawson 2008).

2 A useful place to start on James' views in this regard is his "The Function of Cognition" (1884), reprinted in (James 1978). I offer further analysis with references in O'Shea (2019).

3 There are important alternative pragmatist conceptions of naturalism that remain influential in contemporary discussions, such as those deriving, for example, from Quine (for a recent critique of which, see Westphal 2015) or from Dewey (cf. Kitcher 2018) or from "Columbia Naturalism" (for a recent discussion of the latter in relation to the pragmatism of Joseph Margolis, see Cahoone 2021).

References

Bernstein, Richard. (2010) *The Pragmatic Turn*, Cambridge: Polity Press.
Blackburn, Simon. (1993) *Essays in Quasi-Realism*, Oxford: Oxford University Press.
———. (2013) "Pragmatism: All or Some?" in Price (2013), pp. 67–83.
Brandom, Robert B. (1994) *Making It Explicit: Reasoning, Representing, and Discursive Commitment*, Cambridge, MA and London: Harvard University Press.
———. (2000) *Articulating Reasons: An Introduction to Inferentialism*, Cambridge, MA and London: Harvard University Press.
———. (2011) *Perspectives on Pragmatism: Classical, Recent, and Contemporary*, Cambridge, MA and London: Harvard University Press.
Cahoone, Lawrence. (2021) "Margolis as Columbia Naturalist," *Metaphilosophy* 52: 49–59.
De Caro, Mario and David Macarthur, eds. (2010) *Naturalism and Normativity*, New York: Columbia University Press.
deVries, Willem A. (2023) "Wilfrid Sellars and Pragmatism" In S. Aikin and R. B. Talisse (eds.), *The Routledge Companion to Pragmatism*. New York: Routledge, 63–69.
James, William. (1978) *Pragmatism: A New Name for Some Old Ways of Thinking & The Meaning of Truth: A Sequel to 'Pragmatism'*, Cambridge, MA: Harvard University Press.
Kitcher, Philip. (2018) "Deweyan Naturalism," in Matthew Bagger (ed.) *Pragmatism and Naturalism: Scientific and Social Inquiry after Representationalism*, New York: Columbia University Press, pp. 66–87.
Kraut, Robert. (1990) "Varieties of Pragmatism," *Mind* 99: 157–183.
Kukla, Rebecca and Mark Lance. (2009) *'Yo!' and 'Lo!' The Pragmatic Topography of the Space of Reasons*, Cambridge: Harvard University Press.
Levine, Steven. (2019) *Pragmatism, Objectivity, and Experience*, Cambridge: Cambridge University Press.
Macarthur, David. (2018) "Liberal Naturalism and the Scientific Image of the World," *Inquiry* 62(5): 565–585, DOI: 10.1080/0020174X.2018.1484006.
Misak, Cheryl. (2004) *Truth and the End of Inquiry: A Peircean Account of Truth*, 2nd edition, Oxford: Oxford University Press.
———. (2016) "Peirce, Kant, and What We Must Assume" in Gabriele Gava and Robert Stern (eds.) *Pragmatism, Kant, and Transcendental Philosophy*, London: Routledge, pp. 85–94.
O'Shea, James. (2017) "'Psychological Nominalism' and the Given, from Abstract Entities to Animal Minds," in Patrick J. Reider (ed.) *Wilfrid Sellars, Idealism and Realism: Understanding Psychological Nominalism*, London and New York: Bloomsbury, pp. 19–39.
———. (2019) "James on Percepts, Concepts, and the Function of Cognition," in Alexander Klein (ed.) *Oxford Handbook to William James*, Oxford: Oxford University Press, DOI: 10.1093/oxfordhb/9780199395699.013.15.
———. (2020) "How Pragmatist Was Sellars? Reflections on an Analytic Pragmatism," in Stefan Brandt and Anke Breunig (eds.) *Wilfrid Sellars and Twentieth-Century Philosophy*, London and New York: Routledge, pp. 110–129.
Peirce, Charles S. (1992, 1998). *The Essential Peirce: Volumes 1 and 2*, N. Houser, and C. Kloesel (eds.). Bloomington, IN: Indiana University Press.
Price, H. (2011) *Naturalism Without Mirrors*, Oxford: Oxford University Press.

———. (2013) *Expressivism, Pragmatism, and Representationalism*, Cambridge: Cambridge University Press.
Putnam, H. (1981) *Reason, Truth, and History*, Cambridge: Cambridge University Press.
Quine, W.V.O. (1969) "Epistemology Naturalized," in *Ontological Relativity*, New York: Columbia University Press, pp. 69–89.
Rorty, Richard. (1982) *Consequences of Pragmatism*, Minneapolis: University of Minnesota Press.
Rosenberg, Jay F. (1980) *One World and Our Knowledge of It*, Dordrecht, Holland: D. Reidel Publishing Co.
Rosenthal, Sandra. (1996) "Classical American Pragmatism: The Other Naturalism," *Metaphilosophy* 27: 399–407.
Sellars, Wilfrid. (1963) *Science, Perception and Reality*, Atascadero, CA: Ridgeview Publishing Company.
Shapiro, Lionel. (2022) "Price's Subject Naturalism and Liberal Naturalism," in Mario De Caro and David Macarthur (eds.) *The Routledge Handbook of Liberal Naturalism*, London: Routledge, pp. 152–163.
Stovall, Preston (2022) *The Single-Minded Animal: Shared Intentionality, Normativity, and the Foundations of Discursive Cognition*. New York and London: Routledge.
Talisse, Robert and Scott Aikin. (2008) *Pragmatism: A Guide for the Perplexed*, London: Continuum (Bloomsbury).
Westphal, Kenneth R. (2015) "Conventionalism and the Impoverishment of the Space of Reasons: Carnap, Quine and Sellars," *Journal of the History of Analytic Philosophy* 3(8): 1–66.
Williams, Michael. (2013) "How Pragmatists can be Local Expressivists," in Price (2013), pp. 128–144.
Wolf, Michael P. and Jeremy Randel Koons. (2016) *The Normative and the Natural*, London: Palgrave Macmillan.

39
PRAGMATIST THEORIES OF TRUTH

Cornelis de Waal

The pragmatists' conception of truth is perhaps the most ferociously criticized part of pragmatism.[1] Part of this goes back to William James, who famously equated truth with the cash value of our ideas—a claim that made others respond that surely there is more to truth than that. G.K. Chesterton expressed this quite well when he wrote, "Pragmatism is a matter of human needs; and one of the first human needs is to be something more than a pragmatist" (1908: 64). What inspired this sort of criticism is by and large the purported need to hold on to some form of objectivism, often spurred by a fear of relativism (or even subjectivism) or a concern that without it all our knowing would become instrumentalist, technocratic—a *Weltanschauung* for engineers, as Heidegger once described pragmatism.

The objectivism of modern philosophy was inspired by a worldview that was dominated by the idea of an all-knowing Creator-God. On this view, propositions are true when God assents to them. Thus, "Snow is white" is true when God *agrees* that snow is white. However, after Darwin, when secular explanations for the origin of the universe and its complexity became mainstream, this worldview lost its natural appeal. A common response was to simply evict God. This, however, left a vacancy (Thomas Nagel's famous "view from nowhere"), one that effectively continued to perform the function that was formerly ascribed to God. The possibility of a single, neutral, and complete description of the world (what an all-knowing God would see were he to exist) was taken for granted and continued to act as the benchmark for truth. Pragmatists consider themselves more thorough heirs of Darwin, and this extends to their conception of truth. They refuse to simply take it for granted that a single, neutral, and complete description of the world is possible.

All of this has profound consequences. If there is no longer a God's-eye view to guarantee what is true, then Ivan Karamazov's claim that "if God is dead, everything is permitted" (Dostoevsky 1990: 589) threatens not only morality but also logic and epistemology. Consequently, the challenge for pragmatism becomes that of rejecting an outmoded objectivism without thereby lapsing into relativism, subjectivism, or a directionless instrumentalism. (For the technocrat, ends are merely the means through which to attain further ends so that in the end, there is no purpose.) In this chapter, I cover some of the salient aspects of the resulting notion of truth by looking at several pragmatists.

The official birthplace of pragmatism is a paper by Charles Peirce titled "How to Make Our Ideas Clear."[2] Here Peirce introduces three grades of clearness, the third of which later becomes known as the pragmatic maxim. Peirce argues that what we can mean by philosophic, religious, or scientific conceptions is determined by this maxim. Consequently, if a conception we entertain fails to pass this maxim, then it must be dismissed as meaningless. Now, what happens when we apply this maxim to our conception of truth?

To fully answer this, we must look at all three grades of clearness. Peirce's first grade of clearness is a mere familiarity with the concept—a familiarity that is typically vague and unarticulated. Anyone who asks a question has such a familiarity with truth. This because asking a question typically implies that one believes that there is a true answer to that question (as opposed to answers that are false). It is quite something different, however, to state precisely what this so-called familiarity with truth entails, something that is particularly important when we traverse unfamiliar terrain, say in quantum mechanics. Here Peirce's second grade of clearness comes into play. We reach this second grade when we have formulated abstract criteria that determine unambiguously, and universally, what falls under the conception and what does not. Aristotle's famous definition, "To say of what is that it is, or of what is not that it is not, is true" (*Metaphysics*, 1011b26), is an attempt to do just that. According to Peirce, we cannot stop here, however, as such a definition does not tell us how to apply it. How do we determine whether something that is, is, or something that isn't, isn't? This leads Peirce to a third grade of clearness, which concerns the most general application of such an abstract definition to the world we encounter. This, Peirce argues, can only be done in terms of practical consequences. By this he means "characters that might conceivably influence rational conduct" (R1170).[3] With rational conduct, he means self-controlled conduct—or, more specifically, conduct subject to the normative science of logic. Consequently, insofar as science, religion, and philosophy aim to steer rational conduct, it has to be pragmatic, meaning that, among other things, it must adhere to a pragmatist conception of truth.

Peirce develops the third grade of clearness for truth by explicitly relating the conception of truth as expressed in the first two grades to the process of inquiry rather than to the desire to assuage the skeptics and other peddlers of paper doubt. By inquiry, Peirce means wholly disinterested rational inquiry, done for its own sake, not for some ulterior purpose (including the satisfaction of urgent practical needs). Since the sole motive for such inquiry is to have our questions answered, come what may, truth may be considered the goal of inquiry. Peirce further observes that when we engage in inquiry, we should not do so aimlessly, but we should do it deliberately, meaning that we engage in it with an eye on what the outcome is likely to be. With this Peirce has in mind the conceivable consequences insofar as they may influence future deliberate conduct. The circumstance that such future deliberate conduct may show that answers we now think are true are actually false then opens up the possibility of defining truth in terms of answers that no future inquiry can show to be wrong. Consequently, Peirce writes,

> when I say that I believe that a given assertion is 'true,' what I mean is that I believe that, as regards that particular assertion, [...] sufficiently energetic, searching, and intelligently conducted inquiry,—could a person carry it on endlessly,—would cause him to be fully satisfied with the assertion and never to be shaken from this satisfaction.
>
> *(R655:27)*[4]

In doing so, Peirce argues for a new objectivism, one that does not rely on anything extraneous to the process of inquiry.

William James's pragmatist conception of truth is best looked at against the backdrop of his phenomenology. When a child first opens its eyes, James argues, it experiences "one great blooming, buzzing confusion" (PP1:488).[5] Over time, this confusion gets domesticated. James denies, however, that there is only one right way of doing this, and he sees our practice bearing this out. We commonly find ourselves in multiple universes, each "a consistent system, with definite relations among its own parts" (PP2:292). In fact, even when these universes openly conflict, it doesn't bother us. For instance, most of us happily accept both the sensory and the scientific conception of a table—the former, a solidly filled object with continuous, colored surfaces; the latter, a mostly empty region of space through which numerous, sparsely scattered electric charges traverse at great speed (as noted by Eddington 1928). With the term "universe," James means a way of capturing our experience against the backdrop of which we can declare certain claims to be true or false. As James further explains, each such universe, "*whilst it is attended to* is real after its own fashion; only the reality lapses with the attention" (PP2:293). Returning to the table, we thus occupy a different universe depending on whether we are dining out or doing physics. Put more generally, since each such universe is self-contained, we live in a multiverse, confidently traveling back and forth between different universes even though they do not add up to a consistent whole. Because of this, a single, neutral, and complete description of the world, which can be called "the truth," is not just humanly unattainable but it is also impossible; it rests on a misconception. Instead, the various universes we find ourselves in must prove their worth within experience, and this experience subsequently becomes the reason for accepting some and rejecting others; it is why we accept chemistry while rejecting alchemy. In the end what counts is, can they deliver the goods? What this does, however, is that it directly connects our understanding of the world with our needs and desires—whether it is a desire for food or a desire to understand the behavior of black holes. For James, truth can no longer be considered something we passively encounter, but, as it is a function of how we conceptualize experience, it is something we make. As James insists, "the human serpent is over all"—we cannot neatly separate what we find from what we contribute. This making of truth does not happen wholly at will, and experience can set us straight. The latter happens in terms of future practical consequences. This causes James to remark, using yet another analogy derived from the business world, that truth lives largely on a credit system.

James's view is broadly reflected in how Peirce comes to see the relation between phenomenology, logic, and metaphysics, in Mead's perspective realism (partially inspired by Whitehead's interpretation of Einstein's theory of relativity), and in C.I. Lewis's conceptual pragmatism. The last strain continues through Carnap, Quine, Goodman, and Davidson in a move that comes to significantly undermine empiricism, especially the idea that we can identify certain elements of the world on their own terms and then use the result as the ground for our knowledge.[6]

Mead's biological focus on problematic situations is in line with James's phenomenological approach. For Mead, our conception of the world is an ongoing working hypothesis, one that is shaped—and continually reshaped—by the problems we encounter. Consequently, when we talk of truth, what we are really talking about is problems we have solved (1964: 332). And just as with James earlier, there is no guarantee that everything adds up to a single, consistent worldview. Quite the contrary, Mead interprets James's multiple universes in terms of objective perspectives, which causes him to ascribe to a form of perspective realism (1964: 306–319). Like Dewey, Mead rejects a spectator theory of knowledge. Truth is something we play an active role in and its scope is always relative to a perspective.

It is because of considerations like these that various pragmatists have remarked that they see little use anymore for a notion of truth. What counts instead is that the problems we face are resolved, that the beliefs we have and the claims we make are warranted. This causes Dewey to favor "warranted assertibility" over truth. This approach, however, raises the following issue: If there is no underlying criterion of truth, how can we be justified in calling certain claims warranted? The traditional pragmatist's reply is they are warranted when they result in successful action—that they help us solve the problems we face. The trouble with this response is that not infrequently what appears to be a perfectly good solution later turns out to have been no solution at all. The long practice of bloodletting in medicine is a good example. To this, pragmatists have replied by shifting away from short-term solutions and answers to those that would be universally agreed upon in the long run, as we saw Peirce doing, or under idealized circumstances, as with Hilary Putnam's notion of idealized rational acceptability. Though strictly unattainable, like the physicists' frictionless surfaces, Putnam argues that this ideal is nonetheless sufficiently close to what we *can* attain to make it a reasonable approximation (1981: 55). This move has caused other pragmatists, such as Richard Rorty, to object that people like Peirce and Putnam either tacitly revert back to the notion of truth that they justifiably rejected or that they unwittingly fall back on our current practices of justification, effectively implying that those beliefs are warranted which we and our peers believe are warranted. For Rorty, there is no way out of this. Consequently, for Rorty, the idea of truth is completely outdated; it is an empty compliment "paid to beliefs which we think so well-justified that … further justification is not needed" (2010: 230). In part, what the situation comes down to is a divide within pragmatism, separating those who think that the idea of truth can and needs to be reformed from those who think that it is beyond reform and that we better give up on the idea altogether. We can call them, respectively, the reformists and the rejectionists.

As many observed, there is also a rather uneasy tension between pragmatism and James's argument for the will to believe—a tension that most certainly hampered the early reception of pragmatism, especially within Europe. James's will-to-believe argument was taken to imply that we can make things true simply by believing them. Though this may work sometimes, as with self-fulfilling prophecies, the objection goes that such are exceptions rather than the rule. More generally, James appears to be confusing the practical consequences of a belief being true with the practical consequences of believing that a belief is true. On the latter view, the belief that God exists is "made true" not by what is posited in the belief but by how the belief itself affects the believer. If it makes the believer happier, more content, more at home in the world, then it is true.

Leaving aside the explicit restriction James put on his argument (as, for James, it only applies to genuine options that cannot be decided on intellectual grounds), the problem with many objections to James's will-to-believe argument is that they happily presuppose the conceptions of truth and reality that James, and pragmatists more generally, reject. James explicitly denies that the world is a finished product that can be captured within a single and complete set of (true) statements. Not only are multiple, mutually incompatible descriptions possible but the world is also profoundly unfinished, meaning that we have options, even though, as James also admits "the squeeze is very tight." The result is a shift in focus. Rather than making our ideas conform to the world as much as we can, we should make the world conform to our ideas as much as the world will let us. That there is still room for deliberate action that profoundly affects us is certainly true in our lived experience, where genuine options habitually appear. Moreover, as Dewey and others have keenly realized, this importantly includes socio-political issues as well as the implementation of new technologies. That is to say, to genuinely improve society we first need a vision and we must believe in it, as only then can we make it true. Pragmatists can further add that over the last 200 years we *did* change our world considerably, trying to make it conform to our ideas and our needs with variable success.

The idea that truth and reality are something we make plays a central role also in the pragmatic humanism of F.C.S. Schiller, who extends James's will-to-believe argument to include even the most basic "axioms" of logic (2008). However, as Schiller also points out, this making does not happen in a vacuum. We are rather like artists, struggling to extract a creation (one that conveys meaning) from something that is already there.

Reformist pragmatists are sometimes accused of changing the topic—that what they call truth is not really what normal people mean by it. Consistent with the pragmatic maxim, pragmatists can respond to this by arguing that the meaning of a term is always provisional because it is a product of what we can conceive, and this shifts over time. When inquiry progresses, the meaning of the concepts we employ is not fixed, because we come to know more about what these concepts entail. This is true even for straightforward empirical terms. For instance, Locke's conception of gold in his *Essay Concerning Human Understanding* (1975: II.xxiii.1) pays no attention to such key properties as its exceptionally low electrical resistivity, its extreme ductility, its role in chemical compounds, its use in curing rheumatoid arthritis, etc. The same can be said for our conception of truth. There is no reason to insist that truth today must mean exactly what it did centuries ago. This then raises the question whether "truth" is more like "phlogiston" or more like "gold"—is it a term for which we no longer have any use, as Rorty at times suggests, or is it a term that has grown in meaning (and that still continues to grow), a process during which its character may alter significantly? Or, returning once more to Peirce, under what sort of circumstances do we need to relinquish the abstract definition that informed the second grade of clearness, or even the commonsense understanding that informed the first grade? Briefly put, one can argue that reformist pragmatists are justified in moving away from old-style representationalist or correspondist conceptions of truth, drawing instead a direct connection with the human practice of inquiry, imperfect and fallible as it may be. To acknowledge that our conception of truth is a living notion, is to acknowledge its potential to grow, that when it does its properties are likely to change, and that when you fail to properly nourish it or fail to realize that its surroundings are no longer the same, it will die. In brief, the fact that the notion of truth as developed by pragmatists fails to fully capture what it used to capture is not necessarily a strike against it. This notwithstanding, pragmatists can add that we should not stray too far. We also have a responsibility to honor the received meaning of the terms we employ, something that caused Peirce, for instance, to talk about an ethics of terminology.

That the meaning of our concept of truth has shifted is a possible solution also to the problem of buried secrets, a problem often cited as a refutation of the pragmatist conception of truth. Peirce was the first to raise this objection, but he never seemed particularly bothered by it. The idea is that we can easily envision claims—say that Cleopatra sneezed five times on her third birthday—that may be true even though inquiry could never show them to be true. In that case, they cannot be called true in pragmatic terms. What is more, whether she did sneeze five times that day seems to depend on what happened, not on what future inquiry might be able to say about it. The result is a looming discrepancy between what can be called true and what pragmatists can call true, as such claims can easily be multiplied ad infinitum. However, raising this as an objection misses the point. This because, like the criticism of James's will-to-believe argument, it tacitly relies on a key presupposition that pragmatists reject, namely that a neutral, complete description of the world is possible and that if "Cleopatra sneezed five times on her third birthday" is part of that description, then it is true, and if it is not, then it is false. Put differently, the mere fact that we are able to formulate a random declarative sentence that in virtue of its form must be either true or false is not enough, it either trades heavily on an outmoded metaphysics or it lacks teeth.

Recasting the issue in terms of what someone who would have been present would have observed, as empiricists may be tempted to do, is of little use here, as it would reduce what is true to the fallible product of that observer's actual inquiry. Perhaps she accidentally missed a sneeze or inadvertently misidentified some sound as a sneeze. To fix this by assuming the observer in question to be "without flaws" would beg the question, as it again presupposes the correctness of the discredited objectivism for which reformist pragmatists are seeking to develop an alternative. The pragmatists can further add that the only way to meaningfully talk about truth in situations like these is to assume that the observer found the issue important enough to ensure that the sounds were really sneezes and that no sneezes were missed and that if the inquiry she started by doing this were to be continued by a sufficiently large community of inquirers, this would lead, in the long run, to the agreed-upon opinion that Cleopatra sneezed five times that day (see de Waal 1999).

In the end, it is this that gives meaning to the idea that all declarative sentences are either true or false. The claim, "the observer found what she observed important enough to inquire into it," again drives home the point that we cannot talk about truth outside the scope of (possible) inquiry.

A further criticism of pragmatist theories, one we already touched upon, is that because of their focus on the long run, reformist pragmatists have nothing to offer us today. What good is it for us to "know" that sometime in the long run our opinions would be settled? One way in which pragmatists have responded to this is by making truth a regulative ideal of inquiry—a way of preventing us from prematurely calling out that we already know the truth, and thereby block the road to any further inquiry. Searching for truth thus becomes a moral imperative, one that we are bound by today. But there is more to it. It is also, and more importantly, a presupposition of inquiry. It makes little sense for us to ask questions unless we believe, or at least hope, that those questions can be answered.

Notwithstanding this focus on the long run, however, pragmatists can still maintain that many of our current beliefs are true. What they deny is that we can extend this to any single one of our beliefs. Pragmatists are fallibilists—we may be mistaken about anything, but we certainly cannot be mistaken about everything. Moreover, even if it turns out that we are doomed to remain fallibilists till the bitter end, and that the pragmatist's final opinion is a chimaera, this need not undermine the notion that committing ourselves to the idea of truth as the opinion that would be agreed upon at the end of inquiry is our best bet for ensuring that inquiry remains an open and productive marketplace of ideas.

Finally, the theory of truth described here results from applying Peirce's pragmatic maxim to contemporary notions of truth. It is important to keep in mind, though, that this maxim is to be applied to all key philosophic notions, not just truth. Consequently, the pragmatic theory of truth must not be looked at in isolation but as part of a broader rethinking of how we can understand ourselves and the world.

Notes

1. For a concise introduction to pragmatism, see Cornelis de Waal (2022).
2. Charles Peirce, *The Writings of Charles S. Peirce*, 7 vols. (Bloomington: Indiana University Press, 1984–present), 3:257ff.
3. Unpublished manuscript as identified in Richard Robin, *Annotated Catalogue of the Papers of Charles S. Peirce* (1964).
4. Unpublished manuscript as identified in Richard Robin, *Annotated Catalogue of the Papers of Charles S. Peirce* (1964).
5. References James's *Principles of Psychology* (1981) will be in the format (PPVol#:Page#).
6. For developments of this line of thought, see de Waal (2022: Chapter 9).

Bibliography

Chesterton, Gilbert Keith. (1908). *Orthodoxy*. London: Collins.
de Waal, Cornelis (1999) "Eleven Challenges to the Pragmatic Theory of Truth." *Transactions of the Charles S. Peirce Society* 35.4: 748–766.
de Waal, Cornelis (2022) *Introducing Pragmatism: A Tool for Rethinking Philosophy*. London: Routledge.
Dostoevsky, Fyodor (1990) *The Brothers Karamazov*. San Francisco: North Point Press.
Eddington, Sir Arthur (1928) *The Nature of the Physical World*. Cambridge: Cambridge University Press.
James, William (1981) *Principles of Psychology*, 3 vols. Cambridge: Harvard University Press. Referenced as (PPVolume#: page#).
Locke, John (1975) *An Essay Concerning Human Understanding*. Oxford: Clarendon Press.
Mead, G.H. (1964) *Selected Writings*. Edited by Andrew Reck. Chicago: Chicago University Press.
Peirce, Charles (1984-Present) *The Writings of Charles S. Peirce*, 7 vols. Bloomington: Indiana University Press.

Putnam, Hilary (1981) *Reason Truth and History*. Cambridge: Cambridge University Press.
Robin, Richard Robin (1964) *Annotated Catalogue of the Papers of Charles S. Peirce*. Amherst: University of Massachusetts Press.
Schiller, F.C.S. (2008) "Axioms as Postulates," in *F.C.S. Schiller on Pragmatism and Humanism: Selected Writings, 1891–1939*. Edited by John Shook and Hugh McDonald. Amherst: Humanity Books, 88–102.

40
PRAGMATISM AND INSURRECTIONIST PHILOSOPHY

Lee A. McBride III

Insurrection and transvaluation are challenging notions for pragmatist philosophy. Open-minded pluralism, cooperative intelligence, and piecemeal meliorism seem fundamentally at odds with tenacious calls for wholesale abolition or a radical revolution of values. To shed light on this, I offer a brief interpretation of pragmatism, then attempt to intimate the motivation for an insurrectionist philosophy. To this end, I provide an account of Leonard Harris's idiosyncratic philosophy born of strife and struggle, clarifying the role of Alain Locke's critical pragmatism and the insurrectionist spirit needed to disavow the intervening background assumptions that lurk tacitly behind the dominant order of things.

Pragmatism

Despite Charles Saunders Peirce's disparagement, William James broadened the notion of pragmatism from a method of clarifying the meaning of an idea to a philosophical approach that sits somewhere between arid abstract rationalism and crude positivist empiricism.[1] Pragmatism, on this account, retains both the exercise of our powers of intellectual abstraction and a positive connection with the actual world of finite human lives; it retains both the concrete personal experiences of the street—multitudinous beyond imagination, tangled, muddy, painful, and perplexed—and the explanatory power of the categories and principles of reason—simple, clean, noble, and logically necessary (James 1981: 13–14). José Ortega y Gasset notes that pragmatism attempts to bridge the perennial duality between action and contemplation, *vita activa* and *vita contemplativa*, Martha and Mary (Ortega y Gasset 1960: 45).[2]

> With the amiable cynicism which is characteristic of the Yankees, characteristic of every new people (a new people, just arrived, seems always to be an *enfant terrible*), pragmatism in North America dared to proclaim this thesis—"There is no other truth than success in dealing with things."
>
> *(Ortega y Gasset 1960: 44)*

With pragmatism, we are limited to a mundane and attenuated conception of truth. "The true" is whatever remains stable, whatever is practically satisfying, whatever is useful/productive, whatever bears fidelity and reliability, whatever is verifiable and supported, whatever is consistent and cogent, whatever is legitimate and pervasive, whatever produces real effects (James 1904: 466). As such, pragmatism is trained on efficacious activity and the intellectual insights that help secure valued experiences within the phenomenal realm. That is, pragmatism spurns a priori approaches to philosophy that necessitate antecedent immutable truths in a transcendental realm, a noumenal realm (apart from human experience) known only through intuition or armchair rational inference (Anderson 2015: 25–27). Rather, pragmatism models itself on experimental a posteriori approaches to knowledge production, paying attention to history, context, causal variables, and corrigibility.

> It is the sincere theory in which the cognitive manner of certain sciences is expressed, the way of knowing possessed by those individual sciences that conserve a vestige of that practical attitude which is not a pure zeal for knowing, and, by the same token, is an acceptance of a problem without limits.
> *(Ortega y Gasset 1960: 64)*

Given its affinity toward empirical experimentation and the systematic pattern of inquiry exhibited in laboratories, pragmatism sees no definite distinction between inquiries in the social sciences, the humanities, or ethics. Pragmatic moral inquiry openly makes use of the evidence, theories, and forms of self-understanding developed by humanistic and social scientific inquiries. For pragmatists, "empirically grounded knowledge and forms of understanding bear upon the justification of ethical principles themselves" (Anderson 1998: 16). Compelling ethical justification for our ethical principles comes from actually living out the lives our ethical principles prescribe for us. Pragmatism thus relies upon more than armchair intuition or a priori justification; pragmatism takes into account lived experience from the point of view of agents. Elizabeth Anderson explains:

> Pragmatic ethical justification thus depends on fine-grain, subject-centered descriptions of actual human experience and conduct. People describe what it is like to live their lives in accordance with ethical principles in terms that *matter* to them, in heavily value-laden terms.
> *(Anderson 1998: 17)*

Pragmatism, thus, avowedly recognizes that its inquiries are not value-neutral. Our values, interests, and selective attention influence the terms and categories we deploy in our descriptions of our experiences; by the same token, "empirical inquiry will shape our conception of our interests, and it will also tell us what things fall under the classifications set by our interests" (Anderson 1998: 24). Thus, our inquiries do not rely upon fundamental value-neutral "natural kinds" or self-evident moral laws. The categories and principles we utilize are not given to us a priori; rather they are developed through value-laden a posteriori investigation.

On this account, pragmatist conceptual frameworks are akin to maps. Note, "all maps are simplifications of reality; they leave out most of the features of the world" (Anderson 1998: 34–35). Yet, maps of various detail can be empirically adequate in helping us navigate the concrete world or in helping us secure valued experiences. Maps are empirically inadequate only if they leave out normatively significant features of our social world, or if they misplace or misdescribe those normatively significant features in ways that confuse and mislead us. Thus, given the mutability of our natural and social environments, our maps will need to be updated or revised as normatively important facts change in light of new evidence.

Picturing Paradox, Predicament, and Undue Duress

Charles Sanders Peirce, William James, and John Dewey are commonly depicted as the classical figures of American pragmatism, the distinctive philosophy of the United States. Does this canon skew the portrayal of life in the United States? Does it conceal particular forms of lived experience? Has this depiction limited the discussion within American philosophy to a particular set of philosophical questions? What if the classical figures of American philosophy included David Walker, John Wannuaucon Quinney, Lydia Maria Child, Frederick Douglass, Anna Julia Cooper, Luther Standing Bear, Zitkala-Ša, and Alain Locke (Collins 2019: 176; Harris 2020: 264)?[3] What if American pragmatism centered on the lived experiences of indigenous First Nations peoples, Afrodescendant people, exploited immigrant populations, and women? Would concerns about chattel slavery, indigenous "removal" tactics, racist terrorism/lynching, or the pervasive subjection of women be more salient in American philosophy? Would American pragmatists discuss democracy and the democratic way of life as if *liberté*, *egalite*, and *fraternité* had been achieved and extended to all inhabitants of Turtle Island?[4] Would modern-day pragmatists argue that sympathetic apprehension of the other, thorough social inquiry, and changes in personal habits are sufficient to ameliorate the systemic inequities that attend the hegemonic techno-industrial order of things in the Global North?

In 1829, David Walker wrote, "My object is, if possible, to awaken in the breasts of my afflicted, degraded and slumbering brethren, a spirit of inquiry and investigation respecting our miseries and wretchedness in this *Republican Land of Liberty!!!!!!*" (Walker 1965: 2). Leonard Harris is helpful in drawing out these depictions of misery and wretchedness. Harris suggests that our philosophical intuitions and received moral principles may harbor meanings and valuations that escape our notice. On this account, our norms, standards, and principles will be entangled in a set of intervening background assumptions—an *episteme* (Harris 2020: 108). In any epoch, the dominant vocabulary, categories, and norms will be imbued with tacit valuations. Thus, the dominant definitions, conceptual categories, and normative principles of any epoch do not ensure absolute truth or unimpeachable morality. Within an episteme bearing white supremacist values, Indian assimilationist schools, public lynchings, and imposed sterilization procedures (at detention facilities on the border) can be understood as rational, moral, or pious—in line with valid inferences and time-honored moral principles. Whether we champion Kantian deontology, utilitarianism, virtue ethics, or a pragmatic heuristic proceduralism, the intervening background assumptions of the episteme may tacitly harbor valuations that authorize and reinforce the subordination of targeted groups (McBride 2021b: 45–46). Some populations have been demarcated by some purported distinguishing feature (e.g., religion, phenotypic race, or sex) and systemically devalued. Some groups have been jettisoned from the human family, rendered savage or three-fifths of a person, inherently and permanently subordinate, the improper object of dignity and honor. Some groups have been barred from protection under the law, ownership of land, educational and vocational opportunities, and access to those materials that make good health possible; as such, they are rendered "social eunuchs," a social group stripped of its δύναμις *(dunamis)*, its ability to shape its own future (Harris 2020: 104).[5] Nota bene, much of this remains peripheral or unpictured in classical pragmatist philosophy. The suggestion is that wanton acquisitiveness, Euro-American racial imperialism, and heteropatriarchy may be so thoroughly ingrained in the existing hegemonic techno-industrial episteme that logic and moral reasoning may countenance genocide, apartheid, or rendering populations perpetually subordinate and infantilized (in a purported democratic nation) (McBride 2021a: 20–23).

Harris argues that any viable philosophy should include immiseration and undue duress (Harris 2020: 20). Philosophy should make possible epistemologies, metaphysics, and aesthetics that include the excluded; the wretched should not disappear. Harris advocates a philosophy born of struggle (*philosophia nata ex conatu*) that tarries here in the messy and mutable world in which things come to be and pass away—the world of anomalies, predicaments, and undue duress (Harris 2020: 15). Rather than positing an impartial rational agent pursuing antecedent immutable truths in a transcendental realm, philosophy should emphasize the centrality of human limitation and existential struggle. To this end, philosophy should acknowledge the primacy of corporeal health. Philosophers should not aim to separate their selves/souls from their bodies, feign the aloof and pristine station of a god, or confine their philosophical acumen to arid thought experiments and byzantine cloistered debates. Philosophy should not trivialize or abstract away from the fact that some social groupings are dehumanized, terrorized, immiserated, and robbed of bodily health. Some groups are made to endure degrading insults, imminent threat of brutality, and a barrage of boundaries and impediments that impede the securing of assets, opportunities, and good health. Furthermore, philosophy should recognize the limitations that attend human perception and reasoning. Any philosophical position will come from a perspective, a standpoint, some provincial platform from which we build and create. We cannot assume that our philosophical intuitions are universal; we cannot assume that what seems self-evident to us holds in all cases. Philosophers, thus, should admit their cultural biases, their value-laden commitments, their perspectival valuations. On this account, although we may strive to exceed our accustomed horizons and proclivities, we never escape some set of provincialisms or biases; value-neutrality, at best, is a regulative ideal. We are organic beings struggling for survival; "we can have hope without entrapment in undue delusions" (Harris 2020: 31). We can build traditions without the pretense that we have transcended provincialism once and for all.

Harris suggests that a viable philosophy should provide resources and reasoning methods that make the management of abjection and existential crisis viable, given impossible odds of relief. Philosophy should provide vocabulary, imagery, poetry, concepts, spirit, epideictic rhetoric, and evidential reasoning, resources that make it possible to imagine and articulate a new way forward. Ideally, it should proffer the sort of spirit/*ethos*/animus needed to escape conceptual backwaters and asylums, retain a sense of hope, negate conventional categories and valuations, and redraw the maps of spring in bold and untested ways.[6]

Harris suggests that Alain Locke's iteration of pragmatism—critical pragmatism—proffers conceptual resources for the subjected facing existential crisis (Harris 2020: 189). Like the classical pragmatists, Locke was antifoundational, pluralist, fallibilist, and a proponent of democracy. But, unlike the canonical figures, Locke wrote from the perspective of an African American (closeted gay) man. And, from that perspective, he remained critical of uniformitarianism, proprietary culture, scientism, racial and ethnic stereotypes, racism, colonialism, and the unfinished business of democracy.

Locke raised concerns about the prevalent "logico-experimental slant" in American philosophy circa 1935. He was concerned that philosophy in the United States (including pragmatism) had become too positivistic, too committed to scientific objectivism (Locke 1989: 36–37). Locke emphasized that human behavior is as selectively preferential as it is experimental. To address this, Locke outlines "a descriptive and empirical psychology of valuation" that is meant to offset the one-sidedness of scientistic/reductive positivism (Locke 1989: 38).

It is worth noting that Locke hoped to discern middle ground between anarchic relativism and dogmatic absolutism. He writes, "The effective antidote to value absolutism lies in a systematic and realistic demonstration that values are rooted in attitudes, not in reality, and pertain to ourselves, not to the world" (Locke 1989: 46). Hence, Locke advocates a functional value relativism; his view recognizes cultural incommensurability and the fact of pluralism but remains open to common denominator values germane to humanity (e.g., reciprocity, parity, confraternity, etc.). Locke recognizes that value pluralism and cultural differentiation (and conflict) will remain perennial issues, yet it is hoped that our values and cultural differences "will be less arbitrary, less provincial and less divisive" (Locke 1989: 50).

On this view, all philosophies ultimately derive from life, products of history, place, and situation; each of our philosophies evinces a provincial type (or personality) projected into their systematic rationalizations (Locke 1989: 34). The critical pragmatist tarries in the phenomenal realm; human (physical and cognitive) limitations and predilections are acknowledged. And from this stance, critical pragmatism retains a world of anomalies, incommensurabilities, thick ethical concepts, paradoxes, and awful predicaments (Harris 2020: 193–196). As a value relativist, Locke embraces moral imperatives and finds it exceedingly important that we find a way to establish some normative principles or valid ultimates for values without resorting to arbitrary dogmatism or absolutism. Value ultimates can function as heuristics (or stereotypes) of feeling-attitudes and dispositional imperatives of action-choices, which are only secondarily reinforced by reason and judgment (Locke 1989: 36). Locke makes use of stereotypes and heuristics as inescapable tools of human cognition yet notes the inevitability of fallacious inferences and mis-categorization. Locke is a cosmopolitan that conceives of phenotype-based races as tropes imposed on racialized ethnic groups bearing concrete cultural types and provincial traditions that change over time, and yet, he champions the New Negro and ardently advocates for Afrodescendant peoples in the United States (Carter 2016; Locke 1992). "Locke embraces moral imperatives, utility of stereotypes, and the dilemma of race, facing without pretention the quandaries that attended each" (Harris 2020: 196). In this fashion, critical pragmatism provides vocabulary, depictions, poetry, spirit, epideictic rhetoric, conceptual categories, and evidential reasoning methods as resources that make the management of abjection and existential crisis feasible.

An Ethos of Insurrection, or, Leaping into the Abyss

But Locke's critical pragmatism may not be sufficient. Nothing in critical pragmatism precludes Locke from supporting slave insurrections or disavowing hegemonic norms but what gives the impetus, the animus, the imperative to engage in such perilous endeavors (Harris 2020: 196). If insurrectionists appear downright ridiculous by customary norms, and insurrectionist endeavors often fail to achieve their intended goal (invariably bringing exponential retaliation to oneself and associates), what would give good reason to disobey authority, to disavow norms and traditions, to start anew?

In short, Harris argues that an insurrectionist spirit/ethos is needed to disavow ingrained norms and traditions. The impetus needed is exhibited in outlandish figures such as David Walker, Maria Stewart, Lydia Maria Child, Henry David Thoreau, and Angela Davis (McBride 2013, 2017). Each, in their own way, called for a break with the instantiated norms and intervening background assumptions of the epoch. Each calls for a transvaluation/revolution so contrary to the status quo that it is hard to imagine.

> Since the condition of slaves is such as I have described, are you surprised at occasional insurrections? You may *regret* it most deeply; but *can* you wonder at it. The famous Captain Smith, when he was a slave in Tartary, killed his overseer and made his escape. I never heard him blamed for it—it seems to be universally considered a simple act of self-defence. The same thing has often occurred with regard to white men taken by the Algerines. The Poles have shed Russian "blood enough to float our navy;" and we admire and praise them, because they did it in resistance of oppression. Yet they have suffered less than black slaves, all the world over, are suffering. We honor our forefathers because they rebelled against certain principles dangerous to political freedom; yet from actual, personal tyranny, they suffered nothing: the negro, on the contrary, is suffering all that oppression *can* make human nature suffer. Why do we execrate in one set of men, what we laud so highly in another? I shall be reminded that insurrections and murders are totally at variance with the precepts of our religion; and this is most true. But according to this rule, the Americans, Poles, Parisians, Belgians, and all who have shed blood for the sake of liberty, are more to blame than the negroes; for the former are more enlightened, and can always have access to the fountain of religion; while the latter are

> kept in a state of brutal ignorance—not allowed to read their Bibles—knowing nothing of Christianity, except the examples of their masters, who profess to be governed by its maxims.
>
> *(Child 1996: 184)*

The insurrectionist boldly depicts the impetus needed to challenge the episteme. They compel critical agents to deconstruct and disavow values and norms that mark stigmatized/racialized groups as subhuman, falling outside of the human family. The insurrectionist prods us to give approbation to conceptions of humanity and personhood that maintain robust notions of reciprocity and fraternity, affording dignity and honor to all human beings. The insurrectionist understands groups as anabsolute, porous and variegated, and thus is willing to build transversal coalitions and advocate for the oppressed and immiserated. The insurrectionist gives approbation to insurrectionist character traits, such as tenacity, audacity, irreverence, anger, and guile.

> We can help ourselves; for, if we lay aside abject servility, and be determined to act like men, and not brutes—the murderers among the whites would be afraid to show their cruel heads. But O, my God!—in sorrow I must say it, that my colour, all over the world, have a mean, servile spirit. They yield in a moment to the whites, let them be right or wrong—the reason they are able to keep their feet on our throats. Oh! my coloured brethren, all over the world, when shall we arise from this death-like apathy?—And be men!!
>
> *(Walker 1965: 62)*

"Imagination requires leaving categories given to us by experience" (Harris 2020: 196). Imagination allows for new categories, new valuations—new identities, new communities/groupings, new cultures. Our categories, our conceptual schemes, our vocabularies and habits are largely entrapped in the dominant/intervening background assumptions of this epoch—the *episteme*. But "we are not completely trapped in the current world" (Harris 2020: 197). Leaving will likely mean breaking from the well-worn grooves of habitual conduct, breaking from established norms. Leaving might include discord, negation, separation, or violent rupture—insurrection/transvaluation. Again, on this view, there is no escape from having to deal with dilemmas of this sort—but we can still hope for conditions bearing less misery and wretchedness, less subjection and degradation.

Philosophia nata ex conatu strays from conventional attempts to settle or reveal the forms, natures, or essences of normative traits (Harris 2020: 198). It is well pictured as emergent discourses inclusive of poetry, testimonials, and discordant arguments, the engaging of issues that have their origin in struggles to destroy boundaries. Philosophy arising from struggle foregrounds the voices of those loathed, humiliated, and stripped of honor and assets—it speaks to immiseration and necro-being. Harris clarifies:

> I argue that if a philosophy tells us of a viable mode of inquiry, valuation, living, and relating to others, as well as providing guidance and suggestions of how to conceive reality, it should be of service to a slave, serf, and proletariat, and a resource for the abused, subjugated, or humiliated, and the object of abjection.
>
> *(Harris 2020: 201)*

A philosophy should help the wretched and oppressed to escape concrete experience, keep vain hope, negate conventional categories/valuations, insurrect, and leap into the abyss. "Insurrection is the negation of a world. It is an attempt to create a new world" (Harris 2020: 202). Such a philosophy warrants the process of searching and making difficult decisions under conditions of trauma, insecurity, and fear, especially for the subjugated. It proffers resources to imagine a new social world, radically different from this one—a radical break from the episteme (Harris 2020: 204; cf. 196). Insurrection, then, denotes a rejection of the intervening background assumptions, a seeking for a radical newness, a daring leap into novel, untested conditions. Insurrectionist philosophy should help us to interrogate and deconstruct the episteme, encourage us to break with implicit avaricious, racist, and patriarchal norms and traditions, and prompt us to critically reassess valorized methods of inquiry. Insurrectionist philosophy should help to picture and manage the trauma and duress, the necro-being, the loss of faith endured by corporeal beings.[7] It should help to create alternative categories, values, and maps of (im)possible futures.

Notes

1 See Perry (1996: 280–291).
2 See Luke 10:38–42 and John 11:5–20.
3 See Harris, Pratt, and Waters (2002).
4 "Turtle Island" is a name used by several Native American cultures to refer to North America.
5 The Greek term δύναμις (dunamis) denotes potency, efficacy, or power.
6 See Wynter (1995: 35); see also Césaire (2001: 34).
7 The term "necro-being" refers to a living death—conditions that kill and destroy health/lives.

References

Anderson, Elizabeth. 1998. "Pragmatism, Science, and Moral Inquiry." In *In Face of the Facts*. Edited by R. Wightman Fox and R. Westbrook, 10–39. Washington DC: Woodrow Wilson Center Press.
Anderson, Elizabeth. 2015. "Moral Bias and Corrective Practices: A Pragmatist Perspective," *Proceedings and Addresses of the American Philosophical Association*, Vol. 89 (November 2015), 21–47.
Carter, Jacoby Adeshei. 2016. "'Like Rum in the Punch': The Quest for Cultural Democracy." In *African American Contributions to the Americas' Cultures*. Edited by Jacoby Adeshei Carter, 107–167. New York: Palgrave.
Césaire, Aimé. 2001. *Notebook of A Return to the Native Land*. Translated by Clayton Eshleman and Annette Smith. Middletown: Wesleyan University Press.
Child, Lydia Maria. 1996. *An Appeal in Favor of That Class of Americans Called Africans*. Edited by Carolyn L. Karcher. Amherst: University of Massachusetts Press.
Collins, Patricia Hill. 2019. *Intersectionality as Critical Social Theory*. Durham: Duke University Press.
Harris, Leonard. 2020. *A Philosophy of Struggle: The Leonard Harris Reader*. Edited by Lee A. McBride III. New York: Bloomsbury.
Harris, Leonard, Scott Pratt, and Anne S. Waters (eds.). 2002. *American Philosophies: An Anthology*. Malden: Blackwell.
James, William. 1904. "Humanism and Truth," *Mind*, Vol. 13, No. 52, 457–475.
James, William. 1981. *Pragmatism*. Indianapolis: Hackett.
Locke, Alain. 1989. *The Philosophy of Alain Locke*. Edited by Leonard Harris. Philadelphia: Temple University Press.
Locke, Alain (ed.). 1992. *The New Negro*. New York: Touchstone.
Locke, Alain. 2010. "World Citizenship: Mirage or Reality?" In *Philosophic Values and World Citizenship: Locke to Obama and Beyond*. Edited by Jacoby A. Carter and Leonard Harris, 139–146. Lanham: Lexington Books.
Locke, Alain. 2016. *African American Contributions to the Americas' Cultures*. Edited by Jacoby Adeshei Carter. New York: Palgrave.
McBride, Lee A., III. 2013. "Insurrectionist Ethics and Thoreau," *Transactions of the Charles S. Peirce Society*, Vol. 49, No. 1, pp. 29–45.
McBride, Lee A., III. 2017. "Insurrectionist Ethics and Racism." In *The Oxford Handbook of Race and Philosophy*. Edited by Naomi Zack, 225–234. New York: Oxford University Press.
McBride, Lee A. III. 2021a. "Culture, Acquisitiveness, and Decolonial Philosophy." In *Decolonizing American Philosophy*. Edited by Corey McCall and Phillip McReynolds, 17–35. Albany: SUNY Press.
McBride, Lee A., III. 2021b. *Ethics and Insurrection: A Pragmatism for the Oppressed*. London: Bloomsbury.

Ortega y Gasset, Jose. 1960. *What is Philosophy?* Translated by Mildred Adams. New York: W.W. Norton.
Perry, Ralph Barton. 1996. *The Thought and Character of William James.* Nashville: Vanderbilt University Press.
Walker, David. 1965. *Appeal to the Coloured Citizens of the World, but in Particular, and Very Expressly, to Those of the United States of America.* New York: Hill and Wang.
Wynter, Sylvia. 1995. "The Pope Must Have Been Drunk, The King of Castile a Madman: Culture as Actuality, and The Caribbean Rethinking Modernity." In *The Reordering of Culture: Latin America, The Caribbean and Canada in the Hood.* Edited by Alvina Reprecht and Cecilia Taiana, 17–41. Ottawa: Carleton University Press.

41
LATIN AMERICAN PHILOSOPHY, U.S. LATINX PHILOSOPHY, AND ANGLO-AMERICAN PRAGMATISM

Denise Meda Calderon and Andrea J. Pitts

The histories of Anglo-American pragmatism, Latin American philosophy, and U.S. Latinx philosophy have long been interconnected. While many Latin American and U.S. Latinx authors were readers of the same strands of post-Kantian European philosophy that influenced early Anglo-American pragmatists, there have been other, more direct dialogues among Anglo-American pragmatists and Latin American and U.S. Latinx philosophers that connect these geopolitical traditions. For example, recent research by Paniel Reyes Cárdenas and Daniel R. Herbert (2020) has traced the reception of Charles S. Peirce in Brazil, Argentina, Colombia, and Mexico. Within this collection, Catalina Hynes notes that the first documented citations of Peirce in Argentina occurred in 1910 through the publication of a book titled *El pragmatismo* by Coriolano Alberini (1886–1960) (Cárdenas and Herbert 2020: 37). Other Argentine philosophers had a favorable reception of U.S. pragmatism as well, including José Ingenieros (1877–1925) and Risieri Frondizi (1910–1985). For example, Ingenieros traveled to the United States in 1916 in search of resources on Ralph Waldo Emerson (1803–1882) and, in the following year, published *Hacia una moral sin dogmas* [*Towards a Morality without Dogmas*], a book which offers a direct and sustained dialogue with the writings of Emerson on topics of moral perfectionism, social action, and the perils of human mediocrity (Gomez 2011). Following a very different trajectory, Frondizi studied at Harvard University and the University of Michigan, Ann Arbor, with mentors such as Alfred North Whitehead, Roy Wood Sellars, and Ralph Barton Perry, important figures for differing Anglo-American pragmatist trajectories in the 20th century. Frondizi eventually earned his PhD at the National Autonomous University of Mexico and later returned to Argentina to continue his career at the University of Buenos Aires. These experiences in the United States and across sites of Latin America, Frondizi notes, gave him insights into the philosophical differences between "las dos Américas" ["the two Americas"] of the North and the South. He thus developed a long, rich dialogue on the metaphilosophy of these differing philosophical trajectories in the Western Hemisphere (Cárdenas and Herbert 2020: 40).

Anglo-American pragmatist John Dewey also shared important connections throughout his life with Latin American philosophy, and, more specifically, Mexican philosophy. Dewey conducted several important trips to Mexico during his lifetime, one of which included his high-profile role as the chair of the "Commission of Inquiry into the Charges Made against Leon Trotsky in the Moscow Trials" in Mexico City in 1937 (Pappas 2012: 2). Dewey also mentored Mexican education reformers Moisés Sáenz (1888–1941) and Rafael Ramírez (1885–1959), both of whom played a significant role in the development of rural schools throughout Mexico in the 1920s (Pappas 2011). Lastly, we find Dewey's reception in Mexico through Mexican philosopher

José Vasconcelos (1882–1959), who wrote critically of U.S. pragmatism in general and characterized Dewey's approach as a "gunship philosophy" that was directed at the expansion of U.S. imperial interests in the Americas (MacMullan 2015).

Beyond these Anglo-American authors in the history of pragmatism, a number of authors have also begun to examine the pragmatist trajectories found within African American and Indigenous philosophical traditions. For example, Alain Locke, a founding figure of the Harlem Renaissance, has been interpreted as developing a form of "critical pragmatism" that

> promotes a deep-seated commitment to transforming the world, too often filled with racial hatred and prejudice, through intellectual engagement in ways that do not rely on what [Locke] considered the enemies of cross-cultural communication—absolutism, metaphysics, and treating existing social groups, including any race or nation, as a natural creation rather than as the vagary of human manufacture.
>
> *(Harris 2006: 88)[1]*

In this sense, Mexican American philosopher Grant Silva has read Locke and Vasconcelos together, both as proponents of an "axiological turn" in American philosophy. This axiological turn in Vasconcelos, Silva argues, was shared by his contemporary, Locke in the United States, as both philosophers developed critiques of the forms of rationalism and scientific positivism that impacted their respective professional and personal lives (Silva 2010). Additionally, Scott Pratt argues in *Native Pragmatism: Remaking the Roots of American Pragmatism* (2002) that the central commitments of classical U.S. pragmatists, such as Peirce, James, and Dewey, were actually "prefigured in indigenous thought at a time when European thought in America was marked by a set of contrary commitments" (Pratt 2002: xii). Engaging Ojibwe, Delaware, Haudenosaunee, Narragansett, and Penobscot stories of their respective nation's encounters with community outsiders Scott demonstrates that a number of Indigenous nations have long lived according to principles of pluralism, interaction, and transformation through community growth, all of which are often considered by Anglo-American philosophers to be hallmarks of classical U.S. pragmatism (Pratt 2002).

In this critical trajectory, editors such as Kim Díaz and Matthew Foust (2021), Erin McKenna and Pratt (2015), and Leonard Harris, Pratt, and Anne S. Waters (2002) have re-read the history of Anglo-American pragmatism as but one thread of a broader series of philosophical dialogues and traditions taking place across the Americas. Such threads include, for example, as Díaz and Froust propose, ancient creation stories of Indigenous communities in the Americas such as the *Popul Vuh* of the K'iche Mayan peoples, as well as the work of contemporary Indigenous scholars such as V.F. Cordova (Jicarilla Apache) and Silvia Rivera Cusicanqui (Aymara).[2] Such important research connecting philosophy across the Americas, as we detail below through writings on Latin American and U.S. Latinx engagements with Anglo-American pragmatisms, spans the Western Hemisphere. Such approaches thereby provincialize Anglo-American pragmatist traditions as but one tradition among many that upholds pragmatist values, commitments, and methodologies. In what follows, we offer four interpretive lenses through which to connect debates and dialogues among Latin American and U.S. Latinx philosophical discourses to those of Anglo-American pragmatism: theories of liberation, conceptions of selfhood and identity, questions of pluralism, and the practice of public philosophy. In each section, we highlight ongoing work among philosophers engaging issues relevant to U.S. Latinx and Latin American communities, and that likewise address questions raised within Anglo-American pragmatism. The hope of this chapter, like the projects mentioned above, is to expand the purview of pragmatism by demonstrating the plurality and depth of philosophical thinking and praxis across the Americas.

Theories of Liberation

One commitment shared by both U.S. Latinx philosophy and pragmatism has been to theorize from within contexts in which ordinary people live, think, and experience the world. For Latin American and U.S. Latinx communities, whose lives are often marked by struggles against systemic oppressions (colonial, gender-based, capitalist, racist, ethnocentric), much of U.S. Latinx philosophy responds to the harms faced by everyday peoples and their struggles against oppression. Within this vein, Alexander Stehn notes that U.S. pragmatism and Latin American philosophy of liberation, "two American philosophical traditions ... share a metaphilosophy insofar as they take experience as both the fundamental point of departure and the necessary point of arrival for every philosophy worth its salt" (Stehn 2011: 26). However, what distinguishes Latin American liberation philosophers like Enrique Dussel (b. 1934) from U.S. pragmatism has been the concrete commitment of the former to continue to ask, "Whose experience? [and] Whose concrete life has been, is, and will be taken seriously?" Stehn notes, for example, that Dussel's project critiques U.S. neopragmatist Richard Rorty's rejection of foundational ontologies, including those of Marxism. Dussel notes that for many people facing systemic oppressions, being able to articulate both *that* they are suffering and the *etiology* of that suffering are important pragmatic means for eliminating the conditions that are causing them misery, obstacles, and ongoing hardships (Stehn 2011: 18). With respect to Dewey, Stehn notes that Dewey's political work addresses the need to begin from the experiences of those who are facing systemic exclusion and harm, including Dewey's statement that "the man who wears the shoe knows best that it pinches and where the shoe is pinching" (Dewey 2008: 364). Yet, as Stehn argues, unlike theorists of liberation like Dussel, who begin from specific positionalities within oppressed groups, Dewey rarely names "the empirical fact that the shoe was (and still is) disproportionately pinching, for example, women, poor people, or African Americans" (Stehn 2011: 24).[3]

Alongside these analyses of philosophy of liberation and pragmatism, perhaps the most central philosopher to bring these discourses into dialogue is Gregory Pappas (b. 1960). While Pappas has numerous published works on Dewey, including a monograph on Dewey's ethics (2008), he is also to credit for the first and only edited collection that addresses the scope and influence of classical U.S. American pragmatism across Spain and Latin America (2011). Notably, in his own research, Pappas has analyzed the relationship between pragmatism and Mexican philosopher Luis Villoro (1922–2014). He states that

> Villoro is an exemplar of the view central to American Pragmatism, especially in Jane Addams, that a philosopher must continue to revise or refine his/her theories according to what he/she experiences and learns 'on the ground' and in the struggle to ameliorate injustices.
>
> *(Pappas 2017: 86)*

This critical insight articulated by Pappas also shapes his own work, as a Puerto Rican-born philosopher living and doing politically driven work in central Texas. Pappas writes that his relationship to the institutional dimensions of U.S. pragmatism began with rather isolated racial and cultural experiences through the pragmatist organization the Society for the Advancement of American Philosophy (SAAP). While a predominantly Anglo-American-centric organization for many years, Pappas eventually became the president of SAAP in 2018, and he notes that the efforts of African American philosopher Leonard Harris and many other African American, Indigenous, and U.S. Latinx philosophers are now beginning to transform the organization (Pappas 2021: 1–2).[4] Pappas also builds from a variety of non-Anglo-American cultural resources across the Americas, including, for example, the notion of contrapunteo, a concept within Latin jazz music that refers to "the relationship between voices that are harmonically interdependent (polyphony) yet independent in rhythm and contour" (Pappas 2021: 3). Inter-American philosophy, Pappas proposes, can embody this practice by doing philosophical contrapunteo, "juxtaposing and rubbing [different philosophical traditions, authors, and ideas together, while], respecting their relative independence and differences, with the creative goal of learning from their resonance in order to ameliorate present problems" (Pappas 2021: 3). In this sense, Pappas offers a normative metaphilosophical relationship between philosophies of liberation in the Americas and those of classical American pragmatism.

Selfhood/Identity

U.S. Latinx philosophers, including Pappas (2001), have connected questions of liberation to conceptions of identity and selfhood among pragmatist traditions. Building such connections between U.S. pragmatism and Latina feminist conceptions of identity, a number of authors such as Jacqueline Martinez, Paula Moya, and Linda Martín Alcoff have considerably engaged with the writings of U.S. pragmatist authors such as Peirce, Hilary Putnam, and Rorty (Pitts 2021). Likewise, Mariana Ortega has linked the works of William James and W.E.B. Du Bois to questions regarding the pluralism of experience, arguing that James' writings in psychology and Du Bois' conceptualization of double-consciousness allow for rich phenomenological reflections on the experiences of Latinas (Ortega 2016: 121–122). Building from political theorist Edwina Barvosa's engagement with James in her book *Wealth of Selves: Multiple Identities, Mestiza Consciousness, and the Subject of Politics* (2008), Ortega addresses the desire for integrated aspects of the self among racialized and gendered subjects. While Barvosa calls for a "self-integrative life project," Ortega notes that many people may not desire or be willing to strive for such integration, and she constructs a conception of selfhood that seeks to preserve the existential and phenomenological resonances of living across multiple sites of meaning, history, and stratification within racist and sexist social worlds (Ortega 2016: 199–200).

José Medina offers another perspective of U.S. Latinx multiplicity of the self that engages with U.S. pragmatism. Medina's writings on the pluralism of social spaces, agency, and meaning-making practices explicitly discuss the work of classical U.S. pragmatists such as Addams, Dewey, James, George Herbert Mead, and Peirce. Like Ortega, for example, Medina discusses accounts of epistemic unification regarding theories of truth and knowledge. He notes that Peirce and Mead (on some readings) offered unificatory conceptions of epistemic practice, but James was committed more so, "not of consensus and unification, but ... coordination and cooperation" (Medina 2013: 282–283). For Medina, this is akin to his own view of *melioristic pluralism*, which refers to when "epistemic contestations and negotiations are directed toward improving the objectivity of the different standpoints available, toward correcting their biases and mistakes, and toward maintaining their truth *alive*" (Medina 2013: 283). Such a view, Medina argues, offers an epistemological approach toward the political and moral demands of racialized and gendered communities, including U.S. Latinx and Black American communities, for social transformation and the improvement of the material conditions impacting those communities. Additionally, Medina addresses Addams' call to "expose ourselves, make ourselves vulnerable, and let ourselves be guided by feelings of *perplexity* that can activate frictions and resistances going in many directions" (Medina 2013: 311). Medina notes that social sympathy with people who are very different from oneself can be developed through what Addams described as a form of *perplexity*. Medina argues that the cultivation of perplexity can be read in an epistemic register, and he proposes that the "cultivation of [an] openness to being challenged and affected by other experiential perspectives" is a virtue that aids in addressing the forms of epistemic arrogance that underlie much patterned racism and sexism (Medina 2013: 20).

Lastly, among the U.S. Latinx philosophers who have explicitly adopted pragmatist lenses to address issues of identity, Alcoff has been quite influential. While her early book *Real Knowing:*

New Versions of the Coherence Theory (1996) built explicitly from the writings of Donald Davidson and Hilary Putnam, her writings continue to address questions of epistemic and subjective pluralism through pragmatist sources. For example, Alcoff's *Visible Identities: Race, Gender, and the Self* specifically addresses George Herbert Mead's account of social interdependency and its importance for his account of agency (2006: 117–119).[5] Alcoff argues that despite not offering a theorization of how differing social perspectives are hierarchically arranged, Mead's work on self-consciousness offered valuable insights for exploring how social perceptions that approve or disapprove one's abidance with social norms can be applied to understanding cultural, racial, and gendered social identities. Namely, she writes that "Mead's claim that our self-consciousness is produced through the reactions of others makes sense" of the importance of the need for gender and racial minorities to identify and name their minoritarian status within largely white or predominantly masculine social spaces (Alcoff 2006: 119). They are, as she notes, calling attention to how one's position as "other" in an otherwise racially or gendered homogenous social space is produced through the shared meanings of that space and not through the individual's own actions or self-perceptions (2006: 119).

Pluralism

Pluralism functions as a primary lens through which to see the fullness, diversity, and richness in communities and across spaces and is shared among both U.S. Latinx philosophy and Anglo-American pragmatic traditions. In U.S. Latinx philosophy, scholars tend to approach complex community relations through pluralist approaches, especially when such communities act in resistance to social oppressions. In this vein, José Medina and José Antonio Orosco each develop analyses of pragmatic pluralism that encourage democratic processes while remaining attentive to the power dynamics that fracture and marginalize differing peoples. In their works, we see two distinct yet interrelated projects addressing social issues that involve differently subjugated peoples to promote a pluralism informed by lived experiences, inclusive community practices, and cross-group solidarities. That is, pragmatic pluralism in both Medina's and Orosco's views theorize the realities that inform the lived experiences of U.S. Latinx peoples, particularly with attention to communication, habits, and generative intercultural engagement.

The pragmatic approach promoted by Medina delivers an account of social relationality that is fundamentally defined by a pluralist social identity and the contingencies of histories, spaces, and languages that give rise to complex experiential exchanges (Medina 2011). Medina engages Dewey's critical reconstruction, Locke's account of the self-empowerment of the "New Negro," and Cuban writer José Martí's (1853–1895) focus on Latin American intracultural and intercultural relations, to argue that a critical reconstruction of collective identity can facilitate regenerative interactions among U.S. Latinx peoples. Such interactions include addressing their shared histories and experiences while simultaneously maintaining the complexities and differences that exist across the Americas (Medina 2004).

Medina theorizes a pluralist approach through the concept of polyphonic dialogue. Polyphonic dialogue, Medina explains, resists homogenization and essentializing tendencies that can erase, mask, or marginalize differences among and across groups (Medina 2011). This form of communication is facilitated by an experiential exploration on two levels: intragroup and intergroup. "Intragroup" explorations are a process of "getting to know oneself," the self here referring to one's own social group and one's own knowledge of their own relevant histories, values, and modalities, among other aspects (Medina 2011: 142), while an "intergroup" analysis refers to explorations that deliberately involve groups that are in distinct relation. In intracultural and intercultural types of communication, the recognition of local histories and interests for deep solidarity requires a robust pluralist view of communities to engage different communicative values, tactics, histories, and gestures. Subsequently, as a multidimensional and multivocal practice, polyphonic dialogue enables a social dynamic that supports mutual enrichment among members who are linked together but also often fragmented. Medina's account of communication thus illustrates a pluralist practice that seeks to improve relationships within multiethnic communities by fighting homogenizing and normalizing tendencies that often erase differences. Similarly addressing pluralist social relations, Orosco develops a type of engagement that includes all community members, regardless of their national citizenship, to foster intercultural relationships. Applying a Deweyan idea to develop social virtues, Orosco describes pragmatic pluralism through an interculturalism that promotes "democratic habits of the heart" (Orosco 2016). This sense of interaction establishes relations that are impacted by circumstances of place and the resources that people have access to and which inform their personal experiences (Orosco 2011). Thus, the pluralist approach for Orosco involves recentering community values produced by and through a deliberate and reconstructive process of cultural, ethical, and political ideals.

Orosco also addresses issues of participatory democratic engagement, which conceives of experiential interrelationality as a network generated through interactions in shared places and in which people are involved in establishing a sense of community (Orosco 2011). Orosco explains that this notion of participatory engagement is exemplified in Chicano activist Armando Réndon's conception of a form of neighborhood self-governance called "barrio unions" (Orosco 2007). Promoting a type of interculturalism, Orosco frames barrio unions as forms of neighborly social praxis that can function as a participatory democratic network of relationality. Such networks of relationality also collectively respond to social and political marginalization through nonviolent direct action. For instance, U.S. Latinx neighborhoods, when understood as diverse populations often from differing geopolitical spaces, face injustices in public policy and immigration practices. A barrio union framework involves pragmatic pluralism and participatory democracy, offering U.S. Latinx communities a paradigm to reframe civic participation and engage intercultural amelioration through neighborly relations.

Public Philosophy

U.S. Latinx and Latin American philosophers also commonly cultivate dialogues in the public sphere that critique dominant social values and practices, bringing attention to the lives of socially disenfranchised peoples. Anglo-American philosophers such as Addams and Dewey were likewise known for their metaphilosophical commitments that philosophy should serve socially relevant ends. Thus, here we find another linkage among Anglo-American, Latin American, and U.S. Latinx philosophical traditions. Notably, for U.S. Latinx and Latin American philosophers, public philosophy should offer a contextualist approach that seeks to intervene against exclusionary narratives and fragmenting practices imposed by private and public distinctions. The use of alternative intellectual mediums such as news articles, podcasts, blog posts, and so on has been vital to counter such harmful practices, and such media generate non-dichotomized approaches to social enrichment that transcend traditional public-private domains. Through an approach that aims to expand public philosophy, for example, the works of Kim Díaz and Mariana Alessandri seek to empower U.S. Latinx communities by facilitating critical public discussions of harmful social conventions that lead to fragmentation for Latinx communities. Their works, both academic and nonacademic, exemplify how philosophy, when attentive to concrete lived experiences and pluralist approaches, can support U.S. Latinx communities who often experience fragmentation.

Díaz's approach to public philosophy leads to a crucial and relevant issue that continues to concern us today: the relationships between education policies, language, and social identity. In her essay, "Dewey's and Freire's Pedagogies of Recognition: A Critique of Subtractive Schooling" (2011), Díaz draws attention to U.S. Latinx children and the problem of monolingualism in education. In many U.S. school settings, Díaz explains, monolingual policies isolate non-English languages from public civic spaces and designate such languages to the private realm (Díaz 2018). The private-public dichotomy functions in language policies that endorse the view that only English should be spoken and used at school, while "other" languages should be limited to the home because they inhibit social mobility. Subsequently, for U.S. Latinx children with multilingual abilities, monolingual policies in schools treat non-English languages as inferior, thereby diminishing students' knowledge and compromising their educational experiences.

Pointing to the harmful implications of such restrictive private-public dichotomies, Díaz's approach to language brings attention to the metaphysical aspects of language that contribute to the lived experience of multilingual people (Díaz 2018). Language, in this sense, is related to the histories, cultural significance, and social meanings involved in sharing a language (Díaz and Rendón 2017). As part of their lived experiences, U.S. Latinx children with knowledge of a language other than English might experience a fragmentation of identity upheld by private-public dichotomies in schools. The critique Díaz proposes is thus meant to discourage such pedagogical harms as well as emphasize the significance of language to our senses of self in both the "public" and "private" spheres. As such Díaz exemplifies a type of public philosophy that is pedagogically valuable and committed to lived experience and pluralist communities.

In a related way, Alessandri has developed sustained public dialogue on the importance of recognizing the diverse range of human emotions for facing social challenges. By engaging Spanish philosopher Miguel de Unamuno, Alessandri promotes a philosophy of living and wellness that challenges "Fake it till you make it" narratives by arguing that such views have harmful implications when enforced on emotions (Alessandri 2019a). Alessandri utilizes Unamuno's positive estimation of feelings such as suffering and pessimism to problematize success-driven narratives that glorify positive feelings and suppress the negative feelings that arise from one's lived experiences.

Moreover, Alessandri elaborates that negative feelings that arise from experiences of suffering "have the power to connect us to people, while positivity often distances us from them" (Alessandri 2019b: 128). Specifically, Alessandri encourages individuals to recognize the difficult and negative emotions that exist in our lives and to engage in honest relationships with others that do not require us to mask our perpetual struggle (Alessandri 2014). As a type of connecting experience, Alessandri's approach to emotional well-being engages a kind of social relationality that embraces the complexities of our human emotions. In this way, theorizing diverse emotional landscapes with social relationships generates opportunities to hold spaces for the negative emotions that often accompany facing hardships and to identify the harms that engender such negative emotions. For social groups like U.S. Latinx and Black peoples who deal with systemic oppressions and ongoing issues such as monolingualism and anti-immigration practices, Alessandri's work extends crucial insights into the commonly neglected feelings of one's humanity and urges a shared ability to respond to others, possibly with compassion (Alessandri 2019b). In this respect, Alessandri's intervention on the complexity of our *lived* emotions in public venues such as *The New York Times* offers resources that can generate deep connections with others who are also experiencing multiple oppressions.

Notes

1 See also Chapter 6 by Corey Barnes on the writings of Alain Locke and Chapter 39 by Lee McBride on the writings of Leonard Harris.
2 Other important work in this direction is rereading the history of Latin American philosophy in relation to that of U.S. Latinx philosophers, such as recent research by Margaret Newton (2017) analyzing 17th-century Mexican philosopher Sor Juana Inés de la Cruz alongside 20th-century queer Chicana writer, Gloria Anzaldúa, and their philosophical defense of writing in the epistolary form for feminist pragmatist endeavors.

3 Eduardo Mendieta has conducted additional comparative work on the relationship between Latin American philosophies of liberation and pragmatism. Specifically, Mendieta has offered extensive editorial and translation work on the writings of German pragmatist Karl-Otto Apel (1922–2017). For developments of this expanded tradition, see Díaz (2021), Dussel (1996), Harris (1999), Mendieta (1995), and Orosco (2021).
4 Such transformations of SAAP include hosting its annual conference outside the United States for the first time in 2020 in San Miguel de Allende, Mexico.
5 Many thanks to Gregory Pappas for highlighting to the authors this connection to U.S. pragmatism in *Visible Identities*.

References

Alcoff, L. (1996) *Real knowing: New versions of the coherence theory*. Ithaca: Cornell University Press.
Alcoff, L. (2006) *Visible identities: Race, gender, and the self*. New York: Oxford University Press.
Alessandri, M. (2014) 'Miguel de Unamuno and William James, el gran pensador yanqui,' *Inter-American Journal of Philosophy*, 5(2), pp. 12–30.
Alessandri, M. (2019a) 'Cheerfulness cannot be compulsory, whatever the T-shirts say – Mariana Alessandri,' *Aeon*, 2 May [online]. Available at: https://aeon.co/essays/cheerfulness-cannot-be-compulsory-whatever-the-t-shirts-say [Accessed 6 February 2021].
Alessandri, M. (2019b) 'The redemption of negative feeling: Miguel De Unamuno,' in: A. Malagon and A. Doukhan (eds.), *The religious existentialists and the redemption of feeling*. Lanham: Rowman & Littlefield, pp. 118–130.
Cárdenas, P.R. and Herbert, D.R. (2020) *The reception of Peirce and pragmatism in Latin America: A trilingual collection*. Mexico City: Editorial Torres Asociados.
Dewey, J. (2008) 'Liberalism and social action,' in: J. Boydston (ed.), *The later works of John Dewey volume 11, 1925–1953: Essays, reviews, Trotsky inquiry, miscellany, and Liberalism and Social Action*. Carbonale: Southern Illinois University Press, pp. 1–65.
Díaz, K. (2018) 'A process metaphysics and lived experience analysis of Chicanx, Spanglish, Mexicans, and Mexicanidad,' *Journal of World Philosophies*, 3, pp. 44–52.
Díaz, K. (2021) 'Dewey's and Freire's pedagogies of recognition: A critique of subtractive schooling,' in: G. Pappas (ed.), *Pragmatism in the Americas*. New York: Fordham University Press, pp. 284–295.
Díaz, K. and Foust, M. (2021) *The philosophies of America reader: From the Popol Vuh to the present*. New York: Bloomsbury.
Díaz, K. and Rendón, A. (2017) 'Somos en el Aire Podcast: Kim Diaz,' *Somos el Escrito: The Latino Literary Online Magazine*, 4 November [online]. Available at: https://www.somosenescrito.com/somos-en-el-aire-podcast/somos-en-el-aire-kim-diaz (Accessed 8 April 2021).
Dussel, E. (1996) *The underside of modernity: Apel, Ricoeur, Rorty, Taylor, and the philosophy of liberation*, E. Mendieta (ed. and trans.). Atlantic Highlands: Humanities Press International.
Gomez, M.A. (2011) 'The neglected historical and philosophical connection between José Ingenieros and Ralph Waldo Emerson,' in G. Pappas (ed.), *Pragmatism in the Americas*. New York: Fordham University Press, pp. 91–99.
Harris, L. (1999) *The critical pragmatism of Alain Locke: A reader on value theory, aesthetics, community, culture, race, and education*. Lanham: Rowman and Littlefield.
Harris, L., Pratt, S.L. and Waters, A.S. (2002) *American philosophies: An anthology*. Malden: Blackwell.
MacMullan, T. (2015) 'Pragmatism as gunship philosophy: José Vasconcelos' critique of John Dewey,' *Inter-American Journal of Philosophy* 6(1) [online]. Available at: http://ijp.tamu.edu/blog/wp-content/uploads/2016/03/v6i1-MacMullan.pdf (Accessed 7 April 2021).
McKenna, E. and Pratt, S. (2015) *American philosophy: From Wounded Knee to the present*. New York: Bloomsbury.
Medina, J. (2004) 'Pragmatism and ethnicity: Critique, reconstruction, and the new Hispanic,' *Metaphilosophy*, 35(1–2), pp. 115–146.
Medina, J. (2011) 'Pragmatic pluralism, multiculturalism, and the new Hispanic,' in: G. Pappas (ed.), *Pragmatism in the Americas*. New York: Fordham University Press, pp. 199–226.
Medina, J. (2013) *The epistemology of resistance: Gender and racial oppression, epistemic injustice, and resistant imaginations*. New York: Oxford University Press.
Mendieta, E. (1995) 'Discourse ethics and liberation ethics: At the boundaries of moral theory,' *Philosophy and Social Criticism*, 21(4), pp. 111–126.
Newton, M. (2017) 'Philosophical letter writing: A look at Sor Juana Inés de la Cruz's 'Reply' and Gloria Anzaldúa's 'Speaking in Tongues,' *The Pluralist*, 12(1), pp. 101–109.

Orosco, J. (2007) 'Neighborhood democracy and Chicana/o cultural citizenship in Armando Réndon's *Chicano Manifesto*,' *Ethics, Place & Environment*, 10(2), pp. 121–139.

Orosco, J. (2011) 'Jose Vasconcelos, white supremacy and the silence of American pragmatism,' *Inter-American Journal of Philosophy*, 2(2), pp. 1–13.

Orosco, J. (2016) *Toppling the melting pot*. Bloomington: Indiana University Press.

Orosco, J. (2021) 'Pragmatism, interculturalism, and the transformation of American democracy,' in: G. Pappas (ed.), *Pragmatism in the Americas*. New York: Fordham University Press, pp. 227–244.

Ortega, M. (2016) *In-Between: Latina feminist phenomenology, multiplicity, and the self*. Albany: SUNY Press.

Pappas, G. (2008) *John Dewey's ethics: Democracy as experience*. Bloomington: Indiana University Press.

Pappas, G. (2011) *Pragmatism in the Americas*. New York: Fordham University Press.

Pappas, G. (2012) 'Dewey in Mexico: An introduction,' *Inter-American Journal of Philosophy* 3(2) [online]. Available at: http://ijp.tamu.edu/blog/wp-content/uploads/2016/03/v3i2-Pappas-Introduction.pdf (Accessed 7 April 2021).

Pappas, G. (2017) 'Zapatismo, Luis Villoro, and American pragmatism on democracy, power, and injustice,' *The Pluralist*, 12(1), pp. 85–100.

Pappas, G. (2021) 'Jazz and philosophical contrapunteo: Philosophies of la vida in the Americas on behalf of radical democracy,' *The Pluralist*, 16(1), pp. 1–25.

Pitts, A. (2021) 'Latina feminist engagements with US pragmatism: Interrogating identity, realism, and representationalism,' in: C. McCall and P. McReynolds (eds.), *Decolonizing American Philosophy*. Albany: SUNY Press.

Pratt, S. (2002) *Native pragmatism: Rethinking the roots of American philosophy*. Bloomington: Indiana University Press.

Silva, G. (2010) 'The axiological turn in early twentieth century philosophy: Alain Locke and José Vasconcelos in epistemology, value, and the emotions,' in: J.A. Carter and L. Harris (eds.), *Philosophic values and world citizenship: Locke to Obama and beyond*. Lanham: Rowman & Littlefield, pp. 31–56.

Stehn, A. (2011) 'Toward an inter-American philosophy: Pragmatism and the philosophy of liberation,' *Inter-American Journal of Philosophy* 2(2) [online]. Available at: http://ijp.tamu.edu/blog/wp-content/uploads/2016/03/v2i2-Stehn-Toward_an_Inter-American_Philosophy.pdf (Accessed 13 March 2021).

42
PRAGMATISM AND RACE

Jacoby Adeshei Carter

Race and Pragmatism's History

In seeking to investigate the question of pragmatism in relation to the philosophy of race, some care must be taken to further delineate the subject of inquiry. At least three separate inquiries are suggested by the topic "pragmatism and race." Of course there is the obvious task of defining pragmatism and defining race and seeing what if any relationship obtains between that philosophical tradition and the specified concept. One sees this in the literature as scholars pursue such thread bear topics as pragmatism and truth, or pragmatism and experience, but also in more tenuously speculative contemporary investigations of pragmatism and feminism or pragmatism and social justice, or to wit, *pragmatism and race*. An important question in this regard concerns whether there is a distinctively pragmatist conception of the relevant concept. Arguably, pragmatism offers unique conceptualizations of truth and experience. Dewey and James certainly spilled considerable amounts of ink pursuing just those philosophical quests. Next is the question of whether there are distinctively pragmatist methodologies for a philosophical investigation of race. Here the concern is not with uniquely pragmatist understandings of race but rather with recognizably pragmatist approaches to the subject of race. One means to ascertain whether pragmatists investigate race in novel ways or whether pragmatism provides characteristic methodologies for interrogating race unlike those on display in other philosophical systems. Third (and this by my lights is the most common approach), one can choose among their favorite classical pragmatist philosophers and then either use their rather sparse writings on race as a starting point or one can apply aspects of such figures' work on problems in philosophy not directly related to race to questions that arise in the philosophy of race.

Understanding, criticizing, or seeking a more just transformation of racialization or racial injustice in the United States has never been a primary concern of any racialized white classical pragmatist philosopher.[1] Contestably, Alain Locke was a pragmatist figure who did all of these things. There is a case to be made that Locke did not regard himself as a pragmatist, and a much stronger case to be made that for nearly all of the history of the tradition, no self-identified pragmatist regarded him in that way either. To make such a case requires an argument that Locke's philosophy of race was informed by his pragmatism. That argument would aim to establish that he ran arguments, developed concepts, or advanced claims or principles about race that he would not have, or would have been very unlikely to, but for his commitment to a substantive or methodological pragmatism. Alternatively, following arguments made by Paul C. Taylor (2004) one might argue that Locke's philosophy of race cannot be properly understood and contextualized unless one takes seriously the influence of pragmatism or particular pragmatists on his work. One might even concede such influence on some of Locke's philosophical thought. He acknowledges a certain indebtedness to James and Royce in his work on pluralism and democracy, but unless that indebtedness extends to his philosophy of race, there remains reason to doubt the influence of pragmatism on his thinking about race.

Beginning in the late 1980s, Leonard Harris has made the case for Locke's inclusion in the pantheon of pragmatist philosophers (1987, 1988, 1989, 1999a, 1999b) but to little avail. But even the most ardent proponents of anti-racism within academia can eventually be worn down. The present point is that since 1918, the year Locke earned his Ph.D. from Harvard (notably with Ralph Barton Perry, not Josiah Royce or William James as his supervisor), there has been a simple and obvious way to approach the question of pragmatism's relation to the philosophy of race—simply look to the work of Alain Locke, in all but a few exceptional cases that did not happen then, and does not to any significant degree happen now, in pragmatist circles. It seems implausible that pragmatists were unfamiliar with the work of Locke. More likely there was an intentional avoidance of his work. Perhaps because he was not seen as a pragmatist, or perhaps he was not seen as a philosopher?

The historical and ongoing avoidance of Locke for philosophers who wish to weigh in on philosophical debates concerning race and racism from the perspective of pragmatism, particularly in those cases where a historically important pragmatist's work is used as the starting point for such inquiries, merits careful consideration. It would strain credulity to argue that his work was simply unknown and unavailable. This is not obviously attributable to a deficiency in academic training. It is not simply that pragmatists are poor researchers and so Locke's work was never discovered by them and engaged in a way that would bring it into their philosophical discussions. Harris solved this problem for them. Likely the reason has to do with a deeply imbedded reticence on the part of racialized white pragmatist philosophers to engage substantively with the work of major figures within the African American intellectual tradition. There is perhaps a hesitancy to delve into a culturally distinct intellectual tradition identified with a different racialized population than that of the investigating scholar. Such reticence is not excused in the case of academics racialized as non-white. Diffidence about grappling with the philosophical work of authors who are culturally distinct from themselves is not a luxury that academics of color are afforded. If anything, the imposition runs the other way.

Race and Contemporary Pragmatism

Pragmatism does not, because it cannot, offer any advantages in terms of accuracy, empirical evidence, rigorousness, explanatory or descriptive value, or critical perspective over other philosophical traditions in the study of race or racism. This is not to claim that it cannot offer anything of value to such inquiries. The contention is that what it does offer has none of the sorts of advantages indicated over other schools of Anglo or European philosophy. In comparison to many works and thinkers from marginalized and racialized non-white communities, pragmatist scholarship on race or racism is at a decided disadvantage. At least since the time of Alexander Crummell, W.E.B. Du Bois, Anna Julia Cooper, and Ida B. Wells-Barnett, it has been apparent that pragmatism is not invested in an empirically rigorous or theoretically unbiased examination of racialization.

Neither pragmatism nor American philosophy is characterized by deep criticism of contemporary pragmatist philosophers unfamiliar with Black intellectual history, who opt instead to approach the philosophy of race by extending the thought of revered members of Anglo and European traditions to the subject of race and racism. The lack of expertise in intellectual traditions with the greatest critical insight concerning race and racism, and the practice of circumventing those traditions to center mainstream philosophical views and philosophers in debates where they lack expertise, is the crux of under-specialization in race theory (Curry 2010). More than the privileged infiltration of Anglo and European descendant philosophers into an area of study in which they cannot reasonably be regarded as specialists, the attempt to make the work of such scholars relevant to race theory obscures the inadequacy and irrelevance of their thinking on race and helps to maintain some rather dubious views about race (Curry 2010: 51).

Pragmatist conceptions of race and racism are not novel, prescient, or indispensable for any adequate philosophy of race. Locke, Fanon, and Du Bois, historically, and Harris and Curry contemporarily offer novel conceptions of race or racism unmatched by any contribution from a pragmatist. Frantz Fanon argued convincingly that racialized white theorists tend to study African descendant peoples in terms of racist stereotypes and the theorist's own racial pathologies (1952). Important consequences follow: first, this impedes accurate theorization of African descendant peoples; second, racialized Black people do not enter racialized white philosophies of race; instead, what appears in their theories are non-veridical caricatures of Afrodescendant people. A third implication is that Anglo European philosophies of race are about a chimerical "Black" or "Negro," the sort of culturally embedded theoretical aberration that Locke termed the "Old Negro." Such philosophies are not about real existential beings but instead about the patriarchal ambitions, stereotypes, phobias, misguided beliefs, implicit biases, and nefarious social aims of the racialized white theorists themselves. Perhaps the most insidious consequence of such so-called philosophies of race is that they ontologize these pathologies into forms of existence that concretize oppression and subjugation.

It has been recognized in the Black intellectual tradition in the United States, at least as early as Walker's 1929 Appeal to the Colored Citizens of the World, and subsequently in the works of Alexander Crummell, Anna Julia Cooper, and W.E.B. Du Bois, and more recently in the work of Frantz Fanon, and in contemporary Africana philosophy by Tommy J. Curry and Leonard Harris, that African descendant people are poorly understood and understudied empirically and defy theoretical encapsulation. Looking more closely at recent philosophy spurred by this theoretical insight, we have at our disposal Curry's conception of the Man-Not and Harris's conception of necro-being. These are novel conceptions of racialization and racism quite apart from any pragmatist account operating in the theoretical framework just described.

Curry argues that "[t]he Black man, deprived not only of an identity but also a history and existence that differs from his brute negation, experiences the world as a Man-Not" (Curry 2018: 6). Black men and boys, as reflected in the anthropology of the 19th and 20th centuries, are neither fully human nor truly men. They do not, in fact cannot, participate in masculinity and patriarchy as do racialized white men. Nor it should be noted, on Curry's account, do they aspire to such patriarchal roles. In contemporary philosophy and gender theory, namely that of the late 20th and early 21st centuries, Black men appear only as empirically unfounded stereotypes and the collective pathologies of theorists and white supremacist society. "The Man-Not," Curry writes, "grows from the incongruity ... between what theory claims to explain and the actual existence of Black men and boys—an actual reality that remains excluded from its purview" (Curry 2018: 7). "Simply stated," he continues, "analyzing Black males as the Man-Not is a theoretical formulation that attempts to capture the reality of Black maleness in an anti-Black world," a reality that has evaded a great deal of theory and philosophy (Curry 2018: 7). Drawing on insights from Frantz Fanon, Curry understands Black subjects, particularly Black male subjects, as existing in a zone of non-being. The Black male is a being from nowhere, a negation of absolute negation. The Black male is a chimerical construction, a theoretical Frankenstein, a gruesome amalgamation of white phobias and perversions. These representations of Black men do not track actually existing beings. The objects of theory are not as they purport to be actual, real, existing Black men and boys.

My aim here is not a critical analysis of Curry's book' there is not the space for that here; instead, I offer his work as a contemporary, yet historically informed, example of a genuinely novel and insightful philosophical contribution to the analysis of race and racism. His example is especially illuminating, given that he is trained in the pragmatist tradition and has even written a book on Josiah Royce; yet his theory of the Man-Not is not pragmatist by any stretch of the imagination, and his deepest intellectual indebtedness is to Fanon and other figures in the African American intellectual tradition, and not to any canonical racialized white pragmatist.

Leonard Harris carries the fundamental insight of Fanon regarding the deficiency of white philosophies of race in a different direction. Harris famously defined racism as:

> a polymorphous agent of death, premature births, shortened lives, starving children, debilitating theft, abusive larceny, degrading insults, and insulting stereotypes forcibly imposed. The ability of a population to accumulate wealth and transfer assets to their progeny is stunted by racism. As the bane of honor, respect, and a sense of self-worth, racism, surreptitiously stereotypes. It stereotypes its victims as persons inherently bereft of virtues and incapable of growth. Racism is the agent that creates and sustains a virulent pessimism in its victims. The subtle nuances that encourage granting unmerited and undue status to a racial social kind are the tropes of racism. Racism creates criminals, cruel punishments, and crippling confinement, while the representatives of virtue, profit from sustaining the conditions that ferment crime. Systematic denial of a population's humanity is the hallmark of racism.
>
> *(Harris 1999b: 437)*

Racism for Harris is a multifaceted phenomenon; it takes myriad forms, but in every instance, it causes ill-health and death to its victims. That death is not always physical; it can be a form of living death. "Necro-being," for Harris, "is always that which makes living a kind of death—life that is simultaneously being robbed of its sheer potential physical being as well as non-being, the unborn" (Harris 2020: 70). Nowhere in pragmatism is such a prescient and critically insightful examination of racism articulated. Necro-being, as Harris understands, is not a condition exclusively faced by African descendant peoples, nor is it caused by racism alone. The "situation of necro-being" as he makes clear "is hardly the sole consequence of racism and the situation can exist under conditions effected by, for example, only ethnic or status variables" (Harris 2020: 70). To exist in a state of "[n]ecro-being" is to inhabit the world as a "non-being, categorically impossible beings. Not invisible but non-existing" (Harris 2020: 70).

Necro-being is an anomalous fact of white supremacy and systems of racialization that enables inequitable asset transfers across generations from one population to another. Harris's conceptualization of racism as necro-being goes beyond offering a philosophical account of racism to provide an alternative methodology and praxis in philosophy. His account of racism is not causal, neither is it descriptive; it transcends the typical boundaries of political philosophy and purports to be imaginatively depictive, not merely speculative. It is neither ideal, dystopian nor non-ideal. Harris terms it "actuarial" by which he intends an historically and empirically informed imaginative depiction of racism as an unjust system of asset transference across generations that trades death, cultural destruction, poverty and immiseration, ill-health, and denial of social and physical well-being to a subject population for the advantage of a dominant population. Harris does not assert that such a transaction takes place in factually confirmable ways. He does not posit mechanism, causal relations, or law-like generalizations that evidence such transactions. Instead, he prompts us to shift our current framework and conceive of racism as a transactive phenomenon of wealth and health exchange (Harris 2020: 69–96).

Nowhere does pragmatism offer comparably novel or critically insightful accounts of racism, nor any methodology for analyzing racism that avoids the psychological pitfalls identified by Fanon. More than that, the sorts of contributions to the philosophy of race presently under discussion have potential to inspire even further originality and critical insight in the philosophical examination of race and racism.

Note

1 Josiah Royce still may not fit the bill as his work is not focused on racialization and racial injustice. Arguably, Royce's work supports a system of "white racial empire."

References

Curry, Tommy J. (2010) "Concerning the Underspecialization of Race Theory in American Philosophy: How the Exclusion of Black Sources Affects the Field." *The Pluralist*, 5(1), 44–64.
Curry, Tommy J. (2017) *The Man-Not: Race, Class, Genere, and the Dilemmas of Black Manhood*. Philadelphia, PA: Temple University Press.
Fanon, Frantz (1952) *Black Skin, White Masks*. New York: Grove Press.
Harris, Leonard (1987) "The Legitimation Crisis in American Philosophy: Crisis Resolution from the Standpoint of the Afro-American Tradition of Philosophy." *Social Science Information*, 21(1), 57–73.
Harris, Leonard (1988) "The Characterization of American Philosophy: The African World as a Reality in American Philosophy." *Quest: Philosophical Discussions*, 11(1), 25–36.
Harris, Leonard (1989) "Rendering the Subtext: Subterranean Deconstruction Project," in Leonard Harris, ed., *The Philosophy of Alain Locke*. Philadelphia, PA: Temple University Press, 279–291.
Harris, Leonard (ed.) (1999a) *The Critical Pragmatism of Alain Locke: A Reader on Value Theory, Aesthetics, Community, Culture, Race, And Education*. Lanham, MD: Roman and Littlefield.
Harris, Leonard (1999b) "Alain Locke," in John Stuhr, ed., *Pragmatism and Classical American Philosophy*. London: Oxford University Press, 687–686.
Harris, Leonard. (2018) "Can a Pragmatist Recite a Preface to a Twenty Volume Suicide Note? Or Insurrectionist Challenges to Pragmatism – Walker, Child, and Locke," *The Pluralist* 13 no. 1:1–25.
Harris, Leonard. (2020) "Necro-being: An Actuarial Account of Racism," In *A Philosophy of Struggle: The Leonard Harris Reader*, edited by Lee A McBride III, 69–96. New York, NY: Bloomsbury Academic.
Taylor, Paul C. (2004) "What's the Use of Calling Du Bois a Pragmatist?" *Metaphilosophy*, 35(1), 99–114.

43
MEANING AND INQUIRY IN FEMINIST PRAGMATIST NARRATIVE

Shannon Dea

This chapter offers a narrative about the origins and futures of feminist pragmatism and about the use of narrative within feminist pragmatism. It charts two histories—the more recent history of the first scholars who thought of themselves as feminist pragmatists and the earlier history of the *avant-la-lettre* feminist pragmatists of the Progressive Era. While 20th-century feminist pragmatism arose in philosophy departments, the first wave of feminist pragmatists largely operated outside of academic philosophy. As a consequence, they used methods seldom deployed within philosophy. I attend here, in particular, to their inventive and powerful use of narrative. The chapter concludes with some thoughts on what the future of feminist pragmatism might hold.

In one sense, feminist pragmatism began in the 1980s and 1990s. Then as now, philosophy as a discipline was male-dominated, but by the late 1980s, most philosophy departments were starting to see gradual increases in the number of women students and professors. The same period saw the rise of neo-pragmatism and a corresponding renewed interest in classic pragmatism. Feminist pragmatism was a natural outgrowth of this conjunction.

To be sure, feminist philosophers' interest in pragmatism arose from more than this coincidence. Feminist philosophers were drawn to pragmatism not only because it was gaining popularity but in particular because they saw in pragmatism a mode of pursuing feminist projects. Some early remarks on the topic by Richard Rorty nicely capture this alignment from the pragmatist side.

In a 1990 lecture, Rorty argued that feminism is better served by pragmatism than by what he termed "universalism"—that is, the philosophical approach that regards moral and epistemic norms as fixed, unchanging, and extending indifferently to all persons. According to Rorty's characterization, universalists attribute injustice and disagreement to people's distorted perception of those norms, not to the norms themselves. On this view, moral and epistemic progress is progress toward a less distorted perception. Rorty argued that this account is too weak to capture such radical feminist projects as those of Catherine MacKinnon and Marilyn Frye. MacKinnon and Frye envision no less than a new being for society, and that new being requires not the universalist metaphor of less distorted perception but the pragmatist metaphor of evolutionary development. For the feminist as the pragmatist, the moral and epistemic world grows and changes with us (Rorty 1991).

Rorty's account of pragmatism corresponds to what Robert Talisse and Scott Aikin (2005) have termed "inquiry pragmatism," which they distinguish from "meaning pragmatism." On this distinction, meaning pragmatists focus on the pragmatic elucidation of concepts. That is, they seek to make ideas clear by interpreting concepts in terms of their possible effects in the world. By contrast, inquiry pragmatists are primarily interested in modes of inquiry, occurring not just in the laboratory but across the gamut of human endeavor.

The feminist pragmatism that emerged in the 1990s is deeply animated by this pragmatist conception of inquiry. Some common themes that emerge in that literature are inquiry as a social practice, the entanglement of knowledge with practice, and an emphasis on the perspective of the knower. For many feminist philosophers of the period, pragmatism's conception of knowledge as situated and contingent was more appealing than the universalizing ideal of objectivity. Thus, Rooney sees feminism and pragmatism alike as rejecting a priori and fixed thinking (Rooney 1993), Duran regards both approaches as anti-foundational (Duran 1993), and Gatens-Robinson sees them as challenging the dichotomous thinking of mainstream philosophy (Gatens-Robinson 1991). Duran puts it vividly: "The core area of intersection between pragmatism and feminism, then, seems to be that they both remind us of the ways in which we experienced life before talking about the experience became more important than the having of it" (Duran 1993: 168).

Further, pragmatism's treatment of social action as a kind of inquiry and inquiry as a kind of social action aligns with and supports feminist social justice projects. This trend started to emerge most fully in the late 1990s and the early 2000s. See, for instance, Green's "deep democracy" (Green 1999), McKenna's vision of a feminist and pragmatist utopia (McKenna 2001), and Sullivan's deployment of pragmatism to reveal the workings of white privilege (Sullivan 2006).

The third main line of scholarship within 1980s–early 2000s feminist pragmatism is the revival of historical antecedents. This work took the form of adopting and adapting canonical male pragmatists—especially Dewey—in the service of feminist interpretations and contemporary projects but also of recovering neglected women pragmatists from the "classical pragmatism" period spanning the late 19th and early 20th centuries.[1]

This scholarship reveals that while feminist pragmatism was christened in the 1980s and 1990s, its real origins were a century earlier at the dawn of the Progressive Era. Figures who are increasingly included in this first wave of feminist pragmatism are Anna Julia Cooper, Jane Addams, Charlotte Perkins Gilman, Ella Lyman Cabot, and Mary Parker Follett.

While they predated the term "feminism," these thinkers' attention to gender roles, inequality, families and children, and the private sphere marks them as feminist. For instance, Gilman devoted much of her work to the deleterious effects of domestic labor and confinement on women's mental capacity. Cooper's lifelong advocacy for the education of and rights for Black women makes her one of the most important founders of Black feminism.

It is more challenging to include these figures in the pragmatist canon because they did not typically occupy the same roles or populate the same institutions as canonical male pragmatists. Pragmatism originated in the association of a group of 19th-century white male scholars who shared not only a cluster of viewpoints but also a cluster of academic affiliations—especially with Harvard and Johns Hopkins. Since women were not admitted to Harvard or Johns Hopkins until 1920 and 1970, respectively, women thinkers were absent from the soil from which pragmatism sprouted.

To be sure, most of the first-wave feminist pragmatists I have listed had some connections with classic pragmatists. Cooper corresponded with W.E.B. Du Bois, who studied at Harvard under William James and Josiah Royce, and who is himself increasingly regarded as a canonical pragmatist. Indeed, Du Bois quoted Cooper's work (albeit without attribution) in his own (Moody-Turner 2015: 51).

Jane Addams was a close associate and friend of John Dewey and influence upon his thought. Dewey taught a number of her books in his courses at the University of Chicago.

The self-educated Gilman is the only one of the figures discussed here not to have received a university education. However, she spent many months at Hull House, the settlement house that Jane Addams co-founded, and often spoke and corresponded with Addams.

Having studied at Harvard's sister college, Radcliffe, Cabot was a student and friend of Royce, and both influenced him and was influenced by him. Royce read a draft of Cabot's *Everyday Ethics* and recommended she change the title to *Conduct and Power*, but Cabot retained her intended title in order to make plain "the pragmatic purpose" of her work (Kaag 2008: 148).

Among the early feminist pragmatists, Mary Parker Follett had the fewest connections to canonical pragmatists. Like Cabot, with whom she was friends, she attended Radcliffe. However, unlike Cabot, she seems not to have studied with Royce. Nevertheless, her work was influenced by Royce and James. That said, Follett resisted having her work categorized under any particular school of thought, whether feminist or pragmatist (Whipps 2014: 406).

In short, due to the exclusionary character of the institutions in which classical pragmatism emerged, the genealogical case for counting these proto-feminist thinkers as pragmatists is somewhat tenuous. However, their work shows the unmistakable stamp of pragmatism. Each in their own way, these figures develop and deploy such pragmatist themes as standpoint, evolution, growth, fallibilism, community, and meliorism. Interestingly, due to their distinctive perspectives and roles, they often deploy them in quite different domains than the mainline pragmatists did.

The development and application of pragmatist themes by first-wave feminist pragmatists are distinctive in two main ways. As mentioned, they often focused to a much greater degree than their male counterparts on gender, families and children, and the private sphere. Second, while they were philosophically sophisticated in their formation and methodology, they rarely participated in academic philosophy. All of the male classic pragmatists were philosophy professors of one stripe or another, but none of the first-wave feminist pragmatists were. Thus, while their work was richly informed by philosophy, it took such forms as pedagogy, social work, management studies, and creative writing.

In some ways, the extra-philosophical, multidisciplinary character of early feminist pragmatism has proven an obstacle to its inclusion in the pragmatist canon. Mary Parker Follett, while influential in management schools, is rarely taught in philosophy classes or cited in philosophical scholarship—this despite an astonishing body of work that develops James's and Royce's (via Cabot's) thought in profound and novel ways.

This exclusion of early feminist pragmatists from the philosophical canon is a regrettable (though remediable) loss for academic philosophy. However, these figures' work outside of philosophy produced highly novel, generative work. For example—and this will be my focus for the remainder of this chapter—the first wave of feminist pragmatists were pioneers of narrative methods decades before the so-called narrative turn. To see this, let us consider, each in turn, the use of narrative in Gilman, Cooper, and Addams.

Gilman was a writer and social reformer, who took particular interest in the oppressive effects of the domestic sphere on women. Gilman argued that "it is not feminine qualities which distinguish the minds of women so sharply; it is the quality of domestic labor; they are heavily modified by kitchen service, by parlor imprisonment" (Qtd. in Upin 1993: 50).

She explored this theme throughout her astonishing corpus of scholarly and creative writing, which included a wide range of non-fiction books, poetry, plays, short stories, serials, and novels. She is best known for her 1892 semi-autobiographical short story, "The Yellow Wallpaper," which examines in grisly and unforgettable detail the harmful effects of domestic confinement on women's psyches and well-being.

In the story, a young married couple moves to a country house to allow the wife to undergo the "rest cure" for her "temporary nervous depression" (what we would now call post-partum depression). While the husband goes about his business, the wife's movements are limited to the upstairs nursery, where the confinement—symbolized by the room's yellow wallpaper—leads to the rapid and disastrous deterioration of her mental health. Unable to leave the room for weeks at a time except for brief intervals, the wife/narrator becomes fixated on the oppressive features of the room—such as bars on the windows and a gate at the top of the stairs—but in particular the drab yellow wallpaper. She becomes obsessed with every detail of the wallpaper; those details start to shift and eventually come to life in the form of a creeping female figure who starts to emerge in the pattern of the wallpaper. In the end, the narrator strips the wallpaper in an attempt to liberate the woman she believes to be trapped within it. In the process, she becomes the woman.

"The Yellow Wallpaper" brilliantly uses narrative as a method of pragmatist inquiry by making manifest the devastating consequences of the belief that women are weak and therefore require rest and isolation. To adopt that belief, Gilman's narrative reveals, makes it a reality. That is, the conviction that women are frail and unstable is the very thing that produces their frailty and instability. While Gilman's philosophical disposition is unmistakably pragmatist, both her use of fiction as a mode of philosophical inquiry and the topics to which she applies that lens are largely absent from her male pragmatist counterparts. Thus, Jane Upin characterizes Gilman as "even more pragmatist than Dewey ... because she addressed problems he did not identify—much less confront" (Upin 1993: 38).

If Gilman was more pragmatist than Dewey, Anna Julia Cooper might be said to be more pragmatist than James. Born into slavery, Cooper studied mathematics and theology before pursuing a career as an educator of Black students and academic administrator. Her 1892 essay collection *A Voice from the South* focuses in the first half on the education of African American women and in the second half on the representation of African Americans.

The book ends with "The Gain from a Belief." "Gain" champions belief as the well-spring of action against the skeptical tendency in philosophy, which Cooper regards as making "the universe an automaton, and man's future—a coffin!" (Cooper 1892: 291). Cooper addresses an imagined philosophical interlocutor whom we first encounter in a busy marketplace "watching as from some lonely tower" (286) as around him a throng of people busy themselves in search of "wealth, fame, glory, bread" (286). She enters into an imagined dialogue with the philosopher, who characterizes the universe in cold, mechanistic terms. She urges him to direct his philosophical powers not merely to resisting material temptations but to helping people in need. This dialogue prefaces Cooper's extended plea for the productive function of belief.

In service of her argument, Cooper offers a compelling narrative of an enslaved man drawn to freedom by the North Star. "You may have learned that the pole star is twelve degrees from the pole and forbear to direct your course by it," she writes (303).

> The slave brother, however, from the land of oppression once saw the celestial beacon and dreamed not that it ever deviated from true North. He believed that somewhere under its beckoning light lay a faraway country where a man's a man. He sets out with his heavenly guide before his face—would you tell him he is pursuing a wandering light? Is he the poorer for his ignorant hope? Are you the richer for your enlightened suspicion? (303)

Cooper tells us that there is a "noble work here and now" in helping people to "live into" a "conscious and culturable" existence "beyond our present experience" (303). The better life toward which belief compels us is not an afterlife but a better life on Earth. She quotes Wordsworth's *Prelude*: we find our happiness not in Utopia, "but in this very world, which is the world of all of us" (303).

Four years before William James's "Will to Believe" lecture, Cooper offered a rich argument for the legitimacy and usefulness of belief rather than skepticism as an epistemic starting point. V. Denise James has noted that (William) James's defense of the will to believe focuses on individual action, whereas Cooper's focus is action in community with others: "While James's notion of belief and self-cultivation may work as a road to personal growth, for Cooper, belief and the actions that it entails are primarily social—for others, with others" (James 2013: 43). Cooper's main point in "Gain," argues James, is to call believers to action, which remains a shared commitment of "most of us who call ourselves pragmatists" (37).

The final feminist pragmatist use of narrative we will look at is Addams's. I am not aware of any evidence that Cooper influenced Addams. However, there were deep similarities between them. Both were educators and social reformers, and both were active leaders in the Settlement movement—a late-19th- to early-20th-century movement in which settlement houses were established in poor, urban areas to provide a range of services to community members. Cooper was a trustee of and supervisor with the District of Columbia's Colored Settlement House, where she worked with vulnerable folks in the neighborhood. Jane Addams was the co-founder with Ellen Gates Starr of Chicago's Hull House. One of her most striking uses of narrative derives from her experiences at Hull House and, like "Gain," emphasizes beliefs' usefulness rather than their justification.

Hull House served a poor, immigrant neighborhood. Addams's thought is deeply inflected by her attention to the perspectives of those community members. In April 1916, she published a popular article on the "The Devil-Baby at Hull House." Addams's article relays what happened when a rumor began to circulate that a devil baby had been born at Hull House. For weeks, people lined up around the block, hoping to see the baby (which, of course, did not exist) and eagerly discussing it. In the article, Addams recounts listening in from her office on the conversations of old women in the queue below.

Addams was struck less by the women's credulity in the tale of a devil baby than by the ways in which they made meaning of the devil baby story, meaning that resonated with their own lives. In one particularly poignant example, a domestic violence victim saw in the devil baby a mechanism that helped her make sense of the abuse she had experienced:

> You might say it's a disgrace to have your son beat you up for the sake of a bit of money you've earned by scrubbing—your own man is different—but I haven't the heart to blame the boy for doing what he's seen all his life, his father forever went wild when the drink was in him and struck me to the very day of his death. The ugliness was born in the boy as the marks of the Devil was born in the poor child up-stairs.
>
> *(Addams 1916)*

For Addams, the women's belief in the devil baby revealed their yearning for order and pattern in a hostile world that seemed on the face of it disordered. Their sense of participating in that order—painful as it was—helped to give them equanimity and served as a means of coping.

Addams's account of these conversations models what she called "sympathetic understanding"—the method she developed in order to take seriously the perspectives of the vulnerable and minoritized people Hull House served. Addams was deeply committed to democracy conceived as grounded in identification with the common lot. For Addams, this identification with the common lot requires centering on and taking seriously the perspectives of the community members themselves. Narrative is key to that project.

These feminist pragmatist deployments of narrative are exemplary of both inquiry pragmatism and meaning pragmatism. Through these three striking stories of the confined wife, the enslaved man looking North, and the devil baby at Hull House, we can see the powerful way in which early feminist pragmatists used narrative as a method of engaging in pragmatist inquiry and of inviting the reader to participate in that inquiry for themselves. Through narrative, they thus opened pragmatism to readers and community members outside of the philosophical world even as that world remained largely closed to them.

However, each of these stories is also a project in elucidating meaning. It is a core tenet of pragmatism that the meaning of a concept consists of our conception of the practical consequences of that concept. For Gilman, the conception of women as weak and unstable has the practical consequence of rendering them just that. Cooper forcefully elucidates belief as that upon which one is prepared to act. Addams's interpretations of the Hull House women's beliefs about the devil baby focus on the practical meaning those beliefs hold for the women themselves. Thus, while the feminist pragmatism of the 1980s, 1990s, and 2000s is overwhelmingly inquiry pragmatism, we can discern within the narratives of the first wave of feminist pragmatists an embrace of meaning pragmatism.

This chapter itself offers a narrative—of feminist scholars in the 1980s and 1990s who saw in pragmatism a means of pursuing feminist projects, and of these same scholars' recovery of feminist pragmatist foremothers from a century earlier who had been excluded from both the pragmatist canon and the philosophical canon. What will the next chapter of that story look like? What should it look like?

Since the early 2000s, feminist pragmatists have continued their work on and in social epistemology, philosophy of science, and social justice, with ever broadening domains of application. In recent years, V. Denise James's Black feminist visionary pragmatism (James 2014) and Amrita Banerjee's transnational feminist pragmatism (Banerjee 2012) have charted important new paths for feminist pragmatists. There remains considerable opportunity and need for much more feminist pragmatist work that actively decenters whiteness and the so-called Global North. Celia Bardwell-Jones (2012) draws connections between classical pragmatism and Latina feminist theorists and thereby suggests the possibility of the emergence of a Latina feminist pragmatism, although, as far as I know, such a pragmatism has yet to robustly emerge.

Two areas of contemporary feminist thought from which feminist pragmatism remains comparatively disengaged are transfeminism/trans philosophy and feminist social metaphysics. The reason for this may well be the strong inquiry pragmatism tendency within feminist pragmatism. Much (but not all) trans philosophy and most social metaphysics are concerned less with modes of inquiry than with the meaning of such concepts as sex, gender, man, and woman. Meaning pragmatism offers rich possibilities for trans feminism and feminist metaphysics because it offers a mode of elucidating the meaning of concepts that resists essentialism, focuses on practical consequences in the world, and regards concepts and world as co-evolving.

If, as I hope, feminist pragmatists increasingly take up anti-oppressive scholarship that centers on racialized and trans people, they would do well to echo first-wave feminist pragmatists' use of narrative as a mode of critique, resistance, and emancipation. Now as then, stories provide powerful mechanisms both of inquiry and of elucidating meaning. Now as then, a story's truth consists in its usefulness in helping us both navigate the world and change it.

Acknowledgment

Thanks to Matt Silk and Jay Solanki for their research assistance.

Note

1 Charlotte Haddock Seigfried's work has been especially valuable. See, for instance, Siegfried (1996).

References

Addams, J. 1916. "The Devil-Baby at Hull House." *The Atlantic*. https://www.theatlantic.com/magazine/archive/1916/10/the-devil-baby-at-hull-house/305428/. Accessed July 20, 2021.

Banerjee, A. 2012. "Dynamic Borders, Dynamic Identities: A Pragmatist Ontology of 'Groups' for Critical Multicultural Transnational Feminisms." In M. Hamington and C. Bardwell-Jones, Eds. *Contemporary Feminist Pragmatism*. New York: Routledge, 71–89.

Bardwell-Jones, C. 2012. "Border Communities and Royce: The Problem of Translation and Reinterpreting Feminist Empiricism." In M. Hamington and C. Bardwell-Jones, Eds. *Contemporary Feminist Pragmatism*. New York: Routledge, 57–70.

Cooper, A.J. 1892. "The Gain from a Belief." In *A Voice from the South*. Xenia, OH: The Aldine Printing House, 286–304.

Duran, J. 1993. "The Intersection of Pragmatism and Feminism." *Hypatia* 8(2): 159–171.

Gatens-Robinson, E. 1991. "Dewey and the Feminist Successor Science Project." *Transactions of the Charles S. Peirce Society* 27(4): 417–433.

Gilman, C. 1892. "The Yellow Wallpaper." *New England Magazine*: 647–656. https://www.nlm.nih.gov/exhibition/theliteratureofprescription/exhibitionAssets/digitalDocs/The-Yellow-Wall-Paper.pdf. Accessed July 20, 2021.

Green, J. 1999. *Deep Democracy: Community, Diversity and Transformation*. Lanham, MD: Rowman and Littlefield.

James, V.D. 2013. "Reading Anna J. Cooper with William James: Black Feminist Visionary Pragmatism, Philosophy's Culture of Justification, and Belief." *The Pluralist* 8(3): 32–45.

———. 2014. "Musing: A Black Feminist Philosopher: Is That Possible?" *Hypatia* 29(1): 189–195.

Kaag, J. 2008. "Women and forgotten movements in American philosophy: The work of Ella Lyman Cabot and Mary Parker Follett." *Transactions of the Charles S Peirce Society* 44(1): 134–157.

McKenna, E. 2001. *The Task of Utopia: A Pragmatist and Feminist Perspective*. Lanham, MD: Rowman and Littlefield.

Moody-Turner, S. 2015. "'Dear Doctor Du Bois': Anna Julia Cooper, W.E.B. Du Bois, and the Gender Politics of Black Publishing." *Multi-Ethnic Literature of the United States* 40(3): 47–68.

Rooney, P. 1993. "Feminist-Pragmatist Revisionings of Reason, Knowledge, and Philosophy." *Hypatia* 8(2): 15–37.

Rorty, R. 1991. "Feminism and Pragmatism." *Michigan Quarterly Review* 30(2): 231–258.

Seigfried, C.H. 1996. *Pragmatism and Feminism: Reweaving the Social Fabric*. Chicago, IL: University of Chicago Press.

Sullivan, S. 2006. *Revealing Whiteness: The Unconscious Habits of Racial Privilege*. Bloomington, IN and Indianapolis, IN: Indiana University Press.

Talisse, R. and S. Aikin. 2005. "Why Pragmatists Cannot Be Pluralists." *Transactions of the Charles S. Peirce Society* 41(1): 101–118.

Upin, J. 1993. "Charlotte Perkins Gilman: Instrumentalism beyond Dewey." *Hypatia* 8(2): 38–63.

Whipps, J. 2014. "A Pragmatist Reading of Mary Parker Follett's Integrative Process." *Transactions of the Charles S Peirce Society* 50(3): 405–424.

44

PRAGMATISM AND ENVIRONMENTAL PHILOSOPHY

Evelyn Brister

Environmental philosophy emerged as a subfield of philosophy in the early 1970s, following the mid-century rise to prominence of the science of ecology and the advent of a political movement for environmental protection in the 1960s. Then, in the 1990s, environmental pragmatism became established as a distinctive approach to environmental philosophy. In its early years, environmental philosophy focused almost exclusively on analyzing ethical questions, such as how to assign intrinsic value to the natural environment and the duties that humans owe to natural entities. But the causes and solutions to environmental problems are complex: addressing environmental problems involves not only ethical reasons and commitments but also the development of environmental policy and practices by policymakers, experts, and the public. The defining contributions of pragmatism to environmental philosophy have been (1) its expansion of the application of philosophical resources to environmental issues, drawing on political theory, philosophy of science, epistemology, aesthetics, and philosophy of religion and (2) its strong emphasis on justifying and motivating action to benefit the natural environment.

Environmental pragmatism, like pragmatism generally, is unified by its intellectual heritage and by a set of overlapping, but not essential, methodological and normative commitments. Classical pragmatism, as in the thinking of Peirce, James, Dewey, and their contemporaries, rejects the view that thought is primarily representational and, especially, rejects the idea that philosophy is a form of a priori theorizing that ignores practical implications. Instead, pragmatism stresses the priority of action and holds that the function of ideas is to formulate and test plans of action. The classic pragmatists had little to say explicitly about the natural environment. Nonetheless, their influence has shaped the approach of contemporary environmental pragmatism, alongside the influence of other American thinkers who wrote on nature, especially Henry David Thoreau and John Muir; pragmatists who engaged in democratic practice, including Jane Addams and Mary Parker Follett; and land managers and horticulturalists, such as Aldo Leopold and Liberty Hyde Bailey.

Environmental pragmatism was born of a conviction that prominent debates in environmental ethics provided little in the way of guidance for addressing specific, concrete environmental issues such as pollution and species loss. It issues a distinctive call to make philosophy relevant to environmental policy. Environmental pragmatists have demonstrated their commitment to action by working with policymakers and practitioners to evaluate and reform conservation priorities, agricultural practices, and climate change policy, while at the same time, and again like pragmatism more generally, a turn toward action has not meant turning away from philosophical discourse. In addition to direct involvement with environmental policy and practice, pragmatists articulate and refine the implications of their commitments and attempt to solve or dissolve philosophical problems in philosophical terms.

Commitments of Environmental Pragmatism

Environmental pragmatists draw on the core methodological and normative commitments of pragmatism to theorize about sustainability, urban planning, industrial waste, agricultural issues, animal welfare, climate justice, and conservation practices. Their particular commitments all relate to pragmatism's instrumental character as an active, constructive philosophy that places an emphasis on supporting human agents to confront and collectively resolve the problems they face. Pragmatists do not, however, share a unitary outlook, and some of the tensions among pragmatist commitments have produced disagreements, or at least differences in emphasis, among environmental pragmatists.

Pragmatist epistemology is *empiricist, naturalistic,* and *fallibilistic*. Peirce, James, and Dewey all identified scientific methods, institutions, and practices as having a crucial role in inquiry, and contemporary environmental pragmatists continue to maintain the centrality of scientific knowledge for environmental practice. For pragmatists, the goal of inquiry is not just true beliefs but also successful action: in Dewey's phrase, resolving "problematic situations." Their work accepts scientific knowledge about, for instance, the causes of climate change and biodiversity loss. However, environmental pragmatists also recognize the limits of science and social science and view them as fallible human enterprises that are pursued within a context of human values.

Bryan Norton notes the similarity between pragmatist epistemology's commitment to experimental inquiry and the epistemological basis for adaptive management, an approach to managing natural resources that is now common (Norton 2015). Environmental systems are characterized by complexity and uncertainty, and 20th-century management of natural resources was plagued by rigid management decisions that were often based on outdated or inadequate knowledge about natural resources and processes. Too often, the result was fisheries management that hastened the collapse of fisheries, fire suppression that created the conditions for more destructive wildfires, agricultural practices that degraded water quality and soil resources, and other unintended and undesirable consequences. Adaptive management has been developed as a management response that attempts to identify areas of uncertainty, monitor the outcomes of management actions, learn from failures, and remain sufficiently nimble and flexible to update knowledge and improve decision-making over time. It shares the pragmatist understanding that actions should be based on empirical knowledge and values the importance of experimentation and scientific methods in order to improve that knowledge.

However, environmental pragmatists have also been critical of overdependence on science, especially when a scientific theory does not adequately fit the empirical constraints of actual practice. The economist Daniel Bromley invokes pragmatist commitments when he notes that by and large economists have not successfully used their field to direct public policy and institutions in ways that serve the public good. He cites the example of economic models that fail to explain and predict observed phenomena (such as deforestation) and diagnoses their failure as caused by a dependence on deductive logic rather than Peircean abductive logic and by the refusal of economics as a discipline to respond to criticism (Bromley 2006: Ch. 11).

Environmental pragmatism has also inherited a commitment to *value pluralism* and *methodological pluralism*. Debates about environmental ethics sometimes center on which capacity (such as sentience) or quality (such as naturalness) is most fundamental, a determination that may be forced when values conflict. For pragmatists, however, debates about the ultimate ranking of values are either unnecessary or of secondary importance. Pragmatists have various reasons for supporting value pluralism: some follow James in his belief that no moral theory can encompass all conceptions of the good, while others follow Dewey in seeing values as shifting in response to changing material, social, and environmental conditions. There are also pragmatists who doubt the validity of value pluralism, but even these pragmatists adopt an attitude of methodological pluralism, holding that debates about the ultimate source of value are of secondary importance to finding the best solution to social conflicts arising from value disagreements.

Pluralist commitments have led environmental pragmatists to evaluate the process—and not just the outcome—of environmental decision-making and, indeed, to see that philosophers themselves have a role in environmental decision-making. In this role, environmental pragmatists may assist citizens, stakeholders, experts, and policymakers to clarify and express environmental values and to find sufficient common ground to identify workable policy solutions to environmental problems (Light 2002a). Environmental pragmatists use the conceptual resources that are relevant in a particular context—appealing to rights or to particular values held by stakeholders, whether economic, moral, or aesthetic—rather than seeking a unified theory of natural value. The values actually held by stakeholders and decision-makers, when these have been subjected to ethical reflection and informed by empirical evidence, are those most likely to motivate progress in identifying solutions to environmental problems (Brister 2012).

This shift from investigating the source of environmental value to theorizing and assisting in the process of decision-making marks a significant difference between environmental pragmatism and traditional environmental ethics. It arises from a pragmatist commitment to *democratic deliberation* and an approach that is *procedural* rather than substantive. Political theory is integral to environmental pragmatism, as is an understanding of environmental policy, practice, and law. Environmental pragmatists seek solutions to environmental problems within existing democratic institutions while also critiquing these institutions. Since pragmatists see the goal of philosophy as to encourage social learning, to improve the situation of humans (and non-humans), and to solve problems and prevent moral harm, they engage not just in philosophical debates but also apply philosophy to particular environmental issues. Because environmental pragmatists are more problem-oriented than theory-driven, they also work in related fields, such as agricultural and animal ethics, science and energy policy, land management, sustainability, environmental education, and environmental economics.

Theoretical Debates and Environmental Pragmatism

Andrew Light and Eric Katz have identified four forms of environmental pragmatism (Light and Katz 1996: 5). Two forms further develop the theoretical framework for environmental pragmatism either by articulating connections between classical American pragmatism and contemporary environmental philosophy or by extending normative inquiry guided by pragmatist commitments to pluralism and democratic deliberation. The other two are practical: either they bridge gaps between theory, practice, and the public or they assist policymakers and environmental advocates in articulating normative judgments that can be the basis for concrete action. I'll discuss the two theoretical senses in this section and the two practical senses in the section following.

Pragmatist thinking is anti-dogmatic, a stance that requires pragmatists to be amenable to civil and open-ended disagreement even among themselves. However, environmental pragmatism is cohesive enough—and different enough from other approaches in environmental ethics—that environmental pragmatists have waged a persistent battle to defend their work as a distinctive environmental philosophy rather than merely an unphilosophic prejudice in favor of unreflective action (Pearson 2014; Samuelsson 2010).

This is not a new charge against pragmatism: in 1926, Lewis Mumford leveled a general form of this accusation, charging that pragmatism affirmed, or acquiesced to, prevailing cultural norms (Mumford 1926). Rather than providing substantive criteria for what is good, pragmatism provides procedural guidelines for rational inquiry and democratic decision-making, but Mumford and others have argued that this makes it vulnerable to appropriation by dominant social forces. Although certain pragmatists, like Dewey, have stood for progressive politics, this charge holds that pragmatism's commitments are too weak to counteract powerful non-democratic interests.

During the emergence of environmental philosophy in the 1970s, environmentalists identified a cultural apathy toward the non-human world as a cause of environmental problems. Indifference to nature and non-humans resulted in a blind spot in western philosophy that environmental ethics corrected by theorizing the source of non-anthropocentric value and the nature of such value as intrinsic (see, e.g., Rolston 1989; Routley 1973). From this ethical perspective, as Anthony Weston notes, "'Pragmatism' sounds like just what environmental ethics is against: shortsighted, human-centered instrumentalism" (Weston 1996: 285). However, a fairer description of the pragmatist project is that it proceeds with environmental problem-solving in a piecemeal fashion, working from the assumption that deliberative groups will learn from experience to define and address particular issues (such as pollution or species loss) in context and that cultivating procedures for sincere inquiry aligns with the ends of environmental ethics. Weston holds that the discovery of a knockdown philosophical proof for intrinsic value would do less to repair environmental harm than working through a policy process to implement actions based on the experiences and values of the people involved (Weston 1996). This is not an argument against any particular view in environmental ethics and is not anti-theory; rather, it is a different philosophical strategy. It holds that the best way to address environmental ethics is not to settle disputes over abstract propositions. Rather, the strategy is to identify the ideals that express our valued relationships to the natural world (Lekan 2012). It is a strategy that is oriented toward action.

Nonetheless, environmental pragmatists were pressed to address the question of whether, by focusing on procedures to support ethical deliberation on environmental topics, they were implicitly aligned with the human-centered values that drive environmental harm. In response to this challenge, Bryan Norton developed a distinction between a strong form of anthropocentrism holding that only humans have moral standing and a weak, or enlightened, form of anthropocentrism that aligns with pragmatist attitudes. Norton argued, first, that a theory of nonhuman rights on which the case for non-anthropocentric value rests has not, and likely cannot, be convincingly made (Norton 1982). Second, he argued that it is a mistake to premise environmental action on making a philosophical case for non-anthropocentric value. Since human health and well-being require developing principles of conduct to sustain the integrity of functioning ecosystems, existing public sentiment for environmental protection that is based in human-oriented values and theories of justice should be sufficient ethical grounds for addressing environmental problems (Norton 1984).

Again, not all pragmatists agree with Norton on these points, but many do, and the key point is not a positive argument for enlightened anthropocentrism so much as a strategic point in favor of prioritizing problem-solving over theoretical precision. This lays out a research agenda for environmental pragmatists that starts from real-life conditions—legal, political, scientific, social, and environmental—and uses philosophical inquiry to support democratic deliberation and collective action. Norton's own argument goes further—he hypothesizes that when it comes to identifying appropriate action, enlightened anthropocentrists and non-anthropocentrists will converge on the same environmental policy goals (Norton 1991). The convergence hypothesis, as it is called, is a view intended to provide environmentalists with rational reasons to cooperate with one another rather than squandering time and resources examining differences within the environmental community. In addition to laying out the theoretical implications that follow from value pluralism, it is presented as a testable empirical hypothesis. Devising a definitive test of the hypothesis has proved elusive, however. Various pragmatists and non-pragmatists have evaluated the convergence hypothesis, and the divergence of views is striking: some find it politically essential, others unnecessary; some find evidence to support it, others find counterexamples that throw it into doubt; yet others find it provocative but ambiguous (Minteer 2009).

In the past two decades, environmental pragmatism has developed to be more concerned with addressing the philosophical questions that are involved in pressing problems than in pursuing ultimate answers. This has pushed environmental pragmatism to attend not only to specific policy issues but also to fields that intersect with environmental issues, including economic and political theory. It is in the nature of many environmental issues to raise questions about justice. This is because these questions emerge when the value of natural resources is exploited and the costs externalized such that they are paid by people living downstream, downwind, or in the future. Thus, environmental pragmatism has both drawn on the existing resources of pragmatic political theory and has contributed new insights to its development.

Pragmatism emphasizes the importance of cultivating and strengthening democratic institutions and practices as a foundational requirement of addressing environmental problems. Environmental issues are divisive; in order to produce collective action at the scale of regions, nations, and the globe, it is important to overcome political divisions to build durable coalitions. Pragmatists show how environmental action creates opportunities to promote environmental and democratic citizenship (Light 2002b). The political process of democratic deliberation has value not only when it successfully identifies a solution to an environmental problem but also because the process itself builds community by generating conversations and opportunities for collaborative action. Taking part in such activities is a form of learning and inquiry, and when communities come together to talk about problems and possible solutions, this can lead not only to policy outcomes but also to creating appreciation of environmental processes, motivation to address problems, and an explicit articulation of value commitments (Norton 2005). The deliberative process thus creates political alternatives that transcend a simplistic division between economic exploitation of natural resources and imperious preservation of wild nature. The successful coalition between hunters and conservationists to preserve rich wetland habitats is one example of how former political adversaries have invested in collaboration for the sake of resolving a problem.

However, pragmatists have also been criticized for placing naive trust in existing political institutions and in idealized images of democratic deliberation. The Deweyan model of social inquiry expects inquiry to be sincere and for participants to aim for the public good. But political institutions are subject to capture by vested interests, and public discourse about environmental issues is plagued by ignorance and deliberate misinformation (Brister 2018). When this happens, pragmatists may be hamstrung by their support for value pluralism and their reluctance to make substantive value judgments.

Pragmatism has also been criticized for overlooking injustice (Glaude 2007) and lacking a theory of justice (Talisse 2017). Specifically, pragmatism lacks critical resources to overturn a status quo that distributes political power unequally in the context of environmental decision-making (Eckersley 2002). Environmental pragmatists' expressed commitment to cultivating shared participation in democratic institutions ought to generate not only robust attention to issues of environmental justice but also support for emancipatory projects to enhance the voice of oppressed and marginal groups when addressing environmental problems. Although pragmatists have been active on a number of policy issues, environmental justice has not received much attention. In this context, pragmatists may find (1) their commitment to institutional solutions scrutinized and (2) their allegiance to the convergence hypothesis tested. In relation to the former, democratic institutions may disadvantage environmental advocates to the degree that compromise, incremental policy shifts, and constrained parameters of options are so inadequate that marginalized participants are forced to disengage or enact civil disobedience (see Maboloc 2016 for a case study as applied to climate justice). In relation to the second, the convergence hypothesis assumes that support for human well-being will, in the long run, converge with the interests of the non-human world. However, in times of crisis, a situation may arise where prioritizing human rights requires sacrificing non-human interests. Environmental pragmatists have argued that the solution to this dilemma is to invest strongly in biodiversity preservation and climate change mitigation so as to avoid crisis (Miller et al. 2011). But the point remains that in the short term, when we are in fact facing crises, pragmatists should strengthen and expand their understanding of—and commitment to—environmental justice.

Environmental Pragmatism and Policy Engagement

As we've seen, environmental pragmatism encompasses a range of philosophical positions and methods that address humanity's relationship with the environment. Some environmental pragmatists explicitly draw on the legacy of classical American pragmatists such as James and Dewey; others share commitments with pragmatism and the practical goal of using theoretical inquiry to ground environmental action. In pragmatic fashion these theoretical commitments should have practical, action-oriented outcomes. Otherwise this is pragmatism in name only. As a result, this has led environmental pragmatists to write on specific environmental issues and to work directly with community groups, farmers, professionals whose work impacts the environment, government agencies, and others. They've taken up work related to a wide range of environmental topics, with some of the most influential being sustainability, agricultural practices, land management and conservation planning, and climate policy.

For example, pragmatists were among the first philosophers to address sustainability (Thompson 2017), sustainable agriculture (Thompson 2010), and sustainable land management (Norton 2005, 2015). During the 1990s, environmental thinkers became bogged down thrashing out general definitions of sustainability. This led to disagreements about which definition best supported human rights and which best prioritized nature protection: the sort of disagreements that pragmatists would argue prevent progress on policy issues. Both Paul Thompson and Bryan Norton examined sustainability from the perspective of how the concept works to unify people and direct action. Thompson's work on sustainability has bridged agricultural and environmental ethics. Typically, agriculture is treated as part of the built environment, entirely separate from the issues of natural value that grip environmental philosophers. Thompson, however, emphasizes the ways that agriculture is embedded in social, economic, and environmental systems. For instance, agricultural runoff affects water quality downstream, farm subsidies affect the health and resilience of agricultural communities, and attitudes toward organic, genetically modified, factory-farmed, or processed foods express social status, political beliefs, and concerns with purity. Thompson advocates for examining specific debates in applied agricultural and environmental research in order to identify the values that research programs assume to be correct and subjecting them to analysis. This requires considering how systems actually function—natural systems, political systems, economic systems—and working directly with farmers, business and land owners, researchers, and government agencies to evaluate the ethical implications of policy options. Thompson writes that helping practitioners and policymakers "think through their practice can bring about significant improvements in the lives of the oppressed (including animals)" (Thompson 2020: 294).

On issues of land management, too, environmental pragmatists have drilled down in specific contexts to examine how normative assumptions guide practice and the deliberative processes through which environmental policy is made. Bryan Norton has made the case that the techniques of adaptive management that are now commonly used by foresters and natural resource managers are congruent with Dewey's commitments to democratic participation in decision-making and using empirical data to improve practice (Norton 2005). Ecological restoration is a specific conservation activity that has been defended by environmental pragmatists. Critics fear that ecological restoration puts undue trust in management interventions to repair harm to the environment and that this could in the long run exacerbate environmental problems since it perpetuates a belief that technological fixes are available to repair ecosystem damage and a dominating stance toward natural systems. Andrew Light argues against this, claiming that ecological restorations bring communities together to participate in a common civic activity, increasing public awareness of the value of functioning ecosystems and building community (Light 2000). Pragmatists see community involvement in environmental activities as an opportunity for people to learn about the environment they live in and ecological processes, to reflect on and clarify their own values, and to strengthen social bonds and civic engagement.

Most recently, environmental pragmatists have begun to build a distinctive pragmatist approach to climate ethics, specifically to issues related to mitigation policies (Fesmire 2020). As noted above, pragmatism lacks robust resources to address issues of justice head-on. However, its emphases on moral pluralism, on the value of participatory engagement rather than unilateral decision-making, and on converging on actionable policies that stem from divergent underlying values are relevant to negotiating international climate agreements. Climate change is forcing not only changes in the atmosphere, ocean, and biosphere; it is also forcing changes to political and social institutions. As the climate crisis intensifies, human relationships to the natural environment are shifting—causing us to grapple with the threats of wildfire, drought, floodwaters and storm surges, declines in forest health, changes to land use, and even displacement and refugees. Pragmatist commitments highlight open-mindedness, pluralism, social learning, and willingness to alter values in response to new evidence, and these will be needed as humanity navigates these changes.

Environmental pragmatists have shown that philosophers have a role to play in the translation of theory into policy and practice. Environmental pragmatists have taken on governmental roles, working with the U.S. Environmental Protection Agency (Bryan Norton) and on climate policy with the Obama and Biden administrations (Andrew Light). Other environmental pragmatists have collaborated with scientists, agricultural researchers, biotechnologists, and farmers. Environmental pragmatists advocate working directly on issues that lead to intentional changes in practice and policy to benefit the natural environment, doing so in ways that generate democratic participation. Providing examples of how to undertake that work, identifying its beneficial impacts, and theorizing the relationship between normative inquiry and ethical practice are ongoing projects of pragmatism (Brister and Frodeman 2020).

References

Brister, E., 2012. Distributing epistemic authority: Refining Norton's pragmatist approach to environmental decision-making. *Contemporary Pragmatism* 9(1): 185–203.

Brister, E., 2018. Proceduralism and expertise in local environmental decision-making. *In*: S. Sarkar and B. A. Minteer, eds. *A Sustainable Philosophy—The Work of Bryan Norton*. Cham: Springer, 151–165.

Brister, E. and Frodeman, R., 2020. *A Guide to Field Philosophy: Case Studies and Practical Strategies*. New York: Routledge.

Bromley, D. W., 2006. *Sufficient Reason: Volitional Pragmatism and the Meaning of Economic Institutions*. Princeton: Princeton University Press.

Eckersley, R., 2002. Environmental pragmatism, ecocentrism, and deliberative democracy: Between problem-solving and fundamental critique. *In*: B. A. Minteer and B. P. Taylor, eds. *Democracy and the Claims of Nature*. New York: Rowman and Littlefield, 49–69.

Fesmire, S., 2020. Pragmatist ethics and climate change. *In*: B. Miller and D. Eggleston, eds. *Moral Theory and Climate Change: Ethical Perspectives on a Warming Planet*. New York: Routledge, 215–237.

Glaude, E. S., 2007. *In a Shade of Blue: Pragmatism and the Politics of Black America*. Chicago: University of Chicago Press.

Lekan, T., 2012. A Jamesian approach to environmental ethics. *Contemporary Pragmatism* 9(1): 5–24.

Light, A., 2000. Ecological restoration and the culture of nature: A pragmatic perspective. *In*: R. B. Hull, ed. *Restoring Nature*. Washington, DC: Island Press, 49–70.

Light, A., 2002a. Contemporary environmental ethics: From metaethics to public policy. *Metaphilosophy* 33(4): 426–449.

Light, A., 2002b. Restoring ecological citizenship. *In*: B. A. Minteer and B. P. Taylor, eds. *Democracy and the Claims of Nature*. New York: Rowman and Littlefield, 153–172.

Light, A. and Katz, E., 1996. Introduction: Environmental pragmatism and environmental ethics as contested terrain. *In*: A. Light and E. Katz, eds. *Environmental Pragmatism*. New York: Routledge, 1–18.

Maboloc, C. R., 2016. On the ethical and democratic deficits of environmental pragmatism. *Journal of Human Values* 22(2): 107–114.

Miller, T. R., Minteer, B. A. and Malan, L., 2011. The new conservation debate: The view from practical ethics. *Biological Conservation* 144: 948–957.

Minteer, B., ed. 2009. *Nature in Common?: Environmental Ethics and the Contested Foundations of Environmental Policy*. Philadelphia: Temple University Press.

Mumford, L., 1926. *The Golden Day: A Study in American Experience and Culture*. New York: Horace Liveright.

Norton, B. G., 1982. Environmental ethics and non-human rights. *Environmental Ethics* 4(1): 17–36.

Norton, B. G., 1984. Environmental ethics and weak anthropocentrism. *Environmental Ethics* 6(2): 131–148.

Norton, B. G., 1991. *Toward Unity among Environmentalists*. New York: Oxford University Press.

Norton, B. G., 2005. *Sustainability: A Philosophy of Adaptive Ecosystem Management*. Chicago: University of Chicago Press.

Norton, B. G., 2015. *Sustainable Values, Sustainable Change: A Guide to Environmental Decision Making*. Chicago: University of Chicago Press.

Pearson, C. H., 2014. Does environmental pragmatism shirk philosophical duty? *Environmental Values* 23(3): 335–352.

Rolston, H., 1989. *Philosophy Gone Wild*. New York: Prometheus Books.

Routley, R., 1973. Is there a need for a new, an environmental ethic? *Proceedings of the 15th World Congress of Philosophy* 1: 205–210.

Samuelsson, L., 2010. Environmental pragmatism and environmental philosophy: A bad marriage! *Environmental Ethics* 32(4): 405–415.

Talisse, R. B., 2017. Pragmatism, democracy, and the need for a theory of justice. *In*: S. Dieleman, D. Rondel, and C. Voparil, eds. *Pragmatism and Justice*. New York: Oxford University Press, 281–293.

Thompson, P. B., 2010. *The Agrarian Vision: Sustainability and Environmental Ethics*. Lexington: University Press of Kentucky.

Thompson, P. B., 2017. *The Spirit of the Soil: Agriculture and Environmental Ethics*. 2nd ed. New York: Routledge.

Thompson, P. B., 2020. Field philosophy in an actual field. *In*: E. Brister and R. Frodeman, eds. *A Guide to Field Philosophy: Case Studies and Practical Strategies*. New York: Routledge, 285–297.

Weston, A., 1996. Beyond intrinsic value: Pragmatism in environmental ethics. *In*: A. Light and E. Katz, eds. *Environmental Pragmatism*. New York: Routledge, 285–306.

INDEX

Addams, J. 27, 35–40, 266, 276, 300, 327, 381, 382, 384, 387
Adorno, T. 147, 149
Aesthetics 287–299
Aikin, S. 159, 161, 194, 345, 380
Alberini, C. 366
Alcoff, L. M. 369, 370
Alessandri, M. 371, 372
American Evasion of Philosophy (West) 84–86
An Analysis of Knowledge and Valuation (Lewis) 53, 277
Analytic-synthetic distinction 60–61
Anderson, A. R. 95
Anderson, E. 359
Anscombe, G. E. M. 260
Aristotle 44, 193, 280
Austin, J. L. 139, 220
Ayer, A. J. 220, 332

Bacon, F. 253
Baier, K. 95
Bain, A. 8, 110, 226, 278
Banerjee, A. 385
Bear, L. S. 360
Bell, Clive 288
Berkeley, G. 8, 120
Bernstein, R. 75, 168
Blackburn, S. 139, 291, 295, 347
Bohnert, H. 95
Boodin, J. E. 118
Bradley, F. H. 98
Brandom, R. 64, 75, 101–106, 139, 141, 168, 181, 287, 292, 304, 314, 331, 333, 345, 346, 348
Bromley, D. 388

Cabot, E. L. 277, 282, 381, 382
Calderon, M. 128
Calkins, M. W. 35
Carnap, R. 55, 138, 140, 293, 353

Carter, J. 41
Cavell, S. 152
Cerbone, D. 160
Chekov, A. 152
Chesterton, G. K. 351
Child, L. M. 360, 362
Church, A. 95
Clapp, E. R. 35
Clifford, W. K. 266
cognitive science 239–251
Coleman, O. 292
Collingwood, R. G. 289
A Common Faith (Dewey) 267
Cooper, A. J. 360, 376, 381, 383
Cordova, V. F. 367
critical common-sensism 91–92
Critique of Pure Reason (Kant) 9,
Croce, B. 133
Crummell, A. 386
Curry, T. J. 377

Darwin, C. 10, 27, 71, 109, 206, 209, 227, 265, 351
Davidson, D. 58, 287, 353, 370
Davis, A. 362
Deleuze, G. 155
democracy 32–33, 37–38, 47–49, 153, 300, 305, 391
Democracy and Social Ethics (Addams) 37
Dennett, D. 56, 242
Derrida, J. 84
Descartes, R. 14, 70, 102, 193, 232, 234
Dewey, J. 26–34, 54, 59, 71, 75, 82, 92, 120, 128, 132, 147, 171, 179, 191, 197, 212, 218, 223, 266, 276, 287, 300, 303, 305, 313, 323, 343, 353, 359, 366, 368, 375, 381, 387, 390
Diaz, K. 367, 371
Dickie, G. 289
Douglass, F. 360
Dretske, F. 19
Dreyfus, H. 165

Du Bois, W. E. B. 43, 81, 86, 87, 118, 277, 369, 376, 377, 381
Duchamp, C. 292
Dummett, M. 15, 296
Duran, J. 381
Dussel, E. 368

education 317
Eliot, T. S. 292
embodied cognition 239
Emerson, R. W. 86, 152, 366
"Empiricism and the Philosophy of Mind" (Sellars) 66–67
environmental philosophy 387–394
Epictetus 280
epistemology 252–263
expressivism 139, 170, 331–342, 347

Fanon, F. 377, 378
Feigl, H. 55
feminism 380–386
Fisch, M. 227
Fisher, R. A. 206
"Five Milestones of Empiricism" (Quine) 59
Fodor, J. 19, 235, 240
folk psychology 226
Follett, M. P. 35, 381, 382, 387
Forstenzer, J. 327
Foucault, M. 84, 87
foundherentism 90
Foust, M. 367
Frege, G. 219, 293
Frye, M. 380

Gallagher, S. 161, 164, 235, 240
Gates, H. L., Jr. 82
Gaukroger, S. 253
Gettier, E. 252
Gibson, J. 240
Gilman, C. P. 381, 382
given 66, 121, 292
Godfrey-Smith, P. 214, 235
Goodman, N. 83, 125, 140, 353
Guattari, F. 155
Gramsci, A. 147
Green, J. 382

Haack, S. 75, 89–94
Habermas, J. 99, 143, 153, 268
Harris, L. 358, 360, 361, 376, 377, 378
Hartland-Swann, J. 260
Hegel, G. W. F. 27, 98, 101, 147, 154, 179
Heidegger, M. 71, 83, 191, 287
Hempel, C. G. 95
Heney, D. 327
Herbert, D. 366
Hintikka, J. 139
Hocking, W. E. 118, 120

holism 122
Holmes, Oliver Wendell, Jr. 7, 11, 109
Honneth, A. 148
Hook, S. 47–52, 86, 191, 281
Hookway, C. 18
Horkheimer, M. 147
Hull House 36–37
Hume, D. 8, 19, 170, 193, 291, 347
Husserl, E. 159, 163, 234

idealism 109, 174
inferentialism 104, 182
Ingenieros, J. 366
inquiry 30–32, 110, 147, 171, 182, 187, 206, 278, 302, 305, 323, 381, 388
insurrectionist philosophy 358–365

James, V. D. 383, 385
James, W. 9, 17, 19–15, 27, 29, 38, 54, 59, 75, 82, 89, 96, 109, 110, 118, 120, 128, 130, 142, 159, 191, 195, 210, 229, 265, 276, 300, 313, 326, 339, 343, 351, 353, 354, 358, 375, 381, 387
Jastrow, J. 206
Joas, H. 268
Johns, J. 295
Johnson, M. 243, 245
Jones, C. B. 385

Kallen, H. 119
Kant, I. 29, 101, 120, 126, 132, 170, 201, 265
Kaplan, A. 75
Katz, E. 389
Kearney, R. 270
King, M. L. K., Jr. 85
Kitcher, P. 174, 198, 215
Kripke, S. 20
Kuhn, T. 83, 206, 213

Lakoff, G. 244
language 217–225
Latinx philosophy 366–374
Leibniz, G 95
Leopold, A. 387
Levi, I. 207, 214
Levinson, J. 292
Lewis, C. I. 9, 53–57, 59, 66, 96, 118, 121, 140, 191, 277, 283, 300, 353
Light, A. 389, 393
Lincoln, A. 36
Locke, A. 41–52, 119, 283, 358, 375
Locke, J. 8, 119, 193, 355, 360, 367, 376, 377
logic 179–190
Logic: The Theory of Inquiry (Dewey) 31, 133, 142, 279, 301
logical positivism 55–56, 138, 191, 213, 219
Long Road of Woman's Memory (Addams) 38–39, 277
Lovejoy, A. O. 118
Lyotard J. 150

Index

MacFarlane, J. 293
Mach, E. 20
MacKinnon, C. 380
Maddy, P. 294
Margolis, J. 75
Marx, K. 37, 150
Matrinez, J. 369
McDowell, J. 348
McKenna, E. 367, 381
Mead, G. H. 82, 153, 191, 218, 277, 280, 353, 369, 370
Medina, J. 369, 370
meliorism 196
Menand, L. 7
Merleau-Ponty, M. 235
metaphilosophical creep 312, 324
metaphilosophy 311–320
metaphor 217
Metaphysical Club 7–12, 17, 128, 191, 227, 291
metaphysics 191–205, 290
method 323–330
Miller, Dickenson S. 118
Mills, C. Wright 86
mind 226–238
Mind and the World Order (Lewis) 121
Misak, C. 15, 51, 171, 174, 208, 279, 291, 302, 345
Moore, G. E. 138
moral life 276–286
Morris, G. S. 26
Mounce, H. O. 325
Moya, P. 369
Muir, J. 387
Mumford, L. 389
Munch, E. 292, 294

Nagel, T. 351
naturalism 122, 147, 211, 241, 343–350, 388
New Ideals of Peace (Addams) 38, 278
Nicolls, S. 261
Niebuhr, R. 81, 86, 87
Nietzsche, F. 87, 150
Neurath, O. 138
Norton, B. 388, 390, 392, 393

objectivity 96, 294, 303, 352
Olson, K. 260
Ortega y Gassett, J. 358

Papini, G. 128, 133, 300
Pappas, G. 368
Peirce, Benjamin 13
Peirce, C. S. 8, 10, 13–18, 21, 29, 51, 59, 65, 75, 82, 89, 92, 96, 109, 128, 132, 139, 147, 160, 170, 191, 207, 221, 229, 240, 246, 266, 269, 276, 279, 287, 291, 300, 303, 304, 313, 343, 351, 358, 366, 369, 387
Perry, R. B. 54, 119, 366, 376
phenomenology 159–167

pluralism 76, 89, 99, 134, 210, 268, 307, 334, 358, 369, 370, 388
political philosophy 300–310, 327
pragmatic a priori 53–54
pragmatic maxim 13, 14–16, 140, 179, 193, 208, 266, 288, 292, 313, 318, 324, 339, 343, 352
pragmaticism 135
Pragmatism (James) 28, 60
Pratt, S. 367
Prezzolini, G. 128, 133
Price, H. 64, 75, 139, 170, 198, 331, 339, 347, 348
Principles of Psychology (James) 20, 21, 27, 60, 211, 239, 242
Putnam, H. 58, 75–80, 118, 124, 139, 142, 164, 170, 195, 197, 242, 267, 269, 303, 345, 354, 370
Putnam, R. A. 75

Quest for Certainty (Dewey) 66
Quine, W. V. O. 9, 58–62, 75, 83, 118, 122, 138, 140, 142, 191, 220, 241, 287, 293
Quinney, W. 360

race 42–44, 375–379
Ramberg, B. 242
Ramirez, R. 366
Ramsey, F. 9, 200
Rawls, J. 143, 308
realism 76–77, 90, 97, 124, 139, 198, 199, 208, 210, 295, 326
Reconstruction in Philosophy (Dewey) 148
Reichenbach, H. 55, 138,
religion 264–275
Rescher, N. 75, 95–100
Reyes Cardenas, P. 366
Rooney, P. 381
Rorty, R. 29, 58, 64, 70–74, 75, 78, 81, 83, 96, 126, 139, 142, 149, 153, 191, 194, 267, 287, 294, 300, 303, 304, 307, 314, 331, 346, 354, 380
Rosen, G. 295
Rosenthal, S. 161
Ross, S. 203
Rothko, M. 297
Royce, J. 54, 120, 140, 266, 276, 282, 376, 378, 381, 382
Russell, B. 23, 54, 138, 142, 219
Ryle, G. 220, 255

Sa, Z. 360
Saenz, M. 366
Santayana, G. 277, 282
de Saussure, F. 221
Sayre, K. 258
Schiller, F. C. S. 13, 96, 126, 129–133, 191, 354
Schlick, M. 138
science 206–216
Sellars, R. W. 63, 366
Sellars, W. 57, 63–69, 71, 83, 95, 101, 138, 142, 181, 287, 292, 297, 331, 345

Index

semantics 179–182, 288
semiotics 220–222
Silva, G. 367
Sokolowski, R. 165
"Some Reflections on Language Games" (Sellars) 64
Spencer, H. 20
Spinoza, B. 228
St. John Green, Nicholas 9–10
Stace, W. 98
Starr, E. G. 384
Stehn, A. 368
Steinberg, L. 295, 297
Stewart, M. 362
Stich, S. 261
Strawson, P. F. 139, 348

Talisse, R. 51, 159, 174, 194, 303, 308, 317, 324, 345, 380, 391
Tarski, A. 142
Taylor, C. 165, 268
Taylor, P. C. 375
Thomson, J. J. 292
Thompson, P. 392
Thoreau, H. D. 152, 362, 384
Tolsty, L. 289
truth 90, 112–114, 120, 125, 142, 170, 208, 259, 279, 288, 295, 301, 303, 328, 334, 335–336, 345, 351–357
Twenty Years at Hull House (Addams) 36
Tufts, J. H. 36

Turri, J. 261
Trilling, L. 86

de Unamuno, M. 372
Upin, J. 383

Vailati, G. 128, 133
van Fraassen, B. 206, 208
Varieties of Religious Experience (James) 267, 282
Vasconcelos, J. 367
verificationism 8, 208
Villoro, L. 368

Walker, D. 360, 362, 377
Walters, A. 367
Weinberg, J. 261
Wells-Barnett, I. B. 376
West, C. 81–88
Weston, A. 390
Whitehead, A. N. 54, 120, 139, 353, 366
"Will to Believe" (James) 8, 212, 267, 327, 355, 383
Williams, M. 292
Williamson, T. 254
Wittgenstein, L. 23, 60, 71, 82, 89, 138, 140, 170, 219, 220, 268, 287, 340
Wright, C. 8, 10–11, 109

Xerohemona, K. 90

Zahavi, D. 161